AF817230

FUTURE DIRECTIONS IN ELECTRICAL ENGINEERING

PROCEEDINGS of the

Cornell University Electrical Engineering

Centennial Symposium Series

March 15 - June 12, 1985

Simpson Linke, editor

Promethean Press
Ithaca, New York

PROCEEDINGS OF THE
CORNELL UNIVERSITY ELECTRICAL ENGINEERING
CENTENNIAL SYMPOSIUM SERIES:
FUTURE DIRECTIONS IN ELECTRICAL ENGINEERING

Copyright 1989
by the School of Electrical Engineering
Cornell University, Ithaca, New York
All rights reserved

ISBN: 0-916859-76-2

Published by Promethean Press
Ithaca, New York

Camera-ready copy for this volume was typeset by means of one of the first commercial laser printers, the QMS LASERGRAFIX ™ using QMS Font 104- "Standard Roman Medium". The font has been reduced to 65 percent of the original size. The text is printed on 50# Finch smooth-finish white vellum.

Printed in the United States of America

March, 1989

TO

ESTHER, MARTHA, and LAURA

WHO

APPRECIATE WORDPOWER

PREFACE TO THE SERIES EDITION

This book contains a complete set of the *Proceedings* of the Cornell Electrical Engineering Centennial Celebration Symposium Series. The six previously published paperback books have been bound into this limited casebound edition to serve as a presentation volume for the Symposium participants and to provide a composite record of the entire Symposium Series.

Since each set of the *Proceedings* is a separate entity, I beg the reader's indulgence for the unavoidable repetition that is present in the Editor's "Preface" and Professor Nation's "Introduction" to each component of the volume. The general table of contents on page IX of this introductory section gives the location of the detailed contents of each Symposium. Complete lists of authors and commentators are also included for convenient reference to the contributions of the individual participants. I am grateful for the encouragement and support of Professor John A. Nation, Director of the School of Electrical Engineering, and of Professor Edward D. Wolf, Director of the National Research and Resource Facility for Submicron Structures (now known as the National Nanofabrication Facility) throughout the course of this publication project. The kind cooperation of Mrs. Jean Coonradt in the management and assignment of the secretarial work, and the assistance of Mrs. Cynthia Telage in the necessary financial arrangements, are also acknowledged with thanks. Finally, I thank my family for the patience and understanding that made it possible for me to undertake and execute this task.

Simpson Linke

Ithaca, New York

SYMPOSIUM CONTENTS *

*Detailed contents of each Symposium are listed on the designated roman-numeral pages.

LIST OF AUTHORS

LIST OF COMMENTATORS

"Victor Hugo once consoled a friend who was depressed at growing old by telling him that 'at the age of 50 you are really entering into the youth of old age'. Disciplines like Electrical Engineering age a good deal more gracefully than we do as humans. It is very clear that Electrical Engineering at 100 has all the vitality and vigor of youth. It has only just begun–in fact the record of the past century is breathtaking as one looks back on it–and the record of the last 30 years is even more impressive: from the cellular telephone to microwave ovens, from computer-aided design to robotics, from fiber optics to satellite communications. These and countless new technologies that you have collectively developed, have not only advanced the field of Electrical Engineering but have transformed the lives of all of us".

– Frank H. T. Rhodes, President of Cornell University
From Centennial Banquet Address, June 12, 1985
Ithaca, New York

FUTURE DIRECTIONS IN ELECTRICAL ENGINEERING

MICROELECTRONICS

PROCEEDINGS of the

Cornell Electrical Engineering Centennial Symposium

at Stanford, California---March 15, 1985

Simpson Linke, editor

Promethean Press
Ithaca, New York

PROCEEDINGS OF THE
CORNELL ELECTRICAL ENGINEERING CENTENNIAL SYMPOSIUM:
FUTURE DIRECTIONS IN ELECTRICAL ENGINEERING-
MICROELECTRONICS

Copyright 1986
by the School of Electrical Engineering
Cornell University, Ithaca, New York
All rights reserved

Published by Promethean Press
Ithaca, New York

Printed in the United States of America

September, 1986

CONTENTS

PREFACE

The celebration of the Centennial Year of the School of Electrical Engineering at Cornell University featured a series of symposia based on the central theme of "Future Directions in Electrical Engineering". Each seminar concentrated on a major facet of the discipline, with individual topics presented by outstanding specialists in each field. The symposia were held at sites around the country that are widely known as centers of activity for each particular subject area, thereby providing significant updates on pertinent electrical engineering technologies at appropriate locations, and allowing a large number of Cornell Electrical Engineering alumni to participate in the celebration in a meaningful way.

The ultimate success of the Symposium Series must be credited to the dedicated efforts of the faculty, symposia, and local alumni chairmen who were responsible for planning the programs, engaging the participants, and attending to local arrangements within specific time constraints. The following schedule of events attests to the overall magnitude of the task and also displays the broad range of topics that were presented:

March 15, 1985—Microelectronics——Stanford, Calif.

April 1, 1985——Computers——E. Windsor, N. J.

April 10, 1985—Communications——Boston, Mass.

April 17, 1985—Atmospheric and Space Sciences——St. Louis, Mo.

April 25, 1985—Energy for the Future——Washington, D. C.

June 12, 1985—Societal Impacts and Future Directions——Ithaca, N. Y.

Each session was videotaped to facilitate preparation of the proceedings of each meeting and for possible use of the individual presentations in classroom seminars and elsewhere. The audio portions of the tapes were transcribed into preliminary text and coordinated with figures obtained from the video images and from hard copy provided by some of the authors. Each presentation was edited with a view toward achieving continuity and coherence without

compromising the natural style of the author. In several instances it was necessary to introduce substantial changes in format and phrasing in order to achieve this goal; in others, the hard copy supplied by the authors made the task easier. In all cases, the final versions were sent to the authors for their criticisms, corrections, and final approval. I am grateful to all of the contributors for their excellent cooperation and assistance in bringing their presentations to this present form.

I wish to thank Professor Edward D. Wolf for his enthusiastic interest in this project and for authorizing financial support from the National Research and Resource Facility for Submicron Structures to assist in the publication of the Symposia Series *Proceedings*. Additional funding from the College of Engineering and the School of Electrical Engineering is also gratefully acknowledged.

It would not have been possible to produce this work without the skillful secretarial assistance of Joan Hammond, Catherine Kuhl, Deborah Overslaugh, and Linda Struzinsky who performed the difficult tasks of transcribing the tapes and retyping the edited manuscripts with meticulous attention to detail. The valuable assistance of John Ralston in the preliminary editing of several of the manuscripts is also acknowledged with thanks. The expert guidance provided by Mrs. Diana Nathanielsz on format and production of this volume is very much appreciated.

I am particularly grateful to Catherine Kuhl and Deborah Overslaugh for their diligence and care in forming the camera-ready copies of the manuscripts on that versatile new instrument that is so illustrative of "future directions in electrical engineering" — the laser printer.

Simpson Linke

Ithaca, New York

INTRODUCTION

John A. Nation

Director, School of Electrical Engineering

The celebration of our Centennial Year affords a unique opportunity to review past accomplishments, to evaluate our present circumstances, and to make some reasonable estimates of prospects for the coming years. A glance at the history of the University reveals that the Cornell *Register* for 1883 contained a "Special Notice" from President Andrew Dickson White announcing the establishment of a "Course in Electrical Engineering." President White had convinced the University's Trustee Executive Committee to create the new program "provided such course occasions no additional expense."

The early curriculum was rigorous and in addition to the traditional engineering topics of mathematics, physics, mechanics, mechanical drawing, and shop work, required rhetoric, French and German. The senior year also included courses in several Cornell specialties such as photometry and electric lamps, telegraph instruments, lines and cables, the dynamo, and electric motors. The first baccalaureate degrees in electrical engineering in the United States were awarded by Cornell (as well as by M.I.T.) in 1885, and the first Ph.D. in the new discipline was conferred by President White at Cornell in the same year. Thus we are truly celebrating 100 years of electrical engineering education.

One of the major figures in the E.E. School of those days was Professor Harris J. Ryan who ran the Electrical Engineering Department essentially as a one-man faculty for 17 years and is credited with having trained almost 1000 electrical engineers during that period. Ryan left Cornell in 1905 to become head of the Electrical Engineering Department at Stanford. It is interesting to note that 12 of the first 13 members of the Stanford faculty were educated at Cornell.

This brief summary of the beginning of electrical engineering instruction at Cornell would not be complete without mention of the internationally known Professor Vladimir Karapetoff who joined the faculty following Professor Ryan's departure and made many contributions to early electrical engineering theory and analysis in several textbooks and many technical articles. Something of the flavor of electrical engineering studies of the time is suggested by the following excerpt from Karapetoff's *Electrical Laboratory Notes*, published in 1906:

> "In coming to the laboratory, bring with you a slide rule, an inch rule or tape, a speed counter, a screw

> driver and a pair of plyers [sic]. This will save you time and trouble of looking for them or borrowing them. Do not forget to have a pocket knife for skinning off wire; a bicycle wrench is also sometimes very handy to have."

In these days of hand calculators and personal computers, it is rare indeed to find a student who has even seen a slide rule, much less to have used one.

Classical electrical engineering education at Cornell was devoted principally to electric machinery, power systems, telecommunications, and early electronics. As the technology developed from the 20's through the 40's, the curriculum was, of course, continually upgraded as new and advanced concepts were introduced, but the basic format remained essentially unchanged until the mid-50's when research began to play a central role in the E.E. School, coincident with the move from Franklin Hall to the newly constructed Phillips Hall. The following figure among the highlights of our research development over the years:

- An early solar telescope project initiated an era of Radio Astronomy research that culminated in the establishment of the world's largest radio-radar telescope at Arecibo, Puerto Rico, now known as the National Astronomy and Ionosphere Center.

- A research program in the new field of Physical Electronics gradually expanded into the present National Research and Resource Facility for Submicron Structures housed in the Knight Laboratory adjacent to Phillips Hall.

- The Old High Voltage Laboratory on Mitchell Street, converted to an Energy Laboratory in 1960, served as the staging area for the Cornell Extra-High-Voltage Cable Test Program and housed E.E. research on high-voltage vacuum switches and arc physics until the formation of the interdisciplinary Laboratory of Plasma Studies in 1966. This facility, together with adjacent installations, now contains the largest particle-beam program in any university in the world and has made many important contributions to the understanding of intense electron and ion beams, particularly as they relate to inertial confinement thermonuclear fusion experiments.

- The national energy crisis in the early 70's created renewed interest in power and energy research within the School leading to the combination of power-system elements, control theory, and new computer facilities, that now constitutes the unique Kettering Power Systems

Laboratory for the analysis of modern complex power-generation and transmission networks.

Two recent events attest to the high regard that the University and the E.E. School enjoy in the national research community:

1. The Program in Microscience and Technology was established in 1982 and receives about $2 million annually through the Center of Excellence Program of the Semiconductor Research Corporation.

2. The NSF-sponsored Cornell Center for Theory and Simulation in Science and Engineering, with its associated supercomputer production facility, will augment and expand research capability in the School and will encourage greater interdisciplinary activities with other scientific units on campus.

One of our strengths in recent years, as the Electrical Engineering School has evolved from a purely teaching organization into a teaching and research institution, has been our ability to attract good new faculty. A positive measure of this quality is in our receipt of Presidential Young Investigator Awards from the National Science Foundation in the first year of the establishment of these awards. The E.E. School received 5 of the total of 26 awards made available to electrical engineering departments in the nation. Our success in attracting these high-quality young people bodes well for the continued future growth of the School.

Remarkable changes in electrical engineering education have occurred in the School within the past 25 years as the influence and impact of new faculty members and their research have brought modern concepts to curriculum and laboratories that had been essentially identical to those of World War II days. The advent of computers has been particularly significant. We have been very fortunate in the last decade in the support that we have received from industrial corporations. Their generous donations of equipment including computers, work stations, and state-of-the-art measurement devices have been of great value in upgrading our laboratories to present-day standards.

On the other hand, the current huge interest in computer engineering and electrical engineering has introduced major enrollment problems in the School. Using 1965 as a year of reference, the size of the faculty has remained fixed at about 40 members whereas the number of undergraduate students has tripled. Our senior class in 1985-86 numbered 186 students. We have 265 graduates students in the School with about 145-150 MS/PhD students and the remainder in the Master of Engineering Program, the latter being the second largest professional degree program in the University. This year we graduated

30 MS/PhD students and 88 Master of Engineering degree candidates. An encouraging factor is that 12 of these advanced-degree students were women. Today about 15 percent of the students in the E.E. School are women and 25 percent are minority students.

The increase in student enrollment is coupled with a research effort that is growing at a compound rate in excess of 10 percent per year in constant dollars. Clearly we are approaching the limits of our present capability. It is hard to realize that within 30 years of our initial occupancy of Phillips Hall we are now bursting at the seams. Further, the building which was then state-of-the-art, is now an antiquated facility in terms of present-day technology, with little in the way of the clean-air environments, air conditioning, and other amenities that are necessary for the proper functioning of modern equipment.

Plans are in progress to build a new engineering complex on campus in which electrical engineering will have substantially more space than in the past. When these new facilities are completed, our faculty will be able to work at its full potential. We expect to see a relatively modest growth in the number of faculty members. Since a faculty lifetime post is forty years, and student-interest time cycles are rather shorter than that, the response to large student increases in enrollments and interest in electrical engineering cannot be followed instantly by new faculty positions. Attempts will be made to maintain our undergraduate size at about its present level with some growth allowed in the number of our graduate students. It is certain that there will be many more interactions with industry in the near future, and these will take on a much broader form as the years go by. For example, an exchange program may be arranged with industrial laboratories so that their scientists will come to work in the School for a period while a faculty member will have the opportunity to spend corresponding time in industrial environments.

This brief perspective of past, present, and future encourages us to begin our second century with a true sense of excitement. We may be justly proud of the important contributions that our School of Electrical Engineering has made to many of the developments that are reported in these *Proceedings*. As we celebrate these past achievements, we can look forward with confidence to producing equally significant additions to our challenging discipline.

PART I

MICROELECTRONICS

PROCEEDINGS of the

Cornell Electrical Engineering Centennial Symposium

at Stanford, California –March 15, 1985

"I am pleased to welcome you to this symposium on 'Future Directions in Electrical Engineering - Microelectronics', the first of six symposia on major topics in electrical engineering that are being held across the country this Spring in celebration of the Centennial Year of the School of Electrical Engineering at Cornell University and in the United States. In keeping with the tradition of excellence that has characterized this century of electrical engineering education, it is appropriate to call attention to an article in this morning's *Wall Street Journal* announcing that Cornell has been selected as one of four University Centers in the country to host research in supercomputers. At Cornell this activity will be a multi-disciplinary, university-wide program under the direction of Professor Kenneth G. Wilson, the 1982 recipient of the Nobel Prize in Physics.

As a prelude to presenting our Symposium Chairman, Charlie Sporck, I wish to thank him for his efforts in assembling this outstanding program for today. It is indeed a rare occurrence to have eight such distinguished speakers participating in one symposium.

Introducing Charlie Sporck here in Silicon Valley is somewhat like introducing a person to his family in his own home so I will only summarize his career briefly. After receiving the B.S.M.E. degree from Cornell in 1952, he joined the General Electric Company, advanced through various manufacturing positions until 1959 when he was named General Manager of semiconductor operations at Fairchild Camera and Instrument Corporation, a post he held until his appointment to National Semiconductor Corporation in 1967 where he serves as President and Chief Executive Officer. Under his direction, National has grown from a manufacturer of discrete transistors to a position of technological leadership in the semiconductor industry. As a founding member of the Semiconductor Industry Association, Charlie is known as a champion of industrial strategy, and a proponent of taxation reform for high-tech industries".

— *Professor Edward D. Wolf, Symposium Faculty Chairman*
From Welcoming Remarks, March 15, 1985
Stanford, California

"When I graduated from Cornell 33 years ago, I had little idea that the microelectronics industry would play such an important role in the future of America and of the world. Given a map of this country it is likely that most people now would find it easier to pinpoint Silicon Valley than Ithaca, New York. While it is true that the two are a continent apart – and with respect to this centennial celebration, they are a century apart – still they have a common bond: engineering schools such as the one at Cornell (I admit to a little bias here) have helped to create the high technology of today. By turning out well-educated students, our engineering schools provide the thinkers, doers, and creators of tomorrow with excellent opportunities for prestigious careers, particularly in Silicon Valley.

It is well known, however, that technology changes quickly. Engineering schools must always be alert to adapt curricula and training to meet the new challenges, and it is essential for industry to lend its support. As you listen to the discussions that will be presented here today, I ask you to think about the advances made possible through education, through hard work, and through the dedication of our talented educators and engineers. I urge you to support those programs that encourage our young people to seek engineering careers, for I believe that young engineers, on the road to fame and fortune, provide the true driving forces for the future directions of microelectronics".

– *Charles E. Sporck, Symposium Chairman*
From Introductory Remarks, March 15, 1985
Stanford, California

VLSI LITHOGRAPHY AND NANOLITHOGRAPHY

Alec N. Broers

Alec Broers obtained BSC degrees in physics and electronics at Melbourne University in 1958 and 1959, and BA and Ph.D. degrees from Cambridge University in 1962 and 1965. From 1965 to 1980 he was associated with IBM Thomas J. Watson Research Center where he became manager of electron-beam technology in 1967, manager of photon and electron optics in 1972 and was appointed an IBM Fellow in 1977. In 1980 he joined the IBM East Fishkill Laboratory with responsibility for semiconductor lithography and process development, becoming a member of the IBM Corporate Technical Committee in 1984. In October, 1984 he was appointed Professor of Electrical Engineering and Head of the Electrical Division, Cambridge University. He received the American Institute of Physics Prize for Industrial Application of Physics in 1981, and the IEEE Cledo Brunnetti Award in 1985.

INTRODUCTION

In opening, I would like to thank Cornell, and Ed Wolf in particular, for including me in the EE Centennial Program. Of the many fields of research in which Cornell is prominent, I am most familiar with their work on the fabrication of solid-state devices with dimensions below 1 micron. I have been associated with the National Research and Resource Facility for Sub-micron Structures (NRRFSS) from its early days, and have been fortunate to have a close view of the remarkable progress that has been made by its members. Of all the accomplishments of Cornell's electrical engineers in the last one hundred years, I especially congratulate them on the success of NRRFSS.

In the following, I am going to add to a review of the present state of integrated circuit lithography, a look into the future of lithography; into the field of nanolithography.

Nanolithography is the study of the fabrication of devices with dimensions less than a tenth of a micron. It is a field of more interest to the fabrication of devices for scientific study, than to the problems of fabricating today's semiconductor devices, but it produces an understanding of the limits of the various techniques that is valuable in predicting future trends.

The major driving forces of the semiconductor business are device speed and cost. They push forward both the level of integration, and the miniaturization of devices. In the case of memory chips, for example, present goals are to increase single-chip capacities to several megabits by the end of the '80s, and simultaneously to improve access time. This means that the task of the lithographer is getting more and more difficult; minimum dimensions are shrinking and at the same time chip sizes are expanding (see Figure 1).

For some years I have been studying trends in lithography by analyzing the proceedings of the IEEE International Solid State Circuit Conference. Up to 1985, device dimensions decreased steadily, with the Japanese slowly catching up with, and now equaling the U.S. (see Figure 2). Interestingly, between 1984 and 1985 there was a slight increase in the average size of the Japanese devices. This slowing seems to represent a consolidation rather than a change in trend, however, as there were many papers on complex devices with dimensions close to the present-day practical limit of 1μ. In those proceedings, and in the proceedings of other important conferences this year (IEDM, U.S./Japan VLSI Symposium), Japanese papers on advances in semiconductor technology outnumbered American papers.

Since returning to England I have become aware that the opinions of Americans and English about Japanese science and technology are the same - "that we invent everything, but the Japanese capitalize on the inventions". The difference in England is that Americans are included with the Japanese. This opinion is supported by the comforting thought that it is intellectually superior to invent. I do not think that any comfort should be enjoyed on this account, however, because inventions in technology are most frequently made by those carrying out the most advanced experiments. The inventions are not so much flashes of genius, as combinations of ideas that come out of the search for solutions to problems encountered in attempting the untried. Today, the Japanese are frequently the first to attempt the untried.

OPTICAL LITHOGRAPHY

Looking forward to 1990 we find that lithography requirements will include linewidths of a half a micron, overlays of 0.15μ, linewidth control to within 0.1μ, and processing capabilities for eight-inch wafers. At the same time, the need for high throughput will remain in order that the economic benefits of further miniaturization are realized.

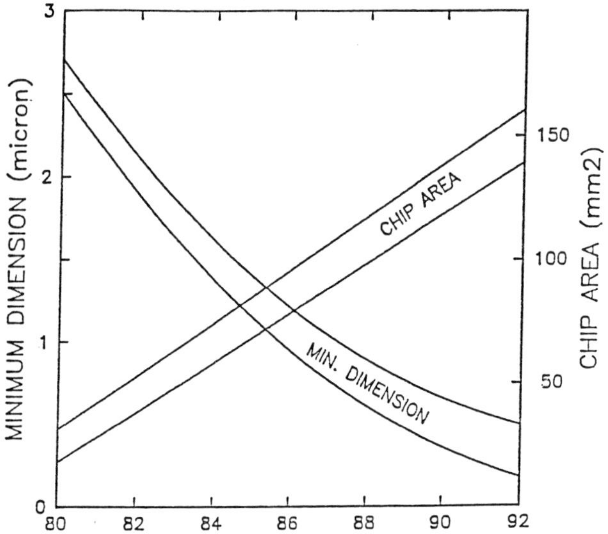

Adapted from Figure 2 by permission from *Solid State Technology*, Vol. 28, No. 6, June 1985, p. 119.

Figure 1. Author's Estimate of the Lithography Requirements Needed to Meet the Performance and Capacity of Memory Chips and Microprocessor Chips.

The work-horses of integrated circuit lithography are the 1:1 full-field scanning optical projection cameras.[1] They use a unique ring-field mirror lens originated by Offner[2] and have a numerical aperture (N.A.) of 0.16, which gives the resolution at different wavelengths shown in Figure 3.

A key advantage of the mirror lens is that exposure can be made over a range of wavelengths. With deep-UV (250nm) illumination, dimensions close to 1μ can be produced. The optics of the newest full-field camera, the Perkin Elmer Model 500,[3] is corrected to the point that the image distortion is less than 0.1μ which is also adequate for 1μ devices.

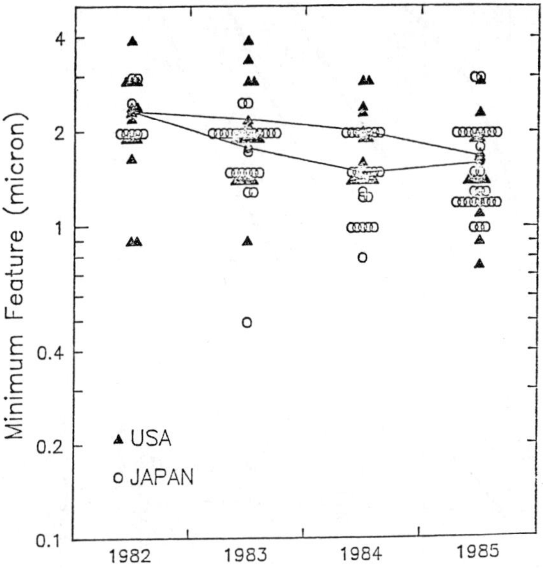

Adapted from Figure 1 by permission from *Solid State Technology*, Vol. 28, No. 6, June 1985, p. 119.

Figure 2. Minimum Dimension of Devices Reported at the 1982-5 IEEE International Solid-State Circuits Conference.

---

[1]Markle, D.A., (1974), *Solid State Technology*, 17:50
[2]Offner, A., (1979), *Photogr. Sci. Eng.*, 23:374.
[3]Greed, J.J. and Markle, D.A., (1984), *Proc. SPIE Conf.*, Santa Clara, CA.

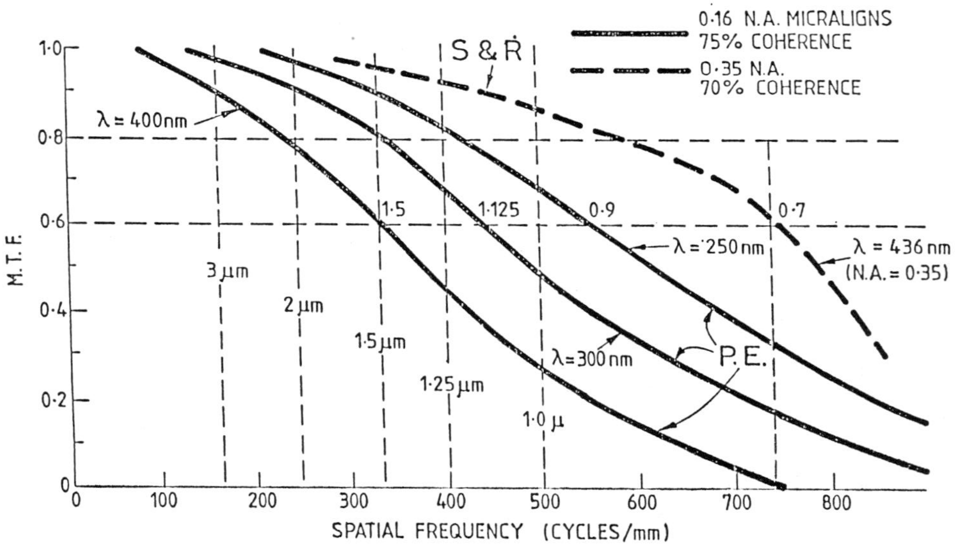

Adapted from Figure 3 by permission from *Solid State Technology*, Vol. 28, No. 6, June 1985, p. 119.

Figure 3. Modulation Transfer Functions for Full-Field Camera with N.A. = 0.16 (Exposure Wavelengths of 250nm, 300nm, and 400nm), and S/R Camera with N.A. = 0.35 (Exposure Wavelength of 436nm).

The main disadvantage of full-field exposure is the need for 1X masks. Masks that are suitable for use with 150nm wafers, but which also have micron-size linewidths, are time-consuming and difficult to make. This is not a problem of fundamentals, rather it is a characteristic of today's electron-beam systems. Nonetheless, it has led many users to switch to demagnifying step-and-repeat cameras, despite the fact that their throughput is only 15-50 wafers/hour compared to >100 six-inch wafers/hour for the full-field cameras.

The present alternative to the full-field camera is the demagnifying (5X-20X) refractive lens step-and-repeat (S/R) camera. Rather than explain in detail how such a camera works, I will restrict my discussion to the performance limits associated with the camera lenses. I will also assume that it will be possible to align the patterns with adequate accuracy, although in practice this is usually as difficult as obtaining the required resolution.

Existing lenses for S/R cameras have more than twice the numerical aperture of the full-field cameras, that is >0.32, and therefore twice the resolution at a given wavelength. The resolution for an N.A. of 0.35 and a wavelength of 405nm is shown in Figure 3. It exceeds that of the full-field scanning camera, even when the latter is operated at 250nm. All refractive lenses cover a much smaller field than the wafer, and step-and-repeat operation is essential. 5X lenses are preferred over 10X lenses because they allow larger fields (up to 2cm x 2cm) and therefore higher throughput.

The penalty of obtaining higher resolution by increasing N.A., rather than reducing wavelength, is that the depth of field is reduced. The small depth of field of the step-and-repeat camera lenses can be tolerated, however, because the exposed area is small, and the sample can be re-leveled and re-focused before each exposure. With full-field cameras, the whole wafer is exposed after a single attempt to level and focus.

Figure 4 summarizes progress in optical lens systems in terms of resolution and field size. Just a few years ago, it was thought impossible to build refractive lenses that would retain diffraction-limited resolution over fields larger than about 10,000 times the resolution limit (0.8λ/N.A.); in effect limiting performance to the region in the upper left of the dotted curve in Figure 4. Lenses used in optical microscopes achieve theoretical resolution for numerical apertures of 0.8 and 0.9, but only for extremely small fields ($<200\mu$ x 200μ). In recent years, however, a series of refractive lenses have been developed that offer diffraction-limited resolution for a numerical aperture of 0.35 ($<1\mu$ linewidth for a wavelength of 365nm) over fields up to 2cm x 2cm.

The scanning cameras have always offered larger field size, but only in one direction; the wafer being covered by scanning the sample mechanically past the optical system. As already mentioned, these cameras have a numerical aperture of only 0.16. Increasing this numerical aperture would require an

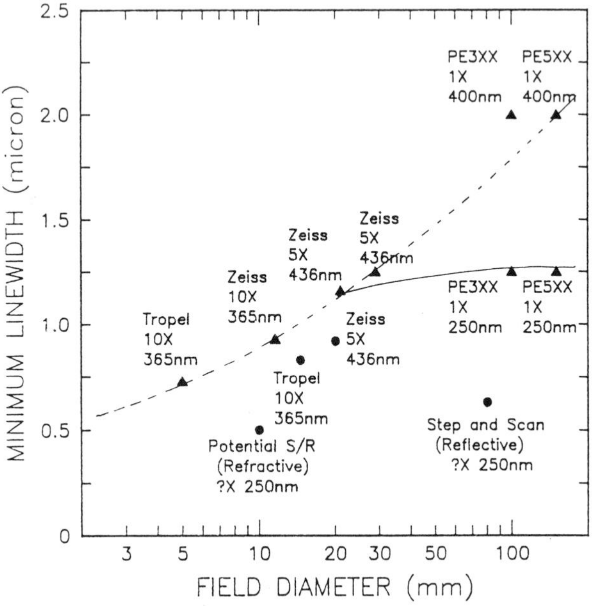

Figure 4. Minimum Line Width vs. Exposure Field Size for Optical Lithography Lenses

increase in the size of the primary mirrors, but these mirrors are already approaching the maximum size practicable from the point of view of fabrication. This has led workers to propose a new camera that retains the ring-field optical system of the full-field scanning cameras but allows the numerical aperture to be doubled.[4] This is accomplished, without the need for impractically large optical elements, by sacrificing the ability to expose a full-wafer in a single scan. The length of the curved slit is reduced from the full wafer diameter to a few centimeters. The wafer is exposed with a series of these narrower scans. The scanned area is about 80mm long and 20mm wide. The name "step-and-scan" has been adopted for obvious reasons. Alignment corrections will be made by continuously 'reading' reference marks in the strips between the scanned areas.

---

[4]Markle, D. A., (1984), *Solid State Technology*, September.

The optical system can be 1:1 or demagnifying and in either case the N.A. can be greater than 0.3. Operation at deep U.V. wavelengths will be readily achieved because it has already been demonstrated with the full-field version of the same type of optical system. At 250nm, the step-and-scan camera should also achieve a resolution of about 0.5μ. Predictions for throughput are similar to those for the full-wafer cameras.

The current major activity for improving the performance of step-and-repeat cameras is the installation of high numerical aperture (>0.35) lenses that operate at 365nm (mercury 'I' line). These lenses offer about a 20% improvement in resolution over those that operate at 436nm, but may introduce problems with image distortion, and lower exposure speed. Image distortion can arise because lens elements absorb more radiation and become heated. Heating may both distort the lens, and change the refractive index of the glass. Exposure time is lower because the output of the mercury arc source is lower at 365nm. These problems will almost certainly be overcome in the near future, however, and the new lenses offer linewidths down to about 0.7μ.

Ultimately, it should be possible to make lenses that operate at wavelengths down to about 200nm with quartz, or with other deep-UV transpar.ent materials if their stability can be improved. Laser illumination would provide such narrow bandwidth (<2nm) that chromatic correction could be ignored. At these wavelengths, a N.A. of 0.4 offers a resolution <0.5μ. Standing-wave interference effects would be exaggerated, but there are processing methods that reduce these effects.

In addition to refractive lenses, one can also build step-and-repeat mirror lens systems that operate at short wavelength (<300nm) and have similar numerical aperture to the refractive lenses. The Ultratech S/R camera[5] is built around such a lens.

All of the optical systems I have mentioned will be difficult to use because of their very small depth of field. We have been accustomed to working with full-field cameras of the Perkin Elmer type that have a numerical aperture of 0.16 and a depth of field of 6μ to 10μ. Step-and-repeat cameras have much smaller depth of field, but work over relatively small fields which alleviates the problem.

When 250nm exposure wavelength is used and linewidths reach 0.5μ, however, the depth of field will fall to about 1μ and it will be necessary to greatly improve resist planarity.

Vertical dimensions of integrated circuit devices will probably not scale with lateral dimensions, so small depth of field will become more and more of a problem even with improved resist processing. At some point between 1μ

[5]Hershel, R., (1982), *Proc. SPIE Conf.*, Santa Clara, CA.

and 0.5μ it will probably become intolerable, and will force optical lithography to give way to electron or x-ray lithography.

SCANNING ELECTRON BEAM LITHOGRAPHY

At first sight, scanning electron-beam lithography is the ideal method for writing integrated circuit patterns. No masks are required and resolution is beyond foreseeable needs (Figure 5). Unfortunately, it is slow and very expensive. Exposure cost is at least an order of magnitude higher than it is with optical exposure, so electron beams are restricted to environments where the high cost can be offset by the savings that result from eliminating masks. In research and development, reduction in device fabrication cycle times also justifies the high cost.

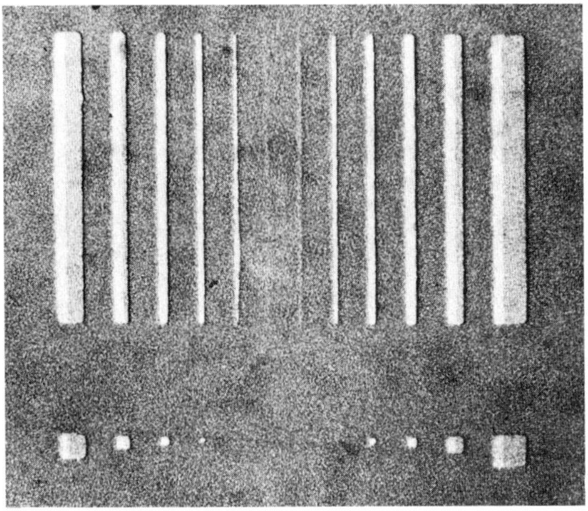

Adapted from Figure 4 by permission from *Solid State Technology*, Vol. 28, No. 6, June 1985, p. 119.

Figure 5. Exposure pattern of extremely small-size linewidths in the standard electron resist, PMMA. The minimum attempted linewidth is 4.4nm and the largest linewidth is 110nm.

Another problem with electron exposure is that of proximity effect – an electron beam writing in one position causes partial exposure in surrounding regions due to back-scattered electrons. However, several different techniques can be used to reduce the problem to an acceptable level.

There are two basic approaches to building faster scanning electron-beam systems. With the first approach, the pattern is written with a small round beam, the size of which corresponds to the required edge definition in the pattern. In order to maximize throughput, an extremely bright electron gun is used so that the maximum current is focused into the small spot. The small round beam exposes a single pixel at each flash. The second approach is to use a variably shaped electron beam that simultaneously exposes hundreds of pixels at each flash.

In addition to the difficulty in obtaining adequate current in the small round beam, the first approach places extreme requirements on the deflection electronics. Incrementing speeds will have to be more than an order of magnitude faster than for a system in which the pattern is written with an electron beam the shape of which is varied to most efficiently fill the areas requiring exposure. To date, the variably shaped beam systems offer highest throughput.

One of the key factors in favor of shaped-beam systems is that throughput does not depend on the address increment of the integrated circuit pattern. Address increment - the accuracy with which pattern features are placed - is frequently much smaller than the linewidth. With a small round beam, the beam diameter must be scaled down with the address increment. The number of beam addresses therefore increase and the throughput is reduced. With shaped beams, the spot-edge definition and the beam position are not related and the number of exposure positions does not increase if the address increment is reduced.

Figures 6 and 7 compare the throughput of vector-scan, EBES, and variable shaped-beam systems at present, and in 1990. The assumptions used in these calculations are discussed in the cited reference.[6] According to these estimates, the throughput in 1990 will be lower than it is today, despite the assumption that the system operating parameters will be upgraded. This is because the increase in wafer size, and the reduction in the pixel size, together produce an unmanageably large number of pixels per wafer, even for a shaped-beam system.

ION-BEAM LITHOGRAPHY

Several semiconductor applications for focused particle beams may be better tackled with ion beams than with electron beams, but at present,

---

[6]Broers, A.N., *Solid State Technology*, June 1985.

lithography is not one of these. The sources and lenses available for focusing and deflecting ions do not produce high enough beam currents to allow adequate throughput. The difficulty is not the lack of source brightness, but the lack of adequate total flux from the source.

The metal and field ion sources that produce high brightness have very small source size ($<0.1\mu$). The ion gun must accelerate the ions away from the tip in such a way that the apparent source size remains unchanged. This

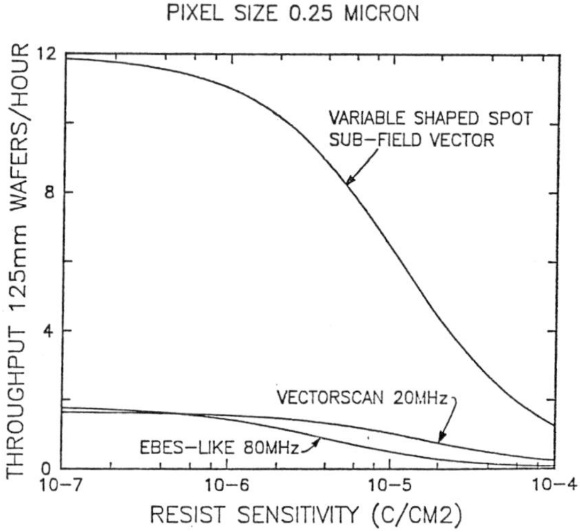

Adapted from Figure 6 by permission from *Solid State Technology*, Vol. 28, No. 6, June 1985, p. 119.

Figure 6. Throughput versus resist sensitivity for present scanning-electron-beam fabrication systems. System operating parameters are given in Table 2. V.S.S. system throughput becomes limited by electronic delay times for resist sensitivities greater than about $5\times 10^{-5}C/cm^2$. At lower resist sensitivity, the vectorscan system is faster than the EBES system because throughput is limited by the beam current/resist sensitivity limitation, and the vectorscan system only has to write 25 percent of the chip area. As the resist sensitivity approaches $10^{-7}C/cm^2$ the EBES system is faster because the beam current/resist sensitivity limitation no longer dominates and the EBES system does not have a chip-to-chip delay.

cannot be done for the several microamps of beam current needed for production throughputs, because the accelerating system must accept current from the source over a relatively large angle ($>>10^{-2}$ radian). In these cases the spherical aberration of the accelerating lens dominates, and the apparent source size is increased. This reduces the brightness of the emerging beam and means that the beam diameter increases rapidly as the beam current increases. This is

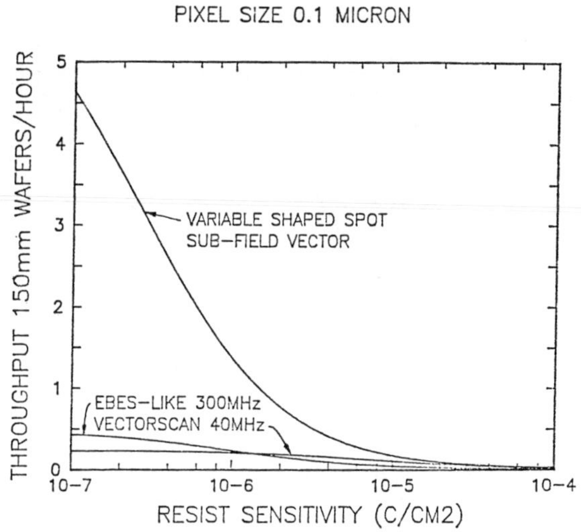

Adapted from Figure 7 by permission from *Solid State Technology*, Vol. 28, No. 6, June 1985, p. 119.

Figure 7. Throughput versus resist sensitivity for future (~1990) scanning-electron-beam fabrication systems. System operating parameters are given in Table 3. V.S.S. system again has the highest throughput, however, even with this system practical throughputs are only realized for resist sensitivities better than 10^{-6}C/cm^2. Electron-electron interactions are the dominant factor limiting current density in the electron beam. For resist exposure smaller than about 5×10^{-7}C/cm^2 (shaded area), the number of electrons required to expose each pixel falls to a level at which unacceptable noise is statistically predictable. (See Everhart, T.E., "Materials in Microlithography," ACS Symposium Series, Ed. Thomason, L.F. et al., American Chemical Soc., Washington, D.C. 1984.)

not the case for thermal electron sources, such as the lanthanum hexaboride source assumed in Figure 8, where the total current is small compared to the total current emitted from the source. In this case the brightness is effectively constant for a large range of beam elements.

Table I compares the throughput to be expected from future electron and ion-beam lithography systems. Despite the greater sensitivity allowed for ion-beam exposure (resulting from the more efficient deposition of the energy of the ions into the resist) the ion-beam system has much lower throughput than even the vector-scan electron-beam system.

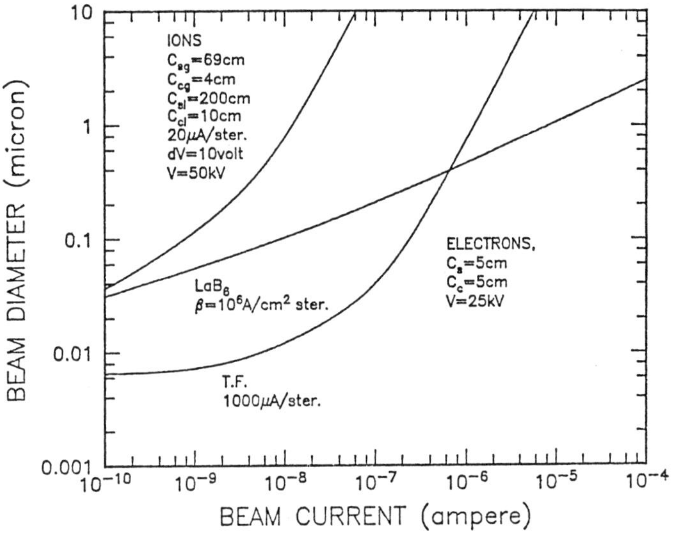

Adapted from Figure 8 by permission from *Solid State Technology*, Vol. 28, No. 6, June 1985, p. 119.

Figure 8. Spot size versus current for typical round beam electron and ion-beam systems. Practical lithography systems require beam currents greater than 1 microamp. Thermal field emitter electron-source parameters: Source current density 1000 microamp/cm², Source diam.=20nm, C_s gun=69cm, C_c gun=4cm, Magnif. gun=4.3, Volt.Ration=5, V=25kV, V=1.5volt, C_s lens=5cm, C_c lens=5cm. Ion-beam parameters: Source-current density=20microamp/cm², Source diam.=30nm, C_s gun=69cm, C_c gun=4cm, Mag. gun=4.3, Volt.Ration=5, V=50kV, V=10volt, C_s lens=200cm, C_c lens=20cm.

~1990

	V.S.S Continuous Table	Vect. 40MHz	IONS Vector 40MHz
PIXEL SIZE	0.1μ	0.1μ	0.1μ
SPOT CURRENT DENSITY	$10A/cm^2$	$300A/cm^2$	$10A/cm^2$
RESIST SENSITIVITY	$10^{-6}C/cm^2$	$10^{-6}C/cm^2$	$3\times10^{-7}C/cm^2$
PIXELS PER SHAPE	30	1	1
PIXELS PER SUB-FIELD	$10^3\times10^3$	–	–
CHIP SIZE	10mm	10mm	10mm
WAFER SIZE	200mm	200mm	200mm
FRACTION EXPOSED	25%	25%	25%
SUB-FIELD TO SUB-FIELD TIME (μsec)	10μs	0	0
SHAPE TO SHAPE TIME (nsec)	25ns	25ns	25ns
CHIP TO CHIP TIME (sec)	0	0.25s	0.25s
THROUGHPUT (200mm WAFERS/HOUR)	1.35	0.2	0.1

Table I. Electron and Ion Beam Throughput

X-RAY LITHOGRAPHY

Of the new methods that offer throughput equivalent to optical lithography, X-ray lithography appears to have the highest promise. The major advantage of x-ray exposure is that the exposing radiation is neither scattered in the resist nor is it diffracted. This results in higher aspect ratio resist patterns than can be produced with optics, and no proximity effect as there is with electrons. Throughput is adequate provided an electron storage ring, or perhaps a pulsed laser plasma source is used to produce the x-rays.

The major difficulties are with the mask which is fragile and is not dimensionally stable. Non-flatness of the mask, or the wafer, also produce overlay errors when exposure is made with the large divergence of conventional electron bombardment x-ray sources. In this case, in order to keep pattern distortion below 0.1μ for a 125mm diameter wafer, and a source-to-wafer distance of 50cm, the mask and wafer must be held flat to better than 1μ. Alternatively, the variations over the gap must be held constant for every exposure level. Wafer buckling that arises during many standard integrated circuit processes makes this very difficult to achieve and this problem combined with mask distortion make it unlikely that full wafer exposure can be accomplished for wafers larger than about 7.5cm. Maximum mask size of 5cm is preferred.

The run-out error could be avoided if resist sensitivity was improved to the point that exposure could be made at a point further removed from the source, but this requires a sensitivity approaching $1mJ/cm^2$. Some resists have this sensitivity but they do not produce high aspect ratios and therefore lack the major advantage of x-ray lithography over optical lithography. High aspect ratio can be obtained by using a tri-layer resist process, but this option is also open to optical lithography.

The run-out error could also be avoided by going to a S/R approach and limiting mask size to a few centimeters. This also alleviates problems due to mask and/or wafer distortions, but exposure times are too long for economical processing with the resists and conventional sources that are available at present . Exposure times of a few seconds would be needed for a source-to-wafer distance of about 50cm which is a similar sensitivity requirement to that for the full-wafer case.

The electron storage ring produces X-ray beams that do not suffer from these deficiencies.[7] The x-ray beams are so intense, and have such small divergence, that the problems of run-out error and long exposure time are removed. Enough x-rays are produced at each of the bending magnets, (slightly more are produced by 'wigglers' or 'undulators'), to expose the same number of

[7]Heuberger, A., *Solid State Technology,* February 1986.

wafers/hour as an optical S/R camera. There are up to 16 such sources in each ring, so provided the cost of the ring remains below about $16M, it is competitive with optical lithography and provides higher resolution and better aspect ratios. Several laboratories are working on designs for 'compact' storage rings that will be even more practicable for production environments.

The resolution of x-ray lithography is more than adequate for integrated-circuit applications, but is ultimately not as good as electron-beam lithography. A combination of diffraction between mask and sample, and the range of the photoelectrons produced by the x-rays, make the method less practicable than electron-beam lithography for dimensions below about 20nm.

NANOLITHOGRAPHY

Nanolithography is the name given to the technology for fabricating devices with dimensions $<0.1\mu$. So far, all of these devices have been made with electron beams, and it has been found that the minimum dimensions that can be produced depend on the interaction range of the electrons in the image-forming layer, and not on the size of the electron beam.

To explore this resolution limit, a series of experiments has been carried out with PMMA,[8] the highest resolution electron resist. A STEM micrograph of a test exposure used in these experiments is shown in Figure 5. The pattern contains lines that range from those much narrower than the expected minimum linewidth, to those considerably wider than the minimum linewidth. Many exposures were made at different doses using an electron beam with a diameter <1nm. From these data the fractional exposure received at the center of each of the lines was calculated and the minimum usable linewidth predicted. This linewidth turns out to be about 10nm, and the contrast for an array of lines and spaces is as shown in Figure 9.

In order to overcome the 10nm limit encountered with conventional electron resists such as PMMA, it is necessary to use a recording medium that is sensitive to the impact of high-energy primary electrons, rather than low-energy secondaries. Low-energy secondaries are produced remotely from the primary electrons and give rise to the delocalized exposure.

Several methods have been discovered which offer this possibility, I will mention just one here. This is the direct sublimation of ionic crystals such as NaCl, MgF_2, and LiF. The high resolution potential of this technique was discovered in the experiment that produced the 5nm-diameter holes shown in Figure 10. These holes were formed in a NaCl crystal by a 1nm-diameter

---

[8]Broers, A.N., *J. Electrochem. Soc.*, 128:166, 1981.

50kV electron beam.[9] The crystal was estimated to be about 0.25μ thick. The convergence half-angle of the beam was 10^{-2} rad., and the beam, therefore, formed a conical shaped hole with the base of the cone being about 5nm in diameter, assuming the beam is focused on one face of the crystal. This explains the discrepancy between the apparent size of the hole and the beam diameter, and suggested that the resolution of the process was much better than 5nm. Isaacson and his co-workers working in the NRRFSS laboratory at Cornell confirmed that higher resolution was possible by writing structures down to 1.5nm in thin NaCl and AlF_3 films.[10]

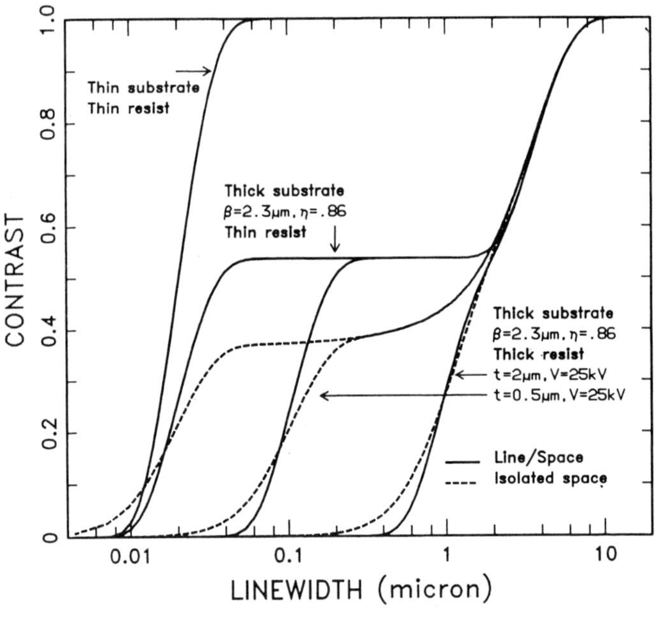

Figure 9. Contrast versus linewidth for electron exposure of PMMA. Details of calculations used to obtain these curves are given in reference 8.

[9]A.N. Broers, J. Cuomo, J. Harper, W. Molzen, R. Labowitz, M. Pomerantz, *Electron Microscop.* 1978, Vol. 111, Ed. J.M. Sturgess, Microscopical Soc. of Canada, Toronto, p. 343-354, 1978.

[10]Isaacson, M. and Murray, A., *J. Vac. Sci. Technol.*, 19:1117, 1981.

SUMMARY

There are two extreme categories of microcircuit components: high-volume memory and microprocessor chips where thousands to millions of chips are required, and custom logic components where only a few hundred chips are needed. For obvious reasons, the two categories require different lithography systems. Figure 11 depicts the continuum of applications between the two extremes, and shows the author's predictions as to where the various lithography approaches fit. The boxes around each of the lithography types (S/R optics, scanning electron beams, etc.) show how the area of application of each of the methods will change with time.

Nanolithography experiments show that it is possible to fabricate structures ten times smaller than solid-state device physicists predict will be useful for conventional semiconductor integrated circuits. This shows that no fabrication difficulties (apart perhaps for cost) will be encountered in reaching these limits. The excitement of nanolithography is that it opens up the possibility for new devices.

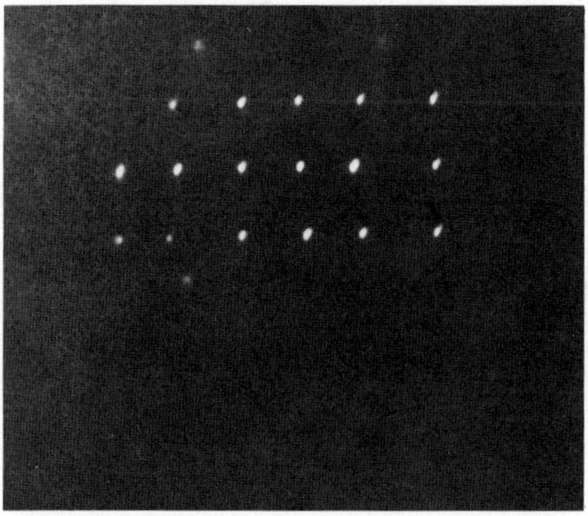

Figure 10. 5nm diameter holes "drilled" through a 250nm thick NaCl crystal with a 0.5nm diameter electron beam[9]. The holes are spaced by approximately 50nm.

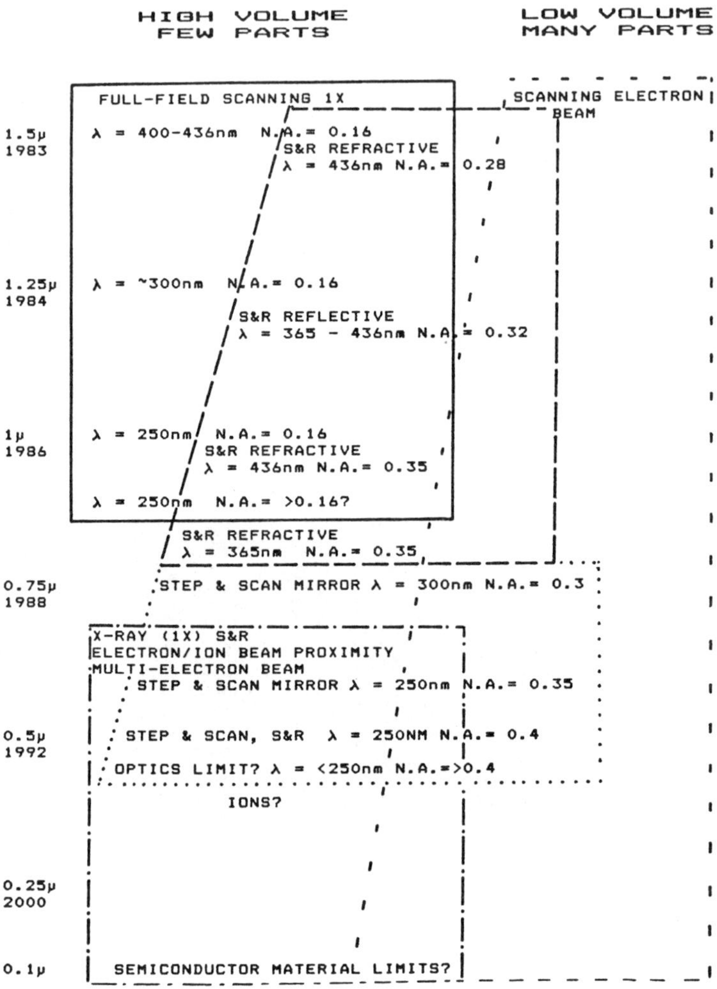

Adapted from Figure 9 by permission from *Solid State Technology*, Vol. 28, No. 6, June 1985, p. 119.

Figure 11. Lithography Flow Chart

COMPOUND SEMICONDUCTOR HIGH-SPEED ELECTRON DEVICES

Lester F. Eastman

Lester F. Eastman obtained the B.S., M.S., and Ph.D. degrees in electrical engineering at Cornell University in 1953, 1955, and 1957, respectively. He has been a member of the Cornell Electrical Engineering faculty since 1957, was appointed the John Laporte Given Professor of Engineering in 1985, and is the Director of the Joint Services Electronics Program. He has been an IEEE Fellow since 1969 and was the IEEE Electron Devices Society National Lecturer in 1983. He is a member of the U.S. Government Advisory Group on Electron Devices, and serves as a consultant on compound semiconductor materials, high-speed switching devices, and circuits with Lincoln Laboratory at MIT and with several industries.

The research reported in this presentation was performed at the National Research and Resource Facility for Submicron Structures in the School of Electrical Engineering at Cornell University. The work is supported by JSEP monitored by AFOSR, DARPA, ONR, ARO, NASA, NRAO, IBM, GE, United Technologies, Ford Microelectronics, McDonnell-Douglas, and other industries.

Fundamental research in compound semiconductors has been conducted for decades with an early objective of finding application in the field of optoelectronics. In recent years, many significant advances have occurred, notably in the realization of the so-called ballistic transistor where average electron velocities have been achieved that allow switching times of the order of 1.5 picoseconds. Development of integrated-circuit elements based on this research will have a profound impact upon many areas of Electrical Engineering, particularly in the computer and communications fields. This presentation summarizes the physical electronics and technology of the short compound-semiconductor transistors that could be used in the future design of circuits for high-speed high-frequency operation.

Compound Semiconductors like Gallium Arsenide have long mean-free-paths ($\approx 0.10\ \mu m$) for electrons. At higher electron energy the mean-free-path lengthens and the scattering angle becomes smaller thereby allowing very high average electron velocity for drift lengths of $0.40\ \mu m$ or longer. Even much longer devices have optimum saturation electron velocities of from $1.2 \times 10^7 cm/s$ to $2.1 \times 10^7 cm/s$ at room temperature, depending on their design. The electronic properties of selected compound semiconductors is given in Table I, including the larger-band-gap materials for electron confinement. Any percentage of aluminum may be substituted for gallium without any serious lattice mismatch.

Compound semiconductor materials must be grown under carefully controlled conditions. Over the past decade the technique known as molecular beam epitaxy (MBE) has been used widely for growing such experimental heterostructures. Figure 1 is a cross-section of such a machine. It is a large vacuum system with heated effusion cells as sources of gallium, arsenic, aluminum and dopant chemicals. The heating substrate is rotated, for uniformity, and about one atomic layer per second is deposited. In-situ measurements of the surface are carried out using attached instruments.

Organometallic vapor phase epitaxy (OMVPE) is another means of growing heterojunction structures. OMVPE has a faster growth rate than MBE, and can be scaled up to produce as many as 8 to 25 wafers. Figure 2 shows a horizontal-flow machine capable of growing many 2" x 4" areas at one time. Vertical-flow machines with a rotating, tapered barrel can be used for uniform, multiwafer growth. This method transports the gallium, aluminum, and arsenic as gases. These gases are dissociated and the desired elements are deposited on the heated substrates on the susceptor, but not on the cold walls.

The forbidden bandgap of GaAs rises as aluminum is substituted for the gallium. Figure 3, from Casey and Panish,[1] shows the key condition-band minima as a function of X, referenced with the valence band. If an abrupt $Al_x Ga_{1-x} As/GaAs$ heterojunction is grown, 3/8 of the increase in the (Γ valley-valence) band gap goes into the valence band potential step and 5/8

Table I

COMPOUND SEMICONDUCTOR ELECTRONIC PROPERTIES

	GaAs	$In_{.53}Ga_{.47}As$	InP
Band Gap (eV)	1.43	0.75	1.33
Mobility (300°K, pure) (cm^2/V-s)	10,000	15,000	6,000
Mobility (300°K, $10^{17}/cm^3$) (cm^2/V-s)	5,000	7,500	3,000
Dielectric Constant	13	13.5	12.4
Effective Mass	0.07	0.045	0.08
Gunn Threshold (V/cm)	3,500	2,800	10,500
$(E_L - E_\Gamma)$ (eV)	0.31	0.55	0.60
Mean Free Path (μm)	0.10	0.125	0.040
Electron Confinement Material	$Al_x GA_{1-x} As$	$In_{0.52}Al_{0.48}As$ or InP	$In_{0.52}Al_{0.48}As$

goes into the Γ valley potential step. Above $X \approx 0.45$ there is a lower energy valley that gives an indirect bandgap.

Figure 4, adapted from Chelikowsky and Cohen,[2] shows the electron energy versus electron wave number for GaAs. At the Γ_6 valley, where electrons reside in n-type material, electrons accelerated to the right are in the [100] cube-edge direction. At left they go in the [111] cube-diagonal direction. The slope of the curve is proportional to the electron group velocity, which maximizes at a value of $9.5 \times 10^7 cm/s$ in the [100] direction. It is 20% lower in other directions being, for example, 45° away in the [110] direction of the crystal. When the electrons reach 0.36 eV or more above the Γ_6 valley, they can transfer to the broad minimum at L_6, where they are at higher mass and very low group velocity.

The frequency response of FETs depends on the time-average electron velocity across the transit region. Specifically, the unity-current-gain frequency, F_T, is proportional to G_M/C which is proportional to V_E. G_M is the transistor mutual transconductance, C is the transistor input capacitance, and V_E is the time-average electron velocity. Table II shows a list of average electron-velocity values obtained, or expected, in short GaAs FETs. Usual MESFETs have a velocity of $1.2 \times 10^7 cm/s$, while the modulation-doped

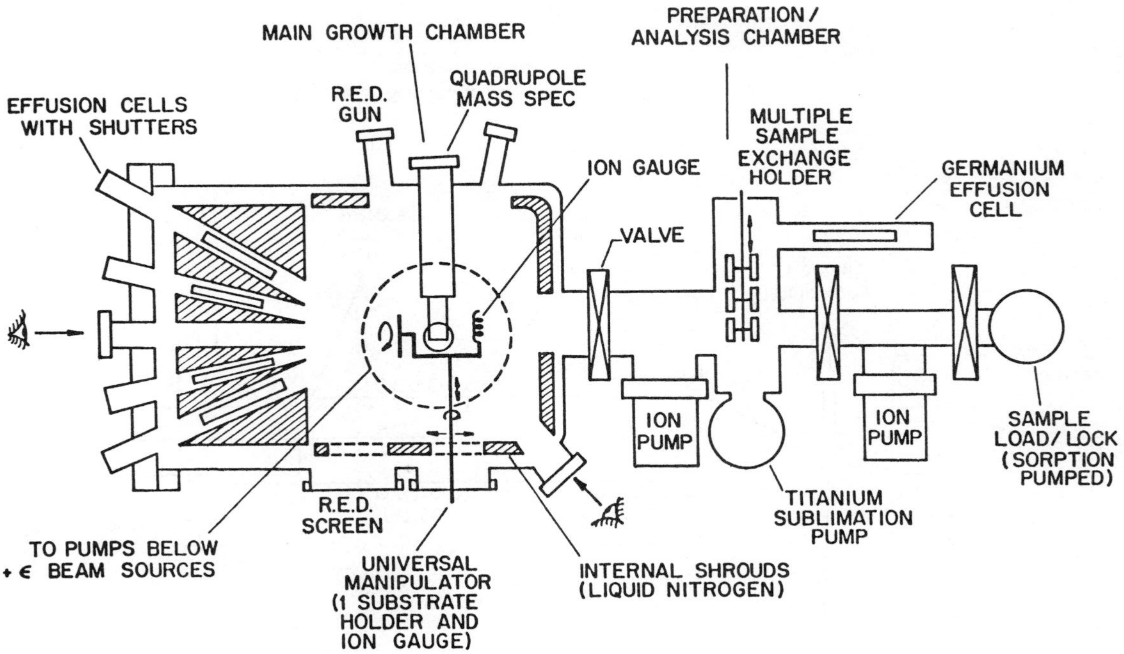

Figure 1. Cross Section of a Molecular Beam Epitaxy (MBE) Machine

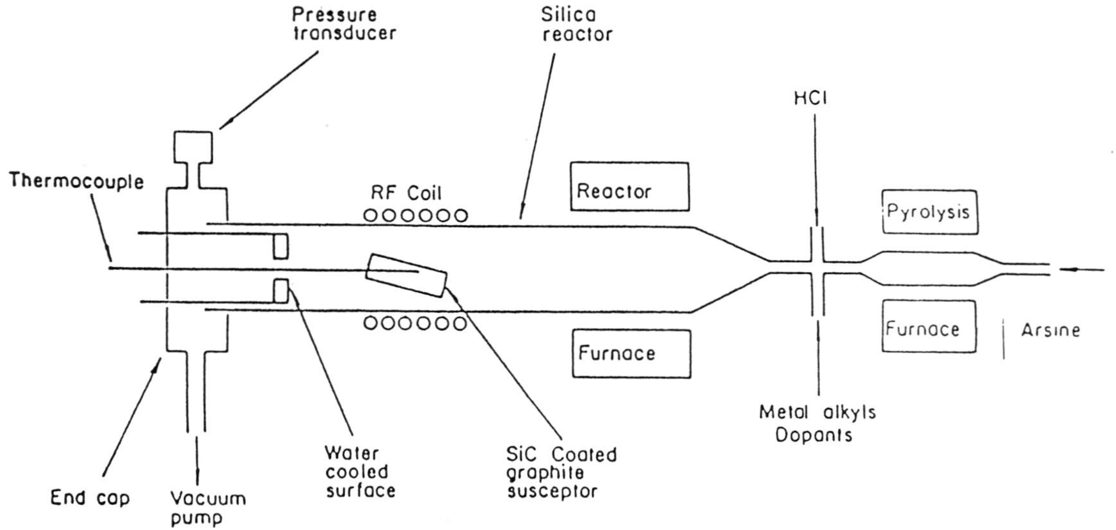

Figure 2. Schematic Diagram of the OMVPE Horizontal Reactor Cell

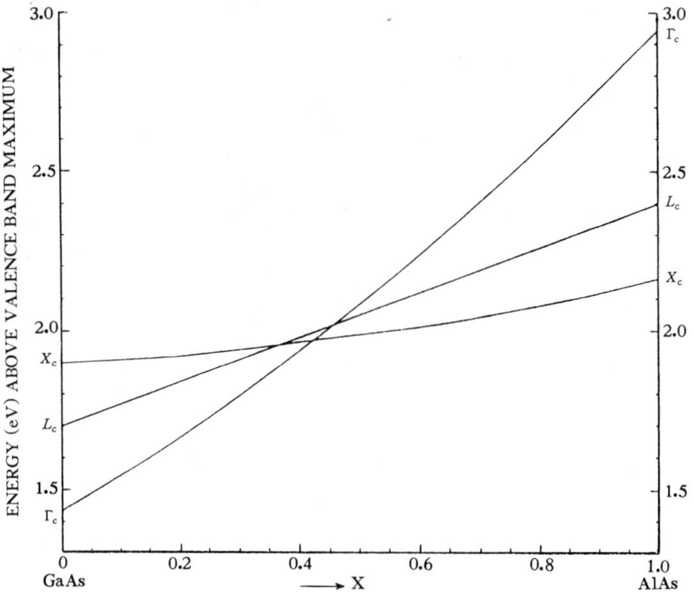

Adapted from Figure 5.3-2 by permission from Academic Press, Inc.

Figure 3. Key Conduction-Band Minima as a Function of X

Al,GaAs/GaAs heterojunction MODFETs have velocities up to $2.1 x 10^7 cm/s$. For short transit lengths < 0.4 μm) and gradual electron acceleration, an average electron velocity of approximately $4 x \cdot 10^7 cm/s$ is expected. For impulse acceleration, (ballistic injection) an average velocity of approximately $8 x 10^7 cm/s$ is expected over short transit lengths (< 0.6 μm). With no collisions, electrons with energies of approximately 0.3 eV traveling in the [100] crystal direction, have a maximum crystal-limited group velocity of $9.5 x 10^7 cm/s$.

Some experimental values of time-average electron velocity versus electrical length are shown in Figure 5 under gradual-acceleration conditions. For comparison, half the theoretical velocity ($4.75 x 10^7 cm/s$) is shown at zero electrical length. A short-pulse optical measurement at Bell Laboratories gave the $4.4 x 10^7 cm/s$ value at 0.22 μm in the [100] direction. The value with $2 x 10^{15}/cm^3$ doping was for a diode with electrons going in the [100] direction.

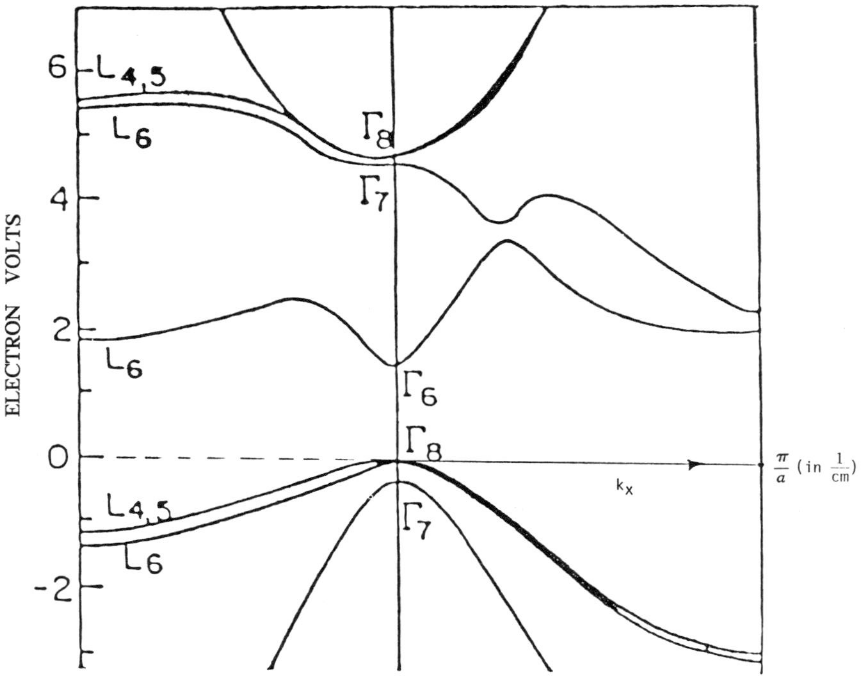

Adapted from Figure 7 (for GaAs) by permission from *Physical Review*, Vol. B-14 No. 2, 15 June 1976, p. 556.

Figure 4. Electron Energy as a Function of Electron Wave Number (k_x) for GaAs

The values with $7\,x\ 10^{15}/cm^3$, $5\,x\ 10^{16}/cm^3$, and with ballistic injection attempted, were for vertical transistors with electrons going in the [100] direction. The other devices, with ion implantation at $9\,x\ 10^{16}/cm^3$, and epitaxy at $1.5\,x\ 10^{15}/cm^3$ and $4\,x\ 10^{17}/cm^3$ were for MESFETs with electrons going in the [110] direction. For the [100] direction in vertical FETs, an average electron velocity of approximately $8\,x\ 10^7 cm/s$ is expected for lengths $< 0.6\ \mu m$.

Modulation-doped Al,GaAs/GaAs FETs have condition-band potential profiles as shown in Figure 6 for normally-off devices. In (a), the MODFET version, the doping is in the $Al_x Ga_{1-x} As$ barrier. X must be $\leqslant 0.235$ in order to have shallow donors. In (b), the MAGFET version, the barrier is undoped, but N^+GaAs on top of it is used to fix its Fermi energy near the conduction band. This latter device is like an M-O-S device, except that here it is metal-Al,GaAs-GaAs instead of Metal-Oxide-Silicon. The degenerately doped GaAs acts like a metal, and has an ohmic contact to deposited metal without alloying. Threshold voltages are easier to reproduce precisely in the MAGFET. Its barrier cannot go above 0.36 eV (obtained for $X = 0.50$) where all conduction-band minima are considered.

If one wishes to limit the space-charge-limited current flowing along fringing fields from source to drain in the material under the two-dimensional electron gas (2DEG), then an $Al_{0.5}Ga_{0.5}As$ buffer layer is required. This technique yields a quantum well, as in Figure 7, which should be as large (~ 160 angstroms) or larger than the thickness of the 2DEG. This quantum-well version helps limit the output conductance of the FET.

Table II

FET FREQUENCY RESPONSE
($f_T\ \alpha\ g_m/C\ \alpha$ Electron Velocity)

ELECTRON VELOCITIES FOR GaAs - 300°K

$1.2\,x\ 10^7 cm/s$ $-N_D \sim 1\,x\ 10^{17}/cm^3$	$L_E \geqslant 0.8\ \mu m$
$1.6\,x\ 10^7 cm/s$ - Pure	$L_E \geqslant 1.0\ \mu m$
$2.1\,x\ 10^7 cm/s$ - MODFET	$L_E \geqslant 1.0\ \mu m$
GRADUAL ACCELERATION - NEAR BALLISTIC	
$\sim 4\,x\ 10^7 cm/s$ - Pure/MODFET	$L_E \geqslant 0.4 \mu m$
IMPULSE ACCELERATION - NEAR BALLISTIC	
$\sim 8\,x\ 10^7 cm/s$ - Pure	$L_E \leqslant 0.6\ \mu m$
CRYSTAL-LIMITED GROUP VELOCITY	
$9.5\,x\ 10^7 cm/s$ [100] direction - 0.3 eV	

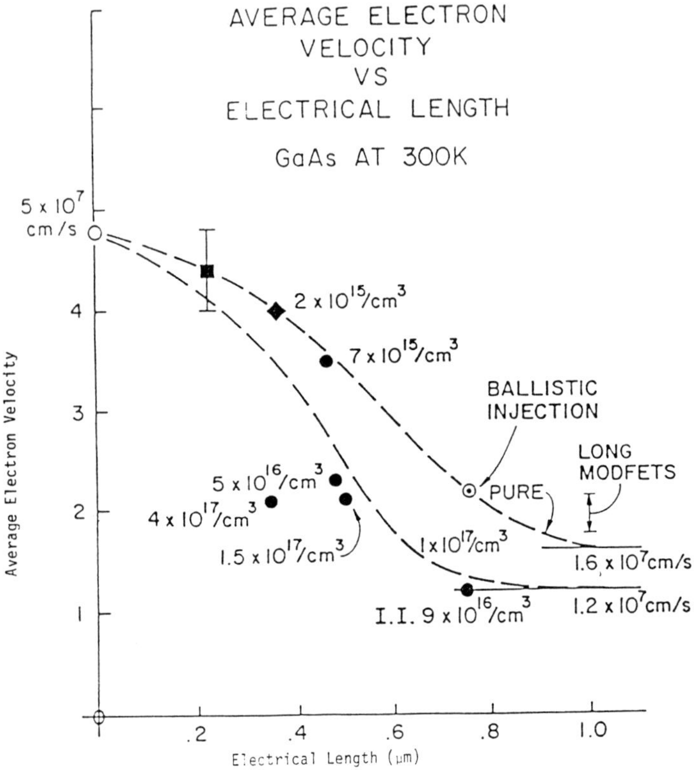

Figure 5. Experimental Values of Time-Average Electron Velocity as a Function of Electrical Length

Both horizontal and vertical short FETs can have submicron active lengths as displayed in Figure 8. For a constant conduction area, space-charge-limited current rises as the square of the reciprocal of this length. The current rises nearly linearly with voltage for a constant length, but normally in FETs, this length rises a little with voltage. In order to analyze the transistor I-V characteristics, one would like a situation with a constant voltage across the submicron active length. This restriction limits space-charge injection, modulation of the length, and limits electron energy.

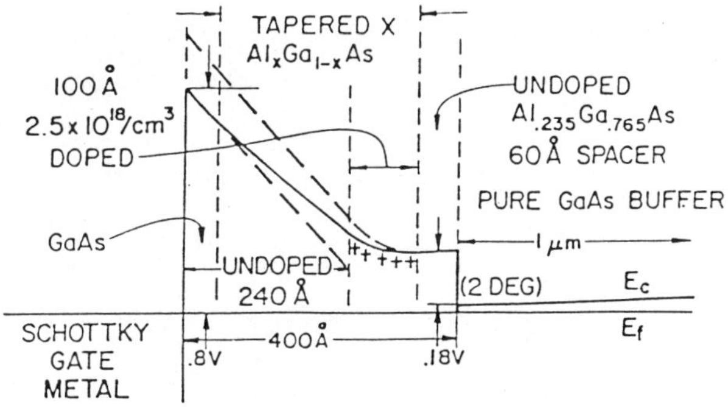

(a) GaAs MODFET

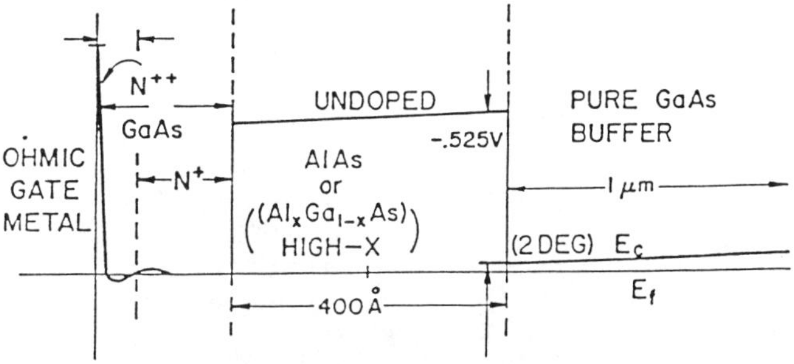

(b) GaAs MAGFET

Figure 6. Conduction-Band Potential Profiles

By subtracting out the voltage drop across the linear resistances (R_s + R_d), a constant voltage, V_A, across the active length can be established. Figure 9 shows the I-V characteristics of a quantum-well MODFET with a graphical approach to obtaining constant V_A. The horizontal distance from the vertical axis to the first sloping line is the voltage drop across the linear resistances (R_s + R_d), and the horizontal distance from the first sloping line to the second sloping line is the 0.5 voltage across the active length. The data from the I-V intercepts are taken, and the current is plotted vs the total gate bias, including the I_{ds} x R_s component as well as the applied V_{gs} voltage. The maximum

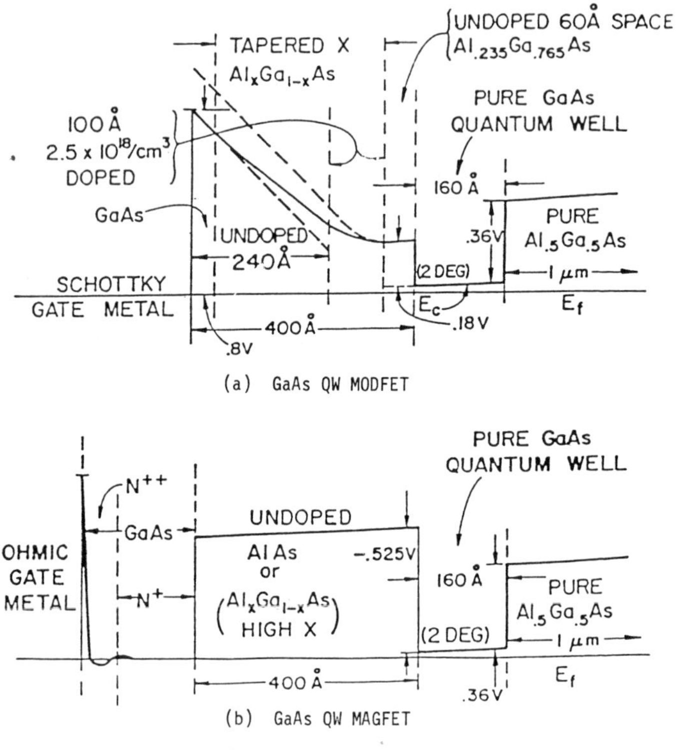

Figure 7. Quantum-Well Result of the
Application of a Buffer Layer

slope of this latter plot yields an intrinsic G_M which equals $(\epsilon W V_E)/\ L_{G-2DEG}$, where ϵ is the permittivity of the Al,GaAs, W is the device periphery, V_E is the time-average electron-transit velocity across the active length, and L_{G-2DEG} is the distance between the gate metal and the center of the 2DEG. The center of the 2DEG is normally 80Å deeper than the key top heterojunction, or half the quantum-well thickness, whichever is least.

Table III lists values for saturation electron velocity for various MODFET structures at 300°K. The top-line structure is a MAGFET made by Ogura of the Japanese Electrotechnical Laboratory. The double H.J. with the top doped is the quantum-well structure of Figure 9. The bottom structure has a 50-angstrom AlAs spacer layer, whose lowest barrier is less than 0.20 eV and is the heavy mass [100] minima at high electron wave number. For comparison, other results are given at the bottom of the table. The proposed explanation for the lower saturation-velocity values for the bottom three lines of MODFET structures is that the $\sim 0.2 eV$ potential barrier does not confine all the electrons at high energy, so some spill into the Al,GaAs where they move much slower, yielding a low time (and space) average electron velocity.

In order to have a reproducible very short structure, it is possible to take a self-aligned approach as shown in Figure 10. Structures like this have been made in MESFETs, in a cooperational Hughes/Cornell program that yielded a 15 psec switching time in January 1983. The overhang of the metal is used as a mask to bring the ion-implanted contacts close to the gate, but not close enough to raise the parasitic capacitances.

If a potential step is used to inject electrons with 0.18 - 0.30 eV energy, the electrons are ballistically launched with velocities of from 8 to $9.5 \times 10^7 cm/s$. With a modest 2 to 3kV/cm electric field, this velocity can be

Table III

SATURATION ELECTRON VELOCITY AT 300°K
IN VARIOUS MODFET STRUCTURES

STRUCTURE	BARRIER	GATE LENGTH	SATURATION VELOCITY
Single H.J.	$x = 0.40$	$2 \mu m$	$2.1 \times 10^7 cm/s$
Single H.J.	$x = 0.23-0.30$	$0.25-1.5 \mu m$	$1.7-2.1 \times 10^7 cm/s$
Double H.J. (Top Doped)	$x = 0.30$	$0.35 \mu m$	$1.4 \times 10^7 cm/s$
Double H.J. (Top/Bottom Doped)	$x = 0.30$	$1-1.7 \mu m$	$0.87-0.90 \times 10^7 cm/s$
Single H.J. (x stepped)	$x = 1.0$ & 0.23	$0.25 \mu m$	$0.79 \times 10^7 cm/s$

Comparisons: 77°K MODFET: $2.6-3.1 \times 10^7 cm/s$ except for double H.J. (Top/Bottom Doped) which yields $1.36-2.53 \times 10^7 cm/s$.

300°K MESFET: $1.2-1.3 \times 10^7 cm/s$ (if $L > 0.5 \mu m$)

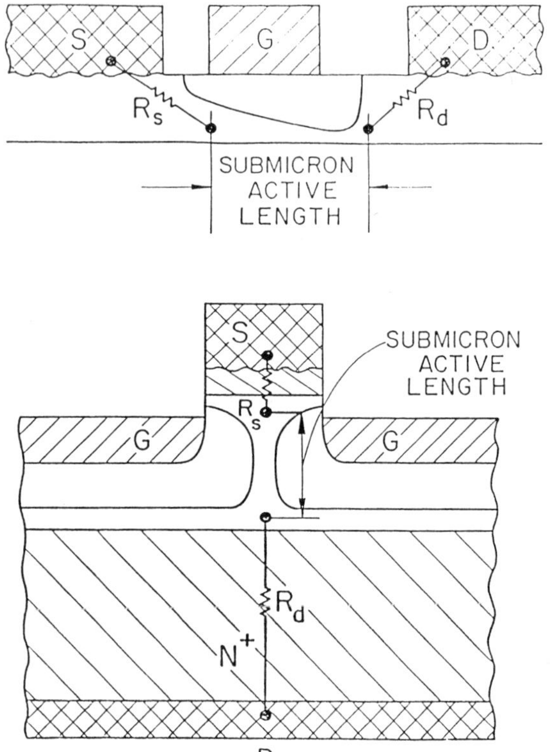

Figure 8. Submicron Active Lengths in Short FETs

nearly maintained across a 0.5 - 0.6 μm active length. Figure 11 shows a ballistic-electron, Schottky-gated transistor structure and its conduction-band profile, based on these physical principles, that was first conceived at Cornell.[3] Figure 12, after Awano *et al.* at the Japanese Electrotechnical Laboratory, shows the Monte Carlo simulation of electrons in a 0.25 μm-long active device of this type.* The average electron velocity versus bias voltage is shown for

*Y. Awano, private communication, January, 1984.

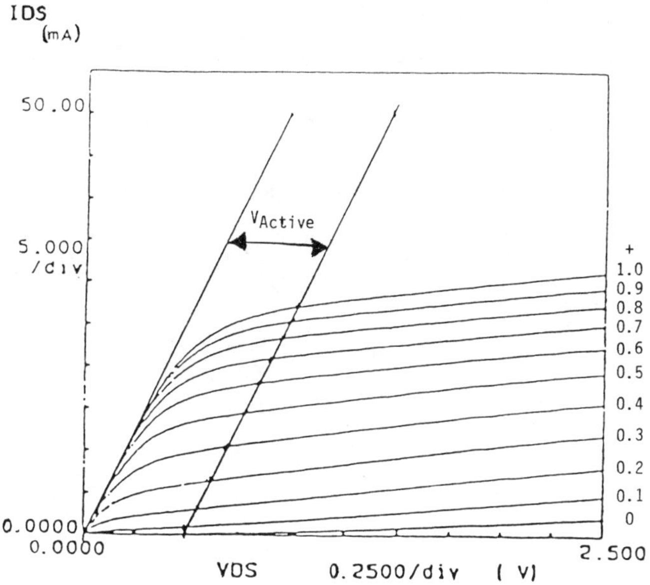

Figure 9. I-V Characteristics of a Quantum-Well MODFET

the case of no doping and for $1 x 10^{16}/cm^3$ doping. The potential step, in approximate volts, is shown as the key parameter. A value near $7.5 x 10^7 cm/s$ is reached with $\sim 0.2V$ launch energy and 0.5 V applied across the active length.

Heterojunction bipolar transistors can also use ballistic-electron transport, although the heavily doped P-type base region causes the ballistic electron momentum to drop rapidly. Also at Cornell,[4] it was found that a built-in electric field can be caused by tapering the aluminum content from emitter to collector across the base, as shown in Figure 13. More or less initial potential step and less or more aluminum content at the emitter side of the base can be experimentally tested. Tables IV and V give the state of the art in GaAs FET microwave and logic devices.

The physical-electronics technology of compound semiconductor, high-speed, field-effect transistors has now reached a level that holds forth great

Table IV

GaAs MICROWAVE TRANSISTOR
STATE OF THE ART

MESFET (300°K)		
0.95 dB NF	12 GHz	0.25 μm gate
6 dB gain	60 GHz	0.25 μm gate
$f_T \leqslant 40$ GHz;	$f_{\max} \leqslant 130$ GHz	0.25 μm gate
MODFET (300°K)		
0.95 dB NF	12 GHz	0.5 μm gate
1.2 dB NF	18 GHz	0.25 μm gate
$f_T \leqslant 70$ GHz;	$f_{\max} \leqslant 120$ GHz	0.25 μm gate
$g_m \leqslant 450$ mS/mm	0.25-0.35 μm gate	
MODFET (12.5°K)		
8.5°K NT	8 GHz	0.25 μm gate
3.5°K NT	3 GHz	0.25 μm gate
H.J. Bipolar (300°K)		
$f_T = 40$ GHz;	$f_{\max} = 25$ GHz	

promise for the future of many aspects of Electrical Engineering. In this review of the state of the art, the major components are:

(1) present-day transistors that have time-average velocities across their active lengths that range from 1.2 to $2.1 \, x \, 10^7 cm/s$ at room temperature for long devices,
(2) projected short devices with electron velocities that are expected to reach $4 \, x \, 10^7 cm/s$ for gradual acceleration across the drift region and $8 \, x \, 10^7 cm/s$ for ballistic injection into the drift space.

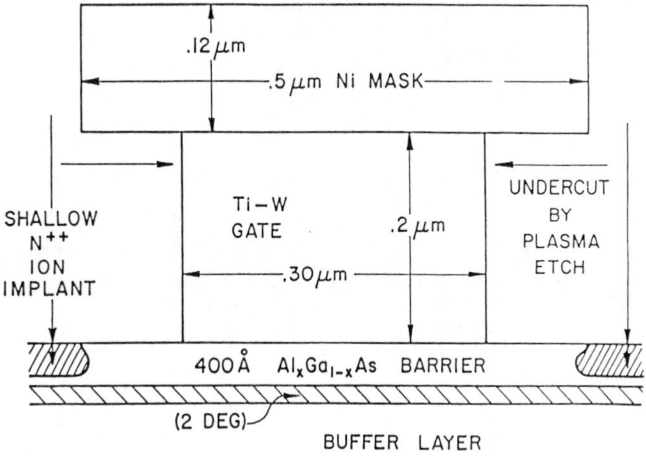

Figure 10. MODFET Self-Aligned Contacts

Table V

GaAs LOGIC TRANSISTOR
STATE OF THE ART

MESFET (300°K)	L_{SD}	L_{GATE} x W_{GATE}
15 ps (or 9fJ)	0.75 μm	0.55 μm x 10 μm
16.7 ps (04 2.4 fJ)	0.5 μm	0.30 μm x 10 μm
13.7 ps		0.5 μm x 30 μm
MODFET (300°K)		
~ 11 ps	~ 0.4 μm	
MODFET (77°K)		
~ 9.5 ps	~ 0.4 μm	
BIPOLAR (300°K)		
25 ps		

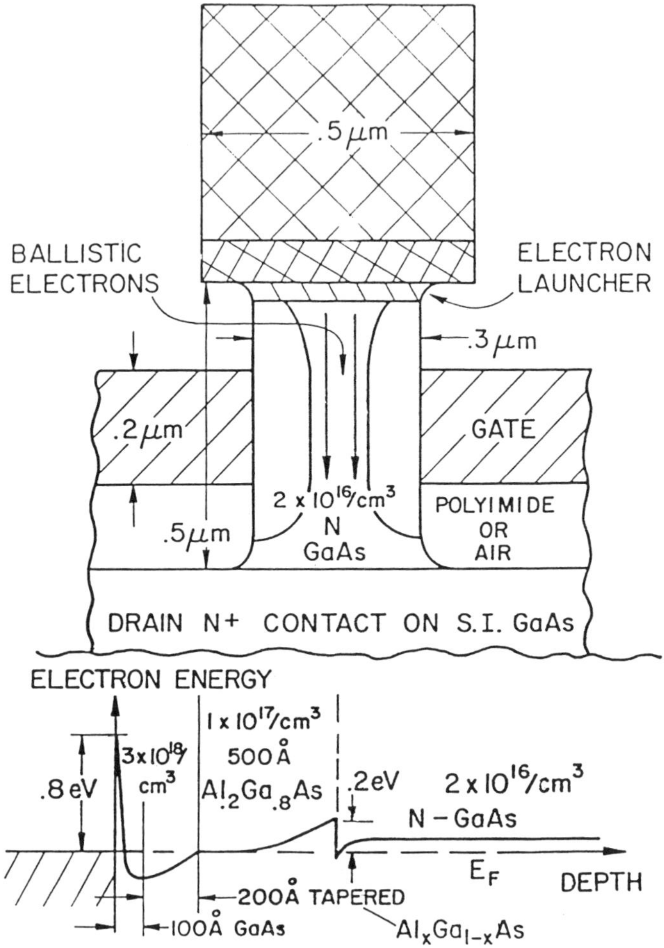

(©1983 IEEE)

Figure 11. Ballistic Electron Schottky-Gated Transistor

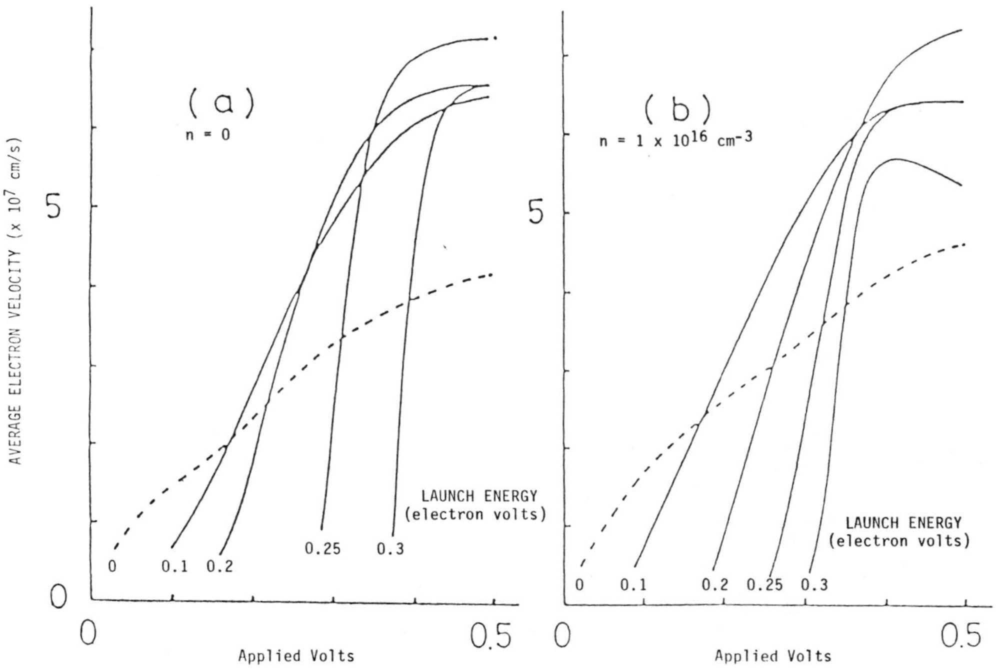

Figure 12. Monte Carlo Simulation of Electrons
in a 0.25 μm-long Active Device

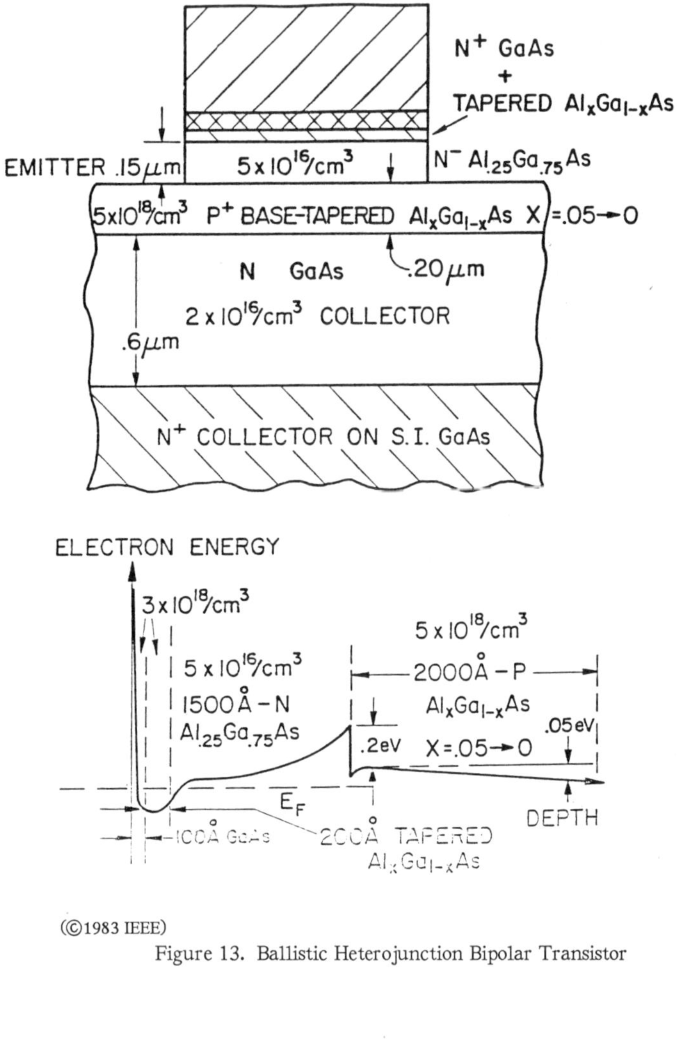

(©1983 IEEE)

Figure 13. Ballistic Heterojunction Bipolar Transistor

(3) expected operating frequencies (for devices with such very high electron velocities) that will be greater than 100 GHz, with switching times that will be well below 10 ps.* The development of such ultra-fast transistors will prompt fundamental revisions in computer architecture and integrated-circuit design, and can be expected to expand communications facilities well beyond present-day capabilities.

*Update:

Following the presentation of this material at the March 15, 1985 Symposium, significant advances have been reported in recent months. At 300°K, MODFETs yielded noise-figure values of 0.85 dB at 18 GHz at Toshiba, and over 80 GHz F_T at 300°K at GE-Syracuse. In a joint Cornell/GE-Syracuse research program, GE experiments on MODFETs at a physical temperature of 12.5°K yielded an 8.5°K noise temperature. At Cornell a high G_m of 450 mS/mm was obtained at room temperature and 580 mS/mm was reached at 77°K. Expected future results are in the range of $F_T \approx$ 250 GHz for near-ballistic ultra-short (0.2 μm active length) quantum-well MODFETs, and $F_T \approx$ 350 GHz for short (0.25 μm active length) ballistic-injection vertical FETs. Hetero-junction (H.J) bipolar transistors should also eventually reach $F_T \geqslant$ 100 GHz. The H.J. bipolar results shown in Table IV were obtained at Rockwell. A switching time of 1.5 ps is expected for unity fan-out in the 0.2 μm quantum-well MODFETs. IBM has made an experimental direct observation of ballistic electron transport in GaAs (reported by Heiblum, M. *et al.*, *Physics Review Letters*, November 1, 1985).

References

1. H. C. Casey, Jr., and M. B. Panish, *Heterostructure Lasers, Part B, Materials and Operating Characteristics*, Academic Press, Inc., New York, N.Y. 1978.

2. J. R. Chelikowsky and M. L. Cohen, "Non-local pseudopotential calculations for the electronic structure of eleven diamond and zinc-blende semiconductors", *The Physical Review*, Vol. B-14, No. 2, 15 July 1976, pp. 556-582.

3. Lester F. Eastman, "Physical Concepts of Some Promising High-Speed High-Frequency Compound Semiconductor Transistors", *Proceedings of IEEE/Cornell Conference on High-Speed Semiconductors*, IEEE Cat. No. 83CH1959-6/83/00000-0010 1983, pp. 10-15.

4. *ibid*. p. 15.

LIMITS ON ULTRA-LARGE-SCALE INTEGRATION

James D. Meindl

James D. Meindl, John M. Fluke Professor of Electrical Engineering at Stanford University, is Director of Stanford Electronics Laboratories, co-director of the Center for Integrated Systems, and Associate Dean for Research, School of Engineering. He received the Ph.D. degree in electrical engineering from Carnegie Mellon University in 1958. From then until 1967 he was on military and civilian duty at the U.S. Army Electronics Command at Fort Monmouth, New Jersey, where he eventually became Director of the Integrated Electronics Division and was recognized by the Arthur S. Flemming Committee as one of the ten outstanding young scientists in government. In 1967 he joined the Stanford faculty and founded Stanford's Integrated Circuits Laboratory in 1969. Dr. Meindl was named a Fellow of the IEEE in 1968, received outstanding paper awards at the 1970, 1975, 1976, 1977, and 1978 IEEE/ISSC Conferences, was elected to the National Academy of Engineers in 1978, and in 1980 received the J. J. Ebers Award of the IEEE Electron Devices Society for outstanding contributions to the field of electron devices. He is the author of a book on micropower circuits and has written over 270 technical papers on integrated circuits, micropower circuits, and medical electronics.

This morning I will discuss the subject of ultra-large-scale integration. To cast it as a question: how many transistors can we expect to fabricate in a single silicon chip that will prove to be useful at some future time? My central thesis is that ultra-large-scale integration is governed by a hierarchy of limits and we can codify these limits as (1) fundamental, (2) material, (3) device, (4) circuit, and (5) system. I would like to propose that at each level there are three kinds of limits: theoretical, practical, and analogical. So that we might then say that ULSI is governed by a hierarchical matrix of limits.

I am going to divide this discussion into four segments. In the first segment I will give you a simple set of examples of theoretical limits that will illustrate each level of the hierarchy. In the second segment I am going to focus on theoretical circuit limits because of the unusual insight they give us into the limiting dimensions of ULSI structures. In the third section I will outline three overall practical limits that indicate the rate at which we have been progressing and are likely to continue to progress in VLSI. In the final section I am going to discuss a new kind of limit, which I have called the analogical limit, and which calls upon history as a guide to the long-term future of ULSI.

We go first to a fundamental limit which is based on thermodynamics. Let us suppose that in Figure 1 we have a node in a complex micro-computer chip and that hanging from that node we have an equivalent resistance of value R. Immediately from thermodynamics we know that there is a mean-square-noise voltage associated with the node given by the first expression in the figure. Circuit theory tells us that if we want to extract the maximum available power from that node we must terminate it in a matching load impedance of value R. Then the available power will be KTB. Now it seems reasonable to assert that if we want to unambiguously represent information at that node, the signal power must be greater than the available noise power associated with the node. I am going to take the factor to be 4 times greater. Furthermore if we want to change the information represented at the node, for example, from a zero to a one, we know that the transition time must be greater than the reciprocal bandwidth associated with the node. Combining these three equations results in Table I. It can be seen that the switching energy required to change information represented at the node must be greater than or equal to 4 kT, where k is Boltzmann's constant and T is absolute temperature. Both of these quantities are independent of the material, the device or the circuit associated with the node. Consequently we have what we might call a fundamental limit (which happens to have a value of about 10^{-20} joules at room temperature) on the energy required to change the information represented at that node. So this is a fundamental limit on ULSI.

In Table II we go to the materials level and consider silicon. We are well informed about its band gap, the intrinsic carrier concentration, the intrinsic

$$\bar{e}_n^2 = 4kTRB$$

$$P_{avail} = \bar{e}_n^2 / 4R$$

$$P_{avail} = kTB$$

Figure 1. A Fundamental Limit from Thermodynamics

TABLE I

P_{avail}	$= kTB$
P_s	$\geqslant 4P_{avail}$
t_s	$\geqslant 1/B$
P_s	$\geqslant 4kT / t_s$
E_s	$\geqslant P_s t_s$
E_s	$\geqslant 4kT$

TABLE II

A Material Limit for Silicon (300° K)

E_g	$= 1.1eV$
n_i	$= 1.45 \times 10^{10}\ cm^{-3}$
μ_i	$= 1350\ cm^2/V_{sec}$ (electrons)
ϵ_c	$= 3 \times 10^5$ V/cm
V_s	$= 8 \times 10^6$ cm/sec (electrons)

mobility, the self-ionizing field strength and the scattering limited carrier velocity.

In the following sketch we have a hypothetical cube of silicon of dimension Δx, with carriers transported through it traveling at the scattering-limited velocity in a field just a bit smaller than the ionizing field strength.

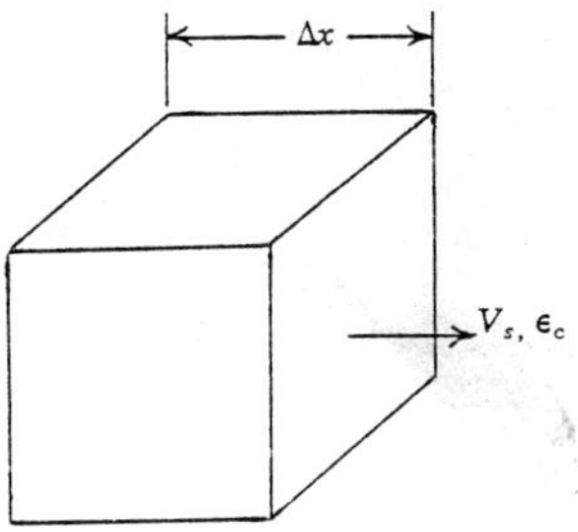

$$\Delta x = V_s \, \Delta t$$

$$\Delta v = \epsilon_c \, \Delta x$$

One can write a simple expression relating the dimension of the cube and the transit time of carriers, and a second one relating the potential drop across the cube and its dimension. Combining these two, we can show that the time required for carriers to pass through a potential difference of one volt, in silicon, is given by the expression:

$$\frac{\Delta t}{\Delta v} = \frac{1}{v_s \, \epsilon_c} = 0.416 \; x \; 10^{-12} sec.$$

This result is a material limit on the switching speed of any bipolar or MOS transistor and it is imposed by the characteristics of the material itself.

In Figure 2 we move to the device level of the hierarchy and consider an MOS transistor. For proper operation of this device, where the gate controls the current flowing from source to drain, we want to have a channel length or separation of the drain and source junctions which is large enough so that their depletion regions do not touch. If they touch we have a punch-through condition and an out-of-control current flowing in the drain. First-order device theory tells us that the separation of the two junctions, the source and drain junctions, is given by the expression:

$$L(min) \geqslant \sqrt{\frac{2\epsilon_{si}(V_D + \phi)}{qN_A}}$$

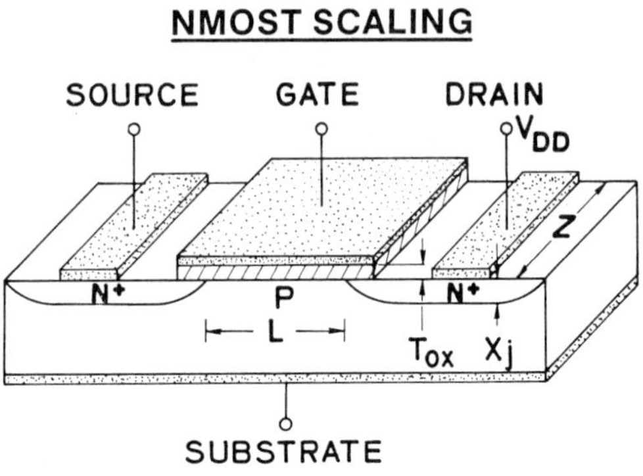

Figure 2. A Device Limit for an MOST

If we put in typical values for substrate doping and drain-supply voltage, we find that a first-order approximation to this minimum channel length is 0.18 microns:

$$N_A = 1.5 \; x \; 10^{17} / \; cm^{-3}, \; V_D = 0.5V$$

$$L(min) = 0.18\mu$$

I would call this a device limit on ULSI imposed by the characteristics of the MOS transistor.

In Figure 3 we move to the circuit level of the hierarchy and consider the complementary MOS structure as drawn. We will use that technology in this thinking experiment to fabricate the complementary MOS inverter of Figure 4 which is excited by a square wave, of period T, energized by the supply voltage and drives the load capacitance C_L. First-order circuit theory tells us that the average power drain is given by the expression at the bottom of the figure which we see approaches a value of zero as the excitation frequency approaches zero. This is then a circuit limit on ULSI imposed by the characteristics of complementary MOS inverters.

Finally, in Figure 5, we come to the system level of the hierarchy and consider a limit due to clock distribution in a synchronous system. We have a clock generator here driving a long-distance interconnect which is really a transmission line. We will designate the near end by N and the far end by F. The square wave of Figure 5 is the ideal clock waveform with period τ. Circuit analysis, as summarized in Figure 6, tells us that the clock waveform will

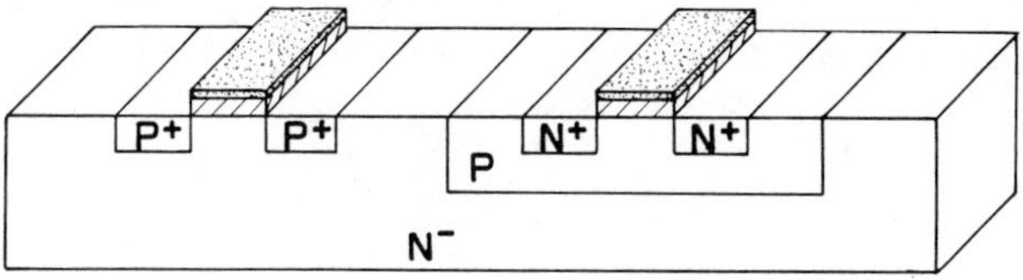

Figure 3. A Circuit Limit for a CMOS Inverter

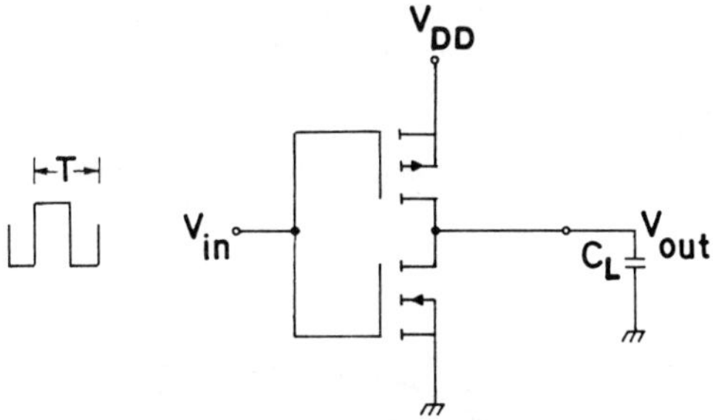

$$P_{AVE} = \frac{E}{T} = C_L V_{DD}^2 f$$

Figure 4. Excitation of a Complementary CMOS Inverter

arrive at the far end of the line at period Δt, after it is available at the near end of the line. This period, Δt, is called "clock skew". We know that the clock period must be greater than twice the value of clock skew in order to have proper operation of a synchronous digital system. So this is then a simple system limit on ULSI.

In review using Table III, we can see that fundamental limits are immutable laws of nature, we cannot do anything to change them. Material limits do depend on composition but they change infrequently. Silicon has been with us for several decades and will continue in that role as the dominant material in ULSI for several more decades as will be seen in the projection I am about to make. Device limits can tell us a lot about minimum allowable

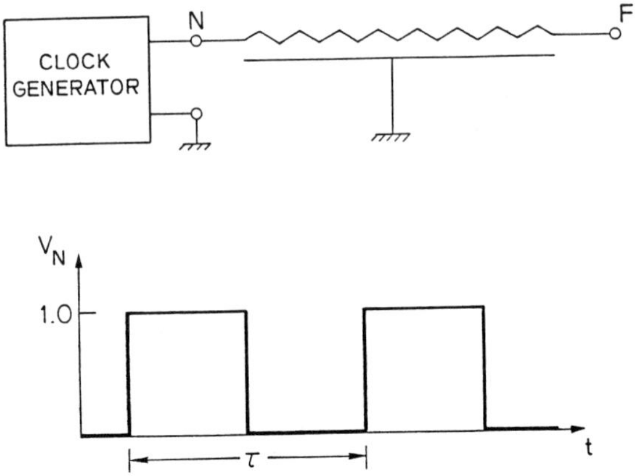

Figure 5. A System Limit due to Clock Distribution

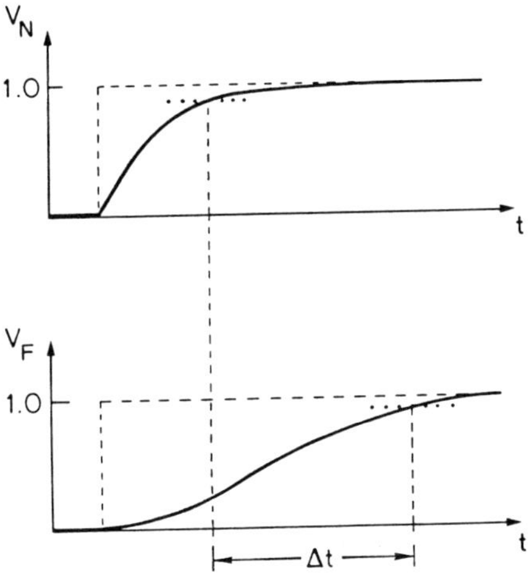

$$\tau > 2\Delta t$$

Figure 6. Circuit Analysis of Clock Distribution System

dimensions in ULSI but they are not as valuable or as helpful as circuit limits for the following reason: the circuit level of the hierarchy is the highest one which retains a complete description of the physical characteristics of the ULSI structure. Furthermore, it is the lowest one in the hierarchy which retains a useful description of the information-processing function that is to be performed by a group of devices. Consequently it is at this level where we have the devices described in a useful operating environment as opposed to sterile isolation. We can therefore project the most accurate values for limiting dimensions.

The second segment of the discussion will deal with some theoretical circuit limits associated with MOS transistors and interconnections because these are the particular elements in a chip that limit its performance. Referring to Figure 2 allows establishment of some pertinent symbols: we will designate the effective channel length as L, the gate-oxide thickness is T_{OX}, junction depth is X_j and the width is Z. The question is how far can we shrink the dimensions of this structure? The following statements summarize ideal scaling theory which helps us to answer that question.

Scaling Factor

$$S > 1$$

Lateral Dimensions

$$L \rightarrow L / S$$

$$Z \rightarrow Z / S$$

Vertical Dimensions

$$T_{OX} \rightarrow T_{OX} / S$$

$$X_j \rightarrow X_j / S$$

Supply Voltage

$$V_{DD} \rightarrow V_{DD} / S$$

We begin in this discipline by defining a device scaling factor, S, which is greater than one. We then proceed to scale down all lateral and vertical dimensions by the same factor $\frac{1}{S}$. In addition, in this ideal approach, we scale down supply voltage by the factor $\frac{1}{S}$. Why do we do that? If we scale a device dimension, Δx, we must scale the corresponding potential drop, Δv, in proportion, if electric field strength, E, is to remain constant, as desired, in order to retain equal safety margins in the operation of the device. If we increase E we tend to degrade reliability. The well-known advantages of doing this kind of scaling are listed in Table IV.

TABLE III

Hierarchy of Limits for ULSI

7. Application/Algorithms
6. Architecture/Instruction Set
5. Register Transfer/Logic
4. Circuit
   A. Physical Characteristics
   B. Information Processing Function
   C. Operating Environment
   D. Limiting Dimensions
3. Device
2. Material
1. Fundamental

TABLE IV

NMOST Device Scaling for Constant Field

DEVICE/CIRCUIT PARAMETER	SCALING FACTOR
Gate Delay	1/S
Capacitance	1/S
Power Dissipation Device	$1/S^2$
Power Density	1
Packing Density	S^2
Speed Power Product/Device	$1/S^3$

We see that the time delay through a gate decreases as $\frac{1}{S}$, the packing density increases as S^2, and the power-speed product in this ideal approach (constant electric-field scaling) tends to decrease rapidly as $1/S^3$. Now with all these good things happening as we increase S, the question of interest is how far can we go, what is the maximum allowable value of S, or putting it somewhat differently, what is the minimum channel length that we can permit in the MOS transistor and still have operation as we know it in the device today? In order to answer that question we have to add one more component to our scaling discipline:

$$N_A \rightarrow SN_A$$

Here we recognize that it is necessary to scale the substrate doping not as $\frac{1}{S}$ but as S to the first power. Why do we do that? We have to bring in another dimension in the structure to answer that question.

The thickness of the depletion region surrounding the source and the drain junctions, is shown in Figure 7 by the symbol d_j. From first-order device theory the expression for the value of the depletion region thickness is:

$$d_j = \frac{\sqrt{2\epsilon_{Si}(\phi_{Bi} + V_{AP})}}{qN_A}$$

If the applied voltage on the junction is large compared to the built-in voltage ($V_{AP} \gg \phi_{Bi}$), then the junction depletion region thickness is given by:

$$d_j \approx \frac{\sqrt{2\epsilon_{Si}\, V_{AP}}}{qN_A}$$

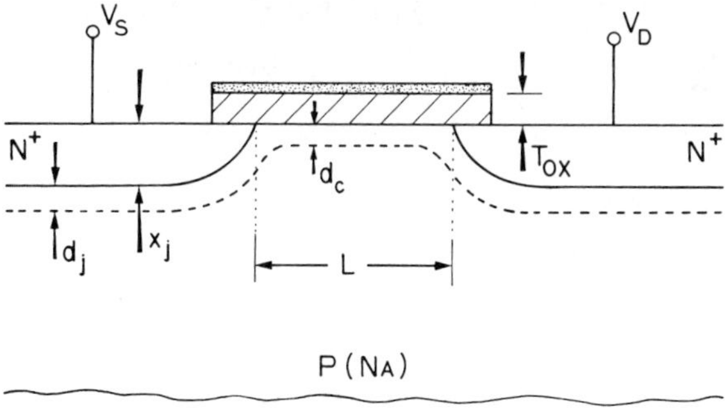

Figure 7. Depletion Region Thickness

If we scale the applied voltage as $\frac{1}{S}$ and the substrate doping as S, we see that indeed the junction depletion region thickness scales as $\frac{1}{S}$ in the same fashion as all the other dimensions of this structure and everything is going well for us as indicated by:

$$d_j \rightarrow d_j / S \approx \frac{\sqrt{2\epsilon_{Si}(V_{AP} / S)}}{q(N_A S)}$$

But now suppose that in the first expression for d_j the applied voltage is not large compared to the built-in voltage. What happens? Looking at the expression for the built-in voltage:

$$\phi_{Bi} = \frac{2kT}{q} Ln(N_A / n_i)$$

we see that indeed it does not scale as $\frac{1}{S}$, in fact it increases slightly as we tend to increase substrate dopings. Furthermore, we often do not have the option of scaling applied voltages as $\frac{1}{S}$. The conclusion is that we do not always have the ability to scale junction depletion region thickness as $\frac{1}{S}$, and this is perhaps the most fundamental reason why we have to put a limit on the value of S.

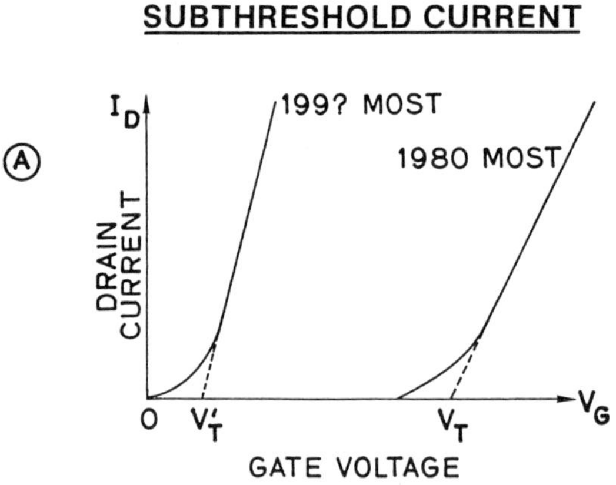

Figure 8. Subthreshold Current Transfer Characteristic

There is a second reason, concerned with subthreshold current, the current that flows when we try to turn off the device. Figure 8 contains transfer characteristics of two MOS transistors showing drain current versus gate voltage. Curve (a) is for an unscaled or 1980 MOS transistor and curve (b) is for a future device that is highly scaled. · Notice the threshold voltage has decreased because we have to scale it along with supply voltage. Figure 9 displays plots of the same variables but now I have used a logarithmic scale for current. Notice that the slope of both curves in the subthreshold region is the same, which means that it is much easier to turn off the unscaled device of curve (a) because the voltage swing will be several times larger than for the scaled device of curve (b) where we have a much smaller voltage swing for the same threshold point. Another way of saying that is to look at the reciprocal slope, α of this curve:

Weak Inversion $V_G < V_T$

$$\alpha \equiv \frac{[\partial(Ln I_D)]^{-1}}{\partial V_G}$$

$$\approx 2.3 \frac{kT}{q}(1 + C_D / C_{OX})$$

$$\approx CONSTANT$$

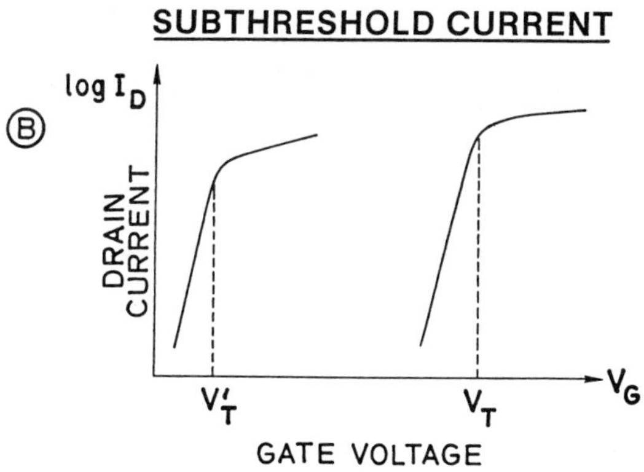

Figure 9. Logarithmic Subthreshold Current Transfer Characteristic

The slope, α, may be regarded as the number of millivolts of gate voltage required to change the drain current by an order of magnitude and this approximate expression for α turns out to be about constant. So the second reason we have to quit scaling is that the swing that we get as we scale down will not allow us to reduce subthreshold current to small enough values.

These features of the MOS transistor have been incorporated in a model for the device which has then been used to analyze three fundamental families of logic circuits that we use in VLSI, as shown in Figure 10. Model (a) is enhancement depletion with NMOS logic. Model (b) is enhancement with CMOS logic. Investigations were also made on Model (c) an unusual third family that has been called "enhancement-depletion CMOS," where a depletion p-channel device is used as the load device for the gates. Typical results of these analyses are illustrated in Figure 11 where we have transfer characteristics, output voltage versus input voltage, for three different inverters, where drain voltage is 2 volts and channel length is 0.15 microns. We see that all three of these are usable characteristics and the best one, of course, is for the enhancement CMOS circuit drawn with the solid line. The net result of investigations of this kind is the projection that we can, in fact, fabricate useful devices and circuits with channel lengths below a quarter of a micron, perhaps even at half that value.

Next I would like to discuss interconnections with the aid of Figure 12. When we examine interconnections, two scaling factors must be brought into play. First is the device-scaling factor, S, as shown previously in Table IV,

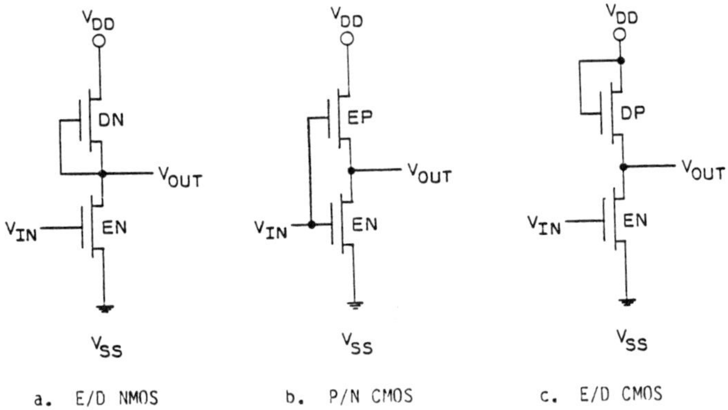

Figure 10. Models for VLSI Logic Circuits

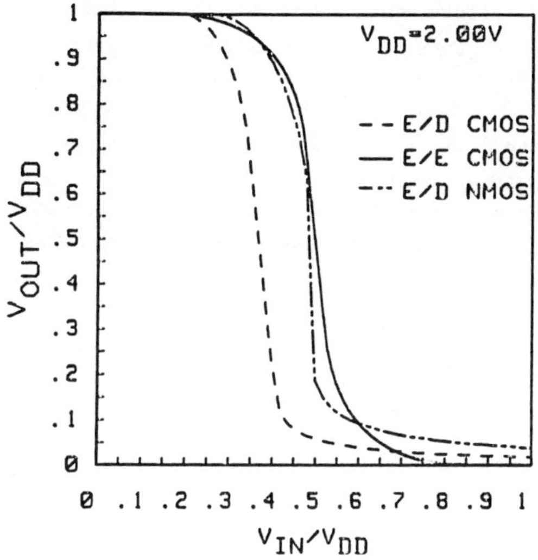

Figure 11. Voltage Transfer Characteristics

where we shrink vertical and lateral dimensions. But we need a second scaling factor, that I will call the chip-scaling factor, S_C, and that is greater than 1 because over time we have tended to increase the size of chips. We have to recognize two different classes of interconnections. The first class I will call

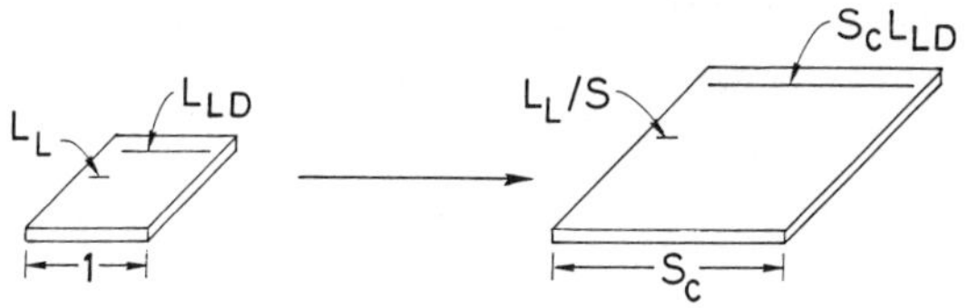

Scaling Factor $S > 1$

Chip Scaling Factor $S_C > 1$

Figure 12. Interconnect Scaling

local interconnections that run within a gate or between adjacent gates. The length of a local interconnect is going to scale as $\frac{1}{S}$ where S is the device-scaling factor. On the other hand, the length of a long-distance interconnect, which runs from corner to corner of a chip, for example, will scale as S_C where S_C is the chip-scaling factor. Now if we use device scaling on local interconnects and scale length, width, and thickness as $\frac{1}{S}$, as listed in Table V, a principal result is that the RC delay time associated with the local interconnect does not scale. It remains constant, as in Table VI, whereas gate delay does scale as $\frac{1}{S}$. This means that local interconnects are placing an increasing

TABLE V

Local Interconnect Scaling (Dimensions)

Length	$L \rightarrow L / S$
Width	$Z_W \rightarrow Z_W / S$
Thickness	$T \rightarrow T / S$
Spacing	$Z_S \rightarrow Z_S / S$
Insulator Thickness	$T_{FOX} \rightarrow T_{FOX} / S$

TABLE VI

Local Interconnect Scaling (Parameters)

Resistance	$R \rightarrow SR$
Capacitance (to substrate)	$C_S \rightarrow C_S / S$
Capacitance (line-to-line)	$C_L \rightarrow C_L / S$
Delay Time	$RC \rightarrow RC$
Voltage Drop	$IR \rightarrow IR$
Current Density	$J \rightarrow SJ$

burden on the speed of circuits as we continue to scale. Looking at long-distance interconnects in Table VII, we are scaling cross-sectional dimensions as $\frac{1}{S}$, but we have no control of that kind over the length, L. It is going to scale up as S_C, the chip-scaling factor. In Table VIII it may be seen that the most important result is that time delay associated with long-distance interconnects scales up rapidly as the square of the product of the device and the chip-scaling factor. This is bad news indeed, and helps to explain why we see long-distance interconnects determining the overall performance of complex chips today. This trend will become even more pronounced in the future unless some unusual steps are taken to effect a change. One of the steps that is important to observe is that we must never let the resistance of an interconnect be larger than about 2.3 times the driver-transistor resistance. Otherwise we suffer a very rapid increase in the delay time associated with long-distance interconnects.

TABLE VII

Long-Distance Interconnect Scaling (Dimensions)

Length	$L \rightarrow S_C L$
Width	$Z_W \rightarrow Z_W / S$
Thickness	$T \rightarrow T / S$
Spacing	$Z_S \rightarrow Z_S / S$
Insulator Thickness	$T_{FOX} \rightarrow T_{FOX} / S$

TABLE VIII

Long-Distance Interconnect Scaling (Parameters)

Resistance	$R \rightarrow S^2 S_C R$
Capacitance (to substrate)	$C \rightarrow S_C C$
Capacitance (line-to-line)	$C \rightarrow S_C C$
Delay Time	$RC \rightarrow S^2 S_C^2 RC$

In summary, looking at the theoretical circuit limits that we have just considered, we can see that it should be possible to reduce the dimensions of transistors to less than a quarter of a micron. In the case of interconnections we certainly are going to be driven towards multilevel metal interconnects in order to reduce the average length of an interconnect and to scale them in different manners rather than simple $\frac{1}{S}$ scaling. In order to keep resistance down we may indeed want to use repeaters on long-distance interconnection lines, and finally, we certainly hope for help at the architecture level, where the layout of a chip will not require so many long-distance interconnects.

The third segment of our discussion is concerned with practical limits that deal directly with the question of how many transistors we can expect to fabricate in a chip that will be useful at some future time. There are three overall factors that determine the number of transistors, N, in a chip and we can include any changes in the technology in these three factors. The first one is the average minimum feature size, F. The smaller it gets, the more transistors we can put in a chip. The second factor is D, the square root of the die area. The larger it gets, the more transistors we have in a chip. The third one is packing efficiency, PE, that is the number of transistors per average minimum feature area. The more efficiently we can pack them the more transistors we get in a chip. The simple relationship between the number of transistors and these three factors is given by:

$$N = F^{-2} \cdot D^2 \cdot (PE)$$

Our history in scaling down transistor-feature size is shown in Figure 13 with average minimum feature size plotted versus calendar year.[1] In 1960 we had about a 25-micron average minimum feature size that dropped to about 2 1/2 microns in 1980. Although it will be difficult to accomplish, if we continue to scale at the same rate, we will arrive at a 1/4-micron average minimum feature size by the year 2000. I will not attempt to elaborate in detail on why we should not expect to do this well in the future, but one of the reasons is that optical lithography is approaching the limits of its capabilities. It is reasonable to project that in the early 1990's we will see some break points in the minimum feature size reduction trend, and that we may well find ourselves on a reduced trajectory somewhere between the two upper dashed lines in the figure. A second factor is plotted in Figure 14 as the square root of chip area versus time. In 1960 we were using about a 1.4-millimeter edge on a die. By 1980, for larger chips, the die edge had increased to about 8 millimeters. Again, if we continue at this same rate, in the year 2000 we will have a 45 to 50-millimeter edge on a die. For a number of reasons, one of them being lithography, again we can project a change in the slope of this curve in the early 1990s so that in the future we may see a slower rate set by the lower dashed line. The third factor of interest is shown in Figure 15 as

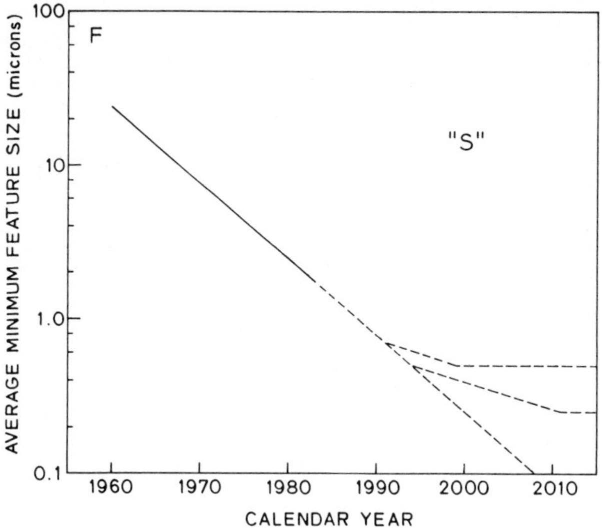

(©1984 IEEE)

Figure 13. History of Feature-Size Sealing

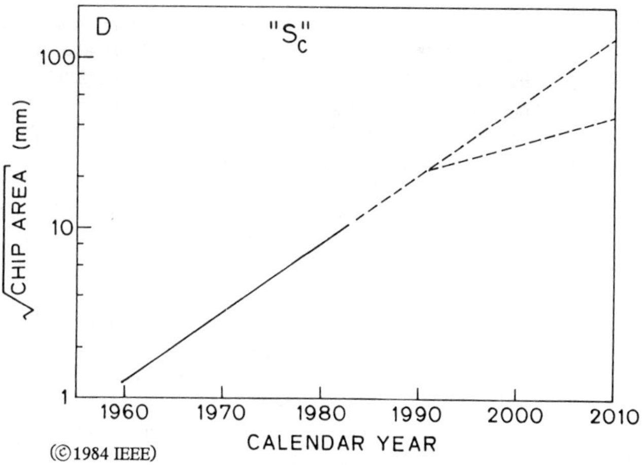

(©1984 IEEE)

Figure 14. History of Chip-Area Scaling

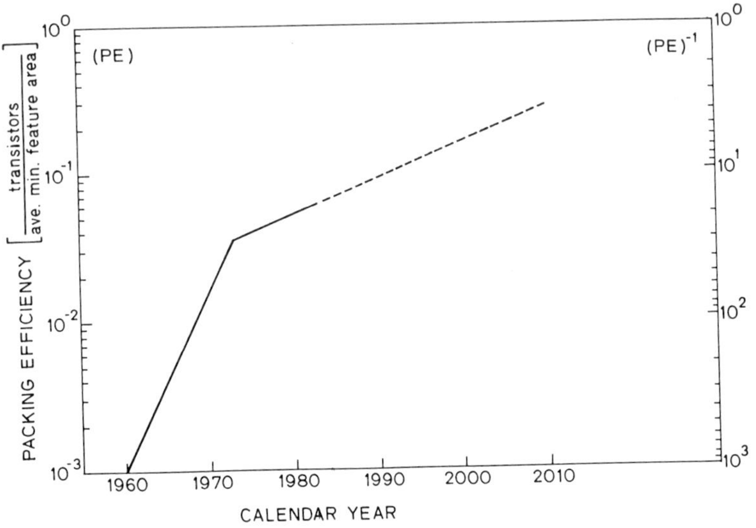

(©1984 IEEE)

Figure 15. History of Packing Efficiency Scaling

packing efficiency versus calendar year, with the right-hand vertical scale plotted as the reciprocal or the number of minimum feature areas per transistor. Packing efficiency increased very rapidly in the 1960's and early 1970's and then suddenly there was a sharp reduction in the slope of this curve because we finally ran out of unused real estate on the chip. At this point we had finally used up all the space, either by filling it with transistors or with interconnects. Since that time we have increased packing efficiency mainly by adding mask levels, in effect by going up in the third dimension. This curve projects that we will be able to continue our trend of about the last ten years or so of a slow rate of increase in packing efficiency.

All of these factors are combined in the curves of Figure 16 that indicate past, present, and future numbers of components per chip versus calendar year. Segment A of this figure was first discussed by Gordon Moore in 1975 when he observed that from 1959 through about 1975 the number of transistors in a chip had doubled annually. Now with ten more years of hindsight to guide us, we can see that indeed there was a change in the slope of this curve about 1972 or so, as in segments B and C. Since that time we have been increasing the number of transistors in a chip by about a factor of 4 every three years. The question of interest is how long will it be before there is

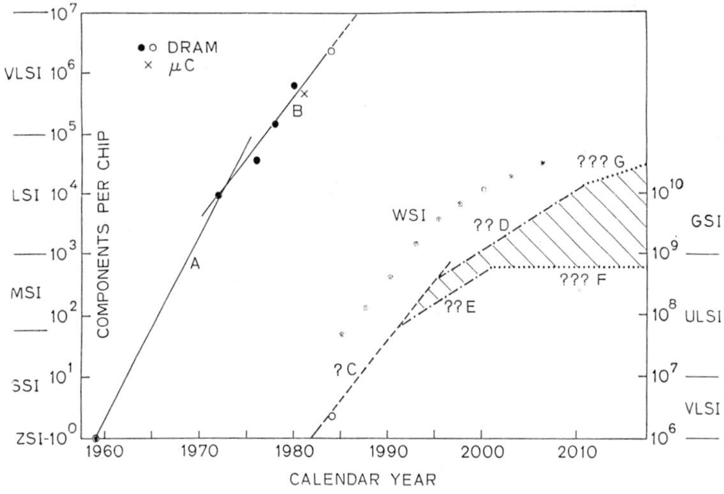

(©1984 IEEE)

Figure 16. Projection of Possible Chip Densities

another breakpoint in this segment? Note that I am using the right-hand vertical scale to do some projecting into the future. I suggest that in the early 1990's we will see a breakpoint that will be caused by a number of factors. First of all, we will be reaching the limits of optical lithography and will have to switch to other technologies if we want smaller dimensions. I am sure that this morning Alec Broers gave you some very interesting comments on the future prospects in this arena. Our inevitable approach to the limits on the dimensions of MOS transistors will be another factor. Further out I think we can see other breakpoints in the curve that will be caused by soft collisions with these limits on the dimensions of transistors. Nevertheless, what is interesting about this projection, even though it is looking at the future in a somewhat realistic way, is that by the year 2000 we should expect to see one billion transistors in a chip of silicon. This prediction represents an increase by a factor of more than one thousand over where we are today. Notice, however, that we have come from one to nearly one million components per chip in the past 25 years.

I will move now to the fourth segment of this discussion which I have called "analogical limits." Historians of technology frequently point out that technology develops in a certain characteristic pattern as shown in the S-shaped curve of Figure 17. Initially the state-of-the-art advances slowly, then when commercial importance is recognized, technology develops very rapidly, followed by eventual saturation of the state-of-the-art and obsolescence of the technology. For example, the efficiency of incandescent light bulbs and the speed of propeller-driven airplanes are among technologies that have followed this pattern. I will try to project the long-term future of ULSI making use of this guide line.

For the remainder of the discussion I will cast an analogy between structural materials as we still use them in the mature industrial revolution, and electronic materials as we are using them in the modern information revolution. I don't think any of us would argue with the assertion that the most important element of the industrial revolution has been iron. Since we would all agree that silicon has been the most important element of the modern information revolution, I suggest that there is an analogy between iron and silicon. We can deepen that analogy if we examine the hierarchy of levels in Table IX that includes compounds, components, and even systems. In the bottom two rows of the table we observe that a steel ingot is doped with different elements such as manganese, nickel, or chromium to obtain special-purpose stainless steels. Similarly, a silicon ingot is doped with boron, phosphorus, or antimony to obtain a special-purpose wafer for a particular type of integrated circuit or discrete device. At the component level we cast pistons

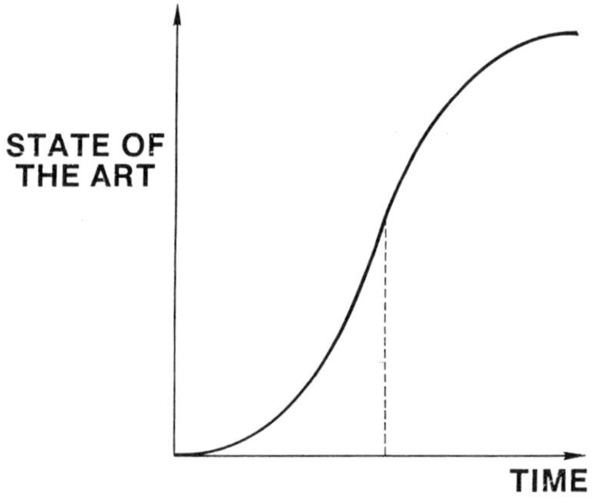

Figure 17. Form of Technology Development

and engine blocks. These are special-purpose mechanical components, just as we build special-purpose integrated circuits, microprocessors, memories, phase-locked loops, etc. Finally at the system level we can cast analogies between, for example, the railroad and the telephone system. The analogy is broader if we recognize that the most important technical figure of merit for a structural material is its strength-to-density ratio, while the most important figure of merit for an electronic material is the reciprocal power-delay product. The larger these two figures of merit become, the better the material is serving us.

These comparisons bring several questions to mind. In certain applications for structural materials, especially in high-speed vehicles such as airplanes where steel has not met the requirements for strength-to-weight ratio, we have turned to aluminum as the material of choice. So my first question is: in electronic systems when silicon has not been fast enough to do the job, what have we done? Well after Les Eastman's talk this morning I am sure everybody here knows the answer: we have turned to gallium arsenide. This comparison suggests that even broader analogies may be drawn between structural and electronic materials. Going one step further, consider titanium. In the most demanding structural applications for high strength-to-weight ratio, especially at high temperatures, designers of jet-engine turbines have turned to titanium as the material of choice. So my second question is: what is the technology that has exhibited the highest reciprocal power-delay product that we know of today? There may be several possible answers but my guess is that it is superconductive niobium Josephson-junction technology. Certainly that technology has given us the highest reciprocal power-delay product of any technology that we have demonstrated today. It is interesting to note that this

TABLE IX

Hierarchy of Analogies

	Structural	Electronic
SYSTEM	Automobile Skyscraper Railroad	Portable Computer Mainframe Computer Telephone System
COMPONENT	Engine Block Piston	Memory IC PLL IC
COMPOUND	Steel Ingot Mn, Ni, Cr	Silicon Ingot B, P, Sb
ELEMENT	Iron	Silicon

technology has not yet emerged from the laboratory into use in practical systems, so we have some exciting prospects for the future in that arena. Moving on to plastics we know that they are a low-cost light-weight, improved, performing substitute for metal in many applications. My next question is: what is the analog of plastics in electronics? Have we identified it as yet? I am not sure that we know the answer but it may be polycrystaline or re-crystalized semiconductors on insulator (SOI) materials which also have not yet emerged from the laboratory for use in practical systems. Certainly there is plenty of effort in this direction at the present time that, in fact, may lead to something akin to the analog of plastics. Finally, if I did not include this one I would expect Les Eastman to do so: the most exotic structural materials that we use today are composites involving crystalline fibers of boron, or carbon, or glass embedded in a binding matrix, giving us the highest strength-to-weight ratios of any materials available today. The question is what can we use as the analog? I would say perhaps super-lattice materials or multi-material wafers, will prove to be the analog of composite materials.

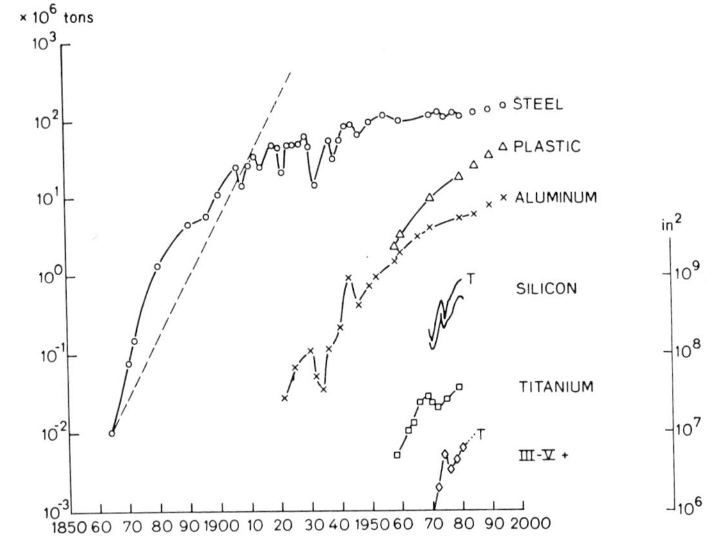

(©1984 IEEE)

Figure 18. Historical Summary of Analogies

Although these speculations are interesting or perhaps even amusing, I would like to cap them with just a bit of hard data. Figure 18 is a summary of the analogies that we have considered. The chart contains a plot of millions of tons of steel or other structural materials produced in the United States versus time, beginning in 1860, as well as similar curves for plastic, aluminum, and titanium. It is interesting to note that from 1860 to 1900 steel production in the United States doubled about every four years, and that almost a century after the curve essentially saturated, steel is still the most widely used structural material in the United States. This trend may suggest the probable future of silicon integrated circuits. Inferences about the future of gallium arsenide or SOI must be made with care, but an interpretation of the other curves on the chart, coupled with some historical sense of what has happened in other technologies, allows us to imagine the possible future of electronic materials.

Summarizing this discussion, I have attempted to establish the point that ULSI is governed by a hierarchy of limits, that the laws of physics and circuit theory tell us that we should be able to reduce the lateral dimensions of transistors to less than a quarter of a micron. But if we look at the rate at which we have been progressing in scaling down transistors, and ask how we are likely to progress in the future, it appears that we will not reach these quarter-micron dimensions until after the year 2000. So there is lots of "gas" left in the silicon tank as well as lots of new "fuels" coming along. It is clear that if we continue to increase the number of transistors in a chip at reasonable levels, we should see a billion transistors in a silicon chip by the year 2000. This conclusion is made without regard to wafer-scale integration and does not include the probable future of gallium arsenide.

Reference

1. James D. Meindl, "Ultra-Large-Scale Integration", *IEEE Transactions on Electron Devices,* Vol. ED-31, No. 11, November 1984, pp. 1555-1561.

FUTURE OF MICROELECTRONIC SYSTEMS

David A. Hodges

David A. Hodges received the B.E.E. from Cornell University and the M.S. and Ph.D. in Electrical Engineering from UC Berkeley. From 1966 to 1970 he was at Bell Telephone Laboratories, first in the components area at Murray Hill, then as Head of the System Elements Research Department at Holmdel, N.J. In 1970 he joined the faculty at the University of California, Berkeley, where he is Professor of Electrical Engineering and Computer Sciences. He is a member of the National Academy of Engineering. Professor Hodges is currently involved with colleagues and students in research on integrated circuit design, VLSI for telecommunications applications, and manufacturing information systems.

It's a pleasure to be here today and see many familiar faces and former professors. This occasion made me think back to my last year at Cornell when I had my first introduction to transistors--that was in 1959 with Professor Paul Ankrum--and I remember very well spending several weeks thoroughly characterizing one alloy junction transistor which had a beta of six. How could we have imagined in those days how far we would have advanced by now? At that time, we had no idea where we were going to be even five years later.

I am convinced that microelectronics technology is nowhere near the end of its development and that a billion-transistor chip will be built. What are we going to do with it? In trying to come up with a set of general principles that could be used to answer this question, I feel that I've failed. The best that I can do is try to outline some of the goals we must pursue in order to achieve further substantial improvements in microelectronics systems.

It is clear that memories and microprocessors will be fabricated with higher densities, better performance, and lower cost, although manufacturing technology will be a limitation. The present manufacturing environment is rather primitive and not highly automated, hence there will certainly be some improvements there. Assuming for the moment that billion-transistor chips can be built, the obvious question is: What functions might such a chip perform? I had difficulty thinking as far ahead as the year 2000, so as a starting point I have listed my expectations for 1990 in Figure 1. Further substantial

	1985	1990	
• MIN. INTERNAL DIMENSIONS	1.5	0.5	MICROMETER
• DEVICES PER CHIP	10^5	10^6	
• PINS PER CHIP	64	184	
• DIGITAL CLOCK FREQUENCY	10	40	MHz
• PRICE PER VLSI CHIP	5	5	$
• MEAN TIME TO FAILURE	10^7	10^8	HOURS

PREDICTION:

SILICON VLSI, MAGNETIC DISKS, CRTs WILL BE THE DOMINANT TECHNOLOGIES FOR INFORMATION PROCESSING, FILE STORAGE, AND DISPLAY FOR MANY YEARS TO COME

Figure 1. Trends in Microelectronics Technology

progress is likely to be made in the component dimensions and the number of devices on a chip, whereas the number of pins per chip and the digital-clock frequency are approaching practical limits, which I do not believe are going to be overcome by improvements in silicon or gallium-arsenide device technologies. In certain specialized applications, it may be possible to operate integrated circuits at very much higher clock frequencies; in complex systems with billions of transistors, I will be surprised if we are able to achieve order-of-magnitude increases in speed. I expect that economic issues are going to limit the very complex circuits to speeds below 100 megahertz. The economics of connecting pins and distributing clocks makes significant advances difficult. The physical limitations imposed by the magnetic permeability and dielectric constant of materials present further obstacles to achieving higher-speed circuit operation.

For systems applications, I expect that silicon VLSI, magnetic disks, and cathode-ray-tube displays will be the dominant technologies for many years to come, although other technologies will not be excluded; this business is far too big for everybody to do everything, and many smaller markets will appear for new technologies.

Figure 2 shows a list of system applications for microelectronic devices; several, in particular toward the bottom of the list, may be able to fruitfully exploit steadily improving device technologies. Of course the microprocessor, microcontroller, microcomputer, and the associated memories in themselves comprise a very major step forward, and there is no question that advances

MICROCONTROLLER, MICROCOMPUTER

TELEPHONE PCM CODEC-FILTER

TOUCH-TONE® RECEIVER

DATA MODEMS—up to 128K baud

ERROR DETECTION AND CORRECTION

LOCAL AREA NETWORKS—10 to 200 Mb/s

HIGH RESOLUTION TELEVISION

FAST FACSIMILE COMMUNICATIONS

SPEECH INPUT

ARTIFICIAL INTELLIGENCE

Figure 2. System-Level Objectives in Chip Design

will continue in those areas. More and more, however, these general-purpose elements are proving to be far from optimum or simply not able to meet the requirements for important system applications such as high-resolution television, fast facsimile communications, speech input, and artificial intelligence. I will discuss these particular applications in more detail.

In addition to the memories and microprocessors, what other system elements are going to be needed? To answer this question, I will pick a few examples from my own recent experience. Among other things, we will require more sophisticated data converters. Figure 3 shows a block diagram of a chip designed by a former graduate student of mine, Hae-Seung Lee,[1] who is now on the faculty at MIT. He pioneered a new concept in self-correcting circuits--a self-calibrating analog-to-digital (A-to-D) converter. The function of an A-to-D converter is to quantize the amplitude of a continuously varying signal (voice or video signals, for example) yielding a binary representation which is suitable for further manipulation by a digital computer. This task must be performed accurately and reproducibly. Microelectronics technology allows us to achieve certain levels of precision fairly readily, but exceeding

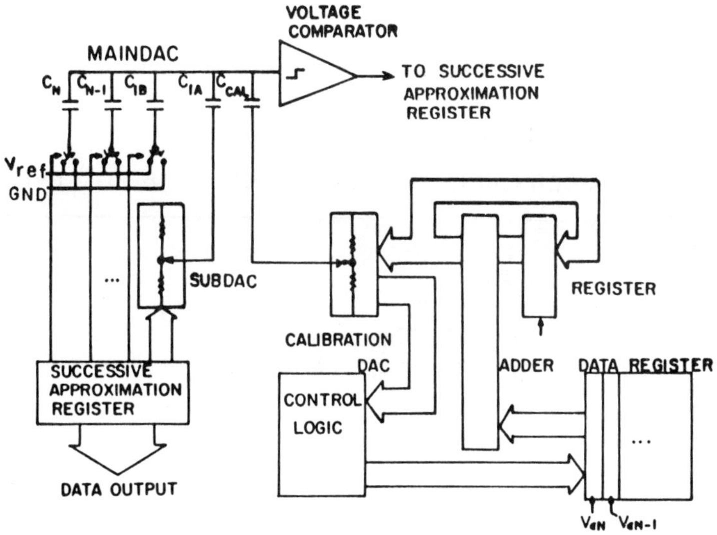

(©1984 IEEE)

Figure 3. Design of a Self-Calibrating A-to-D Converter Chip

that level of precision rapidly becomes much more complex and costly. With the addition of a few extra functional blocks, the performance of the imperfect A-to-D converter can be substantially improved.

Figure 4 shows typical measured errors determined by Lee[2] in the ratios between successive capacitor values which determine bit levels in an uncorrected A-to-D converter. Both systematic and random errors are found in detailed analysis. We have shown how self-correction circuitry can compensate for the inaccuracies introduced into the transfer function of this converter by the undesired capacitor-ratio errors. I will limit my discussion to an outline of the basic concept rather than describing the detailed operation of the circuit. The self-calibration circuitry first determines the capacitor-ratio errors, as shown in Figure 4, with the chip powered on. These data are stored, and during subsequent normal operation, the errors are automatically corrected. Systematic errors can be corrected through design revision, but random errors cannot be eliminated via design changes. The self-correction techniques permit us to go beyond the limitations of imperfect electronic devices and achieve a better result at the system level.

The above self-correction technique finds application in digital audio for instance, where very low distortion is desired at a sampling rate of 50 kilosamples per second, 16 bits per sample. The amount of distortion can be determined via spectrum analysis, as in Figure 5, where the output spectrum of a converter with a 500 Hz sinusoidal input is shown without any correction.

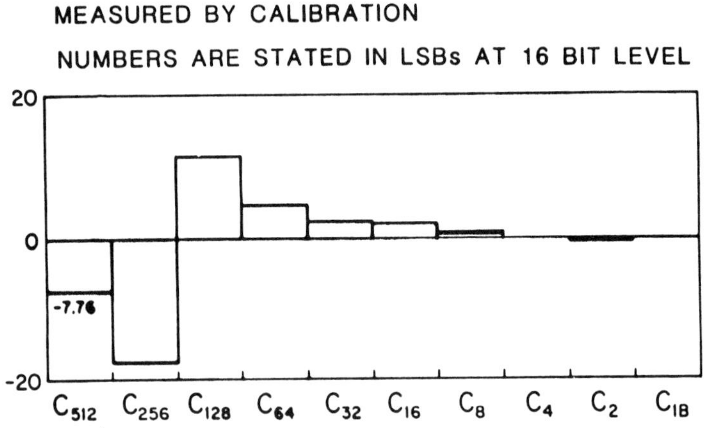

(©1984 IEEE)

Figure 4. Typical Capacitor Ratio Error

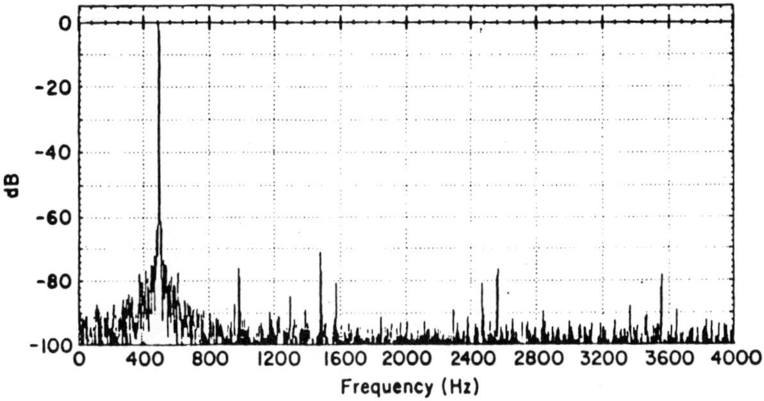

Figure 5. FFT Spectrum for a 12-Bit ADC Sampling a 495 Hz. Sine Wave

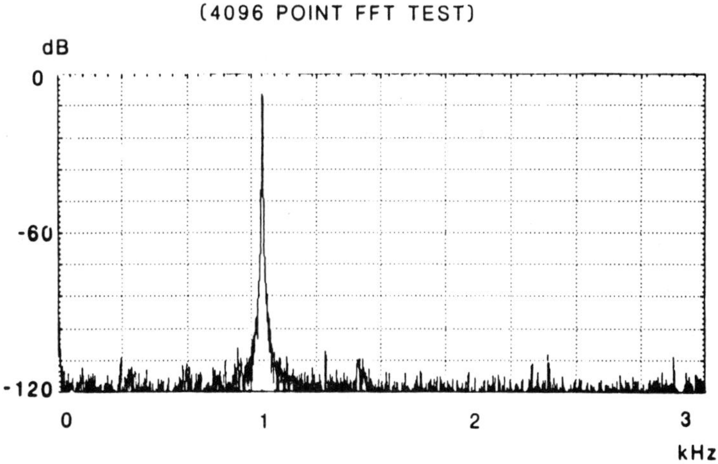

(©1984 IEEE)

Figure 6. FFT Analysis of an Experimental Chip with Self-Correcting Circuitry

The amount of harmonic distortion apparent in the spectrum is more than would be acceptable in a high-quality audio system. When the correction technique is applied, then we can build a much better data converter; and Figure 6 shows a spectral analysis of the experimental chip incorporating self-correcting circuitry.[3] The second harmonic is now virtually invisible, and the third harmonic is suppressed to about 108 dB below the fundamental component–a substantial improvement.

Drawing from the communications arena, I will now discuss several systems applications which may benefit from the use of additional chip integration to compensate for device deficiencies that are hard to overcome with

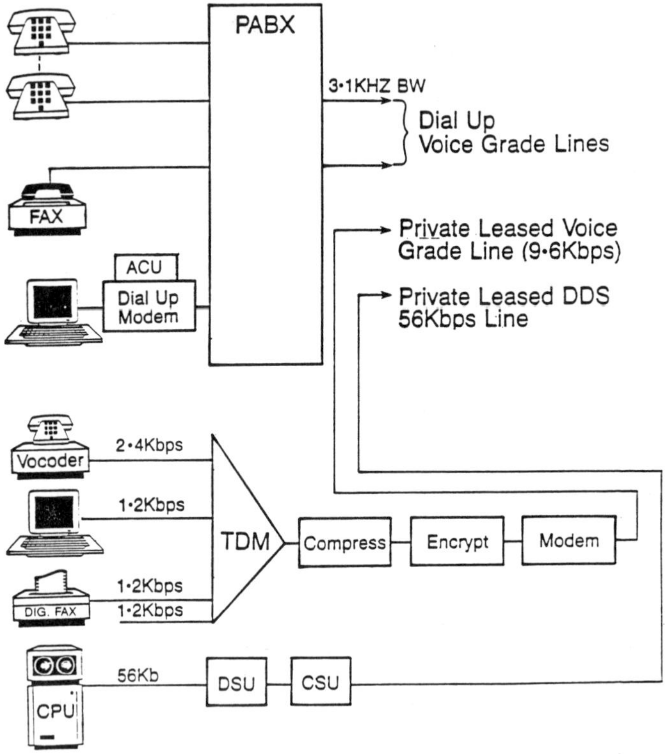

Figure 7. Current VOICE/DATA Networks

brute-force technology. Figure 7 illustrates schematically what our telecommunications environment looks like today, with such typical devices as telephones, private automatic branch exchanges, facsimile machines, personal computers, and mainframe computers all talking to one another over twisted-pair networks with a four-kilohertz analog bandwidth—the old telephone cable that has been installed over the last 50 years. Such cable represents an enormous investment that will not be replaced immediately by fiber optic cables. Many new installations will employ optical fiber, and the higher data capacity (up to hundreds of megabits per second) will certainly be put to good use. It would be to our advantage, however, if we could upgrade that multibillion dollar investment in telephone cable by somehow expanding its bandwidth to permit new services that are being sought for the future. In fact, a great deal of planning has already gone into such an upgrading, and there are plans for an international integrated-services digital network that will provide private switched digital-communication channels between central offices anywhere in the world starting well before 1990. This service won't be available immediately to everybody, but the network will grow much more rapidly than the network of fiber simply because it is much cheaper to add new terminal equipment to existing channels than it is to install new transmission facilities. Currently, the weak link in such a network is that between the customer premises and the first telephone exchange. These subscriber loops still use plain old twisted-pair cable, with three-kilohertz bandwidth, tied into plain old telephone sets. The question is: what can be done to extend the standard 64 kilobit per second single-channel telephone link all the way to subscriber premises? Digital systems permit a multiplicity of channels; multiples of 64 kilobits could be used to accommodate requirements for higher data rates. To take advantage of the above, digital channels with end-to-end digital integrity (bits are not encoded into an analog form at any point) must eventually be employed. The twisted pairs that are already installed are capable of carrying signals of several hundred kilohertz bandwidth (over distances up to about five kilometers or three and a half miles in length). So if we sought to make two-way digital transmission at rates much higher than, say, 9600 bits per second, which is characteristic of the highest speed modems, what techniques could be considered?

Two techniques have been studied: the so-called burst mode, in which transmission takes place alternately in the two directions, and the so-called hybrid mode in which transmission is simultaneous in both directions (see Figure 8). The disadvantage of the burst-mode technique is that, on the average, data are transmitted at less than half the maximum channel capacity. The hybrid mode makes use of the full channel bandwidth, but it is necessary to have some way of separating the transmitted and received signals. Silicon VLSI can provide the necessary functions, but the circuitry becomes fairly

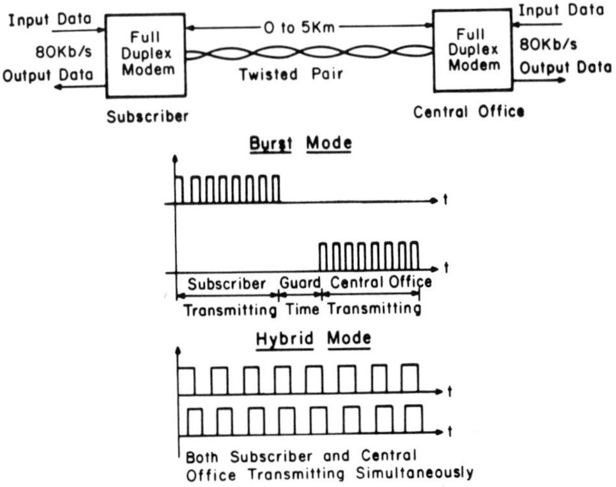

Figure 8. Two-Way Digital Transmission Modes

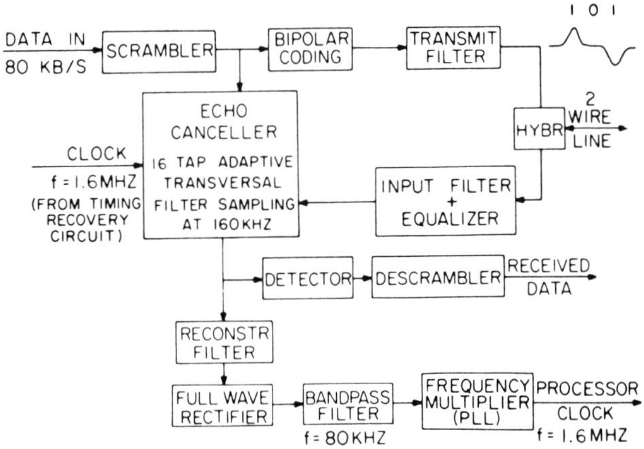

Figure 9. Elements of an End-of-the-Line Two-Signal Separator Circuit

complicated. Figure 9 shows the elements that are required to separate the hybrid-mode signals at the end of the line; included are a hybrid transformer (which may well be an iron-core device with copper windings, or it can be electronic) as well as a number of filters, digital and analog processing blocks, and timing-recovery circuits. Timing recovery requires the use of phase-locked loops, which ten years ago would have been regarded as a prohibitively costly system-level function to provide on a per-subscriber-line basis. Today it is quite feasible to consider implementing such complex devices.

A student of mine, Oscar Agazzi,[4] has studied such systems carefully and concluded that the entire signal-separation function can be performed on a single chip using a rather old LSI technology if nothing better becomes available. I will now present some results from an experimental system which performs this task. Figure 10 shows the transmitted pulse shape; for a variety of reasons, mainly to avoid DC on the line, alternate positive and negative-going pulses are used to represent logical "ones." Of course when data are being transmitted in two directions at once, the two signals become scrambled, as shown in Figure 11. Intersymbol interference arises due to phase distortion in the lines, echoes (reflections) from the ends of the line, and from the so-called bridge taps that are used at each network access port. After applying corrective echo cancellation, we recover the signal shown in Figure 12, which is quite adequate to work with. The traces shown in Figures 10-12 are called "eye" diagrams, and the eye-opening represents how well the logical states can be distinguished. There are three states in this case--the minus, the zero, and the plus state--and in Figure 12 they are very well separated. So with the addition of a few dollars worth of silicon at the end of each link, the investment in the line (several hundred to more than a thousand dollars per subscriber) can be upgraded in value so that instead of being limited to a few kilohertz bandwidth, data can be transmitted at 64 or 128 kilobits/second. In fact the ISDN standard now will be for two voice channels at 64 kilobits/second each (they need not be limited to voice) plus a lower-speed data channel for signaling or other functions. Consider, for example, a home computer with a 1200-baud modem; filling up the screen is a slow process. Now think about the advantages of having a 64-kilobit channel, which is exactly what can be gained simply by upgrading the modem.

Another application which cannot be satisfied with present microprocessors and memories but instead requires some more-sophisticated system engineering techniques is high-resolution television. The present television standard has been in place now for more than twenty years, and it was, of course, defined in a day when there was no VLSI and there was little or no opportunity for sophisticated signal processing in commercial television sets. Now what we would like to implement is an improved grade of television--

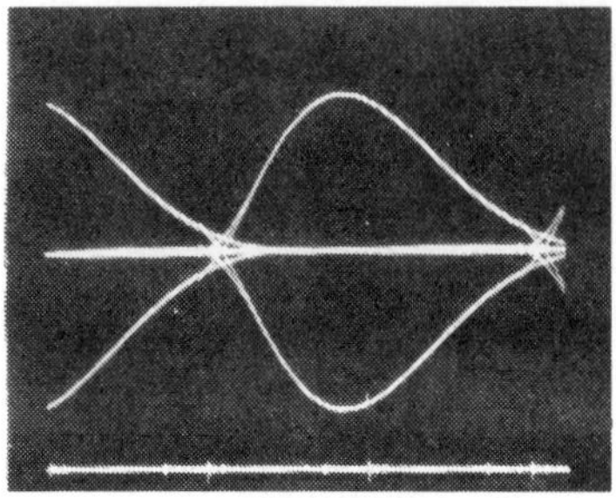

Transmitted pulse shape
200mV/cm
2 μs/cm

(©1982 IEEE)

Figure 10. Eye Diagram at the Output of the Transmitter. Peak Transmitted Signal is 600 mV. (From Reference 4)

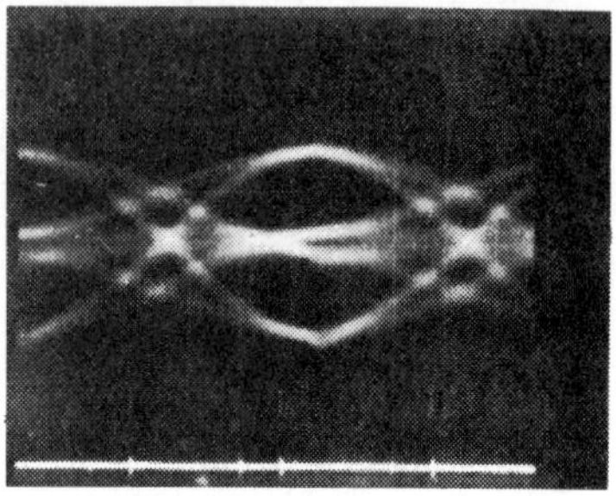

Local transmitter of
20mV/cm
2 μs/cm

(©1982 IEEE)

Figure 11. Eye Diagram at the Input of the Echo Canceller with Local Transmitter off. Scale: 20 mv/cm, 5 μs/cm. (From Reference 4)

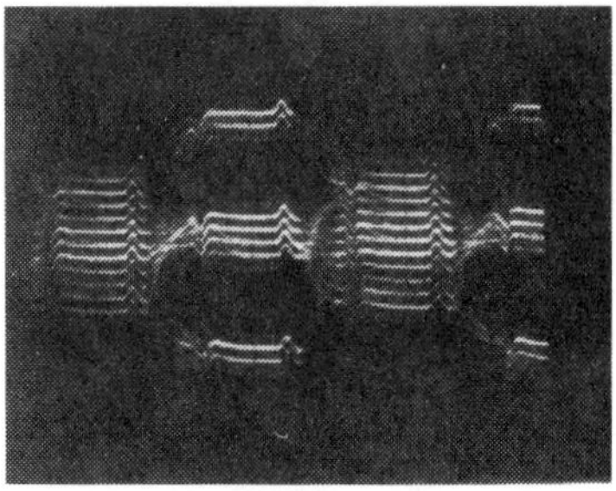

(©1982 IEEE)

Eye Diagram at Output of Echo Canceller
with 40 dB of Line Attenuation.

Figure 12. Eye Diagram at Output Echo Canceller: 2 μs/cm 100 mV/cm. (From Reference 4)

say twice the resolution in the vertical and horizontal directions. Unfortunately, the required transmission bandwidth does not scale linearly with resolution, and with the radio spectrum nearly filled up it seems very unlikely that more than the present four megahertz per channel will be allocated for television broadcasting. What could be done? A close look at today's television sets, with our 575-line standard, will allow you to see the individual lines in the picture.

Any engineer who looks at that picture and thinks about what might be done to improve it within a limited channel capacity would realize that there is a lot of redundancy in the picture; the illumination intensity doesn't change very much from one point to the next along each horizontal line. Similarly, excessive vertical redundancy is also present. Some degree of line-to-line signal comparison could eliminate the need to transmit every single picture element and instead transmit only the changes.

The resulting decrease in bandwidth requirements might possibly allow quadrupling the number of picture elements to be transmitted without occupying any greater bandwidth. What is required to implement such a scheme? Well, you can gain some idea of the complexity involved by noting that proper digital encoding of today's television signals requires about 14 million samples per second (14 megasamples) at 7 or 8 bits per sample. In order to perform the line-to-line correlations necessary to reduce the bandwidth, an entire line of video information must be stored, or 60 microseconds worth of data sampled at 14 million times per second. Even a one-line store requires about 7000 bits of very-high-speed memory. Sophisticated signal-processing circuits are then necessary to reconstruct the picture with a very small time delay—the minimum unavoidable delay in such a system is one line time, which is of no consequence whatsoever. I believe such systems are entirely possible. I was first made aware of this by Dr. Jerry Tiemann of General Electric Research Laboratory. I'm not sure how rapidly the technology is progressing, but it is an attractive prospect and an immediate application for very complex, very-high-performance VLSI.

Another reason why high-resolution video would be attractive is that conventional television screens are not well suited for displaying text; reproduction is fuzzy at best. With higher resolution, text services could be provided over the television with much-improved quality. A related application is facsimile transmission. Facsimile machines take from one to five minutes to send a single sheet of paper containing drawings and other graphic material. Sending straight text is not so much of a problem, but I think facsimile transmission would find a very broad market if it were inexpensive and could produce a page of text about as fast as a low-cost copy machine—say five or ten pages per minute. The necessary resolution would be on the order of 2,000 by 2,000 or more picture elements.

High-speed local subscriber loops could be implemented to provide several different services. For example, a 64-kilobit-per-second channel could be used to send a few megabits of data in considerably less than a minute; with image coding to reduce the redundancy, transmission could be speeded up further. Although these services will require a great deal of complexity, as well as lots of storage and signal processing, I think that they are feasible and attractive applications for microelectronics. I'm on the verge of buying my family a home copier, and I'm looking forward to the day when I can obtain an attachment that allows connection to an information-retrieval service in order to access graphic and textual information over the phone line at the rate of ten pages a minute.

Speech processing is another important application for microelectronic systems engineering, one which has been studied for many years now. I'm grateful to my colleague Bob Brodersen at Berkeley for the loan of some of his materials; Bob has been conducting research in this area with some growing success. Now of course everybody thinks first of all about a dictating machine that has the capability to turn verbal input into a beautiful final copy of a letter. Even before we reach this difficult goal, however, there are a lot of useful applications of speech input and recognition in electronic systems. Figure 13 shows a list of various levels of sophistication and complexity in speech recognition. Present experimental systems are advancing down into the lower

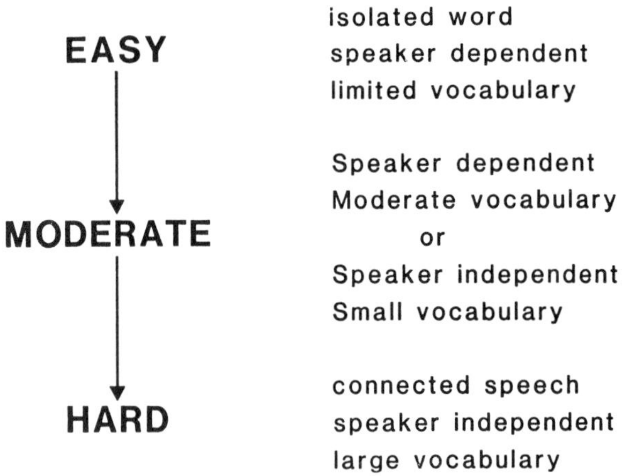

Figure 13. Speech Recognition System Criteria

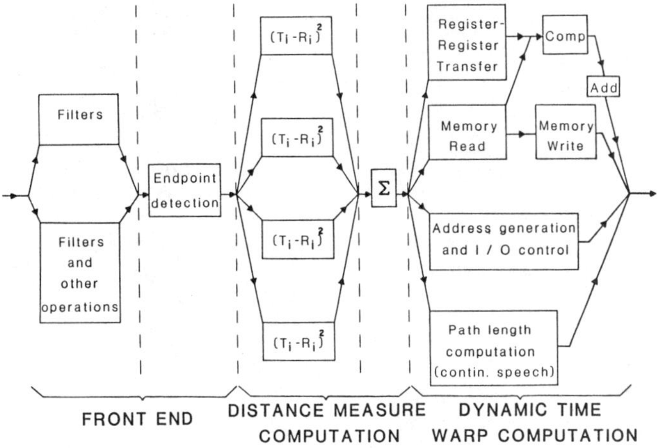

Figure 14. Basic System Architecture

range shown in Figure 13. Operators are not yet able to use a full 100,000-word vocabulary; but thousand-word vocabularies, with some degree of speaker independence, have already been implemented in experimental systems. In most cases, some speaker training of the system can improve performance; but if speaker training is not possible, the systems will still work.

Professor Brodersen and his students have built a number of systems with different architectures. Figure 14 shows an example of a system that has yielded considerable success. A number of custom chips have been used in building this system incorporating several "block functions." End-point detection, for instance, is used to decide when a word has ended; some quite interesting algorithms are used for this purpose. Next the machine must estimate which word in the stored vocabulary is the closest match to the one that it just heard. A "dynamic time warp computation" is required because a speaker might say "moon," for example, or "moooon," and the machine must be able to rescale the time axis appropriately to get the proper match. This amount of computation for a 1,000-word vocabulary is far beyond the capabilities of a standard microprocessor or digital signal processor, but by choice of an appropriate architecture it can be done essentially with a single chip. The heart of such a chip is a special-purpose digital signal processor such as the one shown in Figure 15; this particular chip has been demonstrated to work well even in noisy environments. Figure 16 shows an economic comparison of the

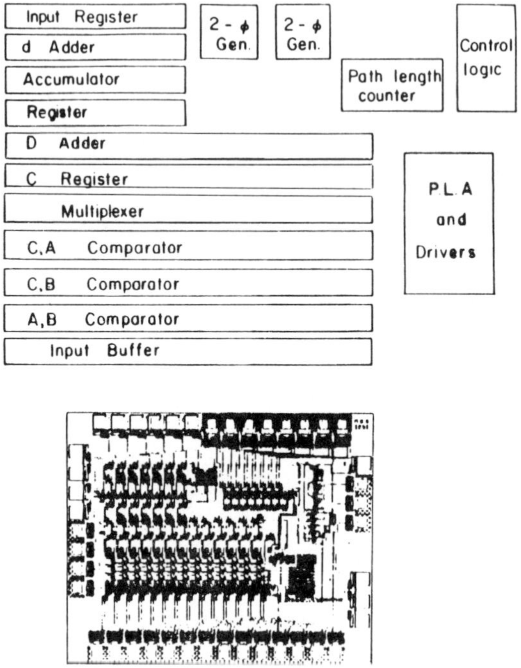

Figure 15. Special-Purpose Digital Signal Processor for Vocabulary Discrimination

"Berkeley" device with those of several other manufacturers. The entire unit consists of a board about the size of those which plug into personal computers and would cost a few hundred dollars. The most important parameter for performance comparisons is the error rate under similar test conditions. Researchers working in this field have standardized some speech-recognition accuracy tests, and the Berkeley system with its specialized digital signal processing performed very favorably in comparison with some of the others listed in Figure 16.

One example of an application of speech-recognition systems can be found close to home. In our laboratories on campus, we lay out a large number of integrated circuits as shown in Figure 17. This task involves manipulating graphic-entry tools such as mice and tablets. Circuit designers do not like to take their hands off the graphic-entry device to type commands at a keyboard,

1. Single Poly N-MOS process 5 micron minimum feature
2. 5 V-single supply 400mW power dissipation
3. Clock frequency for 1000 words recognition: 5 MHz
4. Recognition accuracy

DODDINGTON EXPERIMENT

Manufacturer	Nominal Price	Errors
a	$85,000	10 (0.2%)
Berkeley		45 (0.9%)
b	$65,000	60 (1.2%)
c	$12,000	73 (1.4%)
d	$ 2,400	147 (2.9%)
e	$ 3,300	300 (5.9%)
f	$ 3,500	388 (7.1%)
g	$ 500	646 (12.6%)

Figure 16. Results of Speech-Recognition Comparison Tests of Several Systems

so commands in the graphic-design system now may be inserted through a voice-entry system. Even a limited vocabulary works extremely effectively and does indeed speed up the work of a designer entering a layout design.

- speaker dependent
- isolated word
- ~200 word vocabulary
- now in use

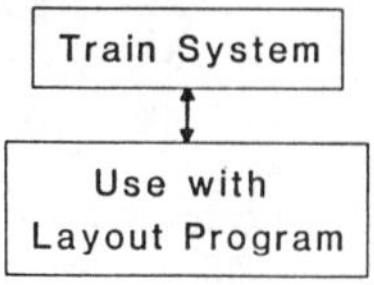

- use voice for commands
- hands free for system operation
- keep eyes on graphics
- satisfied users

Figure 17. Criteria for Speech-Recognition System for Integrated Circuit Layouts

(a)

FUNCTION	OPERATIONS/sec
VISION	10^{10}-10^{11}
NAVIGATION AND MOBILITY	10^{9}-10^{10}
MANIPULATION	10^{8}-10^{9}
SPEECH AND NATURAL LANGUAGE	10^{9}-10^{10}
COMBINATORIAL SEARCH	10^{9}-10^{10}
EXPERT SYSTEMS	10^{12}

(b)

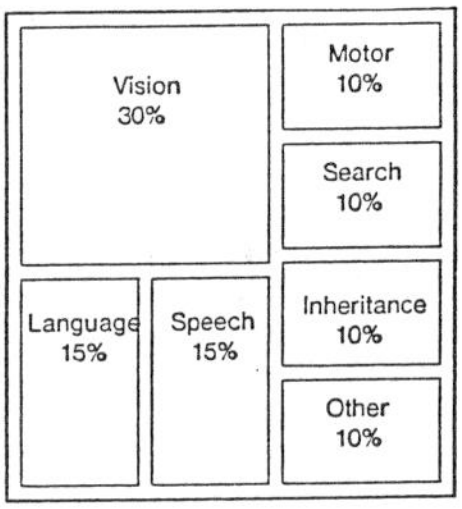

(©1985 IEEE)

Figure 18. Criteria for Artificial-Intelligence Systems

At the recent ISSCC meeting, Professor Raj Reddy[5] of Carnegie Mellon University described in detail how the field of artificial intelligence is ready to consume all of these billion-transistor chips that we're going to be making. I was persuaded that AI certainly can take advantage of the available complexity, although I remember hearing projections of the complexity required to perform some typical artificial-intelligence functions (see Figure 18a) some years ago, and they always seemed to be about two or three orders of magnitude ahead of what we were able to do easily. I do hope that the rate of progress in artificial intelligence and the required complexity of the calculations slows down a little bit so that the technology can catch up. Raj Reddy also foresaw billion-transistor chips, and he has stated that a system comprising something like a hundred of those chips might constitute a useful artificial-intelligence system (see Figure 18b). Such a system might find a very broad market.

In conclusion, my impression is that our technology has progressed very rapidly and perhaps even more rapidly than our skills at exploiting it usefully. It has become clear that we are going to have to go beyond simply doing more of the same things better than we have done before. If not, it will be the system-application skills that limit the applications of microelectronics technology. What we need is some very imaginative thinking about useful applications and about intelligent ways of breaking them down and implementing them with the technology we have available. Very little talk is heard anymore, for instance, about actually duplicating the physical structure of the human brain in answer to the needs of artificial intelligence. I think most people feel by now that there is no particular reason why electronic technology that operates in nanosecond time scales has to be hooked up like biological technology which operates in millisecond time scales. But it is still true that progress has been slower in systems applications than it has been in device technology, and very intensive efforts will be required in order to realize the maximum advantage of microelectronics technology.

References

1. Hae-Sung Lee, David A. Hodges, and Paul R. Gray, "A Self-Calibrating 15 bit CMOS A/O Converter", *IEEE Journal of Solid-State Circuits*, Vol. SC-19 No. 6, Dec. 1984, pp. 813-819.

2. *ibid.* p. 817

3. *ibid.* p. 818.

4. Oscar Agazzi, David A. Hodges, and David G. Messerschmitt, "Large-Scale Integration of Hybrid-Method Digital Subscriber Loops". *IEEE Transactions on Communications*, Vol. COM-30, No. 9, Sept. 1982, pp. 2095-2108.

5. Raj Reddy, "Super Chips for Artificial Intelligence", 1985 IEEE International Solid-State Circuits Conference, *Digest of Technical Papers*, IEEE Publ. 0193-6530/85/0000-0054, pp. 54-55.

YANKEE INGENUITY REVISITED

Robert N. Noyce

Robert N. Noyce received the B.A. degree at Grinnell College in 1949 and the Ph.D. in physical electronics at MIT in 1953. After three years of research at Philco Corporation, he was with Shockley Semiconductor Laboratory in Palo Alto, California until 1957 when he co-founded Fairchild Semiconductor Corporation, Mountain View, California, and remained with that company until 1968 serving as research director, vice president, and general manager. In July 1968, Dr. Noyce co-founded Intel Corporation with Gordon E. Moore, served as President until 1975, and as Chairman of the Board until 1979. He is now Vice President of the Board of Directors. Dr. Noyce is co-inventor of the integrated circuit with Jack Kilby. They have jointly received the Ballantine medal of the Franklin Institute, and the Cledo Brunetti Award of the IEEE for this work. With Gordon Moore he has received the AFIPS Harry Goode award for leadership in computer science. Dr. Noyce was awarded the National Medal of Science and the I.E.E. Faraday Medal in 1979, and the IEEE Medal of Honor in 1978. He is a member of the National Academy of Science, the National Academy of Engineering, the American Academy of Arts and Sciences, and is a Fellow of the IEEE. He holds 16 patents for semiconductor devices, methods and structures.

When asked what I would talk about, I thought long and hard and came up with "Yankee Ingenuity Revisited" as my title. I think you will understand this choice of title as I proceed through my talk; the subject matter is not concerned so much with technology as with some of the other limiting factors in the microelectronics industry. In previous talks at this meeting, you have heard about how we are advancing microelectronics to higher and higher levels of integration. In the past, such advances have typically been achieved by identifying limiting factors and then seeking a way around them. It is my understanding that we may be able to integrate with appropriate technological advances up to a billion transistors on a single chip by the end of the century.

Such new technologies as diffusion, lithography, epitaxy, or computer-aided design all arose from systemic efforts to bypass various road blocks that stood in the way of further progress in microelectronics. Present barriers will be broken down by the introduction of X-rays or electron beams or bigger computers or whatever. I guess that is what I have referred to as Yankee Ingenuity—diligence in finding solutions around technical problems. There are some other characteristics that come to mind when I think of Yankees. Like self reliance. And frugality, saving money and investing it wisely. Such traits have been the source of much success. We have now attained the highest standard of living in the world; and as Jude Wanniski, economist and author says, people are voting with their feet—they want to come to America.

There is, however, an additional limit being imposed on our ability to exploit microelectronics and that is a limit on our ability to compete economically. We are quite adept at solving our technical problems, although I understand that you have heard earlier in this conference that perhaps the U.S. solutions are not necessarily the winning solutions, that indeed we have competitors overseas that can outdo us both in terms of innovation and investment. As we look around, the engine which has generated so many benefits for the population of the U.S. has been slowed down a lot. Perhaps rightfully, we have been concerned with equity versus productivity; we have been concerned with today instead of tomorrow. We have sought a risk-free society and that of course leads to a society that, overall, is profit-free as well. There has been rising concern on this matter, sufficient concern in fact that the Presidential Commission on Industrial Competitiveness was formed in 1983. I served on that commission, chaired by John Young from Hewlett Packard; and its report was issued last month under the title "Global Competition—The New Reality." I can say I contributed that title, but I am not sure that I contributed a great deal more to the report because it was really a consensus of a large number of people. The commission drew its membership from both small and large industries, from academia, from labor, from government—both Democrats and Republicans—and they did reach a surprising degree of consensus. There was a strong feeling that we need a new focus, a new shock, a new sputnik to which we might respond by rekindling our efforts to accomplish the equivalent of landing a man on the moon.

First of all, let's discuss what we mean by competitiveness. There are many definitions, but the one we chose to work with was as follows: competitiveness is the degree to which a nation can, under free-market conditions, produce goods and services meeting the test of international markets while maintaining or improving the real income of its citizens. There remain several difficulties with the above definition. For one, we compete in markets that are not really free. In addition, determining real wages is not all that concise; we must decide whether to measure before or after taxes, whether to consider consumer taxes, etc. It would appear that we as a nation are not going to choose to lower wages in order to compete in the world; I have not heard anyone suggest such a solution, although in all probability it would work. Indeed, if we are not careful, lower wages may be imposed upon us rather than being a course of choice. So our challenge is really to earn the higher wages that we desire to receive in this global economy.

Let me offer a pair of interesting statistics. A quarter of all goods produced in the world cross national borders. Seventy percent of all of the goods produced in the U.S. compete with products manufactured elsewhere. There really is no free lunch to be had here; competitiveness is the currency which pays for whatever we place on our personal or public menu. The question then is whether or not we have a problem. Those of us in the microelectronics industry don't need to think twice in order to answer that question; we are being subjected to severe international competition. Whereas failing competitiveness in the steel and auto industries may have been due to mismanagement or a lack of awareness of what was going on in the rest of the world, that was certainly not the case in microelectronics. This industry was international from its beginning; and most of us in microelectronics have always paid very strict attention to what has been going on overseas, in order that we might remain competitive in the international markets. This international awareness began long before we had significant problems with competition from overseas.

There are many indicators that suggest the presence of a definite problem. In absolute terms, Japan is more productive in autos, steel, electrical and precision machinery; and competition is certainly to be found in these markets. Wages in the microelectronics sector, as well as overall wages in the U.S., have remained essentially stagnant since 1973; at this point, I ask you to remember our definition of international competitiveness. We haven't improved our real wages. The return on investment in the manufacturing industries has dropped sharply to an average of about 4 percent in 1983, down from 14 percent 20 years earlier. There are those who claim that our economy is moving to a service economy and that we need not worry about manufacturing, but the Commission on Industrial Competitiveness concluded that a manufacturing basis was essential to the service industry. After all, who do the consultants work for? Who buys the computers? Another measure of our declining competitiveness is the trade deficit. For the century before 1971, the U.S. had maintained a positive trade balance. Since that time, however, there has been a steady decline to the point where last year saw a deficit of $125 billion. The numbers look almost like an upside-down exponential when plotted over the past few years. One explanation is that the dollar has grown stronger; but in fact, this trend started in the 1970's when it was generally acknowledged that the dollar was weak. A particularly direct indication of our declining fortunes is that since 1965 seven out of ten U.S. high-tech industries have lost market share. In the semiconductor industry, such losses are accelerating rapidly. In the first quarter of 1984, the Japanese held roughly 14 percent of the U.S. market for microelectronic components; by the fourth quarter, the Japanese share had increased to over 18 percent of the U.S. market. During the same period of time, the U.S. share of the Japanese market declined from 11 to 10 percent. Last year the bilateral balance of payments with Japan in electronics products was $15 billion—a figure which exceeds that of the automobile industry. The question, of course, is what can we do? This erosion of U.S. market share has been underway for a couple of decades; we have taken a long time to get into this situation, and it will take us a long time to get out as well. There are no simple solutions.

The Commission agreed on four general areas that need to be addressed. First of all, it was concluded that technology is our strongest advantage and our greatest asset; and yet the U.S. is spending a smaller percentage of its GNP on commercial R&D than is Japan or West Germany. Roughly 50 percent of R&D in this country is supported by the Federal Government, and two-thirds of that is on military and space programs. The fallout to industrial applications from such programs is only incidental, so that if the objective is to improve our industrial technology, military programs do not contribute a great deal. Based on the above conclusion, the Commission called for the formation of a Cabinet-level department of Science and Technology—a suggestion that has been very controversial, particularly among those who receive research contracts from the Federal Government. I think that it is largely because the Devil you know is better than the Devil you don't know. But certainly today technology is more important to our economy than, for example, agriculture or transportation; and there are in existence federal departments in both these areas. It is necessary to improve the form and coordination of the various research programs that are funded by the Federal Government and to aim more specifically at providing the technology needed by American industry to remain competitive. About a year ago, David Packard produced a report on the Federal Laboratories (there are some seven hundred of them) concluding that there were major inefficiencies in these laboratories and that we needed to take steps to create linkages between government laboratories and American industry. One count shows that there are 2,700 different R&D programs that are subjected to line-by-line review by various government agencies. The Commission called for the renewal of the R&D tax credits--something that the Administration has felt inclined to do--and indeed the R&D tax-credit suggestion was included in the Treasury tax bill. There is a lot of risk in doing good research; it does create a public good, and those who pay for the research cannot possibly reap all the benefits. I call to mind the advent of the microprocessor, which certainly contributed many advantages to a broad spectrum of U.S. industries that did not contribute a dime to its development. This fact should be kept in mind this year by those engaged in budget cutting. Various studies indicate that government-funded research in microelectronics will generate revenues exceeding the tax costs, although there is a timing difference--it costs money this year for reaping benefits next year. Unfortunately, such arguments have usually fallen on deaf ears at the Treasury, although we can sometimes convince congressmen that it is indeed true. I might mention that the R&D bills that are currently before Congress provide "zero-based" R&D tax credit for contributions made to university research, and I would encourage those in academics to get solidly behind such bills. At the present time, R&D tax credits are incrementally based; tax credits are allowed for contributions higher than those of the previous year. This scheme was aimed at trying to improve the level of contributions to university research, but it would be far more effective to allow tax credits for the entire contribution rather than for the incremental amount.

The next stage, of course, involves applying the technology that has been created. The Commission found this to be a glaring weakness in the American system today. We are devoting too little attention to the application of new manufacturing technology. It does us very little good to create all of these wonderful new products if the Japanese beat us to the production line. But manufacturing simply has not been stylish with us recently. Within industry, manufacturing managers have been paid less than R&D managers--a state of affairs that seems to neglect the importance of manufacturing to a company's "bottom line." Within our universities, there has been relatively little interest in teaching manufacturing, whether it be manufacturing processes, process technologies, or manufacturing management. Only five or six years ago I asked the administration at one major U.S. business school what courses they had in manufacturing and the answer was none. Even now, the number of universities doing work in this area can be counted on one hand.

Creating and applying technology are just two steps in a competitive strategy. Innovative ideas must be protected from counterfeiting and other forms of misappropriation. Last year the semiconductor industry finally pushed a chip-protection bill through Congress, providing a copyright on chip designs. It looks now as if a similar bill will be passed in the Japanese Diet, although there are some problems with the language of that bill. There remain severe problems elsewhere, however, such as in Korea and Taiwan—countries that are rapidly emerging as significant competitors in the microelectronics industry. Last year the International Trade Commission suggested that counterfeiting alone costs American business $8 billion in sales and over one hundred thousand jobs. If we are going to protect our intellectual property, then we must address these areas. First, the patent laws need a thorough review, both in terms of coverage and in process. Right now seven years are required to get a patent award, whereas technology is changing on a much shorter time scale. Secondly, there has been an abuse of the Freedom of Information Act. The original intent of the Act was most worthy, but it is often subverted to gain access to scientific information that American business is required to supply to the government. Consequently, outside competition has been able to get its hands on such information. Third and most important, we are going to have to negotiate better protection with our trading partners. Many countries, especially the newly industrialized ones, offer little if any protection for intellectual property.

Now let us assume that as a nation, we do a magnificent job in our research, develop a wealth of technology with commercial potential, quickly and broadly apply this technological innovation to create market advantage, and fully protect our intellectual property—what next? All of the above will be for naught if we cannot move the technology to production because of prohibitive costs of domestic capital. To anybody but a banker, the word "capital" doesn't invoke very fond feelings. But capital is really the fuel of our economic machine. There is clearly a correlation between the level of a nation's investment and its improvement in productivity. A ranking of our six major trading partners in terms of capital formation is the same as their ranking in productivity growth. I might mention that the U.S. is at the bottom of that list, and Japan is at the top. Looking into the reasons for the low level of U.S. investment, one of our great surprises was to find that economists from a wide range of backgrounds agree; their consensus was that high capital costs are the major competitive disadvantage for American firms. The cost of capital in America is about twice as high as in Japan. This disparity hurts the ability of U.S. firms to compete in international markets. In fact, some studies have concluded that lower capital costs, and not technological supremacy, was the prime factor behind the Japanese incursion into the U.S. semiconductor industry. If we are going to reduce the cost of capital to American industry, we have to deal with some macroeconomic issues. First, we have to cut the deficit. The government must stop competing with industry for scarce capital resources. Federal borrowing pushes up interest rates and that makes the dollar strong. Since 1980, the value of the dollar has almost doubled against major European currencies. For companies trying to sell in international markets, that means higher export prices and lower import prices. Secondly, our tax system must be restructured. This is certainly the year for tax revision in Washington, and the de facto industrial policy implied by the present tax code discourages savings and encourages consumption. Without savings today, there will not be any capital to invest tomorrow. The tax code also assigns the highest effective tax rates to that sector of the economy most affected by international competition—the manufacturing sector. Rather than wading into the mire of tax reform, the Commission proposed a series of criteria that could be used to judge the competitive consequence of each reform package. Among those would be how far along the various proposals advance toward the goal of a more neutral tax treatment between different industries and between various kinds of assets. We also recommended ways of encouraging investment, such as indexing for capital gains and allowing fuller deduction of capital losses on individual tax returns. There is a third way to lower the cost of capital to American firms and that is to pursue a more stable monetary policy. The final report contains one graph in particular that plots the variation in consumer price index and the prime interest rates since 1971, and it resembles a picture of the Sierras, jumping up and down erratically due to an unstable monetary policy. Those lending money have to protect themselves against potentially higher interest rates and higher inflation rates in the future, forcing them to add risk premiums to their loans. American business thus has a reasonable excuse for its short-term investment practices—it's pretty hard to do long-term planning and investment in a wildly changing business environment. High capital costs force business to look to the short term for quick return on their capital investments.

Capital may be the fuel for our economy, but people are an equally important resource—a subject which I will discuss only briefly this afternoon. We know that everyone involved in a business needs to share in the common success of that business. For Californians, programs like employee stock purchase and profit-sharing plans are commonplace; elsewhere they are still considered somewhat revolutionary. We know that this nation's research universities face severe stress from faculty shortages and equipment obsolescence, so we support such incentives as credits for contributions of equipment to universities or such programs as the American Electronic Associations Education Foundation. We join together to fund collaborative university-industry research such as the Semiconductor Research Corporation or the effort here at Stanford University's Center for Integrated Systems. In mentioning our own human resource initiatives, I certainly don't want to imply that we in this industry have a monopoly on virtue. I also don't think that we can rest on our laurels when it comes to the human side of management; we are still going through severe employment fluctuations in our industry and that is not the proper way to treat our work force. Still, we have earned an enviable reputation and now the challenge is to continue to deserve it.

So far I have talked about three factors that effect our ability to compete—technology, capital, and human resources—and now you should know why I say that improving competitive performance is not a simple task. Let me move on to a fourth area which I mentioned earlier, that being the nature of the international trading environment. The Committee's first conclusion was that international trade has not been a priority item for our nation. We came out of World War II with devastated economies around the world, and our primary objectives were to help rebuild those economies and trade with the affected nations so that they would become stable. That was a long time ago and the situation has changed enormously, but we have not reorganized sufficiently to address the changes. Responsibility for trade policy is highly splintered. In one of the diagrams in our final report, we tried to summarize who makes what decisions necessary to implement various policies affecting trade—the resulting plot resembled a ball of string. Clearly, there is no clear cut responsibility, with the Department of Commerce on one side, the U.S. Trade Representative on the other, and in between, each owning their own piece of the action, are the Departments of Defense, Treasury, Agriculture, and State, along with a host of other executive agencies and congressional committees. The layout of our trade policy formation processes appears more complicated than an integrated-circuit design schematic. The complexity and lack of accountability make it impossible for us to deal effectively with the growing importance of trade and the many accompanying issues that must be resolved. We therefore recommended the creation of a Cabinet-level Department of Trade to provide a single strong voice on trade issues. We were encouraged to hear President Reagan endorse such a move in his State-of-the-Union address, and I am told that the White House has formed a task force for that purpose. If we make trade our national priority and can find effective leadership in addressing the domestic issues that affect our ability to trade, there are several goals we should pursue in order to get our own house in order. First of all, we need a new omnibus trade bill that provides ways to help U.S. industry adjust to international competition before the damage is irreparable (I have likened our current response to the embalming of a corpse). The bill should also clarify how we intend to respond to industrial targeting by our trading partners. Secondly, we need to find a more reasonable approach to export controls. The Commission heard testimony that put U.S. industry cost of current controls at $12 billion a year in sales, which translates into a lot of jobs. We also need to streamline the licensing process for those technologies that we do choose to export. More often than not, the delays in getting export licenses are themselves a major competitive disadvantage; many of us have horror stories to tell of waiting for an export license while the Japanese accepted and delivered the order. It now takes American exporters five or six times as long to get an export license as it does an average European company or the average Japanese company. Third, we need to look for more ways to encourage U.S. exports, including better export financing, better information about foreign markets, and the active support of U.S. industries abroad. Looking at the international trade environment in which American business operates, we observe two simultaneous trends which point in opposite directions. On the one hand, the total volume of world trade has grown enormously, registering a seven-fold increase since 1970. On the other hand, the portion of trade that has been covered by mutually established rules of trade has shrunk dramatically. Trade laws and trade agreements have simply not kept pace with the changing business environment. There is no coverage for trade in services or investments, and there is little provision for agriculture or state-owned industry. While tariffs have come down, the use of non-tariff barriers has increased significantly. As is the case in American trade law, international rules have not responded adequately to government targeting practices and similar non-tariff trade barriers. The newly industrialized nations have the weakest commitment to trade rules. We must strengthen the international trading system by increasing the amount of trade and the kinds of practices it covers, and by ensuring adherence to the rules by our trading partners, especially the newly industrializing countries. Again, President Reagan in his State-of-the-Union address has called for a new round of negotiations on Government Agreements on Trade and Tariff (GATT), which is a step in the right direction.

Let me summarize the major areas that I have discussed. Our goals are first to create, apply, and protect the technology, our greatest competitive advantage and one for which we need to work diligently. Secondly, we need to decrease the cost of capital available for investment in America. Third, we need to develop a more skilled, flexible, and motivated work force. Fourth, and finally, we need to incorporate international trade as a national priority here at home and to strengthen the world trading system in which we operate. It seems to me that all of these represent a return to the values of some years ago that served us very well. I guess in that sense I can be considered an arch conservative, but I think that overall this will be the most liberating action that we can take for the American people.

INTERNATIONAL COMPETITIVENESS IN ELECTRONICS

William G. Howard, Jr.

William G. Howard, Jr. obtained the B.E.E. degree (with distinction) in 1964 and the M.S. degree in electrical engineering in 1965, both at Cornell University, and the Ph.D. in 1967 at the University of California, Berkeley. He was an assistant professor of engineering and computer science at Berkeley until 1969 when he joined Motorola, Inc. where he served in a number of engineering managerial positions related to linear integrated circuits until 1977. In that year Dr. Howard was named director of strategic operations for what then was the integrated circuits division of the Semiconductor Group, later becoming vice president and director of technology and planning for the Semiconductor Products Sector. In 1983 he was appointed to his present position as Senior Vice President and Director of Research and Development. Dr. Howard was named an IEEE Fellow in 1981 and has held a variety of positions with IEEE, including vice president of the Circuits and Systems Society, Chairman of the 1977 International Symposium on Circuits and Systems, and representative for university accreditation and professional development. He holds patents for an improved reference current source, a ladder termination circuit and a zener diode having three terminals. He has published 10 papers on various subjects and is co-author of the book, *Basic Integrated Circuit Engineering*.

Charles Dickens, in his classic "A Tale of Two Cities," characterized the French revolution as "the best of times, the worst of times." That description applies as well to today's electronics industry. 1984 was a year of record sales and profits for many electronics segments, accompanied by continuing concerns over the bookings and increasing competition from abroad.

Electronics technology, as well, is experiencing a "best" and "worst" times phenomenon. Our core technology developments are continuing their forward march, but we in the U.S. have awakened to the fact that our leadership of the parade may be in question. The whole question of our international competitiveness is a vexing one, both in business and technological respects. It seems so abstract, yet it has tremendous impact upon our success in short-term business. Industry is concerned about its competitive position, labor is concerned about it, our government is concerned about it, and if anyone here isn't concerned, you've been living on another planet, or at least have undergone a twenty-year Rip Van Winkle sleep, since the issue has received extensive coverage in all sorts of public forums.

We all have read newspaper articles, seen TV documentaries, heard our national leaders speak, attended courses and seminars, and talked to one another extensively about this problem, how it might affect us, and what we should be doing about it. My company even has assumed an unusually public position on its feelings about its own and our nation's competitive position.

Do we have an international competitiveness problem?

Consider, first, the trend in sales of the world's leading electronics firms outside their home countries as shown in Figure 1. Note that during the four most recent years for which data are available, the international percentage of sales of all but one of the U.S.-based electronics leaders has fallen, while those of non-U.S. companies has remained the same or risen. Clearly, the most impressive gains have been made by Japanese-based competitors. While some of this shift is attributable to international currency exchange rate fluctuation, that factor alone cannot account for the trend depicted here.

The U.S. electronics trade balance shows a similar story. Figure 2 indicates what has happened to the various components of the U.S. Electronics Trade balance over the past five years. Although we remain strong in computers and communications, the trend is downward; and we have slipped to the negative in every other category. The result is a net $6.8 billion trade deficit for 1984, and the American Electronics Association estimates a $12 billion deficit for 1985. These data are representative of those that show a business problem, but we have always regarded American technical leadership as our strong suit. What about that?

Figure 3 illustrates trends in the origin of U.S. patents filed over 14 years. Note that patents of U.S. origin are now being filed at a rate less than 60 percent of their 1971 peak, yet the number of patents of Japanese origin has

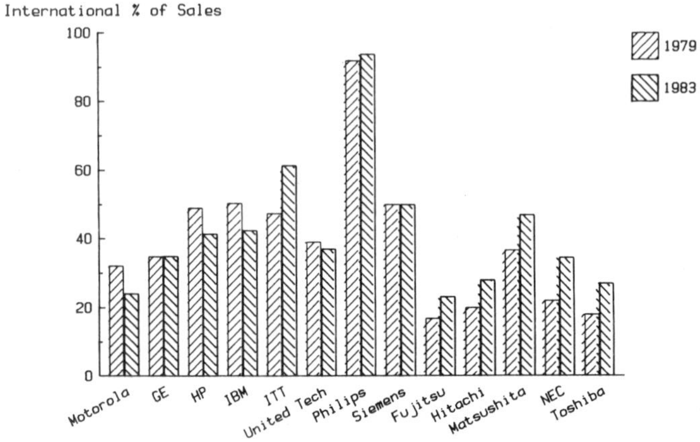

Source: Company Reports

Figure 1. International Business Mix -
Major Electronics Companies - 1979-1983

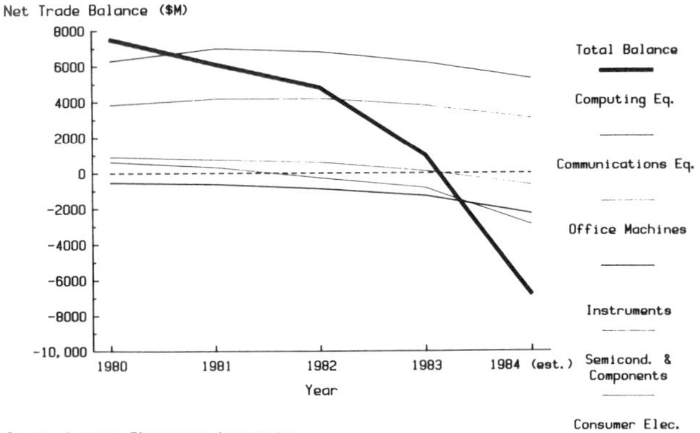

Source: American Electronics Association

Figure 2. U.S. Electronics Trade Balance (total export minus total imports)

climbed steadily. The trend is more disturbing if we restate this in terms of percent, as displayed in Figure 4. Clearly, the Japanese technical commmunity, in spite of its reputation for "me-too" thinking, has been very inventive. Other well-known data comparing origins of conference papers, numbers of technical professionals, progress in the semiconductor and computer fields, and similar indicators illustrate a problem in the technical realm as well as in the business mix.

In cases of this kind, it is often helpful to turn to the thoughts of others who have studied this problem and expressed ideas that might be helpful in the hopes that they may have seen things from a different perspective. I'd like, then, to reflect on a report I found recently of a government study group made up of the best available talent from both public and private sectors which has addressed the international competitiveness problem. The group travelled extensively abroad to study the issue and to make recommendations on means to solve it. The committee was established when it became apparent that there was a serious international competitiveness problem facing the nation. After its extensive travels, discussions, and studies, this group verified the seriousness of the initial concerns and urged that immediate action be taken.

The report notes that foreign competitive goods frequently have better performance, better quality, and cost less than those made at home. Furthermore, as a result of an unusually open attitude on the part of the competition (in fact, the committee was allowed to tour factories, and offers were made by competitors to supply manufacturing equipment), the committee found that those products were made on much more automated machinery. In fact, this study group was awestruck by what they saw in off-shore factories. The facilities, although not as polished as those at home, produced more with the same or poorer resources; and it appeared that the successes realized had been the result of good application of basic principles; diligent management; good, inventive ideas; and hard work. The committee noted that there was far less inventory evident and that the manufacturing process was much better organized than the standards with which they were familiar. The committee went on to note that the off-shore competitor very often used technology invented elsewhere but put it to far more effective use than the inventor.

The report notes with alarm that competitive companies abroad have adopted a philosophy of mass-manufacturing standard products in very large volume with attendant low costs, and that these products were selected to penetrate the export market aggressively. In fact, the lower cost base of foreign goods was allowing the competition to dominate markets worldwide that had previously been served by others. The extreme competitive pricing policy of off-shore manufacturers was seen as the result of illegal, or predatory policies. The report also pointed out what we have observed—that the off-

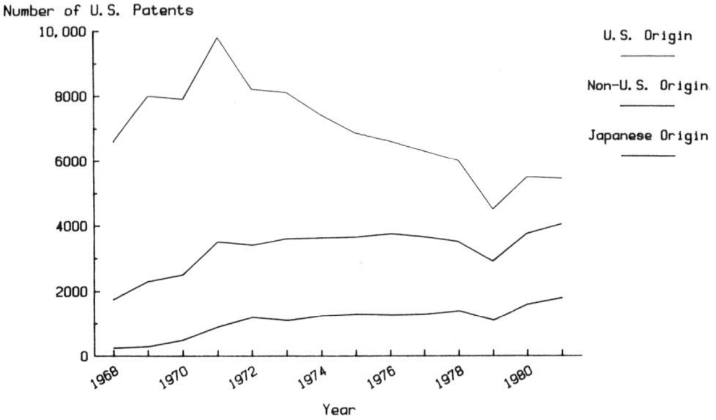

Source: Office of Technology Assessment

Figure 3. U.S. Patents in Information Technology

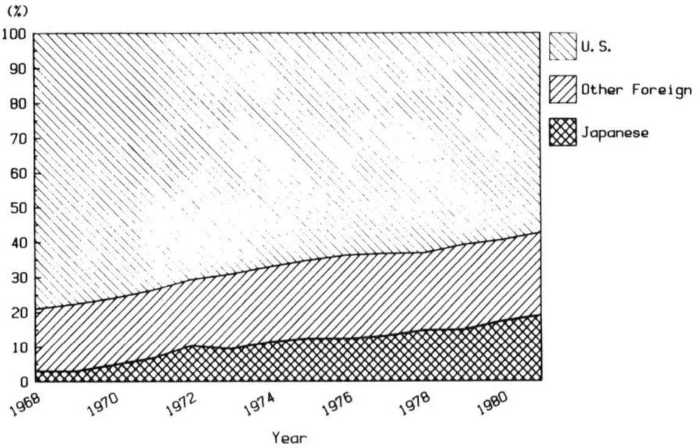

Source: Office of Technology Assessment Data

Figure 4. Percentage of U.S. Information Technology Patents by Origin

shore competitors have an extraordinarily hard-working labor force that is very well educated. It cooperates with its management, rather than resisting. In turn, these competitive companies respond with paternalistic approaches to employees, even housing them, thereby increasing dedication to company goals and objectives. In addition, the committee reports that an unusual spirit of cooperation exists between these foreign competitors and their governments and that these governments seem to side with industry, rather than against it.

Finally, the committee found that the military authorities at home, alarmed with decreased quality and functionality of weapons produced domestically (a situation dramatized by a number of well-publicized confrontations with uncooperative suppliers), were being urged to turn to off-shore sources for technology and arms essential for the country's defense. In fact, the military itself was a contributor to the problem by insisting upon cumbersome purchasing regulations and complex controls for weapons systems procurement and resisting insertion of new technology in weapons systems.

In the committee's opinion, domestic industry was either unaware of the problem, or was stymied in its attempts to counter the threat to its own existence. Much of the domestic industrial base was built during a war forty years before and had not been upgraded since, even though the off-shore competitors' plants were much newer. Many companies were resistive to new ideas, insistent upon continuing to do business in time-honored fashion, and were extremely reluctant to invest in improvements needed to counter the off-shore threat. The report foresees that failure to adopt new approaches would lead to loss of important parts of the national industrial strength. It is also clear that a large part of the problem was managerial. Management of these companies seemed to fail to grasp the extent to which their companies had to make serious changes in their business practices to meet the new threat.

All in all, this report reflects serious alarm about the international competitiveness situation, and the damage being done to the economy, and presents a series of steps to be followed in trying to meet and overcome the challenge. In essence, these recommendations boil down to advising against import and foreign purchasing controls and for aggressive steps to upgrade the country's industrial strength, reasoning that, as the world's technological leader, wealthiest economy, and largest market, the inherent economic strength to accomplish all this is at hand.

This recitation of the findings of a relatively dry document sounds like any of a number of similar calls to action that we have heard over the past several years. So what's the sense in talking about this report now? There *is* something different. This report was not written by our government about the Japanese, or the Koreans, or anyone else; in fact, this report was not written by the U.S. government, it was written by the British—about the

Americans! The report, reprinted in a 1950's book, *American System of Manufacture*, by Stanford's Nathan Rosenberg, was dated 1855.

While the concern of the British was mechanical manufacturing, interchangeable parts, and mass-production technologies developed by the Americans, the parallels between this case and our present electronics world are inescapable. It is clear that the British did not overcome their international competitive problems. The malaise of the mid-nineteenth century festered out of sight for the remainder of the 1800's and burst into full view following World War I. By that time, it was too late to cure the disease which remains evident to this day.

The advice given the British government by the committee was correct. They foresaw the problem, they correctly identified the steps needed to be taken to solve it, and yet the British did not succeed—why? What lessons can we learn about the British experience that could help us today? Others have speculated on government investment and trade policy and industrial management aspects. I would like to comment on thoughts affecting microelectronics technology and technologists. The number of parallels between British history and our own present day are many; some, indeed, may be quite profound. This afternoon, I would like to consider some which may have implications to us today.

Inflexibility lay at the heart of the British inability to cope with the new world-competitive order. That inflexibility manifested itself in numerous ways: The concept of hand fitting lay at the heart of the British system of manufacturing. In each of the industries studied by the committee (machinery, textiles, publishing, metal manufacturing, glass and ceramics, lock making, woodworking, masonry, rubber, railways, and telegraphy), British practice involved high labor content to hand-tune products or services. This was particularly apparent in firearms production, where individual pieces were each hand-fitted by filing. In fact, the equivalent of one file was consumed in the production of each musket. The Americans, on the other hand, had invented (or in many cases adopted from the British) machines for precision machining of metal components, and thereby eliminated the need for fitting. While the underlying technologies for the milling machine and the turret lathe arguably were of British origin, the Americans put them to work in practical machines to accomplish useful ends. Nothing about the new precision methods was secret—they had been openly discussed in the literature for 20 years prior to the awakening of British concern, yet British industry itself had not adopted them. This was partly the result of the structure of the Birmingham metal-working industry, which strongly reflected dedication to hand-production methods. A large part of the problem, however, lay in sheer ignorance. British metal workers, secure in their traditional methods, did not pay attention to developments across the Atlantic, and dismissed references to off-shore progress as not applicable to the British situation. The assumption was that Britain was a sheltered enclave, independent of the rest of the world and unaffected by international developments.

That viewpoint was no more true then than it is today. We live in the midst of the largest, most open, unprotected market the world has ever seen—a veritable American bazaar. During the past four months, I have had the opportunity to extensively tour Europe and Asia to discuss the future of electronics with government officials, other manufacturers, and academics. Everywhere, people are seeking to build their future industry on electronics; and after developing their home market, are planning to sell aggressively in the U.S. I feel as if I live in everyone's target zone, at ground zero. Clearly, developments abroad directly affect us right here in what we in the U.S. have come to regard as our home market. Furthermore, electronics markets abroad have developed to the point where we *must* compete internationally. We can no longer rest assured that leadership in the U.S. market, alone, is sufficient to guarantee a premier world position.

The British did not comprehend the real impact of non-British developments on their own ability to succeed. We must not fail to realize the important implications of international developments on us. In this respect, the 20th century U.S. electronics industry is better off than its 19th century British predecessor.

The British experience illustrates another failing—inability to perceive the world as it is, rather than the way we want it to be. Prior to the 1860's, European shoe styles emphasized the use of buckles to keep shoes closed. A substantial industry had arisen around the making and selling of metal shoe buckles. Again, the Americans arrived on the scene with a new invention—the shoe lace. British industry responded defensively—by starting the rumor that lacing one's shoes was effeminate. They even convinced the royal family to always appear in public in buckled shoes. Alas, the defensive approach did not work. Faced with the ease and flexibility of laced shoe designs, the shoe buckle died a slow, painful death but not before evolving for a short time into buttoned shoes. Nevertheless, the Americans triumphed in the end.

The British simply did not realistically assess the pressure for change brought about by the improved "new technology"—the shoe lace. In the same way, we see American and European businessmen pursuing defensive strategies in the face of what I call the "technological imperative"—the irresistable force of technical advancement. Defensive ploys to keep others from competing cannot overcome the pressure for technical change. Failure to comprehend the strength of new technological developments as forcing elements for change in our society today will increase the eventual upheaval, not lessen it. In our case today, however, the culprit is not the lowly shoe lace, it is manufacturing line automation. The transition from traditional production technology, which emphasized individual islands of mechanization coordinated mainly by manual scheduling techniques, is rapidly giving way to entire production lines which are managed by computers in real time.

The semiconductor industry is a good example of both ends of the manufacturing automation spectrum. Semiconductor manufacturing really consists of two parts: a front-end factory consisting of highly sophisticated, expensive wafer-processing lines, and a back-end, made up of less complex, more labor-intensive packaging, assembly, and test. Front-end wafer-processing areas, in spite of the sophistication of individual pieces of equipment, are good examples of the "island of mechanization" type of line. Successful integration of the entire line into a single, computer-organized entity remains to be practically demonstrated, although major pieces are now emerging. In fact, commercial I/C wafer processing today still reflects its antecedents, as factories still look much like overgrown laboratories. In spite of the technical sophistication of those who have contributed to the advancement of the art, wafer-fabrication lines still leave much to be desired as paragons of manufacturing practice.

We have a real need for engineering professionals who understand both the technology of semiconductor processing and good industrial manufacturing practice (and who are open to new and different methods and ideas). In fact, the back-end portion of the business, in spite of its less glamorous beginnings, is now showing impressive signs of developing into a much more sophisticated manufacturing environment. The combined pressures for effective inventory management and reduction of high labor costs, together with the desire to respond quickly to fluctuations in the incoming order base, have resulted in aggressive plans to automate back-end factories.

Traditionally, the manufacturing management and engineering disciplines have been seen by many to be "second-class" fields—that the true innovation arises from R&D efforts that result in new processes, devices, and products. It now is clear, however, that our principal competitors abroad have placed most emphasis on the efficient manufacturing of high-technology products and have used that expertise to solve other technical problems as well. We Americans must learn from the success of manufacturing-based industries in countries such as Japan that we must master manufacturing technical and management leadership.

Let me return to the British experience for one additional observation: I believe the British suffered one serious structural problem which they share with the French and several other European societies. It is the clear distinction between the annointed scientific elite and the manufacturing engineers. The members of the elite were franchised by virtue of their educational experience—in the science curricula of Oxford, Cambridge, the École Polytechnique, or the École Supérieur or any of a number of prestigious universities.

The science curricula of the best schools in these countries provided superbly educated technical professionals capable of competing on equal footing with the world's best. Their graduates would most likely go on to senior management responsibility in either government or industry. The more ordinary schools produced graduates who performed more mundane technical functions, usually within factory or service industries. These individuals were expected to spend their careers in middle-range positions which require technical background but are not the leadership, creative assignments.

The boundary between the two is clearly drawn, and very difficult to cross. I have found in many cases, even today, where the two types of individuals don't even share the same concepts of what is important or even, for that matter, speak the same technical language. This is sometimes referred to in the U.K. as the Oxbridge effect, although it is as much an industry effect as it is academic. The result of the Oxbridge syndrome is frequently stalemate when it comes to putting new designs and technology into production, and its artificial barriers seriously weaken the entire industrial structure. The hierarchy thus established is very resistant to change.

We must be very careful not to fall into the same trap. I have no compunction against our having technical stars—we rely on them to be the originators of new and innovative ideas, designs, and methods. What I am warning against is having such an elite franchised either by right of birth, alma mater, or degree instead of performance. We, in the U.S., have enjoyed a commonality of outlook between researchers, product developers, and manufacturing engineers. We must continue to emphasize this in order that our society succeed as a competitive industrial power. It is essential that we continue to insist upon evaluation of an individual's future based on his or her performance. To fail to reward performance thusly leads to an unconscionable waste of human resources which we can ill afford. This problem is not solely educational. The Japanese, in spite of a highly structured educational system which is used to differentiate between career advancement paths, have succeeded in overcoming this kind of implied class distinction's impact on industrial competitiveness. It appears that this results, at least in part, from the practice of large Japanese companies of starting new technical employees in manufacturing areas. After an initial assignment as a manufacturing engineer, the journeyman employee moves on to other departments. The employee retains a lingering orientation toward the company's manufacturing needs and is thoroughly steeped in the company's production approach.

We in U.S. industry can learn a number of things from Japanese practices, and this is an area we should carefully examine. I consider this to be a relevant point for this occasion because I sense a split beginning to take place in our technical educational system between the fields of engineering and "engineering technology." We in industry are being asked to differentiate between Bachelor of Engineering and Bachelor of Engineering Technology degrees in our hiring practices. The implication of this differentiation seems to be that one degree endows the graduate with a broad technical outlook, suitable for more developmental, cerebral work; the other for more practicel, "hands-on" efforts. I fear the development here of two classes: one, the certified thinkers--the elite, and the other, the certified doers--the proletariat. Surely, this is not the distinction between the degrees that is explained to applicants and incoming freshman! It is also not the experience of many industrial firms. In many cases, after a few years of industrial experience, a degree is only one of several predictors of ability to contribute to and advance within a company. We must look for ways to be sure that each university-trained engineer has the basic tools with which to operate at any level in the organization, so the tendency to pigeon-hole individuals will be minimized.

The British realization that they had a problem competing with the Americans came a mere 130 years ago. Americans were then experiencing the exhilaration of seemingly limitless horizons of the western frontier. Similarly, over the past thirty years, we have been thrilled by the promise of unbounded horizons on the information-technology frontier. Just as American expansion reached the limits of the continent, we may be seeing the beginnings of somewhat slower growth in some sectors of electronics. Previous speakers have painted optimistic pictures for the device and system technical outlook. Figure 5, however, examines one other aspect. It displays the apparent consumption

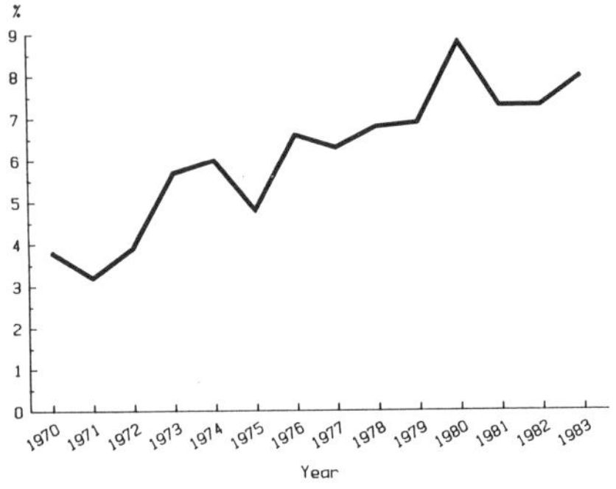

Source: U.S. Department of Commerce Data

Figure 5. Apparent U.S. Consumption of I/C's as a Percentage of U.S. Electronics Equipment Sales

(that's shipments minus exports plus imports) of integrated circuits within the U.S., expressed as a function of U.S. equipment sales. Note that the I/C business captured equipment manufacturers' dollars at a rapid rate during the early 1970's but that the rate of increase now appears to be slowing. This trend indicates that the I/C business growth rate is beginning to match that of equipment-making customers–slower than historical I/C growth. I hasten to add that these data are still new and this conclusion is very speculative. In essence, however, what appears to be happening is that I/C manufacturers are saturating their customers' ability to afford I/C's as they now exist.

As this is taking place, and compounding the competitive problem, the technology and equipment required to manufacture semiconductors has diffused widely throughout the free world, enabling companies and countries with no previous semiconductor experience to believe they can successfully become competitive in world markets after only a few years' experience. Such competition, often unsupported by a native R&D base, and unencumbered by its cost, is creating new competitive pressures in the market.

Both the worldwide diffusion of volume I/C manufacturing technology and potential limiting of the market as we now know it, will force I/C manufacturers to seek a technology advantage–by inventing new ways for customers to use their products. It has become clear that the application-specific I/C concept will help to increase customers' applications of I/C's in electronic equipment in the late eighties and will help establish the advantage. Furthermore, customer design-needs for unique circuits could well make it more difficult for distant, inflexible, high-volume manufacturers to successfully participate. Beyond these near-term product changes (already well under way), however, we are challenged by the need to develop new technology to expand future opportunities. We must think deeply about our long-term technical future. If the I/C business is to continue to grow faster than its customers, it must capture more of the customers' hardware dollars. This line of thought leads to the need for radical changes in software and packaging and interconnection within I/C assemblies, since most remaining customer dollars are there.

The U.S. as a whole has been steadily increasing its commitment of resources to research and development. Since 1965, total national expenditures for R&D have increased more than fourfold, and our commitment to R&D has grown from just under 1.3 percent in 1965 to 1.75 percent of GNP in 1983. Figure 6 shows that over this period, the Federal share of R&D funding has declined steadily, to be picked up by industry. From Figure 7 it appears that on the surface, 70 percent of our research and development is done by industry, implying that the development of new long-term information technology is principally an industrial concern. In Figure 8, however, a closer look shows that industry actually does only 60 percent of applied research in this

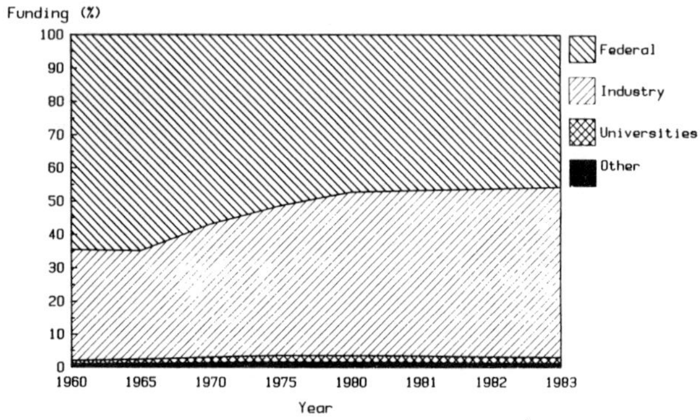

Source: Statistical Abstracts of the U.S.

Figure 6. Sources of R&D Funding

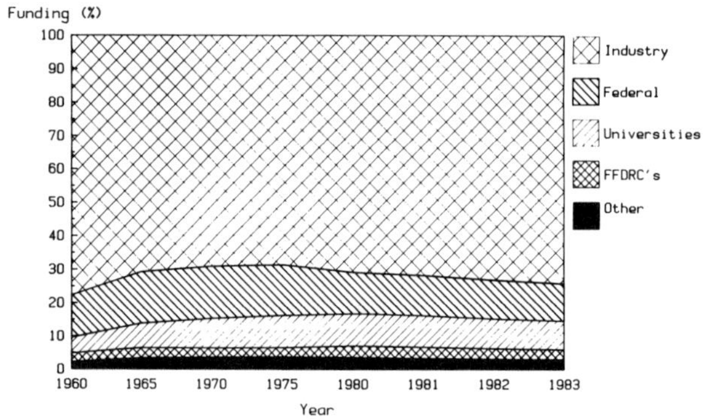

Source: Statistical Abstracts of the U.S.

Figure 7. R&D Performing Sector

country. Figure 9 shows that less than 15 percent of basic research is performed by industry and that universities are the biggest factor in the development of industry's long-term technology.

The problems of our universities in terms of enrollment and faculty recruiting have been well publicized. Research and development capability has had similar serious problems. Figure 10 illustrates the drop in Federal support of university R&D plant that occurred during the late 1960's. The consequences of this reduction have been becoming increasingly evident over the past decade. Since universities perform the bulk of basic research in the U.S., the effect on our long-term industrial world competitiveness is a major concern. Industry and government *must* find ways to help rebuild our universities' R&D capabilities and to more effectively utilize the results of university efforts.

One step, an unusual industry alliance, the Semiconductor Research Corporation, was formed to focus industry attention on helping to support the R&D needs of universities and to couple the academic and industrial technical communities more closely together. The results of the SRC effort have been encouraging and may set an example for other electronics disciplines. The SRC is but one example of the kinds of alliances that can develop between otherwise competing companies to advance their mutual interest.

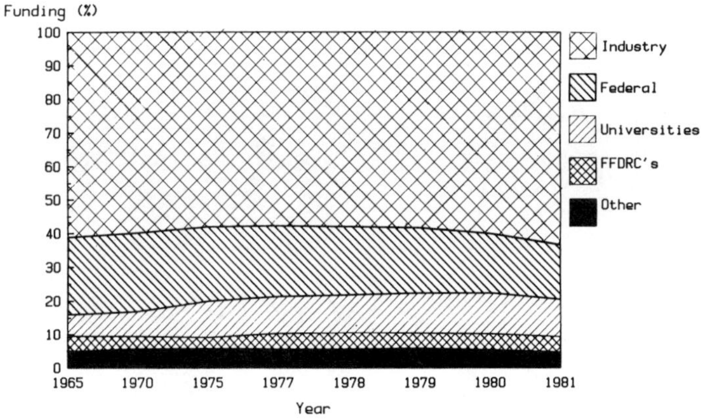

Source: Statisical Abstracts of the U.S.

Figure 8. Applied Research Performing Sector

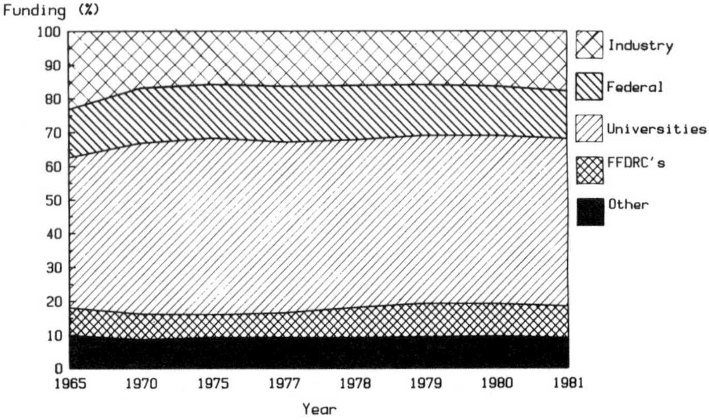

Source: Statisical Abstracts of the U.S.

Figure 9. Basic Research Performing Sector

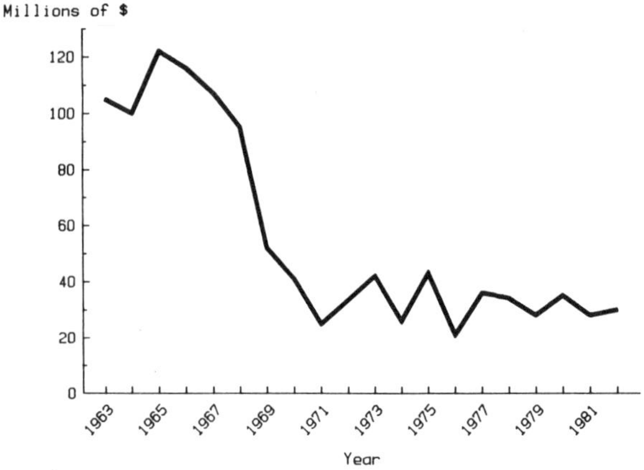

Source: National Science Foundation

Figure 10. Federal Obligations to Universities & Colleges for R&D Plant

The increasing scope and cost of technology development means that single firms can no longer meet all of their R&D needs on their own—something the Japanese discovered some time ago and which we are just now beginning to learn. America's technical competitiveness depends upon the ability to form alliances and to obtain mutual benefit from them—a new behavior pattern! Cooperation between industrial, academic, and as we will hear later this afternoon, governmental interests, is essential if we are to compete effectively in the future. Other alliances, such as the Microelectronic and Computer Technology Corporation, the Microelectronics Center of North Carolina, and various software efforts, as well as the teaming concepts used in the VHSIC program, are other imaginative multi-company solutions to business and technology needs of several companies simultaneously.

I expect we will see still more alliances in the future. If we are to avoid the kinds of problems that beset British industry in its attempts to wrestle with world competition, we must be open to new and different approaches to solving our business and technical problems. The rigid structures and thought patterns that kept the British from succeeding in the new manufacturing approaches of the day must not typify the response of America to the challenge of the 1980's.

In summary, I have touched on various aspects of international competitiveness: realization of the problem, need for manufacturing technology and technologists, increase in the scope of microelectronics products to capture more user dollars, and innovative ways to jointly use all available resources to advance the technology. They're not very exotic ideas—in fact they are mundane—but we must master them if we are to lead in microelectronics. A great deal of flexibility is required. British industry was slow to change—was inflexible—that's still evident 130 years later. We have the resources, the creativity, and the position from which to continue to succeed. If we fail to do so, it can only be because we lacked the willingness to be flexible enough to do so. We do not have 130 years to solve our problem. We must act now.

THE ROLE OF GOVERNMENT IN MICROELECTRONICS

E. D. Maynard, Jr.

Egbert D. (Sonny) Maynard, Jr. received the BSEE and MSEE degrees from the University of Louisville, Kentucky in 1962 and 1977 respectively. In 1977 he joined the Naval Ocean System Center in San Diego, California as an electronics engineer, rising to the position of program manager. In July 1983, he was appointed Director of the VHSIC/Electron Devices Program, under the Office of the Undersecretary of Defense (R&AT), U.S. Department of Defense. In this position he is responsible for the overall management of programs, functions, and budgets for electron-device technology as well as for advanced integrated-circuit technology. He provides for the establishment of plans and the determination of major investment strategies to be followed by the Services and major defense agencies, and coordinates program activities that are interrelated with other Secretary of Defense offices. He is the Department of Defense representative to the International Committee on Technology Liaison for Electron Devices.

For some time I have been working on the new Department of Defense (DoD) Program on Very High Speed Integrated Circuits (VHSIC), so in keeping with the subject of this talk on the role of government in microelectronics I will concentrate on that program. In previous discussions today we have heard about some of the greater problems of our society and the relationship of microelectronics to their solution. I will try to outline my beliefs on how the government can attempt to participate in this work as well as providing some degree of assistance.

One of the first questions that might be asked is "Do we really want any help from the government?" a query that inevitably compares the roles of government and management. Good government and management are sometimes thought of as a nonsequitur. A few years ago Senator Lowell Weicker, in a debate over the national medical program, said that he thought that the last thing he wanted was some bureaucrat deciding or making decisions about his health care. I think we all have similar feelings about some bureaucrat making decisions about our economic affairs. But as we've heard, the interdependence of government and industry is steadily increasing, and there certainly is a long-standing interaction between government and universities. I will try to avoid making any political statements today so my references to the proper role of government will be based on the natural sciences of Archeology, Anthropology, and Ethnology. The study of societies has led to an understanding of the principal roles that society has played in evolution; for example, when a particular group does not attend to education, its next generation will falter because their children will not be able to perform their required tasks. By inference, those of us who do not attend to the Nation's defense will be selected out of the next generation. That is why I work in the Defense Department: I'm working on the number-one priority.

Relative to management, I will not attempt to interpret current or popular management schemes, but any Management 101 course would emphasize that to achieve a successful winning strategy generally oriented to profits in the commercial sector, five components must be present: (1) men, (2) money, (3) materials, (4) time, and (5) authority to make decisions. Since the first four items are already in the public domain, the development of a successful governmental policy depends heavily upon the application of what government generally is best qualified to do, and that is to provide the authority to get the job done—certainly not by arbitrary dictation but rather by motivation of all individuals involved cooperating with one another in order to accomplish the desired tasks. I think Franklin Delano Roosevelt said it best when he was swept into office back in the 30's and began to combat the effects of the Great Depression. He said, "We have everything we need here in this country to solve our problems. What we need is the will to solve them." Later, in World War II, he was proved to be right, particularly since there was strong motivation for all citizens to work together. By the end of the war the economic problem had disappeared because we discovered that one successful approach to boot-strapping the economy is to make lots of expensive things, take them overseas and throw them away. To be facetious for a moment, perhaps when we have trouble selling our products we should follow the same approach: as long as we keep at it, we will be OK, provided we don't look back over our shoulders. Seriously, there is no question that the authoritative aspect of government can lead to successful management programs.

A currently popular book on management has taken a look at the VHSIC program and concluded that it follows some of the rules of good management. Consequently, in this discussion I will present a status report on the VHSIC program, what it is all about and how it was started. I think that the program is a reasonable model for government interactions with microelectronics industries of the future.

In order to establish a solution strategy it is necessary to understand the problem, and the problem in the microelectronics and military weapons systems that developed over the last decade is illustrated in Figure 1.

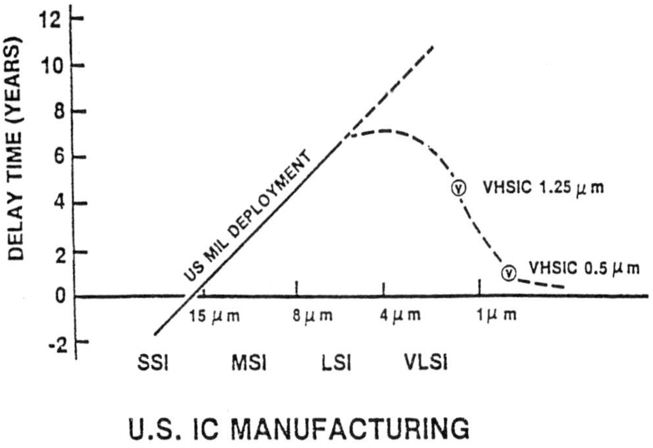

Source: DoD/VHSIC

Figure 1. Military Microelectronics

With the base line representing the gradual development of integrated circuits, the vertical axis indicates the elapsed time between the appearance of a particular product in the commercial market and its successful deployment in a military system. In the 60's, the Department of Defense funded much of this work and had committed itself to the possible risk of concurrent development. Sometimes military integrated-circuit products were matched by equivalent devices that appeared in the commercial market. By the late 70's, we discovered that we were on our way to falling ten years behind the commercial market in terms of putting VLSI into systems that could be used in the field. That situation was unacceptable for a variety of reasons not the least of which was that potential enemies would have plenty of time to insert the same technology into their weapon systems before we could. So the basic strategy for the Very-High-Speed-Integrated-Circuits Program was to return to the state of affairs in the early 60's when we were able to field military hardware containing technology that was equivalent to what the integrated circuits industry was fielding at that time.

We decided to do that in two steps: (1) to recover five of those years by aiming for a clearly definable target, well within lithographic capability at 1.25 microns, where the process would be feasible and the products could be defined; and (2) to intercept the commercial market capability at the half-micron point. The same information is shown in Figure 2, in more familiar form, where feature size is plotted versus time. It is apparent that we

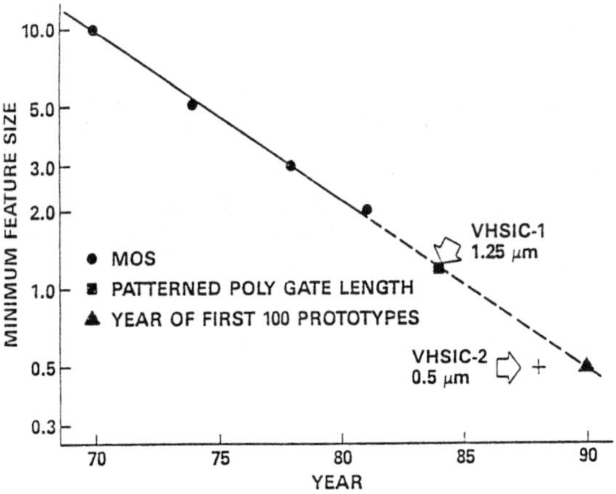

Source: DoD/VHSIC

Figure 2. Historical Trend of Minimum-Feature Size

achieved the 1.25-micron device simultaneously with its development for the commercial market, thus fulfilling the purpose of the program which was to prepare ourselves to acquire the prototypes of unique military products as they become available. On the other hand, for the half-micron device, we are pushing to process and demonstrate a prototype. We believe that we are off the curve (as indicated in Figure 2) and ahead of the probable commercial market position. It is necessary to plan ahead simply because it takes us so long to prepare ourselves for a new product and we want to start early enough to let us deploy it at the same time as it becomes popular in the commercial market.

Figure 3 outlines three critical elements that are essential for putting together a winning strategy. These guidelines may be summarized briefly as: (1) get the message out to everybody who needs to know about it, (2) lay out a specific plan, and (3) carry it through. The following recent comparison shows that we have followed these rules carefully.

BESIDES COMPETENT PEOPLE AND A BIT OF LUCK, THREE SPECIFIC ACTIONS WILL CONTRIBUTE HEAVILY TO SUCCESSFUL EXPLOITATION.

1. CREATION OF SPECIFIC INFORMAL ORGANIZATIONS

   A) WITH EXPLICIT RESPONSIBILITY IN RIDING THROUGH TO SUCCESSFUL PRODUCTIZATION

   B) WITH THE AUTHORITY AND RESOURCES TO MOTIVATE EXISTING LINE ORGANIZATIONS TO PROVIDE THE ATTENTION IT DESERVES.

2. ESTABLISHMENT OF THOROUGH PROCEDURES TO PLAN AND MONITOR THE EFFORT, DOLLAR FLOWS, AND TIMING OF ALL PHASES IN EXPLOITATION

3. THE INTEGRATION OF THESE ORGANIZATIONS AND PROCEDURES TO IMPLEMENT THE CRITICAL STRATEGY WHICH DETERMINES SUCCESS OR FAILURE

Source: DoD/VHSIC

Figure 3. Exploiting the Strategy

Our early plans in the VHSIC program are outlined in Figure 4, abstracted from a set of vugraphs used by Leonard Weisberg to formally announce the program back in November, 1978. We had been working along similar lines for a year or so before that date but this was the first formal announcement. Comparing that early listing with what has been actually accomplished in the program reveals the following information:

- We said it would be a tri-service program that would be managed in the Office of the Secretary of Defense from a corporate standpoint rather than from an industrial standpoint. Such a plan in industry would be equivalent to the unlikely practice of running a major new product development from the corporate staff office, but it worked. Perhaps it is the only program of its kind that could have worked under the circumstances. In fact the program has been used as a model for a number of new programs that have come along.

- We said we were looking for a 100-fold increase in processing speed. The objective was to improve the throughput, not just the clock speed, but the overall process of throughput. We have exceeded that goal and believe we will have demonstrated 100-fold improvement at the end of Phase I, with an expected additional 50 to 100-fold improvement in Phase II. (I should add that the ultimate goal of the entire program is to determine our position by the end of Phase II.)

- We said the budget would be in excess of 150 million dollars for a six-year period, and we certainly have met that goal. Our present overall ten-year program, including some technology insertion on the back end of the program, will cost about one billion dollars, give or take 50 million dollars.

- We said that we wanted to develop these devices specifically for military system needs, and we have established chip sets that are clearly applicable to our military requirements. Certainly these developments will eventually be used for other purposes, but we have definitely satisfied our own requirements.

- We have inserted the new devices into specific equipment that has been demonstrated satisfactorily several times.

- As originally planned, we have put most of the emphasis in the program on signal processing rather than on data processing.

In considering why the VHSIC program should be started, it was clear from the beginning that we were not doing this just to pump some money into the integrated-circuit industry to help it beat the Japanese. The point we were trying to make was that, in fact, we had a problem in military microelectronics and weapon systems that we were trying to solve. I think that generally, the semiconductor industry will agree with this assessment. It was pointed out earlier that much of the R & D that goes on in the

Department of Defense has only incidental application in the civilian sector. Recently somebody from the Congress came over to talk to me about the program and one of the questions he asked was related to the sort of economic impact that VHSIC plans to have on industry. In my cautious reply I pointed out that after 25 years of training in the Department of Defense, I know that the last thing we want to happen is to be caught doing something nice for the people who are doing the work for us. But I believe that there is a permanent change in the way government looks at industry and cooperates with industry. The present philosophy is that there ought to be more of a calculated intent that will result in both economic and militarily expedient approaches to the program.

LEONARD WEISBERG AND LARRY W. SUMNEY
GOMAC-78
14 NOV 78

- TRI-SERVICE PROGRAM, MANAGED BY OUSDRE
- 100-FOLD INCREASE IN PROCESSING SPEED
- BUDGET IN EXCESS OF $150M FOR SIX-YEAR PERIOD
- DEVELOP VHSIC TECHNOLOGY SPECIFIC TO MILITARY SYSTEM NEEDS
- BUILD ADVANCED VHSIC DEMONSTRATION SYSTEMS
- EMPHASIS ON SIGNAL PROCESSING

Source: DoD/VHSIC

Figure 4. The New DoD Program on Very-High-Speed-Integrated Circuits (VHSIC)

From the standpoint of materials, we decided to limit our work to silicon devices for a variety of reasons and continue to work on gallium arsenide outside the program. We have followed that procedure, but there is some serious planning going on at present to determine what the next generation of gallium-arsenide electron devices ought to be after the VHSIC program is completed. Both analog and digital devices are serious contenders for that development.

The VHSIC program was established because it was clear that we would be in deep trouble if we could not solve the problems that we had encountered in the insertion of LSI into military systems. It was also felt that we would probably never solve the even more difficult problems associated with VLSI. The Department of Defense had adopted the hands-off policy of obtaining all electronic technology from the commercial market and that we were going to have to learn to live with it. As with many government policies, this one took a long time to mature. As I mentioned earlier, after 10 to 15 years we discovered that it wasn't working, so we reversed that policy. Our present policy in microelectronics is very strong, the government is conducting a considerable share of the work, and the technology is being inserted successfully into the military systems. In the VHSIC program we are devoting about 200 million dollars out of our budget to specific placement of demonstrator units into military hardware, with another 400 million dollars coming from the services to add to the demonstration program.

Another side effect of the VHSIC program was the reinstatement of the legitimacy of concurrency—representing a unique strategy in the conduct of R & D in the Department of Defense. Several years ago, Dr. Hitch, of the University of California, came to the Department of Defense and introduced a taxonomy of research principles known as the "Hitch Elements", (see Figure 5) that became universally accepted in the Department. VHSIC took the last three stages of this listing and began a concurrent program of simultaneous process development, product definition, and product marketing—a procedure that is quite similar to that of commercial practice.

FUNDAMENTAL RESEARCH SEEKS THE PRINCIPLES AND RELATIONSHIPS UNDERLYING THIS KNOWLEDGE.

APPLIED RESEARCH FURTHER CRYSTALLIZES THE TRUE NATURE OF THE KNOWLEDGE AND DEMONSTRATES ITS POTENTIAL UTILITY THROUGH USE OF BENCHMARK ACHIEVEMENTS.

VHSIC STRATEGY RE-INSTATES CONCURRENCY:

DEVELOPOMENT REDUCES THE KNOWLEDGE TO PRACTICE IN WORKABLE PROTOTYPE FORM.

ENGINEERING REFINES THE KNOWLEDGE FOR COMMERCIAL EXPLOITATION.

OPERATIONS SUCH AS PRODUCTION AND SALES PUT THE TECHNOLOGY FINALLY INTO USE.

Source: DoD/VHSIC

Figure 5. Technology Evolution Stages

The summary shown in Figure 6 provides a general overview for those who are relatively unfamiliar with the VHSIC program. Our plan was to have a Phase I component that would run for about four years, develop 1.25-micron feature size hardware, and demonstrate brassboards of prototype military systems. During the same time frame, we would develop technology for 0.5-micron devices and initiate a submicron program that would carry us on to the late 80's. The table shows that approximately half of the money in the VHSIC program was required for the basic technology developments of Phase I. Figure 6 also displays the diverse set of companies that are involved in the program. One of the things that we had to do in terms of getting the job done, was to combine system-design and fabrication capability with integrated-circuit design fabrication capability. Some companies had both capabilities and others did not, so it was necessary to form teams that could provide the desired combination of skills. I can remember at the time it seemed that there were some unlikely bed fellows in these groups, but throughout the program we have had business as usual in a variety of programs, and team cooperation has become common.

	PHASE 0 1980	PHASE 1 1981-1985	PHASE 0' 1983-1984	PHASE 2 1985-1988	PHASE 3 1980-1984
ARMY	GE HUGHES ROCKWELL	HUGHES TI	ATT HUGHES TI	IBM	89 CONTRACTS (A/N/AF)
NAVY	IBM TRW WESTINGHOUSE	IBM TRW	RCA IBM TRW	TRW	— LITHOGRAPHY — MATERIALS — PACKAGING
AIR FORCE	HONEYWELL RAYTHEON TI	HONNEYWELL WESTINGHOUSE	HARRIS HONEYWELL WESTINGHOUSE	HONEYWELL	— QUALIFICATION — RADIATION HARDENING — VHDL
	$10.3M	$167.4M	$11.2M	$170.3M	$77.8M

Source: DoD/VHSIC

Figure 6. VHSIC Technology-Development Contract Structure

The technical scope of the Phase I effort is outlined in Figure 7. It is clear that we have tried to cover all our bets. All of the major technologies and design philosophies are represented, and the resulting demonstration units throughout the military complex are working satisfactorily. The breakdown of VHSIC Phase I chip fabrication activities by individual companies given in Figure 8 is only partially complete since new items are being added continually. All of the manufacturers have been at least partially successful, and some have produced fully functional designs of all of their chips. Not one of the participants has left the program, and all of them have promised to deliver their products.

- All Major Technologies Represented, Backup Approaches on Bipolar and Bulk CMOS
  - Bipolar
  - CMOS/SOS
  - Bulk CMOS
  - NMOS
- All Major Architectural Approaches Covered
  - Chip Sets
  - Custom
  - Programmable
- Brassboards for Major Systems Applications Will Be Developed
  - Radar
  - Acoustic
  - Electronic Warfare
  - Communications
  - Electro-Optical
  - Missile Control
- Army, Navy and Air Force Systems Requirements Optimally Addressed

Source: DoD/VHSIC

Figure 7. Technical Scope of Phase I Effort

The VHSIC program has signed up military "customers" for these products as shown in the listing of Department of Defense systems in Figure 9. Identification of these acronyms is not significant at this point--rather the listing serves as a histogram that illustrates the spread of applications that have been tied to the program. It is clear from the chart that we are avoiding a mindless long-range implementation of this technology with the goal of eventual insertion into military systems in 15 to 20 years. The lesson we had to teach ourselves in the Department of Defense is that we are going to receive new generations of integrated circuits at a faster and faster rate, and we must learn to take them as they come, use them, and then go on to the next development—applying the latest devices whenever we have the opportunity and the need.

IBM - COMPLEX MULTIPLY AND ACCUMULATE (COMPLETE SET)

TRW/MOTOROLA - MATRIX SWITCH, CONTENT ADDRESSABLE MEMORY, WINDOW ADDRESSABLE MEMORY, FOUR-PORT MEMORY

HUGHES - DIGITAL CORRELATOR

TI - MULTIPATH SWITCH, 10K GATE ARRAY, 72K RAM

WESTINGHOUSE/NATIONAL SEMICONDUCTOR - 64K RAM COMPLEX ARITHMETIC VECTOR PROC

HONEYWELL - SEQUENCER, ARITHMETIC, PARALLEL PIPELINE PROCESSOR

Source: DoD/VHSIC

Figure 8. Chips That have Been Fabricated

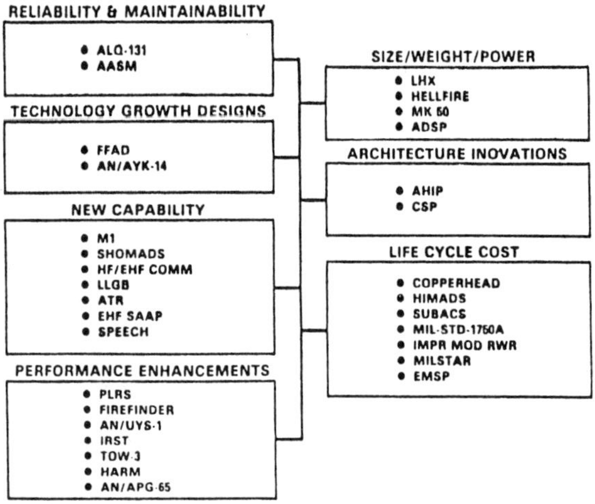

Source: DoD/VHSIC

Figure 9. VHSIC Technology-Insertion Requests
(Areas of Greatest Benefit)

There are about 40 systems signed up to use VHSIC at this point and we have just been going through another exercise in which about 55 proposals have been received for application to this program. Most of the Phase I seed money allotted to technology insertion has been spent, with a small amount left over to allow joint-venture activities with system program officers. This is probably the last year that we will have a formal VHSIC program technology-insertion effort, but the response from the services has been tremendous. They have put up two dollars for every one of ours on these joint ventures and have begun to assume job responsibilities. In fact, we are approaching the point where VHSIC will become "business as usual" within the next three or four years. Overall, Phase I has been successful. It has accomplished essentially all of the goals that were set by Len Weisberg, Larry Sumney, and the host of others who participated in establishing that plan, and Phase II has been launched.

Figure 10 is useful as a reminder of what Phase II means and what we are trying to do with the 0.5-micron feature size. The objective is to attempt to deploy 0.5-micron integrated-circuit technology into military weapons systems at about the same time that the commercial market is going to be selling its first parts. Such a program will take much more effort than was required for

- QUANTUM JUMP IN FUNCTIONAL THROUGHPUT
  - PHASE 1 FTR = 5 x 10^{11} gate-Hz/cm^2
  - PHASE 2 FTR = 100 x 10^{11} gate-Hz/cm^2
- RADIATION HARD:
  - BASIC REQUIREMENT — 50 K RAD TOTAL DOSE
  - ENHANCED REQUIREMENT — 200K RAD TOTAL DOSE
- EXPANDED FAULT TOLERANCE & BUILT-IN TEST
- BROAD CHIP SET APPLICABILITY

Source: DoD/VHSIC

Figure 10. VHSIC Phase II Submicron-Technology Goals

Phase I since it will be necessary to establish a process well ahead of the time when a rational prudent business man would set up a similar production line for the large-volume manufacturing of the product. Consequently, we have expanded some of the initial requirements for Phase II. For example, we have included an R & D option to examine least-space levels of radiation. All three of the contractors that have been selected for Phase II will attempt to meet these enhanced radiation requirements. We have also applied much broader requirements for built-in testing and fault tolerances, as well as greater applicability of the chips to circuit functions.

The general goals for Phase II are shown in Figure 11. The three major contractors are listed with their assigned team projects. Motorola, a funded partner with both Honeywell and TRW, will back up the bipolar process for Honeywell, and is the principal developer of the CMOS process that is included in the TRW project. The range of technology in Figure 11 goes from about 75,000 to 80,000 gates on a relatively small chip of 5.5 to 8 mm, all the way up to several million transistors on a very large chip, where TRW has elected to go with a limited form of wafer-scale integration involving a programmable signal processor approach. We have great hopes that this set of contractors, and the work that they are doing, will give us capabilities for all of our weapon systems that we cannot possibly achieve now. We have made some validation studies on this advanced technology and it is clear that we have requirements that meet or exceed what we will be able to do at the projected level of integration. As in Phase I, we are trying to hedge all our bets. We are not restricting ourselves to any one particular technology--rather we are taking a variety of design approaches and applying them to all of the pertinent major military-system demonstrations.

When VHSIC was first started, we solicited advice from industry on suitable contents and procedures. In addition to receiving many helpful suggestions, we also got a lot of comments on why we should not establish this kind of program. Figure 12 contains a list of these concerns of industry which may be summed up as consisting of two basic complaints: (1) there will be too great an effect on industry, and (2) there will be no effect. In some cases, both viewpoints were expressed by the same critic. I think that we have answered and overcome most of the criticisms that are listed. The principal concern of the first complaint was that the program would absorb much of the best engineering talent in the semiconductor industry. That has not been a problem during the last two or three years. The second complaint was based on a belief that the proposed goals could not be accomplished with the kind of program that the government had assembled. Our success with our "customers" in the military-systems business has put that doubt to rest.

Projected elements of the program that were considered to be of benefit to the microelectronics industry are outlined in Figure 13. In retrospect, it is

CONTRACTOR	TECHNOLOGY	DESIGN APPROACH	BRASSBOARD MODULE APPLICABILITY
Honeywell/ Motorola	CML Bipolar E-Beam Litho Single & Multi-chip Package Radiation Hard to 50K Rads	Cell Library Configurable Gate Array 100% Testability ~80K Equiv. Gates	Radar, EO Systems
IBM	CMOS Bulk E-Beam/U-V Lithography Single & Multi-Chip Radiation Hard to 50K & 200K Rads	Cell Library Master Image Chip Structure ~5K Equiv. Gates	Sonar, EW, Radar Systems
TRW/ Motorola	3-D Bipolar/CMOS E-Beam for 3-D E-Beam/Optical Litho for CMOS 1400-mil Super Chip Radiation Hard to 200K Rads	Cell Library Wafer Scale Integration Via Redundancy ~4M Equiv. Gates	Cruise Missile, Satcom, Avionics Systems

Source: DoD/VHSIC

Figure 11. VHSIC Phase II Submicron Technical Summary

clear that we have followed the line very closely. As part of the development of submicron support technology, we have invested from 25 to 30 million dollars in both E-beam and x-ray lithography and we plan to examine some optical lithography options for half-micron dimensions. We have installed eight communication links around the country that have enabled us to provide considerable feedback to the industry through our evaluation of prototype processing equipment. A number of CAD tools, developed as part of the program, have been placed in the public domain in addition to their being provided to Fort Monmouth. Perhaps the principal benefit of the program to the government is the re-establishment of a spirit of confidence in the ability of the Department of Defense to insert high-technology weapons in the field rapidly. From the standpoint of industry, we have shown that the Department of Defense is an interesting customer as well as a source of creative and useful R & D projects.

- INDUSTRY FEARS VHSIC PROGRAM WILL HAVE TOO GREAT AN EFFECT:

  -WILL PUSH TECHNOLOGY TOO FAST
  -WILL SKEW THE MARKETPLACE TO ONLY LARGE COMPANIES
  -WILL MAKE EQUIPMENT TOO EXPENSIVE
  -WILL CREATE A MANPOWER SHORTAGE
  -WILL DIVERT INDUSTRY TOO MUCH TO DoD's NEEDS
  -WILL CREATE ANTI-TRUST PROBLEMS

- INDUSTRY FEARS VHSIC PROGRAM WILL HAVE TOO LITTLE AN EFFECT:

  -WILL BENEFIT ONLY DoD
  -WILL NOT HELP COMPANIES THAT DO NOT RECEIVE CONTRACTS FROM DoD
  -DoD CANNOT APPRECIABLY AFFECT INDUSTRY SINCE $30M ANNUAL PROGRAM IS ONLY 10% OF THE IC INDUSTRY'S R&D

Source: DoD/VHSIC

Figure 12. VHSIC - Some Industry Concerns

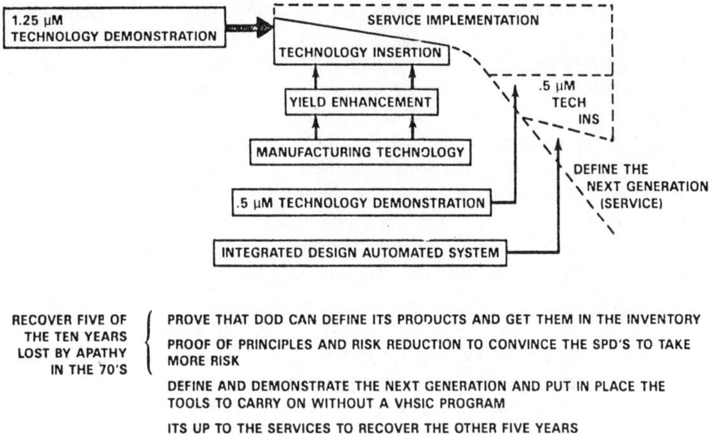

Source: DoD/VHSIC

Figure 13. VHSIC - Fall out to Industry

DEVELOPMENT OF KEY EQUIPMENT
- -E-BEAM AND X-RAY LITHOGRAPHY
- -DRY PLASMA ETCH
- -AUTOMATIC TEST

ADVANCEMENT IN IC DESIGN/ARCHITECTURE AND RELATED TOOLS

TECHNOLOGY DIFFUSION:
- -VERY RAPID IN IC INDUSTRY
- -DoD ROLE WILL BE TO CATALYZE TECHNOLOGY THROUGH FEASIBILITY DEMONSTRATIONS AND ESTABLISHMENT OF A SMALL BUT VIABLE MARKET (SIMILAR TO 1960's)
- -WILL ENCOURAGE REPORTS, WORKSHOPS, AND MEETINGS

Source: DoD/VHSIC

Figure 14. VHSIC Future Strategy

Perhaps the best way to summarize the VHSIC program is to discuss the lessons that we have learned since its inception. We know that we can operate in a business-like fashion, project manufacturing capability for the next five years, predict our requirements, and obtain the desired product. We have learned that the military "customers" want the product. We have demonstrated by proof of principle that we have developed a winning approach in terms of system development and procurement in the Department of Defense. As a measure of our success, a large number of other contractors have made investments of their own to insure themselves of similar capabilities. Strange as it may seem, we have found that the three services *can* work together harmoniously. Furthermore, it was necessary to attend to a number of unanticipated contingencies including (a) funding to buy and/or rent production techniques that would increase silicon yields and move us further up on the learning curve so that manufacturing technology could be advanced enough to satisfy the needs of our Phase I developments, (b) early financial support to encourage system program offices to assume higher risks than normal, and (c) development of fiscal flexibility in terms of being able to pay for the kind of work that was required for the program. We have been pleasantly surprised by the strong support that we have received from the Department of Defense and from Congress. Both feel that VHSIC is a program that is beneficial to the Department and to the Nation. One other lesson we have learned is that there is no such thing as an unimportant detail.

The overall VHSIC Phase II strategy for the rest of this decade is illustrated in Figure 14. In Phase I we have successfully demonstrated that we can acquire the necessary products and insert them into the military system. In Phase II we are trying to establish that the products are feasible and desirable, and that we can design the next generation of integrated-circuit devices. We will not, however, have a specially focused, centrally managed program to do the actual implementation. Our corporate staff is giving that task back to the line organizations. So in the next generation, the individual services will have the responsibility of managing such procedures as technology insertion and yield enhancement. Finally, as a line item in the program, we have added the development of an integrated design-automated system that will be ready at the projected end of the program in the late 80's. This collection of tools will provide our system customers with the capability to interface with the semiconductor industry so as to allow us to step out of the way, and then VHSIC can truly become "business as usual."

TOWARD THE DOMESTICATION OF INSTRUMENTATION*

Joel S. Birnbaum

Joel Birnbaum received a Bachelor of Engineering Physics from Cornell in 1960, and MS and Ph.D. degrees in experimental nuclear physics from Yale in 1961 and 1965 respectively. From 1965 to 1980 he was with IBM Watson Research Center in Yorktown Heights, New York, becoming Director of Computer Science in 1975. He joined the Hewlett-Packard Company in Palo Alto, California, in 1980 as Director of the newly formed Computer Research Center. He is now Vice President and Director of HP Laboratories.

---

*Major portions of this presentation also appear in "Toward the Domestication of Computers," a chapter in *Cohabiting with Computers*, edited by Joseph F. Traub, published by William Kaufmann, Inc., Los Altos, California, 94022, 1985, and in the article "Toward the Domestication of Microelectronics" in the joint issue of the IEEE Computer Society magazine *Computer*, Vol. 18, No. 11, pp. 128-140, and *Communications of the Association for Computing Machinery* (ACM), Vol. 28, No. 11, pp. 1225-1235, November, 1985. Reprinted by permission. A similar version of this article also appears in Part 2 of these *Proceedings* ("Computers").

The French writer Paul Valery once said that "the trouble with our times is that the future is not what it used to be." That seems particularly true right now in the computer and instrument businesses. Modern measurement technology is now inextricably linked to computers--most instruments today contain at least one microprocessor, and many connect to minicomputers and communication networks as well. Concomitantly, an increasing availability of sensors and signal processors for myriad applications of personal computers and workstations in networks is making the distinction among computer-based instruments and instrumented computers merely a semantic one.

The advances of the past thirty-odd years in microelectronics have resulted, in large degree, from the fact that the value of information is independent of the physical size of its container. The number of devices per unit volume has increased about one hundred fold per decade, and this relentless pursuit of smallness has led to a commensurate scaling of speed, power, size, and cost. The next ten years will witness a new type of quest: we are as concerned now with usability and usefulness as we have previously been with price and performance. The Japanese call this the Fifth Generation, intended to signify that the representation of knowledge and its processing in ways that exploit concurrency is an evolutionary step from the miniaturization that has come before. Others talk of Fourth Generation Languages, indicating a higher level of abstraction and purported benefits in our ability to understand and produce computer programs. Similarly, the media are full of articles about Fourth Generation communication systems which can switch digital and multi-modal information. Consumer-driven industries, such as hi-fidelity audio and 35mm photography, announce a new generation in time for each Christmas shopping season.

I do not much care for these advertising-driven characterizations of the march of technology, for they do not provide a useful taxonomy for evaluating the implied evolutionary process. I prefer a model proposed by my colleagues John Doyle and Egon Loebner which, for me, provides a better framework for understanding the impact of microelectronics. In my presentation today, I will try to sketch the outlines of that model and then speculate on two related questions: what will the computer and instrument systems of the future be like, and, more importantly, what will they do?

THE EVOLUTION OF PERVASIVE TECHNOLOGY

Among the ancient records of human life, we find almost 5000-year old reports of the domestication of dogs in Egypt. Pictures of horses drawing vehicles some 3500 years ago have survived. Man's ability to tame wild creatures and to train them as beasts of burden, providers of food, clothing, and companionship was a critical step in the evolution of civilization. As man learned to domesticate animals, he also learned to modify them—that is, to breed them to make them better suited for the purposes to which they were being put. The first known treatise on horse breeding and training is over 3000 years old. The horse of today bears little relationship to that early animal, and the same is true for chickens, cattle, sheep, and many of the other creatures that have had a close relationship with man.

There are now more computers than people in the U.S., and microprocessors are being produced at a rate that probably exceeds one million per day. Although the computer population is growing far more rapidly than the human one, computers are nevertheless far from pervasive in our society in the way that more mature technologies such as the telephone, television, and automobiles are. The mark of a pervasive technology is that its absence is more noticeable than its presence—it has evolved to become part of the very fabric of society. Such a technology is characterized by being generally accessible and by offering a service perceived indispensable by most people.

I hope to convince you today that while computers, and therefore the instruments based upon them, are not yet domesticated; there is every indication that they will become so in the next ten years. It is time for the selective breeding to begin in earnest, and I am going to suggest that many of the *ad hoc* engineering tradeoffs of the last three decades now inherent in personal computer design and application need to be re-examined in the light of new requirements.

Virtually all the pervasive technologies that I can think of seem to have passed through four evolutionary stages. They usually begin as an experimental rarity, a laboratory curiosity, sometimes as an accidental discovery in the pursuit of fundamental science; more often they result from an entrepreneurial attempt to exploit a perceived opportunity. The second stage of evolution usually produces an exotic tool or toy, often solving only a cross-section of a problem and used primarily by a narrow class of experts. In the third stage, the technology has evolved to the point where it is manufactured in quantity, has become well known and commonplace, but is used directly by only a rather small portion of the population. In the fourth, pervasive stage, the technology has become integral to daily life. For example, a home or a modern hotel room without a color television set is unusual since we are in the fourth stage of television technology. Contrast this with the third stage some thirty-five years ago when neighbors would congregate at a pioneer owner's home or in the streets outside store windows to watch the World Series or the Milton Berle Show on the black and white sets of that day. During the third stage of a new technology, there is usually a proliferation of companies seeking an early competitive edge. Product distinctions are clear although there are many look-alikes, and pricing is somewhat arbitrary. A characteristic of the fourth stage is that the businesses consolidate, major innovations are far more difficult because of large in-place investments, and prices stabilize. For example, the 1100 or so American car companies of the third stage have now become three and a half, and prices of comparable (and very similar) models are exceedingly competitive. Personal-computer technology today is in stage three, and we can confidently predict that the several hundred personal-computer companies vying for a share of this rapidly expanding market will shrink to some number about one hundred times smaller, and that the large dynamic range now seen in software pricing, for example, will soon reach a more rational equilibrium between cost and price.

Many studies have been made of these evolutionary processes of technological change; and while we tend to think of them as continuous, in fact they proceed in fits and starts. Social and economic events, such as wars and depressions, as well as major technical innovations, can produce an acceleration or deceleration in the rate of change of the substitution of a new technology for an old for a while, but then the original pattern is resumed. It doesn't seem to matter very much what these technologies are; once started the replacement process continues to completion in a strangely predictable way, whether we are talking about the replacement of sailing ships with steamships, the substitution of airline travel for rail travel or the introduction of plastic for glass in consumer bottles. The rate of dispersion of a particular technology is controlled by several factors; probably the strongest force is the development of a community of common interest. For example, the automobile as a technology for personal transportation could not have achieved stage four without the public's support of the development of a highway network. Similarly, the acceptance of the computer as a serious alternative to the mails or the telephone and telegraph systems will require the creation of extensive data networks, and that will require the public's conviction that this is worth doing.

While the intensity of this common interest is building, invention and entrepreneurship are extremely rapid and exhibit a form of positive feedback in which each advance feeds on those before, often without the underpinnings of fundamental science or understanding. Remember that most of the industrial revolution was based on early theories of electricity, magnetism, and Newtonian mechanics. Quantum mechanics and electrodynamics were not needed or used for this first tier of invention. This has also been the case in computing, where the hazy outlines of theoretical understanding are only now beginning to emerge. It is not an exaggeration to say that the personal computers of today, while useful, are primitive, unadapted versions of what their domesticated descendants will look like within a few years. The effects on the cost, appearance, and function of instruments can be expected to be profound.

THE DOMESTICATION OF COMPUTERS

The airline magazines and other learned journals would have us believe that the so-called "user-friendly" computers of today are already domesticated. I couldn't disagree more strongly; the average American today thinks about programming a personal computer in Basic or Pascal with about the same relish as he would an invitation to climb into a cage with a circus lion and for the same reason. Neither the computer nor the lion are really tame; they violate the rule of general access that is required for true domestication, for neither a lion tamer nor a professional programmer qualify as paradigms of general society.

The test for domestication that I like best is the one proposed by R.M. Pirzig in his book, *Zen and the Art of Motorcycle Maintenance*: "The material objects . . . can't be right or wrong. They don't have any ethical codes to follow except those people give them. The test of the machine is the satisfaction it gives you. There isn't any other test. If the machine produces tranquility, it's right. If it disturbs you, it's wrong until either the machine or your mind is changed." The media are full of examples which tell us unequivocally that we have gotten it all wrong. For example, Timex/Sinclair, in advertising their Model 1000 computer, ran a very successful campaign with the slogan "Why spend thousands to find out you hate computing when you can find out for $49.95?" The cover of a national training magazine featured a photograph of a blue-suited executive cowering before a personal computer, a braid of garlic around his neck, and a silver cross held fearfully before him.

I do not think this is all mindless paranoia. The average person begins with a distaste for computers which may be rooted in sentimentality for the technology being replaced (the same phenomenon that causes romantic attachments to steam trains and sailing vessels) or is the result of the uneasy perception that the computer may in some way become a substitute for the very faculties of mind that distinguish us from the beasts. Science fiction often portrays this in a threatening way, and the computer scientists have produced a guaranteed reinforcement of this malaise by making machines that are so complicated to use.

Advertising to the contrary, using most personal computers today requires arcane knowledge of specific and unforgiving sequences of commands. The computers' designs are optimized by engineers for cost and performance and not for natural usage by ordinary people. Truly friendly machines must be designed so that the particular idiosyncrasies of a user are accommodated in a way which supports his mental abstraction of the process taking place in the machine. Our software today is rude; it does not recognize individuals.

Ease of use, of course, ultimately depends on the programs that specify the machine's actions. A key factor is the degree to which these programs can be made to hide the details of operation from the users. It is one of our popular fantasies that some people can learn how to use machines rapidly no matter how complicated they are. Consider James Bond rushing into the atomic reactor or the space station that he has never seen before, quickly examining the bewildering maze of dials, buttons, flashing screens and then pressing precisely the right switches in the proper sequence to save the day just as time runs out. We all know that this is science fiction, but we like to think that somehow it should be possible to fathom the inner workings of complex machinery. The computer scientists have not really concentrated on this very much until recently; I believe that the industry is at an inflection point now, for we have finally realized that it is the human resource that must be optimized and not the cost/performance of the machines.

One of the great advances that I foresee is that computers will handle these complicated control sequences for us; only then will we be able to substitute simple abstractions for specialized knowledge, just as we now do in using many domesticated technologies. The final stage of technology evolution is often one of transparent complexity. Consider one button automatic tuning on a modern television set, for example, compared to the complex set of knobs to adjust the hue, intensity, fine tuning, brightness, contrast, vertical and horizontal hold of just a few years ago. It is possible to rent an American automobile of any model, any manufacturer, and drive away within a few minutes, having understood the location and function of the primary controls. Automatic transmission has made it unnecessary for people to understand the detailed workings of a gear box. The electric starter has produced a crankless, horseless carriage, and we no longer think at all about what happens when we turn the ignition key since automatic choking mechanisms or fuel injection have replaced some of the specific sequences that used to be required. Any user, even professional managers, can very rapidly grasp the concept behind a lever which is labeled "neutral," "drive," and "reverse," or a button labeled "start." The challenge is to produce this type of mental model to drive computer and instrument systems and then to represent it consistently to the user. Why shouldn't an office workstation have a button labeled "MAIL" or "Dow-Jones," and an instrument one called "determine resonant frequency?"

Hiding this complexity from the user while retaining a rich function set will require computer programs far more sophisticated than those we now write. The computer power required to execute them will be staggering. Simply increasing the speed of a computing cycle, and decreasing the cost and access time of memory by evolutionary advances in hardware technology that result from the application of scaling laws to switching devices, will fall far short of what is required for this task. Computer scientists must reorganize the internal structure of the computers, and modify the system architecture to increase greatly the number of operations that can be carried out in parallel. I think this will lead to computers that are organized as sets of special-purpose processors, each optimized for a particular class of tasks, such as extracting the patterns of speech or handwriting, or rapidly searching large databases, or managing the information on a display screen or doing a particular type of simulation or computation, such as a Fourier transform. These will be coordinated by very-high-performance, mass-produced, yet simple, general-purpose controllers. We will need systems thousands of times more powerful than the primitive general-purpose serial engines of today. Early examples of such machines are now beginning to appear in research laboratories, but there are many problems that relate to their interconnection and coordination. The programming style will be object-oriented--that is, control will be passed in the form of messages among self-descriptive entities. Although less efficient in terms of executed instruction path length, structural networks of objects help a lot with the two greatest limitations to the development of applications by professional programmers: reusability of programming solutions, and the difficulties of maintaining and modifying code.

Software must be designed to optimize not the way that the machines work, but the way that people think; and we can go a long way even with our present machines. Integrated programming environments are beginning to mask details from the user and to provide consistent access to services; natural language, touch screens, and speech are being combined to replace and augment the formalized keyboard interaction. In the new world of domesticated computers, slow, surly, uncooperative software will not survive. I believe that as we enter the fourth stage of computer technology, the successful companies will be the ones that recognize that in a rapidly expanding industry most of the user are beginners. The cumbersome instruction manuals and training classes of today must give way to an intuitive form of learning by doing the things that seem natural. The video-games people have stumbled on to the right experiential model—domesticated computers and instruments should be self-evident and self-documenting, and they should offer strong positive reinforcement almost immediately. Most users should never need to write a program in the way that we now think of that job. Professional computer scientists and programmers will survive, I hope, but increasingly their jobs will be to evolve the high-performance architectures and interfaces to shield users from the internal complexity and to develop tools so that users can build and maintain applications without becoming programmers.

Remember that while a congenial user interface may be a necessary condition for reaching the fourth stage, it is not sufficient. The degree of penetration will ultimately be determined by the value of the technology, as perceived by the people it serves; therefore, the critical question becomes what these pervasive computers, and the instrument systems based on them, will actually do.

We can think of technologies in a general way as amplifiers of human capabilities. So, for example, the telephone has widened the range of human speech and hearing, the television has expanded the range of our sight, and the automobile has served to increase our speed of foot and our ability to carry weight. The great promise of computing technology is that it will expand human memory, augment human reasoning, and facilitate human communication. Similarly, instruments can be thought of as enhancing the senses we already have or adding faculties to those not provided by nature. If we judge current computer and instrument systems by these lofty ideals, we must consider our progress at this time embryonic at best.

Computers in the workplace today are used largely to automate those tasks we have always done. Many are used to improve the performance of job-specific functions like designing circuits, or analyzing financial spreadsheets, or typing manuscripts. The augmented speed and memory capacity adds value, and the price of the machine is determined by the increase in productivity or improvement in quality of the product or service rendered. Professionals performing these functions will invest the requisite time and effort in learning to use the machines, however unpleasant that may be. Unquestionably, applying computer systems to these specific competence functions has made many previously intractable tasks possible; however, by and large, this is a difference in degree, and not in kind.

More interesting are the so-called computer-aided work functions which most people do, and which are largely independent of industry or profession. These can be thought of as the creation and distribution of "documents," electronic or otherwise; the storage, retrieval, and analysis of information, and user-to-user communication. Technologies usually evolve by replacement, and that we can think of this substitution of a new technology for an older one as an x-less-y phenomenon (for example, the automobile as a horseless carriage, the radio as a wireless telephone). We can recognize the computer-aided work functions as replacing a secretary and a typewriter, a filing clerk, support staff or librarian, and substituting for conventional telephones, telegraphs, or facsimile machines. In fact, these are the fundamental amplifications of memory, reasoning, and communication alluded to before, even if they are rather mundane.

Acceptance of these advanced systems has been much slower than predicted, due, I think, to how hard they have been to use and because the functions being replaced have not been significantly improved. This is now changing; the office-systems industry is early in stage three now, with inventions coming furiously and huge companies jockeying for position. Electronic voice mail, for example, with useful store-and-forward capabilities that eliminate the missed phone call and the barriers of international time zones will be widespread shortly. Soon even the simple home machines using TV sets for displays and telephone lines for communication will have easy access to large and diverse databases—the world's cookbook collection; or ten different articles from ten different encyclopedias on a topic of immediate importance; world-wide local weather or news; repair manuals for discontinued products—a seemingly endless set of attractive global services and information, hard to come by today, selected by simple menu-like commands and purchased on a pay-as-you-go basis. There are few, if any, technical impediments to this vision, and prototype systems have been working for several years now.

When these low-cost combinations of telephone network, television set, and personal computer are complete, I think that you will all want one in your home or office. You will be able to send and receive messages instantaneously to and from any connected television set. The world's libraries will be at your disposal, and intelligent, centralized reference services will aid in the collection and updating of information. The spread of cable TV and video-disk technology will add color video and sound to these capabilities—imagine a pronouncing dictionary, the Schwann record catalog on demand (we'll be able to compare, for example, different performances of the same piece of music), encyclopedias with film clips instead of still pictures, access to facsimiles of rare documents, interactive bedtime stories for children in which they'll be able to select their own endings. Driven by the economics of electronic communication, banking, shopping, and publishing, myriad new personal applications as yet unthought of will evolve. Once these services are easy to use from a consistent user interface, they will become the commonplace facilities of stage four. They will not need to be resident on the personal computer itself but will be features provided by the network of information servers.

The computer stores would have us believe that America's key priority is now checkbook balancing, with home recipe filing and word processing running close seconds. Don't believe it! We are on the eve of a great intellectual revolution, and it has nothing to do with bankbook-less banking or postman-less mail. The evidence is all around us, but because much of it is now directed at the largely mindless video-game market, it is perhaps hidden from the general adult view.

What is happening is that software is emerging as a unique medium, a very hot, interactive medium. A culture of media stars is developing—many of them teenagers and college students—who sign their programs, have a following, and hire software agents to represent them. The programs are beginning to be distributed by book publishers in bookstores, including some of our oldest and most respected firms. Genius programmers appear on TV talk shows. There is a software top-ten best seller list, and sub-industries like software distribution and software packaging have emerged. Similarities to the record and book industries are obvious, and problems of piracy, copyright, and compatibility are already troublesome. Software review magazines have appeared, and the *Whole Earth Software Catalog* has been published this year. Within this framework, a new type of literature is being written. Some have said that this is the literature of the future, but software and literature are alike only in that they are both media to present cool, static, one-way broadcasts. They evolved from the technology available at the time of their conception. Software is hot, interactive, dynamic. It is unlike anything we have seen before in its ability to draw people into the environment that it has created. It is expressive, but for now it can only be manipulated by the media stars. Computer science's principal task in this decade will be to provide the tools to eliminate this barrier, and once we have done this I believe a flood of creativity will occur and that it will come mostly from the large grass-roots population now excluded from such interaction. Interacting with the new software will be as different from watching TV or doing word processing or reserving hotel rooms electronically as can be imagined. We will need a new verb like "microwaving" to describe it. We have seen only the pop music side of the medium so far—the classical versions are coming now, and they are not going to be aimed at developing the hand-eye coordination necessary to repel an alien invasion.

THE DOMESTICATION OF INSTRUMENTS

For now, I'd like to assume that domesticated computers have come into existence and speculate about some of the effects that specific technology could produce on future instruments. I am not going to discuss today, for want of time, the tremendous progress in analog-to-digital and digital-to-analog converters, critical to success in so many of the applications. Suffice it to say that sampling rates in excess of one billion per second have already been achieved in laboratories, and that the resolution/speed/cost tradeoffs are increasingly attractive in commercial products. I'll also not discuss today the important area of sensors, although it is still the bottleneck in many situations. Innovation in this area is steady, but slow; cost, reliability, performance, and size of sensors are often limiters to the application of digital technology to measurement and test applications.

Let us first consider the impact of computers and digital signal processing on precision measurements. Since 1967, the world's time has been regulated by commercially available cesium-beam atomic clocks to an accuracy of 1 second in 100,000 years, or about 3-5 parts in 10^{12}. Incredible as this seems, it is not good enough for many modern navigation, communication, and astronomical applications. Len Cutler and his group at HP Laboratories, who developed the HP cesium standard in wide use throughout the world, have demonstrated an elegant new technique, which is about 10 times more accurate, based on the hyperfine transitions of trapped mercury ions. Ideally, an atomic frequency standard is based on the electric or magnetic dipole transition of a free atom observed for an infinite amount of time. In the cesium standard, the ideal of the free atom is realized to a high degree, but the observation time (that is, the time of flight through a microwave cavity) is fairly short.

Figure 1 shows the experimental apparatus designed to overcome this limitation. Mercury 199 ions are trapped for long periods in a quadrupole configuration driven by a precisely calculated radio-frequency field. The ions are pumped to an excited state using Mercury 202 light, and the ions are then irradiated with the microwave frequency tuned first to one side of the resonance at 40.5 GHz and then the other. The counter signals derived from the fluorescence which results are differenced and digitally integrated by the computer, using second differencing methods, and then converted to an analog signal. If the mean frequency of the microwave source is not at the line center, the error signal can be fed back, after a second integration, to a voltage-controlled crystal oscillator, which in turn regulates the frequency source. Complex control sequences minimize the noise due to secondary effects, and careful attention must be paid to other noise sources, such as second-order Doppler shift. Figure 2 indicates the results: the transition has been measured to an accuracy of 2 parts in 10^{11}, with a line width of 0.85 Hz at 40.5 GHz. If the full width at half-maximum of this transition is depicted as 1 inch wide, then the origin is three times the distance to the moon! The servo signal derived from this system yields an absolute accuracy conservatively estimated as 3 parts in 10^{13}, or about 1 second per million years. It is remarkable that today only a garden variety microprocessor and high-speed microelectronics are all that are needed to make such elegant instrumentation practical. While the precision here is an extreme case, many modern instruments are based on this kind of digital control and processing capabilities, unthinkable just a few decades ago.

We can find some far less exotic examples in the nearest computer store. A best-selling program, Typing Master, available now for home computers, analyzes the user's performance on some typing assignments and provides immediate feedback on improvements of a very specific sort. Even lifelong professional typists learn immediately and interactively how to improve their speed or accuracy or both. Extension to many other skills by replacing the keyboard with a sensor is possible. For example, one-chip signal processors could soon provide useful biofeedback to people learning to sing or to play musical instruments, helping with the difficult-to-teach relationships between anatomical control and auditory phenomena. We will be able to see on a screen in real time the spectrum by octaves of the sound which is produced. Visual feedback of auditory phenomena can show dynamically the relationship of harmonics to the fundamental or the transitions between head and

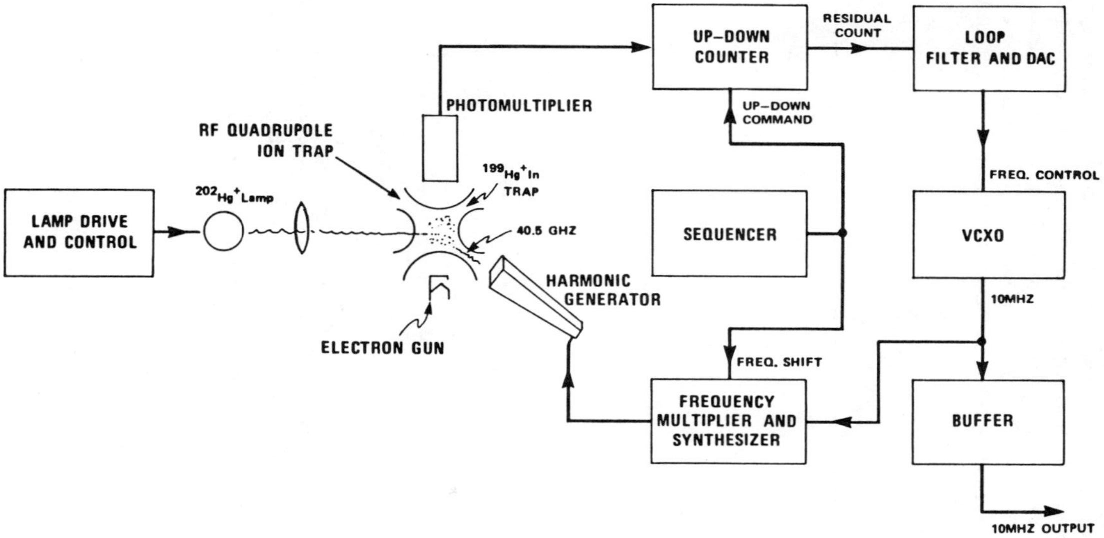

(©1985 ACM)

Figure 1. Trapped Mercury-Ion Frequency Standard

chest resonances, for example, in a way that our ears, which are not Fourier's children, cannot. Such equipment will enable singers and musicians to compare sound patterns and rhythms with stored templates produced by master musicians or with their own earlier efforts. The cost should be low enough to teach school children in the home. Programs to analyze speech accents and cadences are further in the future but certainly possible. Similar techniques with different sensors will permit the semi-automatic tuning of instruments like pianos and Porsches.

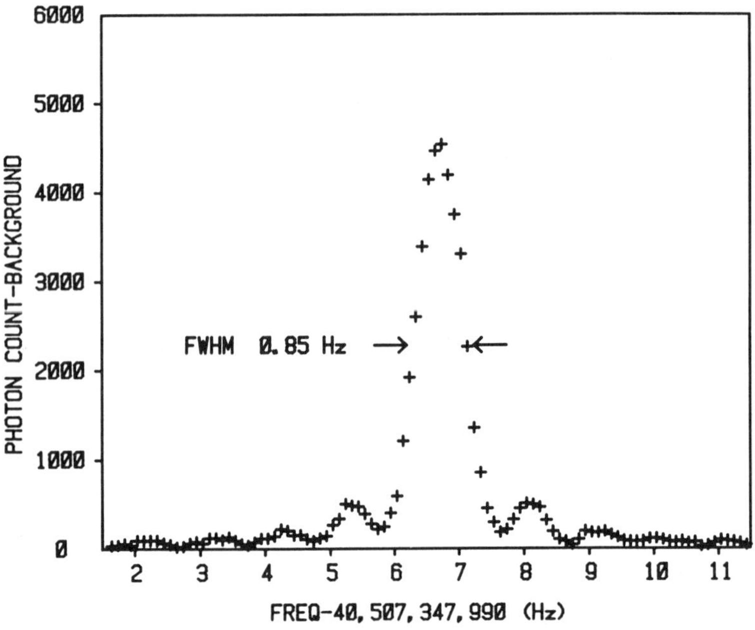

(©1985 ACM)

Figure 2. 199_{Hg}+ Hyperfine Resonance

Another type of personal computer program now becoming popular is the kit, like the Pinball Construction Set. Using simple pictorial interfaces and pointers, objects (in this case flippers, bumpers, spinners, bonus lights, etc.) are selected by the user from a pictorial menu, moved about on a screen and assembled according to the rules which have been previously imbedded in the kit by its creator. Many parameters in the pinball program can be changed—the location and elasticity of the bumpers and flippers, bonus point combinations, etc. Items difficult to change on real pinball machines could also be modifiable: non-physical gravitational fields, balls of arbitrary cross section, etc. After the almost infinitely variable game has been designed *without writing a single line of code*, it can be played immediately and further modified by the user. The output of the program is the pinball game. A later version of this idea involves a music construction kit. The objects now are notes, score markings, tempi, etc. We can create melodies, snip them with iconic scissors, combine them with pre-programmed melodies, transpose to new keys, and so forth with no programming. The application is music composition, and the output of the program will be the score of the music itself. We will be able to play the results on conventional instruments or directly on electronic instruments if we desire.

This entire notion of kits seems to me a very powerful one. If the appropriate sets of rules of combination, the constraints, can be imbedded in the machine by experts, then design can become possible based only on function and aesthetics for a vastly greater proportion of our population than is now possible. Consider an amateur architect designing an addition to his house using such techniques. The cost, the structural factors for different materials, the connection rules for electrical and plumbing fixtures will be preprogrammed in the machine, and the computer will monitor these constraints of the real world, physical or economic, as the design is produced. Cost estimates, blueprints, and material lists will be the output. It is reasonable to consider an electronic test construction kit. Sensors, analytical or quantitative measurement tools, and statistical techniques now constitute the objects; and the constraints are determined by the application domain. A test engineer need only configure the control and data flow on a screen; the program would be developed automatically, with analytical and statistical functions called into play as appropriate. Professional mechanical and electrical engineers and architects are already using early versions of such tools for computer-aided design. In ten years, I think there will be hundreds of variations available for everything from landscape architecture to factory simulation to cash management—people will be able to concentrate on the problem and not on the routine drudgery, and "computer-aided design" and "computer-aided test" will be important new ways of solving problems in many fields.

It is ironic that computers themselves have been so poorly instrumented and, until the last decade, have not been effectively analyzed with respect to internal performance. There are at least three reasons for this. First, making such measurements has been expensive because the computers themselves must be dedicated to the task in much the way that the biologist sacrifices the frog, and most manufacturers and users were satisfied with gross characterizations, such as millions of instructions per second (Mips) or transactions per second. Second, the early attempts at computer performance analysis were so crude and imposed such overhead that they perturbed the measurement, in the Heisenberg uncertainty-principle sense, in precisely those cases of heavy load and congestion that are of the most interest. Third, our ability to make mathematically tractable, predictive models of computer behavior has been limited, and so the measurements have been little used during design. This is changing now with rapid progress in all three areas due to both decreasing costs and better understanding. One result has been the emergence of the so-called reduced instruction set computers (RISC), a poor generic name for computers that base engineering tradeoffs with respect to instruction sets, storage hierarchies, and input/output design on the results of frequency-of-use measurements. These deceptively "simple" machines are, in fact, far more elegant in matching real requirements than their ancestors. G.K. Chesterton expressed this nicely, I think, in the aphorism: "There is more simplicity in the man who eats caviar on impulse than in the man who eats grapenuts on principle."

Microelectronics has also now made it possible to build less invasive, more sophisticated instrumentation into the initial design of a computer system; and the benefits of these built-in hardware and software monitors in terms of reliability, availability, and configuration management for improved price/performance will be large.

An experimental voltmeter with a cathode-ray "faceplate," that we have built at HP Laboratories, illustrates this approach. The techniques of computer graphics and touch screens or other pointing devices combine well with instruments based on computers, since information can be presented numerically or graphically, and the results of theory, or simulation can often be compared in non-physical coordinate spaces during the acquisition of the data to insure validity or to direct experimental strategy. Instructions to the user can be placed on the screen, as can calibration and repair procedures. Data can be probed in real time, as for example, in a mass spectrograph display. Windowing techniques can be used to combine the results of different instruments, as might be required in the display of a potential speech analyzer.

The emergence of the videodisk technology will also have profound effects on the human interface to instruments. At HP Laboratories, a video screen is used to show the layout of the components on an experimental system that blends video information from the disk with conventionally generated computer graphics. In such a system, icons can now be photographs, useful perhaps for parts in a catalog or people in a phone book. Motion and sound can now be mixed freely and inexpensively with conventional text and data. The techniques developed by the TV and movie industries over the last quarter century for dealing with transformations of vast amounts of visual information are being adapted for use on personal workstations. Self-documenting instruments and integral training manuals are sure to result. Prototypes have been developed for such a multi-media document for IC process description which allow the insertion of moving pictures and sound in the computer-generated text and graphics. Soon, locally rewritable disks will be available, and then a new style of lab notebook using video cameras will be possible; and an interesting base for interactive teaching will be available.

A new breed of videogame is emerging now which, like the early sword and sorcerer games such as Adventure, contains a world model and basic reasoning capability. The computer, in a sense, acts as a type of expert combining algorithmic and heuristic information. This, of course, represents the great promise of artificial-intelligence technology. We can think of a calculator as an arithmetic apprentice (how many of you can still take a cube root by hand?) and be confident that soon portable personal algebra helpers will be commonplace. We can now purchase spelling assistants, and one day grammar tutors will appear to help us with the mechanical tasks of writing so that we can concentrate our energies on thinking and style. Many companies are now working on these knowledge-based expert systems; they will be used to fix machines, to help diagnose illness, to translate languages, to locate oil; and they will help to codify human wisdom so that knowledge becomes reusable across a wide range of only superficially different problems.

As always with emerging incomplete technologies, the claims are overblown, the time scales are too aggressive, and many of the applications proposed by the myriad artificial-intelligence (AI) start-up companies will probe non-viable. Nevertheless, this kind of technology is becoming practical on home computers and instrument controllers. It will, for example, enable personalized filters which can eliminate electronic junk mail and sort the remainder according to content and priority. Intelligent database agents will provide a greatly enhanced ability to search global databases to find desired information without providing such detailed parameters for the search.

At HP Laboratories we have developed a high-performance technology (based on the LISP language) to represent knowledge and have applied it to wide range of applications. We think that the appearance of this type of technology on moderate priced, high-performance symbolic processors with a stable, industrial-strength software system, and interfaces that make it accessible to the non-expert user will lead to a change in the way we think about instrument and computer systems. What are now tools will become advisors and apprentices. These intelligent agents will indeed add reasoning capability and expand our memories, and may help us communicate with each other. For example, new classes of smart, composite instruments could come into being. A "hyphenated" instrument, such as a gas-chromatograph mass spectrometer combines complementary information from previously distinct measurements. Expert-systems technology allows us to synthesize diffuse human wisdom, unlikely to be found in a single individual and to reuse it across different domains. Many new combinations of instruments, perhaps even triplets, will evolve. The knowledge base will often not reside locally but will be in a network server, offering economies of scale, security, easy updating, and local machine independence. There are myriad opportunities for such smart servers in instrument interpretation as well as trouble-shooting or diagnostic applications, for either sick people or sick machines. Such servers can maintain up-to-date symptom and repair histories, as well as the latest information from the R&D labs. They will be usable during product development as well as after installation. I can image technology advancing to permit quality verification analyzers to be integral to the products themselves, useful during manufacture, and key to preventive maintenance and repair after installation.

At HP Labs, we have prototyped a manufacturing server on one of our experimental IC lines. It analyzes problems in yield in one of our complex processes and makes the expertise of one or two specialists far more widely available. That it does this twenty-four hours a day and on weekends has made it popular, even with the experts who were previously on call. The inputs to such manufacturing servers could come from humans, design files, or sensors. Modern fiber-optics communication technology makes the notion of remote sensors feeding an assemblage of expert servers pragmatic. A great appeal of AI technology is that it should help us to reuse solutions, since it clearly separates the data, or knowledge, from the algorithms, or rules, that are used to process it.

One of the problems about projecting the future is that it is hard to be outrageous enough about the consequences of the mature versions of the technology. Could Gutenberg have been expected to foresee the greeting card industry? Did the Model T portend the lunar rover? We are in the infancy of computer-based instrument technology, and as we look ahead to stage four, it is hard not to be too timid in our predictions. It is a challenge I cannot resist, so here are two outrageous ideas that I hope will provoke some of you.

First, consider the notion of the shirt-pocket instrument, made feasible through on-going cost reduction and miniaturization of sensors and microelectronics. We usually think of smarter and more accurate measurements resulting from technology evolution, but there might be a vast market for low-cost, personal, portable, low-resolution instruments, just as there is a computation market for handheld calculators. The digital thermometer, the wall-stud locator are examples of such instruments today. Imagine a pocket mass spectrometer used by factory foremen to determine whether a wall is painted with lead or alkyd, or a personal instrument to determine the composition of effluents, a poison analyzer and antidote determiner for the home, a pipe finder for factories. Convenience, immediacy and price, not precision, will determine the viability of such devices.

The last speculation I'd like to subject you to today concerns what I think of as "virtual instruments"; simulations of instruments, themselves acting on simulations of physical systems. The hardware VLSI logic simulators reported on by several companies are early examples of this concept that demonstrate the enormous leverage attainable through specialized architectures and modest concurrency when a task has micro-parallelism inherent in it. Even these early machines can achieve three orders of magnitude speed-up compared to large conventional computers built in the same technology and performing the same computations in normal sequential software. Figure 3 is a simplified view of one such machine that we have built at HP Labs. A designer inputs a

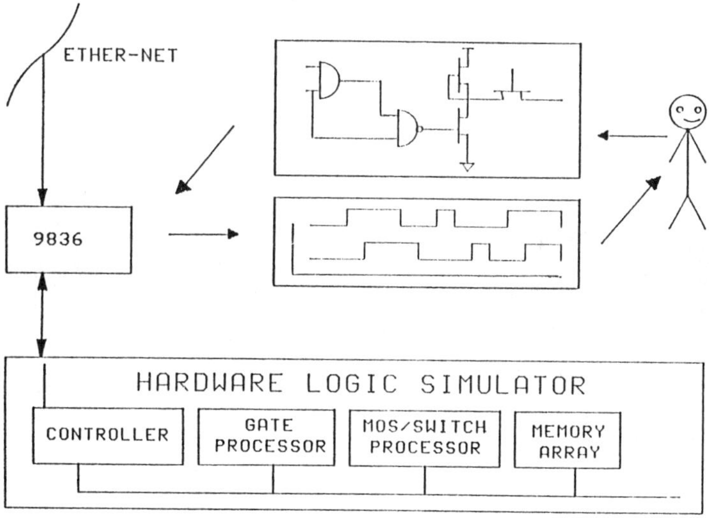

(©1985 ACM)

Figure 3. An Example of a "Virtual Instrument" Simulation

logical representation of the machine being designed in a conventional way on a standard design workstation and receives output, in close to real time, that is equivalent to probing a real machine with a very powerful and flexible logic analyzer. Thus, we are using a virtual instrument to examine a not-yet-built virtual machine. Design become iterative and interactive. The machines' three specialized processors and the custom memory partitions they address are so organized that entire operating systems or applications can run and be debugged on the machine under development before any silicon fabrication is attempted.

Figure 4 shows how the machine works. Conceptually, think of a simulated timing wheel continuously rotating. Each tick of the virtual clock

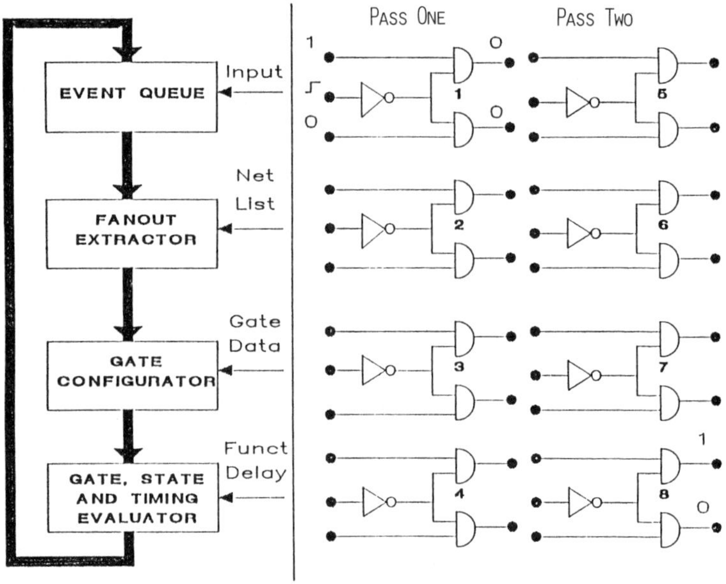

(©1985 ACM)

Figure 4. Details of HP Logic Simulator

represents a state of the design under investigation. Each of the four stages in the pipeline performs a dedicated function at each virtual cycle. The first stage, after initialization, represents those logic elements that have changed since the last tick and is one of the innovations of this particular design, since the machine concerns itself only with these elements. The second stage contains the topology of the machine or the net list, and, when a change occurs, points to those other components in the memory that are affected by the fanout. The third stage is the gate processor which references the gate library to determine the logical characteristics of a particular element. The last stage adds timing and other state information and updates the event queue (the timing wheel) for the next pass. The machine is designed very modularly. The gate capacity is extensible, parallelism can be improved by providing multiple processors (for example, multiple paths of the fanout can be examined simultaneously). Modest versions of the simulator can serve as co-processors on workstations. As indicated in Figure 5, larger versions are capable of running multiple designs simultaneously in much the way that multiple student projects are now fabricated on single wafers of silicon, and can be used to validate an entire operating system or compiler. These machines can be resident on a server – perhaps accessed via a network.

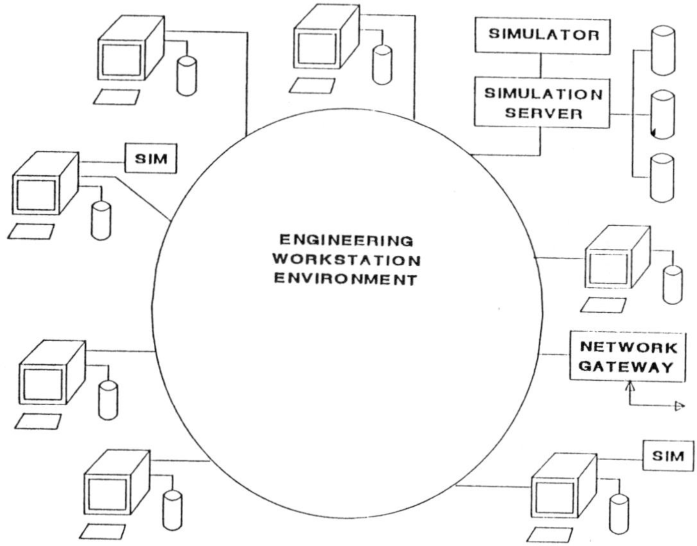

Figure 5. Example of Simultaneous Multiple-Design Simulation

Figure 6 shows the situation in micro-electronic system fabrication today.

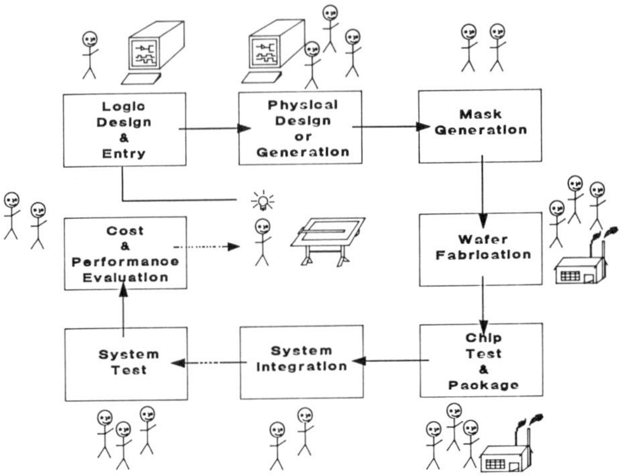

Figure 6. Steps in Micro-Electronic Fabrication

Each of the steps shown is necessary, and the process takes a great deal of time and money. Many types of expertise are involved and, typically, at each stage a new team of specialists is needed. The logic simulator represents just one early step in a prototyping system which could substitute mathematical modeling and testing of designs for much of the manufacturing work that we now do. Consider a collection of special-purpose systems, as displayed in Figure 7, each containing the expertise of one of the specific teams involved. Such an expert system would, as our knowledge of process models and device characteristics improved, begin to replace the large expenditures of time and dollars which are now necessary. It is, perhaps, outrageous today to imagine developing a process, producing the logical and physical designs, and then verifying a complete computer architecture using these virtual instruments before fabricating a single chip. Nevertheless, rudimentary versions of many of these subsystem components now exist in many laboratories, and the progress in mathematical modeling of devices, expert systems to capture process knowledge, and the massive application of specialized computers seem to me a combination that puts this in the realm of eventual possibility.

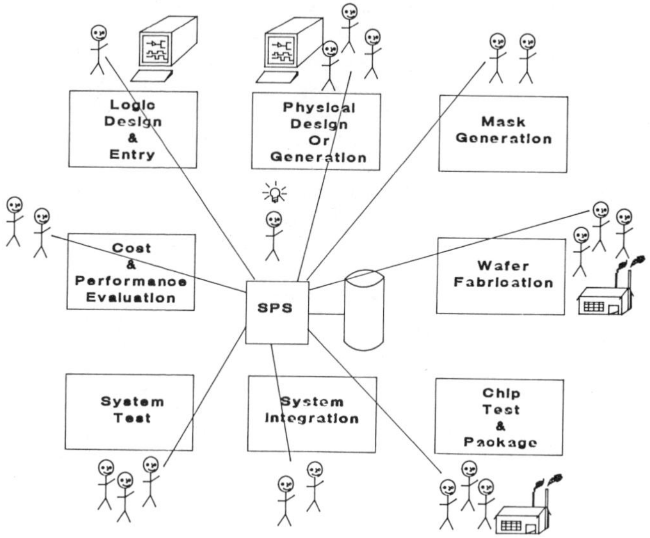

Figure 7. Group of Coordinated Special-Purpose Systems

THE NEXT CENTURY

When I was studying to be an engineer, I used to think how wonderful it would have been to have lived one hundred years ago, when the great mechanical and electrical inventions of the Industrial Revolution were being produced at a dizzying rate. Later, when I studied physics, I regretted not being alive in the period of the great discoveries of relativity and quantum mechanics. I have changed my mind. I am very glad that I am around now, because I think that the next ten years will be more exciting than anything we have ever seen before, because domesticated computers will allow so many more people to participate in the invention process. We have at last arrived at an understanding of the real potential of microelectronic technology; we are finally concentrating on the right problems—making machines usable by everyone and doing things with them that amplify our faculties in new ways. Computers, long pilloried as a dehumanizing force, will become the most powerful means of personal creative expression and communication ever known.

I have talked a lot about the domestication of technology and what I think that it could come to mean, and I would like to end by reading a short passage to you from *The Little Prince* by Antoine de Saint Exupéry. In that book, you may remember, the Little Prince, in love with what he thinks is a unique flower on his tiny planet but exasperated by her pettiness and idiosyncrasies, has wandered from asteroid to asteroid observing the strange ways of men. At last he reaches Earth where he is dismayed to learn that his flower is nothing more than a common rose and that he has wasted so much time catering to her whims. He is disconsolate when, in a vast desert, he meets a wild fox. The fox says to the Little Prince, "To me, you are still nothing more than a little boy who is just like a hundred thousand other little boys, and I have no need of you, and you, on your part, have no need of me. To you I am nothing more than a fox, like a hundred thousand other foxes, but if you tame me, then we shall need each other. To me you will be unique in all the world. To you I shall be unique in all the world. One only understands the things that one tames . . . you become responsible, forever, for what you have tamed . . . if you want a friend, tame me . . ." Later, after the Little Prince has tamed the fox, they come to a sad farewell. "Goodbye," said the fox, "and now here is my secret–a very simple secret. It is only with the heart that one can see rightly. What is essential is invisible to the eye . . . it is the time you have wasted for your rose that makes your rose so important."

The fox has given you my prescription for domesticating technology: our success should be measured by how much we come to care about the results. I hope that our descendants at the Bicentennial Symposium for Electrical Engineering at Cornell will care very much indeed.

FUTURE DIRECTIONS IN ELECTRICAL ENGINEERING

COMPUTERS

PROCEEDINGS of the

Cornell Electrical Engineering Centennial Symposium

at East Windsor, New Jersey—April 1, 1985

Simpson Linke, editor

Promethean Press
Ithaca, New York

PROCEEDINGS OF THE
CORNELL ELECTRICAL ENGINEERING CENTENNIAL SYMPOSIUM:
FUTURE DIRECTIONS IN ELECTRICAL ENGINEERING-
COMPUTERS

Copyright 1987
by the School of Electrical Engineering
Cornell University, Ithaca, New York
All rights reserved

Published by Promethean Press
Ithaca, New York

Printed in the United States of America

February, 1987

CONTENTS

PREFACE

The celebration of the Centennial Year of the School of Electrical Engineering at Cornell University featured a series of symposia based on the central theme of "Future Directions in Electrical Engineering". Each seminar concentrated on a major facet of the discipline, with individual topics presented by outstanding specialists in each field. The symposia were held at sites around the country that are widely known as centers of activity for each particular subject area, thereby providing significant updates on pertinent electrical engineering technologies at appropriate locations, and allowing a large number of Cornell Electrical Engineering alumni to participate in the celebration in a meaningful way.

The ultimate success of the Symposium Series must be credited to the dedicated efforts of the faculty, symposia, and local alumni chairmen who were responsible for planning the programs, engaging the participants, and attending to local arrangements within specific time constraints. The following schedule of events attests to the overall magnitude of the task and also displays the broad range of topics that were presented:

March 15, 1985	Microelectronics	Stanford, Calif.
April 1, 1985	Computers	E. Windsor, N. J.
April 10, 1985	Communications	Boston, Mass.
April 17, 1985	Atmospheric and Space Sciences	St. Louis, Mo.
April 25, 1985	Energy for the Future	Washington, D. C.
June 12, 1985	Societal Impacts and Future Directions	Ithaca, N. Y.

Each session was videotaped to facilitate preparation of the proceedings of each meeting and for possible use of the individual presentations in classroom seminars and elsewhere. The audio portions of the tapes were transcribed into preliminary text and coordinated with figures obtained from the video images and from hard copy provided by some of the authors. Each presentation was edited with a view toward achieving continuity and coherence without

compromising the natural style of the author. In several instances it was necessary to introduce substantial changes in format and phrasing in order to achieve this goal; in others, the hard copy supplied by the authors made the task easier. In all cases, the final versions were sent to the authors for their criticisms, corrections, and final approval. I am grateful to all of the contributors for their excellent cooperation and assistance in bringing their presentations to this present form.

I wish to thank Professor Edward D. Wolf for his enthusiastic interest in this project and for authorizing financial support from the National Research and Resource Facility for Submicron Structures to assist in the publication of the Symposia Series *Proceedings*. Additional funding from the College of Engineering and the School of Electrical Engineering is also gratefully acknowledged.

It would not have been possible to produce this work without the skillful secretarial assistance of Joan Hammond, Catherine Kuhl, Deborah Overslaugh, and Linda Struzinsky who performed the difficult tasks of transcribing the tapes and retyping the edited manuscripts with meticulous attention to detail. The valuable assistance of John Ralston in the preliminary editing of several of the manuscripts is also acknowledged with thanks. The expert guidance provided by Mrs. Diana Nathanielsz on format and production of this volume is very much appreciated.

I am particularly grateful to Catherine Kuhl and Deborah Overslaugh for their diligence and care in forming the camera-ready copies of the manuscripts on that versatile new instrument that is so illustrative of "future directions in electrical engineering" — the laser printer.

Simpson Linke

Ithaca, New York

INTRODUCTION

John A. Nation

Director, School of Electrical Engineering

The celebration of our Centennial Year affords a unique opportunity to review past accomplishments, to evaluate our present circumstances, and to make some reasonable estimates of prospects for the coming years. A glance at the history of the University reveals that the Cornell *Register* for 1883 contained a "Special Notice" from President Andrew Dickson White announcing the establishment of a "Course in Electrical Engineering." President White had convinced the University's Trustee Executive Committee to create the new program "provided such course occasions no additional expense."

The early curriculum was rigorous and in addition to the traditional engineering topics of mathematics, physics, mechanics, mechanical drawing, and shop work, required rhetoric, French and German. The senior year also included courses in several Cornell specialties such as photometry and electric lamps, telegraph instruments, lines and cables, the dynamo, and electric motors. The first baccalaureate degrees in electrical engineering in the United States were awarded by Cornell (as well as by M.I.T.) in 1885, and the first Ph.D. in the new discipline was conferred by President White at Cornell in the same year. Thus we are truly celebrating 100 years of electrical engineering education.

One of the major figures in the E.E. School of those days was Professor Harris J. Ryan who ran the Electrical Engineering Department essentially as a one-man faculty for 17 years and is credited with having trained almost 1000 electrical engineers during that period. Ryan left Cornell in 1905 to become head of the Electrical Engineering Department at Stanford. It is interesting to note that 12 of the first 13 members of the Stanford faculty were educated at Cornell.

This brief summary of the beginning of electrical engineering instruction at Cornell would not be complete without mention of the internationally known Professor Vladimir Karapetoff who joined the faculty following Professor Ryan's departure and made many contributions to early electrical engineering theory and analysis in several textbooks and many technical articles. Something of the flavor of electrical engineering studies of the time is suggested by the following excerpt from Karapetoff's *Electrical Laboratory Notes*, published in 1906:

> "In coming to the laboratory, bring with you a slide rule, an inch rule or tape, a speed counter, a screw

driver and a pair of plyers [sic]. This will save you time and trouble of looking for them or borrowing them. Do not forget to have a pocket knife for skinning off wire; a bicycle wrench is also sometimes very handy to have."

In these days of hand calculators and personal computers, it is rare indeed to find a student who has even seen a slide rule, much less to have used one.

Classical electrical engineering education at Cornell was devoted principally to electric machinery, power systems, telecommunications, and early electronics. As the technology developed from the 20's through the 40's, the curriculum was, of course, continually upgraded as new and advanced concepts were introduced, but the basic format remained essentially unchanged until the mid-50's when research began to play a central role in the E.E. School, coincident with the move from Franklin Hall to the newly constructed Phillips Hall. The following figure among the highlights of our research development over the years:

- An early solar telescope project initiated an era of Radio Astronomy research that culminated in the establishment of the world's largest radio-radar telescope at Arecibo, Puerto Rico, now known as the National Astronomy and Ionosphere Center.

- A research program in the new field of Physical Electronics gradually expanded into the present National Research and Resource Facility for Submicron Structures housed in the Knight Laboratory adjacent to Phillips Hall.

- The Old High Voltage Laboratory on Mitchell Street, converted to an Energy Laboratory in 1960, served as the staging area for the Cornell Extra-High-Voltage Cable Test Program and housed E.E. research on high-voltage vacuum switches and arc physics until the formation of the interdisciplinary Laboratory of Plasma Studies in 1966. This facility, together with adjacent installations, now contains the largest particle-beam program in any university in the world and has made many important contributions to the understanding of intense electron and ion beams, particularly as they relate to inertial confinement thermonuclear fusion experiments.

- The national energy crisis in the early 70's created renewed interest in power and energy research within the School leading to the combination of power-system elements, control theory, and new computer facilities, that now constitutes the unique Kettering Power Systems

Laboratory for the analysis of modern complex power-generation and transmission networks.

Two recent events attest to the high regard that the University and the E.E. School enjoy in the national research community:

1. The Program in Microscience and Technology was established in 1982 and receives about $2 million annually through the Center of Excellence Program of the Semiconductor Research Corporation.

2. The NSF-sponsored Cornell Center for Theory and Simulation in Science and Engineering, with its associated supercomputer production facility, will augment and expand research capability in the School and will encourage greater interdisciplinary activities with other scientific units on campus.

One of our strengths in recent years, as the Electrical Engineering School has evolved from a purely teaching organization into a teaching and research institution, has been our ability to attract good new faculty. A positive measure of this quality is in our receipt of Presidential Young Investigator Awards from the National Science Foundation in the first year of the establishment of these awards. The E.E. School received 5 of the total of 26 awards made available to electrical engineering departments in the nation. Our success in attracting these high-quality young people bodes well for the continued future growth of the School.

Remarkable changes in electrical engineering education have occurred in the School within the past 25 years as the influence and impact of new faculty members and their research have brought modern concepts to curriculum and laboratories that had been essentially identical to those of World War II days. The advent of computers has been particularly significant. We have been very fortunate in the last decade in the support that we have received from industrial corporations. Their generous donations of equipment including computers, work stations, and state-of-the-art measurement devices have been of great value in upgrading our laboratories to present-day standards.

On the other hand, the current huge interest in computer engineering and electrical engineering has introduced major enrollment problems in the School. Using 1965 as a year of reference, the size of the faculty has remained fixed at about 40 members whereas the number of undergraduate students has tripled. Our senior class in 1985-86 numbered 186 students. We have 265 graduates students in the School with about 145-150 MS/PhD students and the remainder in the Master of Engineering Program, the latter being the second largest professional degree program in the University. This year we graduated

30 MS/PhD students and 88 Master of Engineering degree candidates. An encouraging factor is that 12 of these advanced-degree students were women. Today about 15 percent of the students in the E.E. School are women and 25 percent are minority students.

The increase in student enrollment is coupled with a research effort that is growing at a compound rate in excess of 10 percent per year in constant dollars. Clearly we are approaching the limits of our present capability. It is hard to realize that within 30 years of our initial occupancy of Phillips Hall we are now bursting at the seams. Further, the building which was then state-of-the-art, is now an antiquated facility in terms of present-day technology, with little in the way of the clean-air environments, air conditioning, and other amenities that are necessary for the proper functioning of modern equipment.

Plans are in progress to build a new engineering complex on campus in which electrical engineering will have substantially more space than in the past. When these new facilities are completed, our faculty will be able to work at its full potential. We expect to see a relatively modest growth in the number of faculty members. Since a faculty lifetime post is forty years, and student-interest time cycles are rather shorter than that, the response to large student increases in enrollments and interest in electrical engineering cannot be followed instantly by new faculty positions. Attempts will be made to maintain our undergraduate size at about its present level with some growth allowed in the number of our graduate students. It is certain that there will be many more interactions with industry in the near future, and these will take on a much broader form as the years go by. For example, an exchange program may be arranged with industrial laboratories so that their scientists will come to work in the School for a period while a faculty member will have the opportunity to spend corresponding time in industrial environments.

This brief perspective of past, present, and future encourages us to begin our second century with a true sense of excitement. We may be justly proud of the important contributions that our School of Electrical Engineering has made to many of the developments that are reported in these *Proceedings*. As we celebrate these past achievements, we can look forward with confidence to producing equally significant additions to our challenging discipline.

PART 2

COMPUTERS

PROCEEDINGS of the

Cornell Electrical Engineering Centennial Symposium

at East Windsor, New Jersey—April 1, 1985

"I take great pleasure in welcoming you to this Symposium on 'Future Directions in Electrical Engineering - Computers', the second of six cross-country symposia on major topics in electrical engineering that are being conducted this Spring in celebration of the Centennial Year of the Cornell School of Electrical Engineering. Our speakers today have all had significant impact on various aspects of the computer field. Their sphere of influence includes the design of products of major computer companies, computer-engineering education, and extensive contributions to the computer literature. Most of us are familiar with their texts and articles, and of course, we are constant users of the pertinent computer equipment. While we may have occasional complaints about those machines, the significant point is that we want very much to use them, even as we expect them to continue their steady improvement.

I want to take this opportunity to express my thanks and appreciation to Jack Scanlon, our Symposium Chairman, and to Roger Berman, our Local Arrangements Chairman, for their excellent work in making this Symposium possible. We can expect that our distinguished panel will present us with a number of provocative viewpoints on our rapidly growing field.

Our Chairman, Jack Scanlon, obtained the Bachelor of Science degree from the University of Toronto in 1964 and the Master of Science in Electrical Engineering from Cornell University in 1965. He joined the staff at Bell Laboratories in that year. From 1968 to 1974 he was supervisor of the Computer Science Group, served as Department Head of the Electronic Switching System Design Division from 1974 through 1977, and was Director of the Software and System Design Laboratory until 1979, when he left Bell Laboratories to become Vice President of Processor and Software Systems of the Western Electronics Company. The active pursuit of this distinguished career has allowed Jack to play a leading role in the design, development, and marketing of computers".

– Professor H. C. Torng, Symposium Faculty Chairman
From Welcoming Remarks, April 1, 1985
East Windsor, New Jersey

Editor's Note: In August 1985, Mr. Scanlon joined A.T. & T. as Group Vice President, Product Development, his present position.

"In the process of preparing my introductions for the eminent members of today's panel, and while reviewing their accomplishments, I could not help being impressed by the enormous progress that has occurred in the field of computer engineering since I completed my graduate studies in Electrical Engineering at Cornell about 20 years ago. My major was known then as "Switching Theory" and I did some research related to the design of a certain class of binary adders. No doubt our speakers will describe how we moved from those relatively primitive days to our present advanced status, and perhaps they will also dare to predict what we can expect to see in the next 20 years of this exciting and dynamic field".

– *Jack M. Scanlon, Symposium Chairman*
From Introductory Remarks, April 1, 1985
East Windsor, New Jersey

THE MINI AND MICRO GENERATIONS: INDUSTRY, TECHNOLOGY AND USE

C. Gordon Bell

C. Gordon Bell received the Bachelor of Science and the Master of Science degrees in Electrical Engineering from the Massachusetts Institute of Technology in 1956 and 1957 respectively. Following several years as a research engineer at the MIT Electronics Research Laboratory, he joined Digital Equipment Corporation (DEC) in 1960 as Manager of Computer Design. In 1966 he was appointed Professor of Computer Science at Carnegie-Mellon University and became Professor of Electrical Engineering and Computer Science in 1978. Starting in 1966 he was a consultant with DEC and served as their Vice President for Engineering from 1972 to 1983. In 1984 he became Chief Technical Officer of Encore Computer Corporation. In June 1986 he joined the National Science Foundation as Assistant Director of the Directorate for Computer and Information Science and Engineering (CISE). Mr. Bell was named a Fellow of the IEEE in 1974 and became a member of the National Academy of Engineering in 1977. He received the Mellon Institute Award in 1973, the McDowell Award in 1975, and the Echert-Mauchly Award in 1982. He is widely published in the fields of computer architecture and computer design.

Last year I wrote an article[1] that appeared in *Computer* magazine in October, 1984. My discussion today is partly a paraphrase of that article coupled with some additional information that I have gleaned from a study of the history of computing. I strongly recommend a visit to the Computer Museum, now open in Boston on Museum Wharf, where the past, present, and in some cases, the future of computing is presented very effectively.

COMPUTER GENERATIONS

My personal model of computing, and, in fact, of most technology, is similar to the evolutionary cycle that characterized the development of present-day particle accelerators from the original cyclotron (see Figure 1). There is a main loop that includes design of the elements, construction, programming the system, using the system, evaluating the design, all followed by a redesign. Fundamental scientific knowledge and accelerator technology, combined with information from other disciplines and from related industries, allowed the evolution of the accelerators. But, of course, the principal element that really caused the particle accelerators to come to fruition was a perceived need driving a resource generator that supplied the necessarily large amounts of financial and auxiliary support. Bear in mind, however, that this concept does not apply to the acceleration of a single particle but rather to many particles, some having low-energy and some having high-energy levels, with all the particles being at different states at a given time. Similarly with computers, there are many states in each generation, and the computers that form each generation all seem to have followed the same evolutionary path: learning from past mistakes and improving their functions along the way.

A generation for a technological system depends upon the convergence of needs, with the greatest "need" probably being a reaction to the threat of annihilation. If we look at the history of computing we see that the basic motivation for its development has been the very real fear of physical and/or industrial destruction that will result from the failure to build a better device. Greed has been a related factor, stemming from the desire to capitalize on an obviously popular product. We must also consider the natural desire of technologists to release or expand current resources against the wishes of the controlling organization. This attitude creates a state of frustration with existing institutions that inevitably causes an exodus of personnel to some other group that holds forth the promise of interesting new developments. Another "need" is to recognize that in evaluating a specific generation of technology

[1]Bell, C. Gordon, "The Mini and Micro Industries", *Computer*, Vol. 17, No. 10, October, 1984, pp. 14-30.

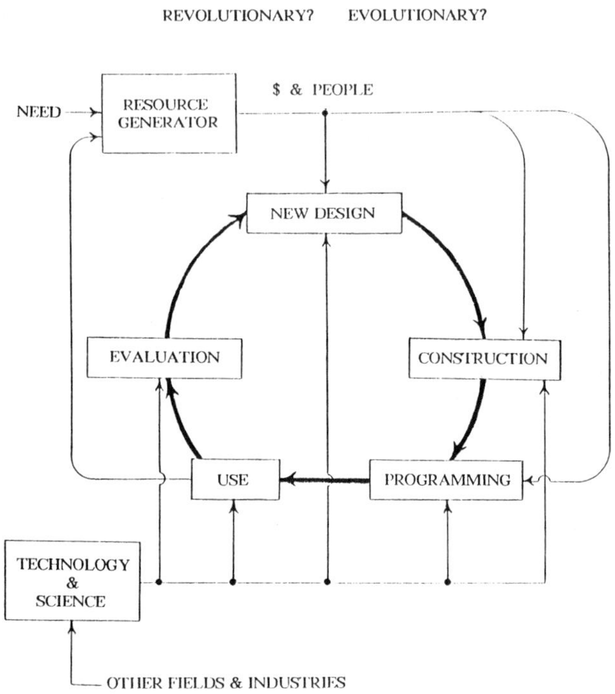

Figure 1. Cycle of Computer Generations

development it is not unusual to find that a mature generation does not have all of the properties that were expected initially.

There is an inherent need for science and technology to provide the basis for design and construction of all machines, but this requirement is influenced by the virtually universal belief that only brand-new organizations can develop completely new machines, and specifically new computer structures. Old organizations can evolve machines but they cannot originate new ones. They may suggest ideas for new machines but often fail to follow through and take them to the market, responding only after some other company has achieved a success. For example, Xerox originated the human-interface features in their ALTO and STAR models but those concepts were exploited by APPLE. We see APPLE introducing the PC, basically an evolutionary machine, only to have it exploited by IBM. When APPLE came out with their personal computer, many PC's were probably already on hand within IBM, each one undoubtedly superior to the APPLE version. In that large organization, with so many creative people available, there were at least a dozen models, but there were also at least a hundred times that many committees ready to kill anything that looked interesting or that would perturb an existing base.

The evolution of the computer is outlined briefly in Figure 2. Earlier devices, such as the abacus, served as convenient aids to computation, but at about the year 1600, when Napier developed Napier's bones (an early version of the slide rule) and the Log Tables, a generation of manual computation began that was based essentially on moving numbers around but with the

PRE-COMPUTER AGE:

Period	Relative Period	Mode
1600 - 1800	3	Manual
1800 - 1890	2	Mechanical
1890 - 1950	1	Electro-mechanical

MODERN COMPUTER AGE:

Period	Generation	Mode
1950 - 1960	1st	Electronic
1960 - 1972	2nd	Transistor
1968	3rd	I/C
to	4th	VLSI
Present	5th	ULSI
?	6th	Optical

Figure 2. Evolution of Computer Generations

added feature of allowing primitive forms of measurement. With the coming of the Industrial Revolution in 1800, mechanical aids to computation appeared. It is interesting to note from Figure 2 that the relative rate of technical improvement increases. In 1890, the invention of the Hollerith tabs allowed the introduction and exploitation of punched cards. This basic electromechanical technique was fully realized by 1950, with the full development time being about half that of the previous generation.

Electronic computers made their appearance in 1950. Since the necessary experimentation was done in the early 40's, this first generation of modern computers covered a period of ten years. Transistorized computers were being built until 1972, representing another ten-year generation. The third-generation I/C computer was started in 1968 and has evolved to the present fourth-generation VLSI and the fifth-generation ULSI designs. Although we have no firm information on what will happen by the year 2000, it is certainly in order to project the sixth generation as being based on optical and particle physics.

STANDARDS

The computer industry today is unlike any previous generation because of the existence of standards. It has become a product-segmented industry, organized by levels of integration that form industrial strata. It is fueled by entrepreneurial energy which releases venture capital funds, but the new products are really formed by software. The strata are displayed in Figure 3, adapted from my previously mentioned article in *Computer* magazine.[2] I have not given these levels an ordinal number but it is evident that we must start with the earth as the source of iron and sand: hence magnetics and semiconductors form the ground plane. We then move up to the silicon wafers that are currently bipolar but competitive with emerging CMOS technologies. The standard chip is at the next level where we see that chips and memory are evolving together, starting with the 8080 that utilized a 4-K memory chip. That system was replaced by the 6502 that contained a 16-K chip, followed by the 8086 with a 64-K chip, and so on. Since memories determine the level of improvement, we observe that major gains occur about every three years, exactly on memory boundaries, but not on particular processor boundaries.

These developments, when coupled with the invention of the floppy disk, the Winchester disk, and other pertinent electromechanical contrivances, have made it possible to form new microstructures that have contributed to the evolution of the supercomputer. Returning momentarily to our accelerator

---

[2]*ibid*. pp. 20-21.

DISCIPLINE AND PROFESSION-SPECIFIC VERTICAL APPLICATIONS
[CAD for logic and circuit design and small-business accounting]

GENERIC APPLICATIONS (Word processing, electronic mail, spread sheets)

THIRD-GENERATION PROGRAMMING LANGUAGES AND DATABASES
[FORTRAN, BASIC → PASCAL → C (evolutionary)]

OPERATING SYSTEMS (Base systems, communication gateways, databases)
[Integrated BASIC → CP/M → MS/DOS → UNIX (evolutionary)]

PRINTED-CIRCUIT BOARDS (Busses synchronized to micro and memory intros)
[S100 → PC Bus, Multibus → Multibus II and VME]

ELECTROMECHANICAL SYSTEMS (Disks, monitors, power supplies, enclosures)[8" → 5" → 3" (?) floppy, 5" Winchester (evolutionary)]

STANDARD CHIPS (Micros, micro-peripherals, and memories)
[Intel and Motorola architectures synchronized to the evolution of memory chip sizes: 8080{S100}(4K) → Z80, 6502(16K) → 8086{Multibus, PC Bus} and 68000{VME}(64K) → 286{Multibus II}, 68020 and NS32032(256K)]

SILICON WAFERS (Bipolar and evolving CMOS technologies)
[Proprietary, corporate-process standards .... require formalization in order to realize a silicon foundry-based industry]

EARTH (Iron and sand)

(© 1984 IEEE)

Figure 3. Levels of Integration Forming Industrial Strata

model, we may think of the supercomputer concept as another "particle" in the ring, operating at a much higher rate of speed, going its own way, and guided by a different set of physical constraints. When individual microprocessors are connected together by collections of printed-circuit boards to make arbitrary systems, it is possible to achieve a variety of machine classes, each with its own evolutionary cycle.

Continuing up the ladder in Figure 3, we observe an evolution in the operating-system/language interface, moving from integrated BASIC, momentarily through CP/M, then to MS/DOS, and finally to UNIX. All of this restructuring is the result of the application of unifying standards. A similar evolution has occurred in the conventional third generation of practical languages that are useful in our current collection of generic applications. Finally, on the top rung, we find a set of highly structured vertical architectural packages.

An interesting conclusion that may be drawn from this summary is that, contrary to general belief, each of these strata contains a highly segmented spread-sheet industry. For example, a massive word-processing industry exists, even for UNIX (although in my opinion a good word processor for UNIX has not emerged as yet). The main point, however, is that there is an intense competition in each major area, unlike the situation in earlier generations when one manufacturer, using a systems approach, would attempt to provide all levels of hardware, software, and applications, with relatively unsatisfactory results.

ENTREPRENEURIAL ENERGY

The computer industry formation process, as I see it today, is outlined in the "entrepreneur venture-cycle program" shown in Figure 4. The program is expressed in a style similar to a PASCAL language dialect, and also appears in

```
                    ENTREPRENEUR_VENTURE_CYCLE

Procedure:
   begin
   while Frustration > Reward {Push from Old_company} and
         Greed       > Fear   {Pull to New_company} do
         begin
         get  (PC, Lotus 1-2-3);
         IF System_company then
              write (Beat_VAX_Business_Plan, New_company)
              else
              write (Business_Plan, New_company);
         get (Venture_capital); {from Old_Venture_company}
         exit {job}; start (New_company);
         get  (VAX, product_development_tools);
         build (product); sell (product);
         sell (New_company); {@ 100 x sales}
         Venture_funds : = Company_sale;
         start (New_Venture_company);
         end
   end

(© 1984 C. Gordon Bell)
```

Figure 4. The Entrepreneurial-Driven, Venture-Capital Financed, High-Tech Company Formation Process

my October, 1984 article in *Computer* magazine.[3] The entrepreneurial cycle is initiated by a frustration/reward syndrome evoked by the "push" of an old company and the "pull" of a new company or a new product idea. It is certain too that much of the motivation for start-up is due to the ever-present greed factor. A PC that runs LOTUS 1-2-3 is required to write two business plans, separated by an IF clause. The first one may be called the "kill VAX business plan" followed by a plan to obtain the necessary capital from one or more venture capital companies. The next step is to make a fast exit from the old company before the new one is started (perhaps this operation can be done in about one day). The new company then proceeds to obtain a VAX to help develop a product that will compete with the VAX (I call this maneuver the "VAX sandwich"). When the new product is built, marketed successfully, and produces a profit, the new company is sold and a new venture cycle is started. This highly volatile behavior is typical of present-day computer business activities.

THE MINICOMPUTER GENERATION

The effect of these entrepreneurial enterprises can be determined by an examination of the recent history of computer development. Figure 5 contains an outline[4] of a new technology, known as small-scale integrated (SSI) circuits, that was applied to many computers. Using this technique, it was possible to start a company without a profound knowledge of computer circuitry. All that was needed to make a new computer was to plug a bunch of chips together in accordance with a limited amount of logic. Power supplies, core memories, semiconductors, disks, and tapes could be bought and combined into appropriate packages. A fairly large number of skills were still required, however, since it was necessary to design a CPU in memory together with some peripheral controllers, build an operating system, add a pertinent language, and integrate the system for specific applications.

A survey[5] of actual minicomputer development from 1968 to 1972 is given in Figure 6. Of the 49 starting companies that I have classified as having retained their autonomy, Data General and Prime grew at significant rates and continue to grow. Eight of the listed companies in this group grew at diminished rates and 39 other organizations have ceased operations. Ten firms have merged with larger companies but only two, Interdata and SEL, have ended up with continued growth. Eight existing computer companies entered the minicomputer field but only two, Digital Equipment Corporation (DEC)

---

[3]*ibid*. p. 19.
[4]*ibid*. p. 14.
[5]*ibid*. p. 16.

BASIC COMPONENT INDUSTRIES	MINICOMPUTER SYSTEM COMPANIES
Power Supplies	Optional
Packaging	Essential
Core Memory	Optional
Semiconductors (MSI)	CPU and Memories
Disks and Tapes	Peripheral Controllers
Terminals	-
	Operating Systems
	Languages
	Applications
	System Integration

(©1984 IEEE)

Figure 5. Minicomputer Technology (circa 1970)

and IBM, continue to build successful minis. In a sense, DEC originated the minicomputer business with its PDP/80, a machine based on a different technology but one that illustrated a possible approach to the design of machines of this class. The IBM machines were more expensive but one could argue that their successful Series 1 and Series 3 units also belong in this period. Then there were 25 non-computer manufacturers who felt that minicomputers were quite trivial and relatively simple to build so they also got into the act for various reasons. Among these was HP who acquired a company called DYMEC and formed a highly successful supplier division.

Summarizing, it appears that the overall success rate has been about 20 percent with a continued success rate of roughly 7 percent. The survival rate is also at about 20 percent. Many useful conclusions may be drawn from this survey. If you are about to start a company I strongly suggest looking up that article in *Computer* magazine and reading my ideas on why some of these events have occurred.

49 STARTED UP AND RETAINED AUTONOMY

- 2 Grew at significant rates and continue to grow
  [Data General and Prime]
- 8 Grew at diminished or declining rates or found small niches
  [Adage, Basic 4, Computer Automation, Four Phase, General Automation, Macrodata, Microdata, Modcomp]
- 39 Ceased manufacture and failed

10 STARTED UP AND MERGED WITH LARGER COMPANIES

- 2 Grew at significant rates and continue to grow
  [Interdata → Perkin Elmer, SEL → Gould]
- 2 Continued and now manufacture niche products
  [Compten → NCR, Datacraft → Harris]
- 6 Stopped manufacturing minicomputers in the merged division

8 EXISTING COMPUTER COMPANIES BUILT MINICOMPUTERS

- 2 Made successful minicomputers and grew at a high rate
  [Digital Equipment Corporation, and IBM]
- 2 Continued with dimishing success with minis
  [Bunker-Ramo, CDC]
- 4 Stopped manufacturing minicomputers
  [GE, Packard-Bell, Recomp, Victor]

25 EXISTING NON-COMPUTER COMPANIES BUILT MINICOMPUTERS FOR BACKWARD INTEGRATION OR SPECIAL-SYSTEM NICHES

- 1 Company acquired an embryonic company in the design stage and formed a division to become a highly successful supplier
  [HP acquired DYMEC]
- 3 Companies continued to build and supply minicomputers for niche markets
  [Hughes, Raytheon, Texas Instruments]
- 21 Companies discontinued building minicomputers

(©1984 IEEE)

Figure 6. U.S. Minicomputer Company Attempts (circa 1968-1972)

THE MICROPROCESSOR GENERATION

The first microcomputer appeared in 1971 with the introduction of the 4004. By 1978, most of the glaring mistakes had been removed, thereby allowing an acceptable micro that actually could be programmed in the general sense by computer users. At the present time, however, I am not very optimistic about the prospects for major improvements in microcomputers in the near future. An examination of the listing in Figure 7 indicates that today's technology requirements are not very imposing when compared with the complexity of the engineering needed to produce early mainframes and minis, as outlined in Figure 4. All that seems to be required to build a system is to combine a number of individual components into a working unit. I know of one company that started up without anyone on the staff whom I would call an engineer. They had an assembler who connected modules together in a nice rack. It is not clear if the system even passed FCC

BASIC COMPONENT INDUSTRIES	MINICOMPUTER SYSTEM COMPANIES
Power Supplies	Optional
Packaging	Optional
Semiconductors (MSI)	-
(Micros, memory, & peripherals)	
CRT's and Terminals	-
Disks and Tapes	-
Board Options (e.g. displays)	Optional
UNIX & Diagnostics	Optional
Languages & Databases	Optional
LAN's & Communication	Optional
Applications	System Integration

(©1984 IEEE)

Figure 7. Microcomputer-Based Technology (circa 1978-)

regulations, but it survived long enough for several units to be marketed before the company discovered that there is a little more to computing than being in the assembly business. That effort really did not bring anything new to computer technology.

Another reason for my pessimistic view is based on the relatively imperfect training of our engineering people. For example, as a user of micros I complain about the designers of instruction sets. Since I have designed several myself, I cannot understand why anyone would have difficulty, particularly since we have about five thousand different architectures that should provide sufficient information for the production of good designs. I suspect that much of the blame can be placed on current textbooks that tend to present bizarre approaches rather than really good engineering ones that work. In educating our students we apparently do not tell them that the interesting examples in the texts are to be used for purposes of analysis only, and that in the real world of computing they will have to learn to devise their own solutions.

These deficiencies in design should cause us particular concern since microcomputers have already reached a substantial level of application that shows every sign of continued expansion. They are certainly prevalent at the desk level in the form of personal computers, smart telephones, word processors, workstation terminals, home appliances, video games, and in portable PC's. At the cabinet level, many individual micros are interconnected to form relatively large computing units.

MACHINE CLASSES

The mini and micro computers that I have just described are best classified on the basis of price, as displayed in Figure 8, again adapted from the October 1984 *Computer* article.[6] I should mention that all of the details of the evolution of these machine classes are not totally revealed in the *Computer* article. Additional information may be found in the publication[7] *Computer Engineering* that contains a review of the 25 years of computer development at DEC through the year 1978. In Figure 8, the machines within the area bounded by the broken lines are all micros, including the desk-level types as well as some of the units, designated as shared systems, that serve a number of users. The so-called "super-micro" is a machine that looks like a mini but is actually based on micros. This region of the chart also includes a machine class that I think is particularly interesting and on which I am working personally. It is

[6]*ibid.* p. 15.
[7]C. Gordon Bell, J. Craig Mudge, John E. McNamara, *Computer Engineering, A DEC View of Hardware Systems Design,* Digital Press, Bedford, MA, 1978.

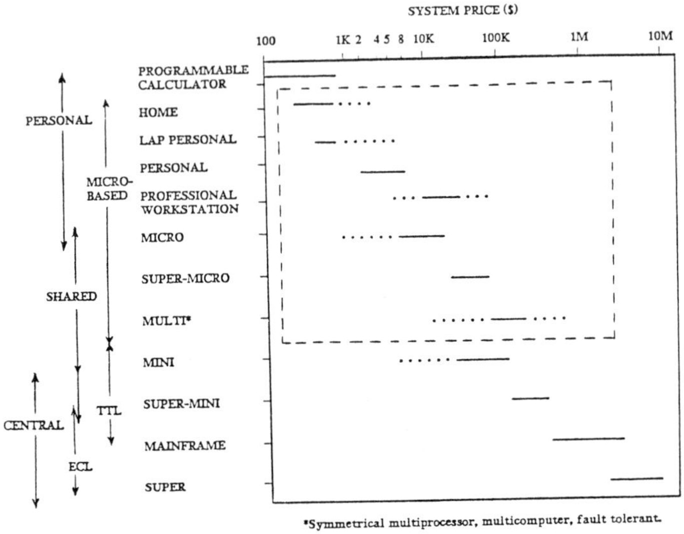

(©1984 IEEE)

Figure 8. 1984 system price versus machine class. The dots on the ends of the lines signify the uncertainty of price range. Because these classes are relatively new, prices are changing rapidly. Also the class has a broad definition; that is, a number of products of varying complexity can go by the same name. Products within a class can be anything from boards to complete systems.

called the "multi" and is based on the idea of putting many micros together and using them to produce a multitude of cycles. The challenge here is to make these cycles available, either to many users or ultimately to a single job. Although this projected task is quite difficult, I believe that it will be accomplished eventually.

Another view of the entire evolutionary process is summarized in Figure 9, a useful chart that will appear soon in an article I have written for *Science*.[8] I have observed and studied this process over the last 15 years. Some of

[8]C. Gordon Bell, "Multis: A New Class of Multiprocessor Computers", *Science*, Vol. 228, No. 4698, 26 April, 1985, pp. 462-466.

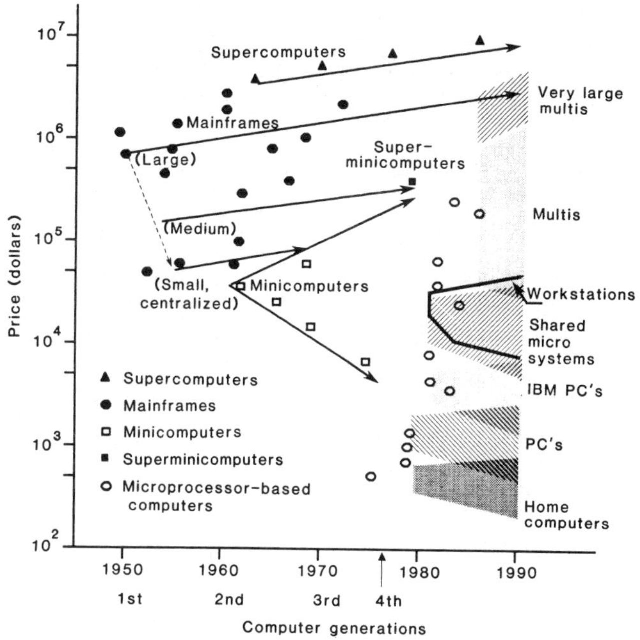

(©1984 AAAS)

Figure 9. The price of classes of computers plotted against time illustrates the clustering of classes such as mainframes, supercomputers, minicomputers and superminicomputers, personal and home computers, shared micros, workstations, amd multis. The emergence of new classes of computers coincides with the advent of new technologies, as marked by the generations of technologies. Shaded areas identify microprocessor-based systems.

the early basic ideas are still applicable but certainly many of the notions about the evolution of computers have changed. It is clear that a major reason for the emergence of these machine classes is the influence of organizational practices. Large business firms characteristically have their computing facilities incorporated into a "mainframe". Even smaller firms, with a ten or twenty-thousand dollar computer, like the idea of having a mainframe. Such facilities, however, provide services that are independent of the actual user, the units require a fair amount of tender loving care, and the user-complaint level is high. By contrast, a mini computer has virtually the same operating characteristics but the users belong to the same organization as the caretakers so the amount of grumbling is greatly reduced. Then there are those who prefer to be their own operators; those who do not mind taking care of all those floppy disks, and, of course, if something goes wrong they have no one to blame but themselves. When these diverse user attitudes are coupled with the advances in technology that affect the price levels, the formation of a variety of machine classes is inevitable.

The shaded areas in the chart indicate that the microprocessor, based on VLSI technology, will have a major, near-term impact upon all machine classes, extending even to the minis and the very large multis. Meanwhile a very interesting machine class has appeared that is characterized by either constant performance and lower price or constant price and improved performance. This so-called minisupercomputer results from attempts of developers to move out of the very costly supercomputer market into a more reasonable price range. Convex, for example, has produced a $500,000, 60-megaflop machine in this class. I know of only ten competitors who have this product in various stages of startup. There are probably ten more companies in the process of formation as the venture capitalists exercise their apparent need to be represented in every new advance in the industry.

In considering the present overall competitiveness in the industry, I am concerned that we have too much venture capital churning a good base of technical ideas. Currently overfunded by perhaps two orders of magnitude, the base is simply too small to justify such large expenditures. For example, there is probably enough room for only two to four good workstation companies instead of the 123 firms that exist at present. This situation is no doubt aggravated by the attitudes of large organizations such as IBM and DEC who view everything that computes as their property. Even AT&T is taking the same approach, although in my opinion that company should concentrate on improving communications systems through the application of computer technology rather than entering the mainstream of computer production.

Returning now to our original model of the computer-generation cycle, as shown in Figure 1, perhaps we should add a block labelled "Entrepreneurship" that in turn is modified by a control block designated as "Marketing". History records that more than half of the companies that are developed in any field ultimately do not succeed. I have outlined the formation of computing machinery over time through the present-day micros. Application of this modified computer-generation cycle to the advanced computer developments of the future may help us to determine if history will repeat itself.

SOME OPPORTUNITIES IN COMPUTING*

Joel S. Birnbaum

Joel Birnbaum received a Bachelor of Engineering Physics from Cornell in 1960, and MS and Ph.D. degrees in experimental nuclear physics from Yale in 1961 and 1965 respectively. From 1965 to 1980 he was with IBM Watson Research Center in Yorktown Heights, New York, becoming Director of Computer Science in 1975. He joined the Hewlett-Packard Company in Palo Alto, California, in 1980 as Director of the newly formed Computer Research Center. He is now Vice President and Director of HP Laboratories.

---

*Major portions of this presentation also appear in "Toward the Domestication of Computers," a chapter in *Cohabiting with Computers*, edited by Joseph F. Traub, published by William Kaufmann, Inc., Los Altos, California, 94022, 1985, and in the article "Toward the Domestication of Microelectronics" in the joint issue of the IEEE Computer Society magazine *Computer*, Vol. 18, No. 11, pp. 128-140, and *Communications of the Association for Computing Machinery* (ACM), Vol. 28, No. 11, pp. 1225-1235, November, 1985. Reprinted by permission. A similar version of this article also appears in Part 1 of these *Proceedings* ("Microelectronics").

The French writer Paul Valery once said that "the trouble with our times is that the future is not what it used to be." That seems particularly true right now in the computer and instrument businesses. Modern measurement technology is now inextricably linked to computers—most instruments today contain at least one microprocessor, and many connect to minicomputers and communication networks as well. Concomitantly, an increasing availability of sensors and signal processors for myriad applications of personal computers and workstations in networks is making the distinction among computer-based instruments and instrumented computers merely a semantic one.

The advances of the past thirty-odd years in microelectronics have resulted, in large degree, from the fact that the value of information is independent of the physical size of its container. The number of devices per unit volume has increased about one hundred fold per decade, and this relentless pursuit of smallness has led to a commensurate scaling of speed, power, size, and cost. The next ten years will witness a new type of quest: we are as concerned now with usability and usefulness as we have previously been with price and performance. The Japanese call this the Fifth Generation, intended to signify that the representation of knowledge and its processing in ways that exploit concurrency is an evolutionary step from the miniaturization that has come before. Others talk of Fourth Generation Languages, indicating a higher level of abstraction and purported benefits in our ability to understand and produce computer programs. Similarly, the media are full of articles about Fourth Generation communication systems which can switch digital and multi-modal information. Consumer-driven industries, such as hi-fidelity audio and 35mm photography, announce a new generation in time for each Christmas shopping season.

I do not much care for these advertising-driven characterizations of the march of technology, for they do not provide a useful taxonomy for evaluating the implied evolutionary process. I prefer a model proposed by my colleagues John Doyle and Egon Loebner which, for me, provides a better framework for understanding the impact of microelectronics. In my presentation today, I will try to sketch the outlines of that model and then speculate on two related questions: what will the computer and instrument systems of the future be like, and, more importantly, what will they do?

THE EVOLUTION OF PERVASIVE TECHNOLOGY

Among the ancient records of human life, we find almost 5000-year old reports of the domestication of dogs in Egypt. Pictures of horses drawing vehicles some 3500 years ago have survived. Man's ability to tame wild creatures and to train them as beasts of burden, providers of food, clothing, and companionship was a critical step in the evolution of civilization. As man learned to domesticate animals, he also learned to modify them—that is, to breed them to make them better suited for the purposes to which they were being put. The first known treatise on horse breeding and training is over 3000 years old. The horse of today bears little relationship to that early animal, and the same is true for chickens, cattle, sheep, and many of the other creatures that have had a close relationship with man.

There are now more computers than people in the U.S., and microprocessors are being produced at a rate that probably exceeds one million per day. Although the computer population is growing far more rapidly than the human one, computers are nevertheless far from pervasive in our society in the way that more mature technologies such as the telephone, television, and automobiles are. The mark of a pervasive technology is that its absence is more noticeable than its presence—it has evolved to become part of the very fabric of society. Such a technology is characterized by being generally accessible and by offering a service perceived indispensable by most people.

I hope to convince you today that while computers, and therefore the instruments based upon them, are not yet domesticated; there is every indication that they will become so in the next ten years. It is time for the selective breeding to begin in earnest, and I am going to suggest that many of the *ad hoc* engineering tradeoffs of the last three decades now inherent in personal computer design and application need to be re-examined in the light of new requirements.

Virtually all the pervasive technologies that I can think of seem to have passed through four evolutionary stages. They usually begin as an experimental rarity, a laboratory curiosity, sometimes as an accidental discovery in the pursuit of fundamental science; more often they result from an entrepreneurial attempt to exploit a perceived opportunity. The second stage of evolution usually produces an exotic tool or toy, often solving only a cross-section of a problem and used primarily by a narrow class of experts. In the third stage, the technology has evolved to the point where it is manufactured in quantity, has become well known and commonplace, but is used directly by only a rather small portion of the population. In the fourth, pervasive stage, the technology has become integral to daily life. For example, a home or a modern hotel room without a color television set is unusual since we are in the fourth stage of television technology. Contrast this with the third stage some thirty-five years ago when neighbors would congregate at a pioneer owner's home or in the streets outside store windows to watch the World Series or the Milton Berle Show on the black and white sets of that day. During the third stage of a new technology, there is usually a proliferation of companies seeking an early competitive edge. Product distinctions are clear although there are many look-alikes, and pricing is somewhat arbitrary. A characteristic of the fourth stage is that the businesses consolidate, major innovations are far more difficult because of large in-place investments, and prices stabilize. For example, the 1100 or so American car companies of the third stage have now become three and a half, and prices of comparable (and very similar) models are exceedingly competitive. Personal-computer technology today is in stage three, and we can confidently predict that the several hundred personal-computer companies vying for a share of this rapidly expanding market will shrink to some number about one hundred times smaller, and that the large dynamic range now seen in software pricing, for example, will soon reach a more rational equilibrium between cost and price.

Many studies have been made of these evolutionary processes of technological change; and while we tend to think of them as continuous, in fact they proceed in fits and starts. Social and economic events, such as wars and depressions, as well as major technical innovations, can produce an acceleration or deceleration in the rate of change of the substitution of a new technology for an old for a while, but then the original pattern is resumed. It doesn't seem to matter very much what these technologies are; once started the replacement process continues to completion in a strangely predictable way, whether we are talking about the replacement of sailing ships with steamships, the substitution of airline travel for rail travel or the introduction of plastic for glass in consumer bottles. The rate of dispersion of a particular technology is controlled by several factors; probably the strongest force is the development of a community of common interest. For example, the automobile as a technology for personal transportation could not have achieved stage four without the public's support of the development of a highway network. Similarly, the acceptance of the computer as a serious alternative to the mails or the telephone and telegraph systems will require the creation of extensive data networks, and that will require the public's conviction that this is worth doing.

While the intensity of this common interest is building, invention and entrepreneurship are extremely rapid and exhibit a form of positive feedback in which each advance feeds on those before, often without the underpinnings of fundamental science or understanding. Remember that most of the industrial revolution was based on early theories of electricity, magnetism, and Newtonian mechanics. Quantum mechanics and electrodynamics were not needed or used for this first tier of invention. This has also been the case in computing, where the hazy outlines of theoretical understanding are only now beginning to emerge. It is not an exaggeration to say that the personal computers of today, while useful, are primitive, unadapted versions of what their domesticated descendants will look like within a few years. The effects on the cost, appearance, and function of instruments can be expected to be profound.

THE DOMESTICATION OF COMPUTERS

The airline magazines and other learned journals would have us believe that the so-called "user-friendly" computers of today are already domesticated. I couldn't disagree more strongly; the average American today thinks about programming a personal computer in Basic or Pascal with about the same relish as he would an invitation to climb into a cage with a circus lion and for the same reason. Neither the computer nor the lion are really tame; they violate the rule of general access that is required for true domestication, for neither a lion tamer nor a professional programmer qualify as paradigms of general society.

The test for domestication that I like best is the one proposed by R.M. Pirzig in his book, *Zen and the Art of Motorcycle Maintenance*: "The material objects . . . can't be right or wrong. They don't have any ethical codes to follow except those people give them. The test of the machine is the satisfaction it gives you. There isn't any other test. If the machine produces tranquility, it's right. If it disturbs you, it's wrong until either the machine or your mind is changed." The media are full of examples which tell us unequivocally that we have gotten it all wrong. For example, Timex/Sinclair, in advertising their Model 1000 computer, ran a very successful campaign with the slogan "Why spend thousands to find out you hate computing when you can find out for $49.95?" The cover of a national training magazine featured a photograph of a blue-suited executive cowering before a personal computer, a braid of garlic around his neck, and a silver cross held fearfully before him.

I do not think this is all mindless paranoia. The average person begins with a distaste for computers which may be rooted in sentimentality for the technology being replaced (the same phenomenon that causes romantic attachments to steam trains and sailing vessels) or is the result of the uneasy perception that the computer may in some way become a substitute for the very faculties of mind that distinguish us from the beasts. Science fiction often portrays this in a threatening way, and the computer scientists have produced a guaranteed reinforcement of this malaise by making machines that are so complicated to use.

Advertising to the contrary, using most personal computers today requires arcane knowledge of specific and unforgiving sequences of commands. The computers' designs are optimized by engineers for cost and performance and not for natural usage by ordinary people. Truly friendly machines must be designed so that the particular idiosyncrasies of a user are accommodated in a way which supports his mental abstraction of the process taking place in the machine. Our software today is rude; it does not recognize individuals.

Ease of use, of course, ultimately depends on the programs that specify the machine's actions. A key factor is the degree to which these programs can be made to hide the details of operation from the users. It is one of our popular fantasies that some people can learn how to use machines rapidly no matter how complicated they are. Consider James Bond rushing into the atomic reactor or the space station that he has never seen before, quickly examining the bewildering maze of dials, buttons, flashing screens and then pressing precisely the right switches in the proper sequence to save the day just as time runs out. We all know that this is science fiction, but we like to think that somehow it should be possible to fathom the inner workings of complex machinery. The computer scientists have not really concentrated on this very much until recently; I believe that the industry is at an inflection point now, for we have finally realized that it is the human resource that must be optimized and not the cost/performance of the machines.

One of the great advances that I foresee is that computers will handle these complicated control sequences for us; only then will we be able to substitute simple abstractions for specialized knowledge, just as we now do in using many domesticated technologies. The final stage of technology evolution is often one of transparent complexity. Consider one button automatic tuning on a modern television set, for example, compared to the complex set of knobs to adjust the hue, intensity, fine tuning, brightness, contrast, vertical and horizontal hold of just a few years ago. It is possible to rent an American automobile of any model, any manufacturer, and drive away within a few minutes, having understood the location and function of the primary controls. Automatic transmission has made it unnecessary for people to understand the detailed workings of a gear box. The electric starter has produced a crankless, horseless carriage, and we no longer think at all about what happens when we turn the ignition key since automatic choking mechanisms or fuel injection have replaced some of the specific sequences that used to be required. Any user, even professional managers, can very rapidly grasp the concept behind a lever which is labeled "neutral," "drive," and "reverse," or a button labeled "start." The challenge is to produce this type of mental model to drive computer and instrument systems and then to represent it consistently to the user. Why shouldn't an office workstation have a button labeled "MAIL" or "Dow-Jones," and an instrument one called "determine resonant frequency?"

Hiding this complexity from the user while retaining a rich function set will require computer programs far more sophisticated than those we now write. The computer power required to execute them will be staggering. Simply increasing the speed of a computing cycle, and decreasing the cost and access time of memory by evolutionary advances in hardware technology that result from the application of scaling laws to switching devices, will fall far short of what is required for this task. Computer scientists must reorganize the internal structure of the computers, and modify the system architecture to increase greatly the number of operations that can be carried out in parallel. I think this will lead to computers that are organized as sets of special-purpose processors, each optimized for a particular class of tasks, such as extracting the patterns of speech or handwriting, or rapidly searching large databases, or managing the information on a display screen or doing a particular type of simulation or computation, such as a Fourier transform. These will be coordinated by very-high-performance, mass-produced, yet simple, general-purpose controllers. We will need systems thousands of times more powerful than the primitive general-purpose serial engines of today. Early examples of such machines are now beginning to appear in research laboratories, but there are many problems that relate to their interconnection and coordination. The programming style will be object-oriented--that is, control will be passed in the form of messages among self-descriptive entities. Although less efficient in terms of executed instruction path length, structural networks of objects help a lot with the two greatest limitations to the development of applications by professional programmers: reusability of programming solutions, and the difficulties of maintaining and modifying code.

Software must be designed to optimize not the way that the machines work, but the way that people think; and we can go a long way even with our present machines. Integrated programming environments are beginning to mask details from the user and to provide consistent access to services; natural language, touch screens, and speech are being combined to replace and augment the formalized keyboard interaction. In the new world of domesticated computers, slow, surly, uncooperative software will not survive. I believe that as we enter the fourth stage of computer technology, the successful companies will be the ones that recognize that in a rapidly expanding industry most of the user are beginners. The cumbersome instruction manuals and training classes of today must give way to an intuitive form of learning by doing the things that seem natural. The video-games people have stumbled on to the right experiential model--domesticated computers and instruments should be self-evident and self-documenting, and they should offer strong positive reinforcement almost immediately. Most users should never need to write a program in the way that we now think of that job. Professional computer scientists and programmers will survive, I hope, but increasingly their jobs will be to evolve the high-performance architectures and interfaces to shield users from the internal complexity and to develop tools so that users can build and maintain applications without becoming programmers.

Remember that while a congenial user interface may be a necessary condition for reaching the fourth stage, it is not sufficient. The degree of penetration will ultimately be determined by the value of the technology, as perceived by the people it serves; therefore, the critical question becomes what these pervasive computers, and the instrument systems based on them, will actually do.

We can think of technologies in a general way as amplifiers of human capabilities. So, for example, the telephone has widened the range of human speech and hearing, the television has expanded the range of our sight, and the automobile has served to increase our speed of foot and our ability to carry weight. The great promise of computing technology is that it will expand human memory, augment human reasoning, and facilitate human communication. Similarly, instruments can be thought of as enhancing the senses we already have or adding faculties to those not provided by nature. If we judge current computer and instrument systems by these lofty ideals, we must consider our progress at this time embryonic at best.

Computers in the workplace today are used largely to automate those tasks we have always done. Many are used to improve the performance of job-specific functions like designing circuits, or analyzing financial spreadsheets, or typing manuscripts. The augmented speed and memory capacity adds value, and the price of the machine is determined by the increase in productivity or improvement in quality of the product or service rendered. Professionals performing these functions will invest the requisite time and effort in learning to use the machines, however unpleasant that may be. Unquestionably, applying computer systems to these specific competence functions has made many previously intractable tasks possible; however, by and large, this is a difference in degree, and not in kind.

More interesting are the so-called computer-aided work functions which most people do, and which are largely independent of industry or profession. These can be thought of as the creation and distribution of "documents," electronic or otherwise; the storage, retrieval, and analysis of information, and user-to-user communication. Technologies usually evolve by replacement, and that we can think of this substitution of a new technology for an older one as an x-less-y phenomenon (for example, the automobile as a horseless carriage, the radio as a wireless telephone). We can recognize the computer-aided work functions as replacing a secretary and a typewriter, a filing clerk, support staff or librarian, and substituting for conventional telephones, telegraphs, or facsimile machines. In fact, these are the fundamental amplifications of memory, reasoning, and communication alluded to before, even if they are rather mundane.

Acceptance of these advanced systems has been much slower than predicted, due, I think, to how hard they have been to use and because the functions being replaced have not been significantly improved. This is now changing; the office-systems industry is early in stage three now, with inventions coming furiously and huge companies jockeying for position. Electronic voice mail, for example, with useful store-and-forward capabilities that eliminate the missed phone call and the barriers of international time zones will be widespread shortly. Soon even the simple home machines using TV sets for displays and telephone lines for communication will have easy access to large and diverse databases—the world's cookbook collection; or ten different articles from ten different encyclopedias on a topic of immediate importance; worldwide local weather or news; repair manuals for discontinued products—a seemingly endless set of attractive global services and information, hard to come by today, selected by simple menu-like commands and purchased on a pay-as-you-go basis. There are few, if any, technical impediments to this vision, and prototype systems have been working for several years now.

When these low-cost combinations of telephone network, television set, and personal computer are complete, I think that you will all want one in your home or office. You will be able to send and receive messages instantaneously to and from any connected television set. The world's libraries will be at your disposal, and intelligent, centralized reference services will aid in the collection and updating of information. The spread of cable TV and videodisk technology will add color video and sound to these capabilities—imagine a pronouncing dictionary, the Schwann record catalog on demand (we'll be able to compare, for example, different performances of the same piece of music), encyclopedias with film clips instead of still pictures, access to facsimiles of rare documents, interactive bedtime stories for children in which they'll be able to select their own endings. Driven by the economics of electronic communication, banking, shopping, and publishing, myriad new personal applications as yet unthought of will evolve. Once these services are easy to use from a consistent user interface, they will become the commonplace facilities of stage four. They will not need to be resident on the personal computer itself but will be features provided by the network of information servers.

The computer stores would have us believe that America's key priority is now checkbook balancing, with home recipe filing and word processing running close seconds. Don't believe it! We are on the eve of a great intellectual revolution, and it has nothing to do with bankbook-less banking or postmanless mail. The evidence is all around us, but because much of it is now directed at the largely mindless video-game market, it is perhaps hidden from the general adult view.

What is happening is that software is emerging as a unique medium, a very hot, interactive medium. A culture of media stars is developing—many of them teenagers and college students—who sign their programs, have a following, and hire software agents to represent them. The programs are beginning to be distributed by book publishers in bookstores, including some of our oldest and most respected firms. Genius programmers appear on TV talk shows. There is a software top-ten best seller list, and sub-industries like software distribution and software packaging have emerged. Similarities to the record and book industries are obvious, and problems of piracy, copyright, and compatibility are already troublesome. Software review magazines have appeared, and the *Whole Earth Software Catalog* has been published this year. Within this framework, a new type of literature is being written. Some have said that this is the literature of the future, but software and literature are alike only in that they are both media to present cool, static, one-way broadcasts. They evolved from the technology available at the time of their conception. Software is hot, interactive, dynamic. It is unlike anything we have seen before in its ability to draw people into the environment that it has created. It is expressive, but for now it can only be manipulated by the media stars. Computer science's principal task in this decade will be to provide the tools to eliminate this barrier, and once we have done this I believe a flood of creativity will occur and that it will come mostly from the large grass-roots population now excluded from such interaction. Interacting with the new software will be as different from watching TV or doing word processing or reserving hotel rooms electronically as can be imagined. We will need a new verb like "microwaving" to describe it. We have seen only the pop music side of the medium so far—the classical versions are coming now, and they are not going to be aimed at developing the hand-eye coordination necessary to repel an alien invasion.

THE DOMESTICATION OF INSTRUMENTS

For now, I'd like to assume that domesticated computers have come into existence and speculate about some of the effects that specific technology could produce on future instruments. I am not going to discuss today, for want of time, the tremendous progress in analog-to-digital and digital-to-analog converters, critical to success in so many of the applications. Suffice it to say that sampling rates in excess of one billion per second have already been achieved in laboratories, and that the resolution/speed/cost tradeoffs are increasingly attractive in commercial products. I'll also not discuss today the important area of sensors, although it is still the bottleneck in many situations. Innovation in this area is steady, but slow; cost, reliability, performance, and size of sensors are often limiters to the application of digital technology to measurement and test applications.

Let us first consider the impact of computers and digital signal processing on precision measurements. Since 1967, the world's time has been regulated by commercially available cesium-beam atomic clocks to an accuracy of 1 second in 100,000 years, or about 3-5 parts in 10^{12}. Incredible as this seems, it is not good enough for many modern navigation, communication, and astronomical applications. Len Cutler and his group at HP Laboratories, who developed the HP cesium standard in wide use throughout the world, have demonstrated an elegant new technique, which is about 10 times more accurate, based on the hyperfine transitions of trapped mercury ions. Ideally, an atomic frequency standard is based on the electric or magnetic dipole transition of a free atom observed for an infinite amount of time. In the cesium standard, the ideal of the free atom is realized to a high degree, but the observation time (that is, the time of flight through a microwave cavity) is fairly short.

Figure 1 shows the experimental apparatus designed to overcome this limitation. Mercury 199 ions are trapped for long periods in a quadrupole configuration driven by a precisely calculated radio-frequency field. The ions are pumped to an excited state using Mercury 202 light, and the ions are then irradiated with the microwave frequency tuned first to one side of the resonance at 40.5 GHz and then the other. The counter signals derived from the fluorescence which results are differenced and digitally integrated by the computer, using second differencing methods, and then converted to an analog signal. If the mean frequency of the microwave source is not at the line center, the error signal can be fed back, after a second integration, to a voltage-controlled crystal oscillator, which in turn regulates the frequency source. Complex control sequences minimize the noise due to secondary effects, and careful attention must be paid to other noise sources, such as second-order Doppler shift. Figure 2 indicates the results: the transition has been measured to an accuracy of 2 parts in 10^{11}, with a line width of 0.85 Hz at 40.5 GHz. If the full width at half-maximum of this transition is depicted as 1 inch wide, then the origin is three times the distance to the moon! The servo signal derived from this system yields an absolute accuracy conservatively estimated as 3 parts in 10^{13}, or about 1 second per million years. It is remarkable that today only a garden variety microprocessor and high-speed microelectronics are all that are needed to make such elegant instrumentation practical. While the precision here is an extreme case, many modern instruments are based on this kind of digital control and processing capabilities, unthinkable just a few decades ago.

We can find some far less exotic examples in the nearest computer store. A best-selling program, Typing Master, available now for home computers, analyzes the user's performance on some typing assignments and provides immediate feedback on improvements of a very specific sort. Even lifelong professional typists learn immediately and interactively how to improve their speed or accuracy or both. Extension to many other skills by replacing the keyboard with a sensor is possible. For example, one-chip signal processors could soon provide useful biofeedback to people learning to sing or to play musical instruments, helping with the difficult-to-teach relationships between anatomical control and auditory phenomena. We will be able to see on a screen in real time the spectrum by octaves of the sound which is produced. Visual feedback of auditory phenomena can show dynamically the relationship of harmonics to the fundamental or the transitions between head and

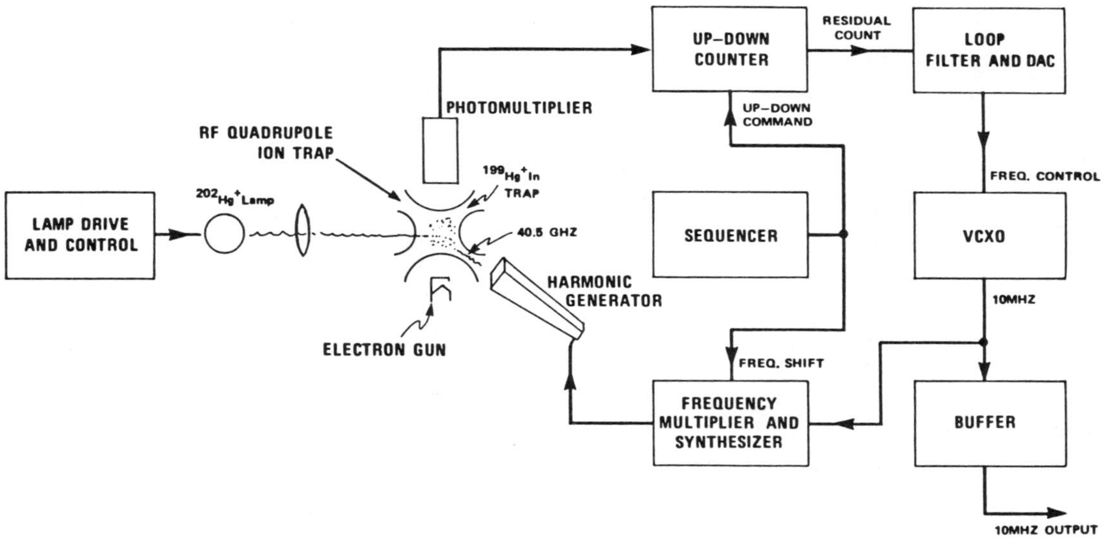

(©1985 ACM)

Figure 1. Trapped Mercury-Ion Frequency Standard

chest resonances, for example, in a way that our ears, which are not Fourier's children, cannot. Such equipment will enable singers and musicians to compare sound patterns and rhythms with stored templates produced by master musicians or with their own earlier efforts. The cost should be low enough to teach school children in the home. Programs to analyze speech accents and cadences are further in the future but certainly possible. Similar techniques with different sensors will permit the semi-automatic tuning of instruments like pianos and Porsches.

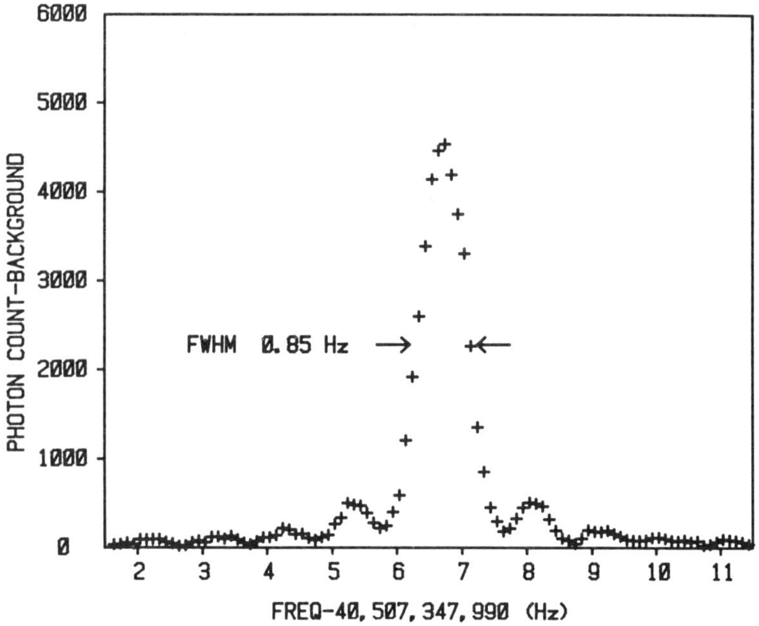

(©1985 ACM)

Figure 2. $199_{Hg}+$ Hyperfine Resonance

Another type of personal computer program now becoming popular is the kit, like the Pinball Construction Set. Using simple pictorial interfaces and pointers, objects (in this case flippers, bumpers, spinners, bonus lights, etc.) are selected by the user from a pictorial menu, moved about on a screen and assembled according to the rules which have been previously imbedded in the kit by its creator. Many parameters in the pinball program can be changed—the location and elasticity of the bumpers and flippers, bonus point combinations, etc. Items difficult to change on real pinball machines could also be modifiable: non-physical gravitational fields, balls of arbitrary cross section, etc. After the almost infinitely variable game has been designed *without writing a single line of code*, it can be played immediately and further modified by the user. The output of the program is the pinball game. A later version of this idea involves a music construction kit. The objects now are notes, score markings, tempi, etc. We can create melodies, snip them with iconic scissors, combine them with pre-programmed melodies, transpose to new keys, and so forth with no programming. The application is music composition, and the output of the program will be the score of the music itself. We will be able to play the results on conventional instruments or directly on electronic instruments if we desire.

This entire notion of kits seems to me a very powerful one. If the appropriate sets of rules of combination, the constraints, can be imbedded in the machine by experts, then design can become possible based only on function and aesthetics for a vastly greater proportion of our population than is now possible. Consider an amateur architect designing an addition to his house using such techniques. The cost, the structural factors for different materials, the connection rules for electrical and plumbing fixtures will be preprogrammed in the machine, and the computer will monitor these constraints of the real world, physical or economic, as the design is produced. Cost estimates, blueprints, and material lists will be the output. It is reasonable to consider an electronic test construction kit. Sensors, analytical or quantitative measurement tools, and statistical techniques now constitute the objects; and the constraints are determined by the application domain. A test engineer need only configure the control and data flow on a screen; the program would be developed automatically, with analytical and statistical functions called into play as appropriate. Professional mechanical and electrical engineers and architects are already using early versions of such tools for computer-aided design. In ten years, I think there will be hundreds of variations available for everything from landscape architecture to factory simulation to cash management—people will be able to concentrate on the problem and not on the routine drudgery, and "computer-aided design" and "computer-aided test" will be important new ways of solving problems in many fields.

It is ironic that computers themselves have been so poorly instrumented and, until the last decade, have not been effectively analyzed with respect to internal performance. There are at least three reasons for this. First, making such measurements has been expensive because the computers themselves must be dedicated to the task in much the way that the biologist sacrifices the frog, and most manufacturers and users were satisfied with gross characterizations, such as millions of instructions per second (Mips) or transactions per second. Second, the early attempts at computer performance analysis were so crude and imposed such overhead that they perturbed the measurement, in the Heisenberg uncertainty-principle sense, in precisely those cases of heavy load and congestion that are of the most interest. Third, our ability to make mathematically tractable, predictive models of computer behavior has been limited, and so the measurements have been little used during design. This is changing now with rapid progress in all three areas due to both decreasing costs and better understanding. One result has been the emergence of the so-called reduced instruction set computers (RISC), a poor generic name for computers that base engineering tradeoffs with respect to instruction sets, storage hierarchies, and input/output design on the results of frequency-of-use measurements. These deceptively "simple" machines are, in fact, far more elegant in matching real requirements than their ancestors. G.K. Chesterton expressed this nicely, I think, in the aphorism: "There is more simplicity in the man who eats caviar on impulse than in the man who eats grapenuts on principle."

Microelectronics has also now made it possible to build less invasive, more sophisticated instrumentation into the initial design of a computer system; and the benefits of these built-in hardware and software monitors in terms of reliability, availability, and configuration management for improved price/performance will be large.

An experimental voltmeter with a cathode-ray "faceplate," that we have built at HP Laboratories, illustrates this approach. The techniques of computer graphics and touch screens or other pointing devices combine well with instruments based on computers, since information can be presented numerically or graphically, and the results of theory, or simulation can often be compared in non-physical coordinate spaces during the acquisition of the data to insure validity or to direct experimental strategy. Instructions to the user can be placed on the screen, as can calibration and repair procedures. Data can be probed in real time, as for example, in a mass spectrograph display. Windowing techniques can be used to combine the results of different instruments, as might be required in the display of a potential speech analyzer.

The emergence of the videodisk technology will also have profound effects on the human interface to instruments. At HP Laboratories, a video screen is used to show the layout of the components on an experimental system that blends video information from the disk with conventionally generated computer graphics. In such a system, icons can now be photographs, useful perhaps for parts in a catalog or people in a phone book. Motion and sound can now be mixed freely and inexpensively with conventional text and data. The techniques developed by the TV and movie industries over the last quarter century for dealing with transformations of vast amounts of visual information are being adapted for use on personal workstations. Self-documenting instruments and integral training manuals are sure to result. Prototypes have been developed for such a multi-media document for IC process description which allow the insertion of moving pictures and sound in the computer-generated text and graphics. Soon, locally rewritable disks will be available, and then a new style of lab notebook using video cameras will be possible; and an interesting base for interactive teaching will be available.

A new breed of videogame is emerging now which, like the early sword and sorcerer games such as Adventure, contains a world model and basic reasoning capability. The computer, in a sense, acts as a type of expert combining algorithmic and heuristic information. This, of course, represents the great promise of artificial-intelligence technology. We can think of a calculator as an arithmetic apprentice (how many of you can still take a cube root by hand?) and be confident that soon portable personal algebra helpers will be commonplace. We can now purchase spelling assistants, and one day grammar tutors will appear to help us with the mechanical tasks of writing so that we can concentrate our energies on thinking and style. Many companies are now working on these knowledge-based expert systems; they will be used to fix machines, to help diagnose illness, to translate languages, to locate oil; and they will help to codify human wisdom so that knowledge becomes reusable across a wide range of only superficially different problems.

As always with emerging incomplete technologies, the claims are overblown, the time scales are too aggressive, and many of the applications proposed by the myriad artificial-intelligence (AI) start-up companies will probe non-viable. Nevertheless, this kind of technology is becoming practical on home computers and instrument controllers. It will, for example, enable personalized filters which can eliminate electronic junk mail and sort the remainder according to content and priority. Intelligent database agents will provide a greatly enhanced ability to search global databases to find desired information without providing such detailed parameters for the search.

At HP Laboratories we have developed a high-performance technology (based on the LISP language) to represent knowledge and have applied it to wide range of applications. We think that the appearance of this type of technology on moderate priced, high-performance symbolic processors with a stable, industrial-strength software system, and interfaces that make it accessible to the non-expert user will lead to a change in the way we think about instrument and computer systems. What are now tools will become advisors and apprentices. These intelligent agents will indeed add reasoning capability and expand our memories, and may help us communicate with each other. For example, new classes of smart, composite instruments could come into being. A "hyphenated" instrument, such as a gas-chromatograph mass spectrometer combines complementary information from previously distinct measurements. Expert-systems technology allows us to synthesize diffuse human wisdom, unlikely to be found in a single individual and to reuse it across different domains. Many new combinations of instruments, perhaps even triplets, will evolve. The knowledge base will often not reside locally but will be in a network server, offering economies of scale, security, easy updating, and local machine independence. There are myriad opportunities for such smart servers in instrument interpretation as well as trouble-shooting or diagnostic applications, for either sick people or sick machines. Such servers can maintain up-to-date symptom and repair histories, as well as the latest information from the R&D labs. They will be usable during product development as well as after installation. I can image technology advancing to permit quality verification analyzers to be integral to the products themselves, useful during manufacture, and key to preventive maintenance and repair after installation.

At HP Labs, we have prototyped a manufacturing server on one of our experimental IC lines. It analyzes problems in yield in one of our complex processes and makes the expertise of one or two specialists far more widely available. That it does this twenty-four hours a day and on weekends has made it popular, even with the experts who were previously on call. The inputs to such manufacturing servers could come from humans, design files, or sensors. Modern fiber-optics communication technology makes the notion of remote sensors feeding an assemblage of expert servers pragmatic. A great appeal of AI technology is that it should help us to reuse solutions, since it clearly separates the data, or knowledge, from the algorithms, or rules, that are used to process it.

One of the problems about projecting the future is that it is hard to be outrageous enough about the consequences of the mature versions of the technology. Could Gutenberg have been expected to foresee the greeting card industry? Did the Model T portend the lunar rover? We are in the infancy of computer-based instrument technology, and as we look ahead to stage four, it is hard not to be too timid in our predictions. It is a challenge I cannot resist, so here are two outrageous ideas that I hope will provoke some of you.

First, consider the notion of the shirt-pocket instrument, made feasible through on-going cost reduction and miniaturization of sensors and microelectronics. We usually think of smarter and more accurate measurements resulting from technology evolution, but there might be a vast market for low-cost, personal, portable, low-resolution instruments, just as there is a computation market for handheld calculators. The digital thermometer, the wall-stud locator are examples of such instruments today. Imagine a pocket mass spectrometer used by factory foremen to determine whether a wall is painted with lead or alkyd, or a personal instrument to determine the composition of effluents, a poison analyzer and antidote determiner for the home, a pipe finder for factories. Convenience, immediacy and price, not precision, will determine the viability of such devices.

The last speculation I'd like to subject you to today concerns what I think of as "virtual instruments"; simulations of instruments, themselves acting on simulations of physical systems. The hardware VLSI logic simulators reported on by several companies are early examples of this concept that demonstrate the enormous leverage attainable through specialized architectures and modest concurrency when a task has micro-parallelism inherent in it. Even these early machines can achieve three orders of magnitude speed-up compared to large conventional computers built in the same technology and performing the same computations in normal sequential software. Figure 3 is a simplified view of one such machine that we have built at HP Labs. A designer inputs a

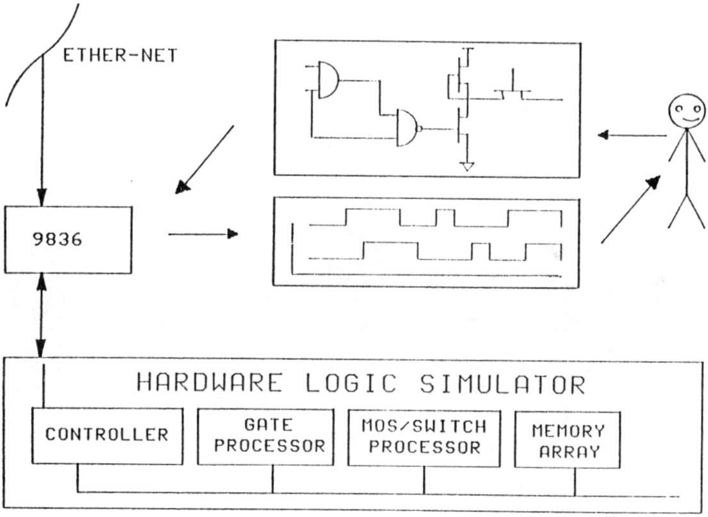

(© 1985 ACM)

Figure 3. An Example of a "Virtual Instrument" Simulation

logical representation of the machine being designed in a conventional way on a standard design workstation and receives output, in close to real time, that is equivalent to probing a real machine with a very powerful and flexible logic analyzer. Thus, we are using a virtual instrument to examine a not-yet-built virtual machine. Design become iterative and interactive. The machines' three specialized processors and the custom memory partitions they address are so organized that entire operating systems or applications can run and be debugged on the machine under development before any silicon fabrication is attempted.

Figure 4 shows how the machine works. Conceptually, think of a simulated timing wheel continuously rotating. Each tick of the virtual clock

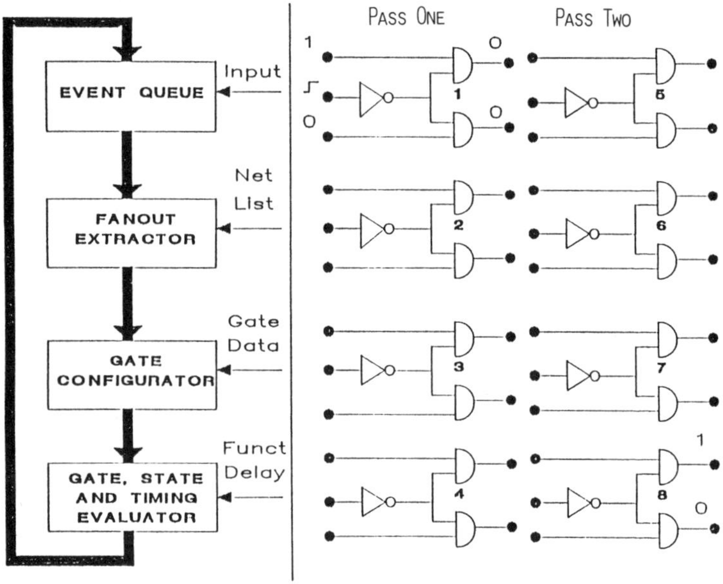

(©1985 ACM)

Figure 4. Details of HP Logic Simulator

represents a state of the design under investigation. Each of the four stages in the pipeline performs a dedicated function at each virtual cycle. The first stage, after initialization, represents those logic elements that have changed since the last tick and is one of the innovations of this particular design, since the machine concerns itself only with these elements. The second stage contains the topology of the machine or the net list, and, when a change occurs, points to those other components in the memory that are affected by the fanout. The third stage is the gate processor which references the gate library to determine the logical characteristics of a particular element. The last stage adds timing and other state information and updates the event queue (the timing wheel) for the next pass. The machine is designed very modularly. The gate capacity is extensible, parallelism can be improved by providing multiple processors (for example, multiple paths of the fanout can be examined simultaneously). Modest versions of the simulator can serve as co-processors on workstations. As indicated in Figure 5, larger versions are capable of running multiple designs simultaneously in much the way that multiple student projects are now fabricated on single wafers of silicon, and can be used to validate an entire operating system or compiler. These machines can be resident on a server – perhaps accessed via a network.

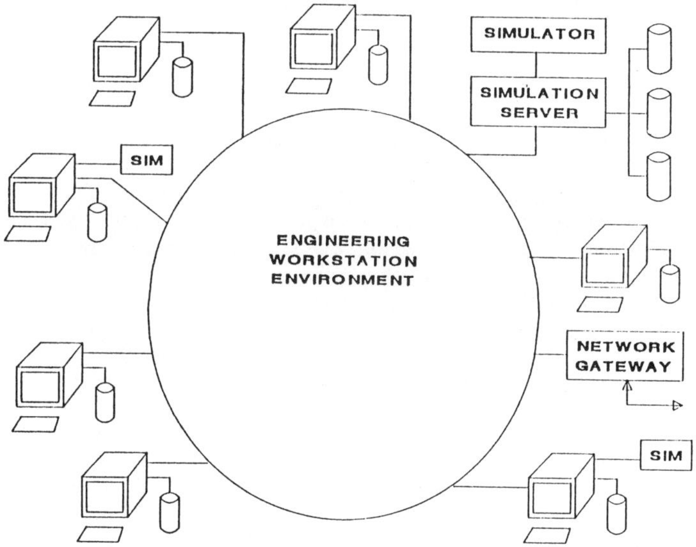

Figure 5. Example of Simultaneous Multiple-Design Simulation

Figure 6 shows the situation in micro-electronic system fabrication today.

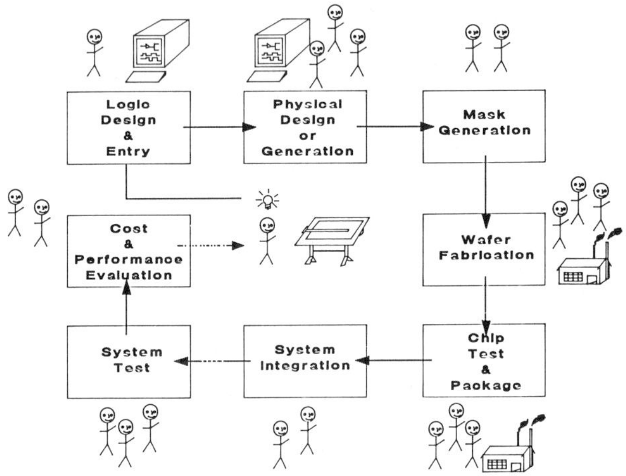

Figure 6. Steps in Micro-Electronic Fabrication

Each of the steps shown is necessary, and the process takes a great deal of time and money. Many types of expertise are involved and, typically, at each stage a new team of specialists is needed. The logic simulator represents just one early step in a prototyping system which could substitute mathematical modeling and testing of designs for much of the manufacturing work that we now do. Consider a collection of special-purpose systems, as displayed in Figure 7, each containing the expertise of one of the specific teams involved. Such an expert system would, as our knowledge of process models and device characteristics improved, begin to replace the large expenditures of time and dollars which are now necessary. It is, perhaps, outrageous today to imagine developing a process, producing the logical and physical designs, and then verifying a complete computer architecture using these virtual instruments before fabricating a single chip. Nevertheless, rudimentary versions of many of these subsystem components now exist in many laboratories, and the progress in mathematical modeling of devices, expert systems to capture process knowledge, and the massive application of specialized computers seem to me a combination that puts this in the realm of eventual possibility.

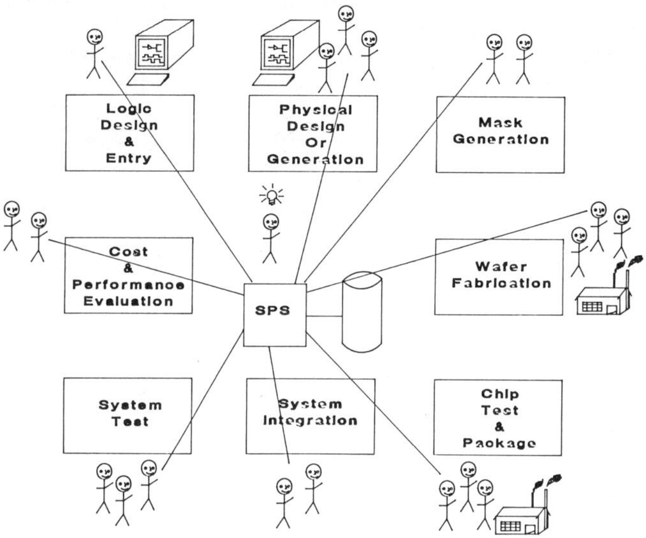

Figure 7. Group of Coordinated Special-Purpose Systems

THE NEXT CENTURY

When I was studying to be an engineer, I used to think how wonderful it would have been to have lived one hundred years ago, when the great mechanical and electrical inventions of the Industrial Revolution were being produced at a dizzying rate. Later, when I studied physics, I regretted not being alive in the period of the great discoveries of relativity and quantum mechanics. I have changed my mind. I am very glad that I am around now, because I think that the next ten years will be more exciting than anything we have ever seen before, because domesticated computers will allow so many more people to participate in the invention process. We have at last arrived at an understanding of the real potential of microelectronic technology; we are finally concentrating on the right problems–making machines usable by everyone and doing things with them that amplify our faculties in new ways. Computers, long pilloried as a dehumanizing force, will become the most powerful means of personal creative expression and communication ever known.

I have talked a lot about the domestication of technology and what I think that it could come to mean. I believe that our future success should be measured by how much we come to care about the results. I hope that you will indeed care about this technology so that our descendants at the celebration of the Bicentennial of the School of Electrical Engineering at Cornell will have many things to talk about that are very difficult for us to project today. I hope that they, in turn, will exercise the same degree of concern as well.

RISC-TYPE MACHINES

John Cocke

John Cocke obtained the B.S. degree in Mechanical Engineering and the PhD. degree in Mathematics at Duke University in 1946 and 1956, respectively. In 1956 he joined the Research Division of IBM as a Research Staff Member and in 1972 was appointed to his present position of IBM Fellow. He was a visiting professor at MIT in 1962 and a visiting professor at the Courant Institute of Mathematics at New York University in 1967 and 1969. He is a member of the IEEE Computer Society and is the author of numerous articles in the computer field. His main research interest has been in systems architecture, particularly hardware design and program organization, with major contributions to the STRETCH Project and the reduced-instruction-set computer (RISC). In 1985 he was elected to membership in the National Academy of Engineering.

Editor's Note: Due to unforeseen circumstances, a manuscript of Dr. Cocke's address to the Symposium is unavailable. In his presentation, Dr. Cocke reviewed the principles of the reduced-instruction-set computer (RISC) and emphasized that the RISC is based on standard old architecture that has been given additional thought. He discussed logic-circuit magnitudes, fan-in and fan-out limits, opcode execution, and cache concepts with special attention to the reasons why cache memories can perform as fast as the internal logic of a machine. He stated his belief that the RISC holds forth promise of better economic performance but suggested that many technical problems must still be solved before we will see a financial revolution in this aspect of the computer industry.

COMPUTING IN THE NEXT COMMUNUCATIONS ERA

Eric Nussbaum

Eric Nussbaum received the B.S. and the M.S. degrees in Electrical Engineering at Columbia University in 1955 and 1956, respectively. He held the Samuel Willard Bridgham Fellowship throughout his academic career. Following three years with General Precision Laboratory, Inc., where he was engaged in work on airborne navigation and air-traffic control equipment, he joined Bell Laboratories in 1959. Over the years he rose to the position of Director of the Advanced Switching Technology Laboratory. On January 1, 1984, after the Bell Company divestiture, he assumed his present position with Bell Communications Research as Assistant Vice President of the Network Systems and Service Research Laboratory. His responsibilities and publications have been concerned with switching systems including recent exploratory studies emphasizing speech processing technologies, voice and data-packet switching, and wideband (video) network architectures and services.

Editor's Note: In January, 1986, Mr Nussbaum was named a Fellow of the IEEE "For leadership in the development of modern digital switching architectures, systems, and services".

In this presentation I will survey some of the important trends in both the computing and communications industries with the hope of pointing out some related problems to the computer-minded people who are present. In many ways, computers have had a major impact on the communications business, but the reverse effect has been minimal. As described earlier by Gordon Bell, the communications industry has been seen largely as a bottleneck to computer progress rather than as an integral part of the development process. Following a review of some of the technological directions in computing configurations, I will comment on the trends in communications technology, emphasizing our expectations as we approach much higher bandwidths while making major qualitative changes. I will then attempt to relate these estimates to the possible impact on computer architectures of the future.

TECHNOLOGY TRENDS AND COMPUTING CONFIGURATIONS

Many of us talk about technology trend curves recognizing, however, that we may choose from a number of examples. The particular ones shown in Figure 1 are concerned with storage. They are typical S-shaped curves indicating that the capacity/price ratio of a new technology may improve rapidly following its introduction. The older technologies fight back valiantly but at some point they give up and die, and the new ones take over. It is necessary to be careful, however, since every new idea that appears does not necessarily survive and flourish—not unlike new computer companies. Magnetic-bubble memories, for example, held forth considerable promise but never really matched the performance of the silicon memories that came along at a later time. The key feature of Figure 1 is that at some point in the progress of a new technology, not only does it become possible to think of it as a one-for-one substitution for an older one, at a better price, but it is sufficiently better so that a qualitative change occurs in the way that the new technique can be used. Things can be done with it that really could not be done before.

Most of our speakers today have felt the need to give a history lesson so I will follow suit except that my version of the history of computing has a somewhat different twist with respect to where some of the elements of change have occurred. As indicated in Figure 2, in the early years, that is in the immediate post-war era of the 50's, computing was essentially confined to number crunching for both scientific and accounting applications. A computer was a precious resource that consisted of a central mainframe dedicated to batch operation. Access to a machine, information transfer between machines, and resulting output were performed by means of a rather non-electronic form of communication that consisted of punched cards and tape.

As we moved toward the 1960's, and as the technology progressed, the machines became faster and more reliable so that it became possible to do on-line and control applications. At that time, I was concerned with electronic switching where we had on-line logic control for voice communication networks using digital computers. Many other examples, such as the SABRE airline reservation system, existed during that period. The essential element of this activity was the development of some ability to time share a resource that was rapidly becoming indispensable. Relatively effective computer time sharing could be accomplished with local terminals that could be accessed at rates of less than a kilobit per second; often as low as 300 bits per second.

In the 1970's, memory and processing power improved sufficiently to allow actual text processing since it was then possible to deal equally well with both numerics and non-numerics. That ability, I believe, was really the foundation of the data-transmission revolution because transmittal of data was no longer confined to the transfer of large chunks of numbers, but now involved the far larger base of textual information. These improved capabilities were enhanced by the advent of the dedicated minicomputer and the introduction of a somewhat faster transmission pace with intermachine data links operating at 56 kilobits per second.

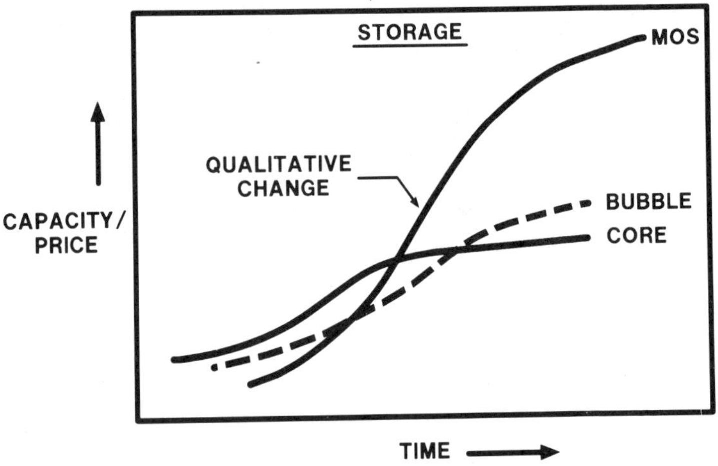

Source: Bellcore

Figure 1. Technology Trend Interpretations

PERIOD	APPLICATIONS INTRODUCED	CONFIGURATIONS INTRODUCED	ACCESS & INTERMACHINE COMMUNICATION
EARLY YEARS	• SCIENTIFIC • ACCOUNTING	CENTRALIZED MAIN FRAME BATCH	CARDS AND TAPE
1960s	• ON LINE • CONTROL	TIME SHARING	LOCAL TERMINAL ACESS AT <1 Kb/s
1970s	• TEXT PROCESSING	DEDICATED MINI	56 Kb/s INTERMACHINE DATA LINKS
1980s	• END USER COMPUTING • DATA BASE ACCESS • VOICE PROCESSING	• DEDICATED PC's • LOCAL NETWORKS	1 Mb/s LOCAL ACCESS >10 Mb/s INTERMACHINE COMMUNICATION 64 Kb/s REMOTE ACCESS < 10 Kb/s PUBLIC NET
1990s	• END USER COMPUTING WITH DISTRIBUTED RESOURCES • IMAGE PROCESSING	• INTERCONNECTED NETWORKS • DISTRIBUTED COMPUTER	100's OF Mb/s LOCAL & REMOTE 64 Kb/s --> >10 Mb/s IN PUBLIC NETWORK

Source: Bellcore

Figure 2. Evolution of Computing

Looking into the current era and examining the personal-computer (PC) world that has claimed so much of our attention, we find that we have obviously upgraded end-user computing, placed much greater emphasis on database access from remote locations, and increased the power of machines so that we are reaching the point where such things as real-time speech processing are a reality in this decade. The configurations, of course, are dedicated PC's, but we recognize that an isolated dedicated PC is of much less value than a PC that is interconnected via some form of local network. With a relatively small number of terminals in a local network it is possible to have a megabit per second of local access, with perhaps 10 megabits per second, or more, for machine communication. In contrast, for remote access on some limited basis, the network rate may be only 64 kilobits per second, with perhaps less than 10 kilobits per second if it is desired to switch in a public network configuration.

Moving ahead to the 1990's, the main theme that I would like to emphasize is that the next big qualitative change we can expect will be in the communications arena where the use of fiber optics will eliminate the distance and bandwidth limitations of today. End-user computing will have distributed resources without the limitation of today's data transmission rates. The faster computers, composed in many cases of special-purpose processors rather than just greater numbers of mixed components, will be processing and moving images as well as alphanumerics. With image processing, the need to interchange data increases enormously. Thus, the interconnected local-area networks will lead to far greater forms of distributed computers. Consequently, the remainder of my discussion today will concentrate on the bottom left-hand corner of Figure 2 in order to explain what is happening in the communications area that may cause some of these qualitative changes to occur.

An examination of the key technology triggers, as outlined in Figure 3, reveals that the transfer from vacuum tubes to transistors created the qualitative change that made widespread commercial computing possible. The advance from tapes to disks was the principal element that allowed us to go from batch to on-line performance. The introduction of high levels of integration stimulated development of the minicomputers, microprocessors, and smart consumer electronics that let us move away from an almost total dependence on shared mainframes. The change in memory components from magnetic to semiconductor materials (from iron to sand as suggested in one of Gordon Bell's earlier figures) has resulted in much larger memories at lower cost for both mainframes and minicomputers, while LSI, of course, has brought us the personal computer.

In the communications area, the list of technology triggers is much smaller, being almost non-existent for public networks. Still, we have gone from wire-pairs to coaxial cable which has given us reasonably high bit rate

OLD	NEW	QUALITATIVE CHANGE
• VACUUM TUBE	TRANSISTOR	WIDESPREAD COMMERCIAL AVAILABILITY OF COMPUTING
• TAPE	DISK	BATCH TO ON-LINE
• SINGLE TRANSISTOR	SSI, MSI	MINICOMPUTER
• DISCRETE LOGIC	μ-PROCESSOR	SMART CONSUMER ELECTRONICS
• CORE	MOS	MAINFRAMES: 1 Mb — 32 Mbyte MINIS: 64 Kb — 10 Mbyte TERMINALS: DUMB — 1 Mbyte, BIT-MAPPED
• MSI	LSI	PERSONAL COMPUTER
• WIRE PAIR	COAXIAL CABLE	HIGH BIT RATE LOCAL DATA COMMUNICATIONS
• COAXIAL CABLE	FIBER OPTICS	HIGH BIT RATE, LONG DISTANCES, MULTIPLE MEDIA

Source: Bellcore

Figure 3. Key Technology Triggers

local-data communications in the various forms. As we substitute fiber optics for coaxial cables, however, the potential for achieving high bit rates at long distances will give us the ability to merge multiple media into a digital format. The resulting impact on the industry will be substantial. I think that communications will then be seen as the wave of the future rather than as a bottleneck to progress.

COMMUNICATION TECHNOLOGY TRENDS

When we examine communication technology trends it is also necessary to consider the corresponding markets that will emerge, as displayed in Figure 4. In the beginning there was "POTS", which for the uninitiated, refers to "Plain Old Telephone Service". POTS could be designated as a "3-3-3 network", that is, it was designed around three kHz of bandwidth, three minutes of holding time, and about three CCS (100 call-seconds) of traffic. Attempts to move too far from these parameters for other types of traffic usually had limited viability. People in private networks who needed data did not wait, however, and came up with specialized systems to meet their requirements. In many ways, this approach has been the driving force behind the so-called public integrated-services digital networks (ISDN) that tend to digitize and integrate

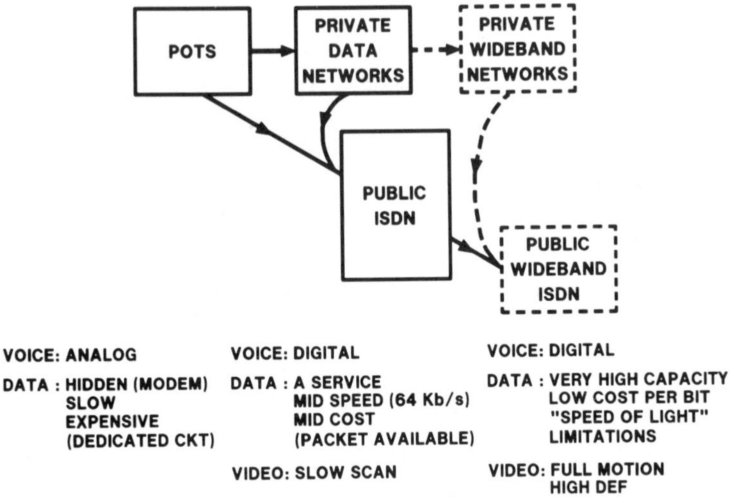

Source: Bellcore

Figure 4. Network Evolution

their facilities and make data transmission more than just a service. Furthermore, the current development of the leading-edge corporation/university networks will require much wider band-area configurations, thereby providing still another driying force for widespread public use of such systems.

In the initial POTS environment, voice was in analog form so that data transfer was essentially a hidden service. Tones were meant to simulate voice and were to be transmitted at whatever speed was possible. As the digital network came into more widespread use in the early 1980's, voices were digitized, and data transfer became a specific service of the new technique. We were able to packet-switch information and to transmit data at higher speeds without the limitations of voice encoding. In fact, even at the 64-kilobit network level, it is possible to begin information processing for slow-scan video, for possible application to video conferences. A projection of wideband network development into the next decade suggests that voice will remain digital and that data transfer will be possible at very high capacities and at very low cost per bit. Also, it is not unreasonable to predict that these future networks will ultimately have full-motion, high-definition video capabilities.

The historical display of communication trends contained in Figure 5 indicates that lead times are significant. In the late 1940's, the transistor was

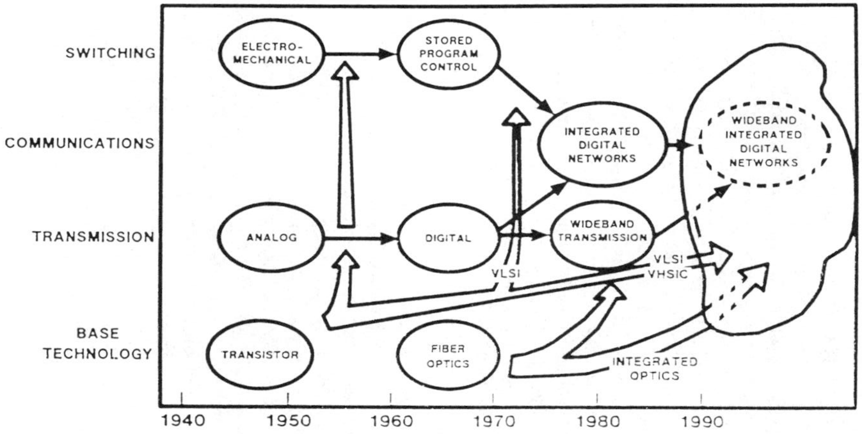

Source: Bellcore

Figure 5. Flow of Communications Technology

invented at the time when transmission was analog and switching was electromechanical. About 20 years passed before the communications industry began to deploy switching that was computer program controlled, and to move toward transmission that was digitized. As the silicon art improved and higher levels of integration were attained, further advances were made that allowed switching and transmission to merge into integrated digital networks. The first papers suggesting fiber-optics communication were also written in the mid 1960's. Even with concomitant breakthroughs in transmitter and receiver design, improvements in the purity of glass, and related technologies, it still took another fifteen to twenty years to achieve the first simple one-on-one replacements for copper as the medium for wideband transmission. The promise of the projected higher levels of integrated optics, combined with high-speed silicon, allows us to predict that really high-speed integrated digital networks will become available in the 1990's.

What do we mean by "high speed"? First of all, the fiber-optics technology is in its infancy, as may be inferred from the learning curve of Figure 6. One of the best measures of performance is by means of the upper band of curves containing the dots. The curves show progress in terms of the number of bits that can be transmitted per second over a distance expressed in kilometers, without an electronic repeater. It is evident that improvement is occurring at a rate of about a factor of two every year. A recent announcement of progress early this year showed a continuation of this trend with a bit rate of four gigabits per second at somewhat over 100 kilometers. The typical time frame from such a "laboratory experiment" to deployment is about five to six years. For an estimate of the straight bit rate, the lower band of curves (with the triangle markers) indicates progress at the rate of 50 percent per year.

Additional information about "high-speed" networks may be obtained from the integrated-electronics learning curve of Figure 7. The curve has hardly stopped but it has leveled off to some extent. It is clear that we are moving toward the megabit chip and beyond. The speed curve at the bottom of Figure 7 indicates another very important trend in high-speed communications technology. When we talk about gigabits in fiber-optics systems, we may very well ask what we will do with all of those photons that are pouring out of the end of the fiber. It is certainly necessary to have some electronics that can handle those high rates of data transmission. Fortunately, the recent developments in micron feature lengths in CMOS and in gallium arsenide show great promise and should provide the required circuits in the late 1980's and early 1990's.

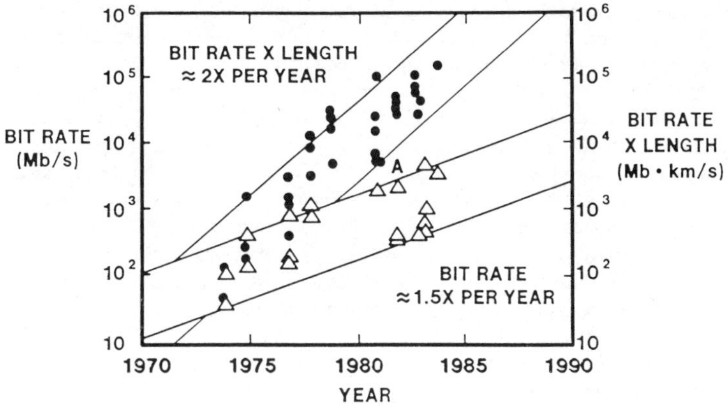

Source: Bellcore

Figure 6. Progress in Performance of Lightwave Technology

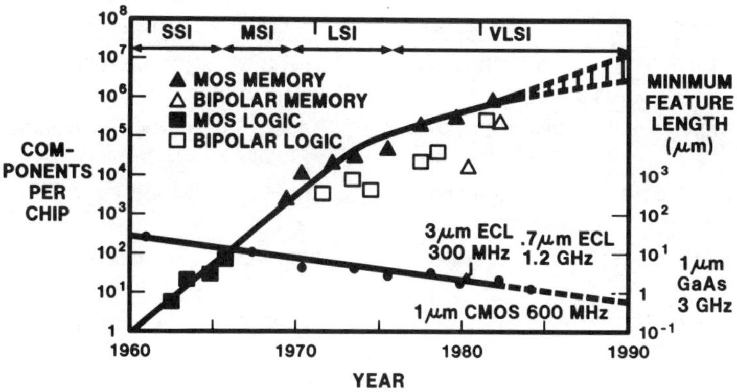

Source: Bellcore

Figure 7. The Integrated Electronics Learning Curve

Since advanced communication technology is making such good progress in the laboratory, both in photonics and in electronics, the natural question that arises is "how are we going to use these new developments to the best advantage?" The initial uses of things often represent one-for-one replacements of the old techniques. For example, in the first simple applications, the transistor was substituted for a mechanical relay or for a vacuum tube. As we become more knowledgeable, however, our applications become far more sophisticated, so we developed computer technology rather than blindly building transistorized cross-bar machines. Similarly, the first uses of fiber optics tend to be one-for-one replacements for narrow bandwidth copper media. But in the same sense as the transistor is superior to a relay or a vacuum tube, fiber optics can give us several orders of magnitude of improvement in bandwidth capability over copper, at comparable costs.

The table in Figure 8 outlines the various forms of information that appear frequently in today's world. We observe that the principal mode of voice transmission today is in 3-kHz analog form which we tend to store on tape. Text material is probably stored most frequently on disk and, depending on the modem, the typical speed may range from one to three kilobits per second. The most common form of transmission of high-fidelity audio these days is by FM stereo which uses a modulator operating at 38 kHz. Still images

FORMS	TRANSDUCER	STRONG	TRANSMISSION
VOICE	TELEPHONE SET	TAPE	3 kHz ANALOG
TEXT	COMPUTER TERMINAL	DISK	"1 Kb/s +" MODEM
HI FI AUDIO	FM RECEIVER	TAPE	38 kHz FM STEREO
STILL IMAGE	FACSIMILE	PAPER, FILM	MODEMS
VIDEO	TV SET	TAPE	4.2 MHz NTSC

DIVERSE ANALOG MEDIA

Source: Bellcore

Figure 8. Forms and Formats of Information (I)

use mostly facsimile techniques with modems, and the storage medium is either paper or film. Video, of course, uses its own peculiar form of modulation, as specified several decades ago by the U.S. National Television Standards Committee (NTSC), at the 4-MHz level.

As indicated in Figure 9, we can begin to see the possibilities of digitizing all of these diverse analog media as we move toward the world of high-speed transmission and high-speed electronics. Transfer of voice at 64 kilobits per second has already happened to a large degree. Text, in packetized form from one kilobit per second on up, is no longer in a hidden modem form. The introduction of the compact disk (CD) has created a completely separate music recording industry devoted to digitization of high-fidelity audio, at 1.2 megabits per second, to produce very high-quality stereo channels. New high-speed facsimile machines now run at about the megabit per second rate to deliver still images at tremendous resolution and very fast transmittal times. Depending upon the coding algorithms, and excluding very heavily processed work, the NTSC video that we have today can transmit high-quality, full-motion images at rates ranging from 45 to 140 megabits per second. Extrapolating this capability by a factor of two or three, even the proposed Japanese high-definition TV can be digitized to produce 35-millimeter-quality pictures. Clearly, the major message is that the communications world is moving to the task of processing all forms of information in a single digital bit stream.

A number of possible architectures exist to help us meet this challenge. Since the current world of high-bandwidth communications in the local areas depends principally on the use of coaxial cable, it is natural to inquire how the translation from coaxial to fiber transmission can be accomplished, with even higher bandwidths. As shown in Figure 10, even with a bandwidth-rich fiber at about one gigabit per second, limitations still exist on the coaxial-cable concept of a local network based on a shared-media bandwidth. Active repeaters and taps would be necessary as no really promising technology exists in the passive-tap area. High-speed and relatively high-cost termination electronics is generally required to trap those bits coming out of the channel. Furthermore, there is a certain amount of limited bandwidth growability because of the inherent media sharing. A gigabit per second is great for many terminals used for data transmission; it is still acceptable for voice transmission, but starts to become inadequate when the information medium is video. The obvious conclusion is that effective transfer of local-area networks to fiber technology is limited by specific communication needs.

At the other end of the spectrum, we can try to upgrade the classic star architecture of today's telephone networks with fiber. Figure 11 shows a typical double-star structure designed to reduce transmission requirements by having remote units that are connected separately to individual terminations. For this arrangement it is possible to upgrade the bandwidth as required, but

FORMS	TRANSDUCER	STORAGE	TRANSMISSION		
VOICE	TELEPHONE SET	TAPE	3 KHz ANALOG	→	64 Kb/s
TEXT	COMPUTER TERMINAL	DISK	"1 Kb/s +" MODEM	→	1 Kb/s + PACKET
HI FI AUDIO	FM RECEIVER	TAPE	38 KHz FM STEREO	→	1.2 Mb/s CD
STILL IMAGE	FACSIMILE	PAPER, FILM	MODEMS	→	64 Kb/s > 1 Mb/s
VIDEO	TV SET	TAPE	4.2 MHz NTSC	→	45 TO 140 Mb/s

DIVERSE ANALOG MEDIA → SINGLE DIGITAL BIT STREAM

Source: Bellcore

Figure 9. Forms and Formats of Information (II)

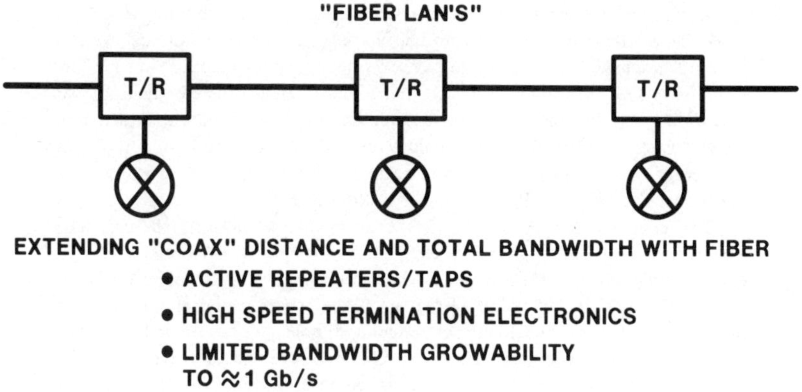

Source: Bellcore

Figure 10. High-Bandwidth Architectural Research Alternatives ("Fiber Large-Area Networks")

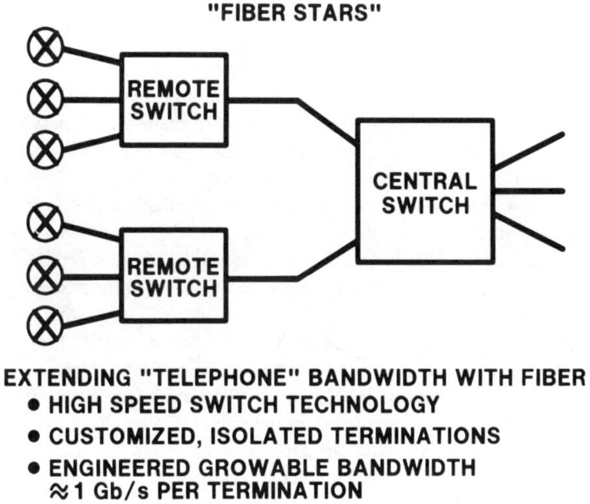

Source: Bellcore

Figure 11. High-Bandwidth Architectural Research Alternatives ("Fiber-Stars")

very-high-speed switch technology is needed since virtual switching would no longer be available in the shared media. For effective operation of this system, it would no longer be necessary to have identical gigabit capability in each termination. The terminals can be customized and isolated, and global bandwidth can be engineered as a function of total bandwidth demand. Today's voice-communication technology uses these techniques, and it is not completely obvious how the two architectures will ultimately merge, but it is likely that the future will have some attributes of both.

The chart in Figure 12 suggests that a channelization structure will present a basic problem in the new high-bandwidth world. We used to live in a very simple world. POTS operated at three kHz as a single channel that constituted the total network. Now as we move towards digital operation in ISDN, with perhaps 100 kilobits per second available, it is not obvious whether the customers will want two 64-kilobit channels, or possibly three 32-kilobit channels. We have gone through a slow, agonizing, give-and-take process of standardization. If we take the system up to the 1.5-megabit level, there are 89 possible arrangements of legitimate standards. If we extrapolate the concept of preassigned channel structures into an optical fiber that has a capability of, say, 0.5 gigabits per second, we could wind up with literally millions of possible combinations of fixed channels. Since it is not possible to preassign channels, there is a need for some form of packetization or dynamic bandwidth management, even in star structures, similar to the procedures that are used in shared-bandwidth media structures.

Those present who are familiar with computer networks in local areas may be interested in the problem of large metropolitan area networks with many interconnected points. Figure 13 is a small picture of the region where my laboratory is located in northern New Jersey. The area covers about 1000 square miles with about a third of a million terminals and subscribers that home-in on some 23 central offices. The "spaghetti" in the diagram represents 2000 separate 1.5-megabit per second transmission spans that attempt to interconnect the largely voice traffic in this arena. This system is an example of the kind of network that should not be replaced item for item with fiber optics. It is necessary, therefore, to envision area network topologies that would be substantially different when fiber transmission links are employed.

A number of alternative topologies may be considered, as described in Figure 14. The "fully interconnected network" of Figure 13 was really only about half interconnected, with an average of 11 spans per office to interconnect all of the other offices. With fiber optics we can go to virtual networks that spin them all back and, in fact, reswitch them through a central hub. Another possible approach is to try some kind of local metropolitan network. Some interesting numbers become evident very quickly from those 2000 separate 1.5-megabit per second lines. For example, all of that traffic could be

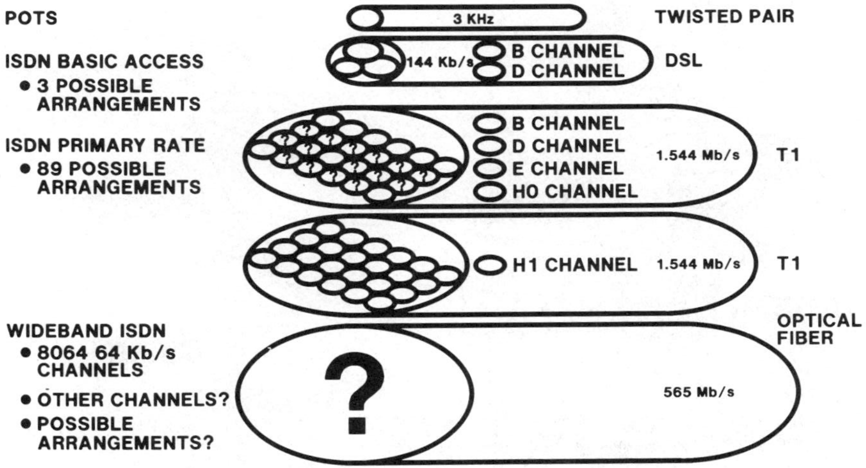

Source: Bellcore

Figure 12. Trends in Channelization of Subscriber Terminations

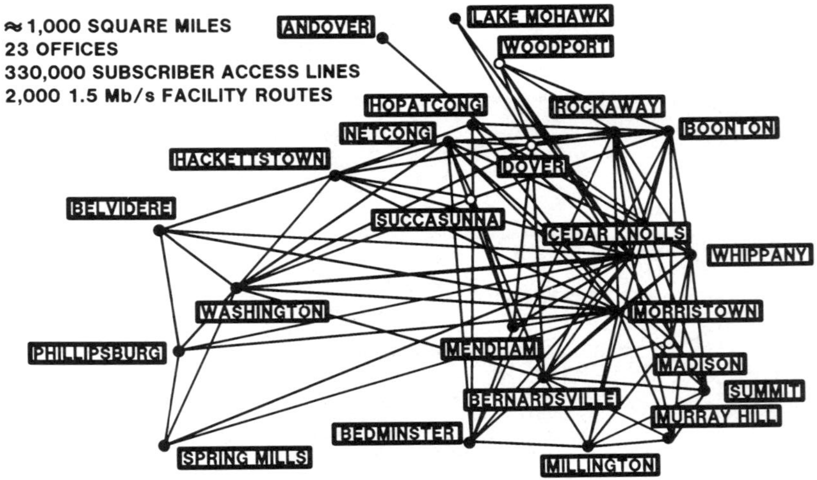

Source: Bellcore

Figure 13. Large-Area Network Topologies

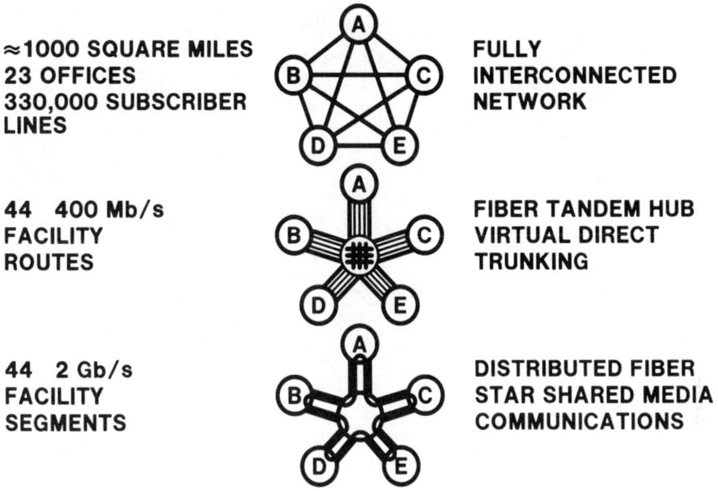

Source: Bellcore

Figure 14. Alternate Large-Area Topologies

handled on 2-gigabit per second ring running around all 23 offices. When all of the available bandwidth has been used, the next 2-gigabit per second ring is added. The trouble lies in the fact that we run out of bandwidth very rapidly. Alternatively, if specific segments are switched in the middle, some spare capacity becomes available when standard 400-megabit per second links are used. The message here is that with fiber optics, we will be re-examining the entire hierarchy and topology of communications networks, and, in the process we will create networks that will have capabilities ranging between 10 and 100 megabits per second, for costs less than those of today's much lower-bandwidth networks.

The potential of these advanced facilities is illustrated in the high-speed distributed multi-media communications architecture sketched in Figure 15. The suggested network would interconnect terminals with multiple distributed data bases containing text, image, and voice information. These data bases would be accessible from very high-power workstations fitted with bit-map graphics, substantial processing power, and perhaps 10 megabytes of memory. Such a communications system would have access and browsing capabilities that would allow the assembly, for example, of a bit-mapped color display that might require 20 megabits. It is necessary to have a high-speed network

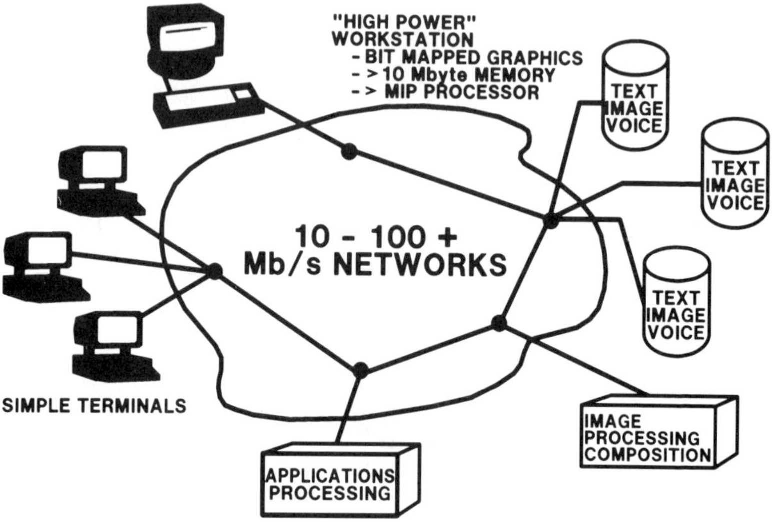

Source: Bellcore

Figure 15. High-Speed Distributed Multi-Media Communication

of this kind in order to be able to browse through pages on bit-mapped color displays by referring to remote data bases much as we now flip through the pages of journals. Examination of the details of a particular item would require occasional browsing stops. Similarly, it might be suitable to employ less powerful processor workstations with relatively simple terminals, a procedure that would require placing image processing and application processing functions as nodes in the network. Whether the work is done by the communications carrier or by some other vendor is not the issue. The key point is that we would have a very highly interconnected, very-high-bandwidth network that would allow us to store all of these aspects of information in a common digital form and to access them in a browsing manner. Perhaps these capabilities will, in time, improve the usability of these machines.

One of the examples that I like to give when I talk about this system is how I read a newspaper. I would not read my *New York Times* in the morning by going through perhaps a 10-kilobit section with a tree structure trying to find what I wanted, as I would today using VIDETEXT. When I read the *Times* in the morning, I am really scanning. I am looking at a page that contains a couple of megabits. Perhaps my eye will catch an article, a headline, or a picture, and I might stop on that item. If my eye does not catch anything,

I'll go on to the next page. That is the point that I want to emphasize with respect to a communications network with browsing capability.

IMPACT ON FUTURE COMPUTING CONFIGURATIONS

How do my previous comments relate to the present and future status of computers in communications? Some insight may be gained from the comparisons made in Figure 16. It appears that today's computer architectures are based either on short distances with very high-speed communications, or on long distances with very low-speed communications. We are dealing basically with alphanumerics and the manipulation of simple textual and graphic data. The future will be based on the applications of fiber-optics technology that, to a large extent, will provide distance-independent, high-speed communication that will allow flexible placement of data-storage resources. One immediate benefit would be the removal of the current necessity to have a disk physically associated with a machine. The stored information could be at a remote location if that is a better way of administering the disk. Furthermore, the digital treatment of all forms of multi-media information would allow the network to become an integral part of the communications system, as opposed to the present bottleneck that computer engineers must attempt to overcome in their architectural designs.

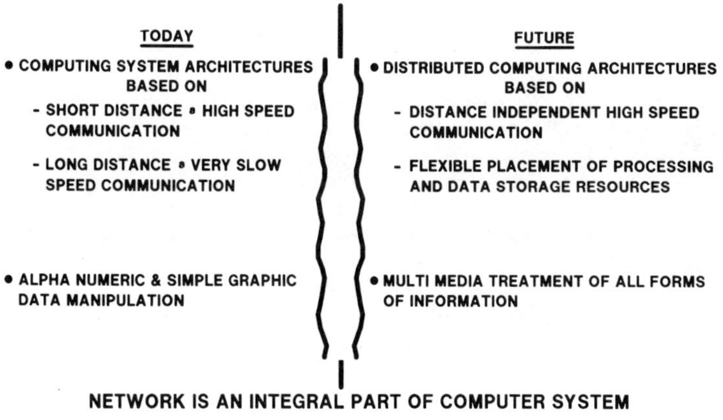

Source: Bellcore

Figure 16. High-Bandwidth Communication

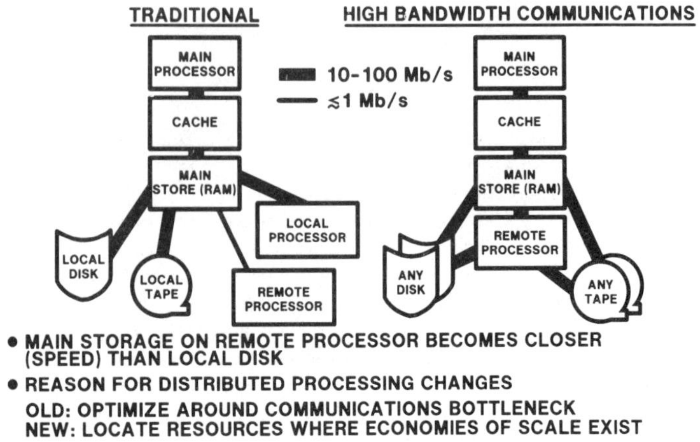

Source: Bellcore

Figure 17. Machine-to-Machine Communication (Memory Hierarchies)

The comparison between memory hierarchies in Figure 17 gives us an example of remote-access capability. The traditional architectures of today offer high-bandwidth paths between disks, tapes, local processors, and caches, while low-bandwidth paths are available to any remote path, process, or other element with which we may wish to interact. If we now move toward higher-bandwidth communications, the remote processor will be able to respond more rapidly than the local disk because the latter must contend with the "latency" of access, that is, the inherent total turn-around time. Consequently, the overall structure becomes far more flexible since the operating procedure changes from attempting to optimize distributed processing around a communications bottleneck, to locating the resources where economics of scale exist.

A similar comparison between traditional and high-bandwidth communications may be made at the user level, as illustrated in Figure 18. If we are concerned with a local-area network, we can get perhaps a megabit per second into our terminal. If we are working into a public network, we are usually limited to 10 kilobits per second at most. Today's terminals tend to be relatively simple with alphanumeric screens and low processing power. As we move toward future workstations with access to peak rates between 10 and 100 megabits per second, via the network, I believe that the previously mentioned downloaded-software and high-speed data-browsing examples can

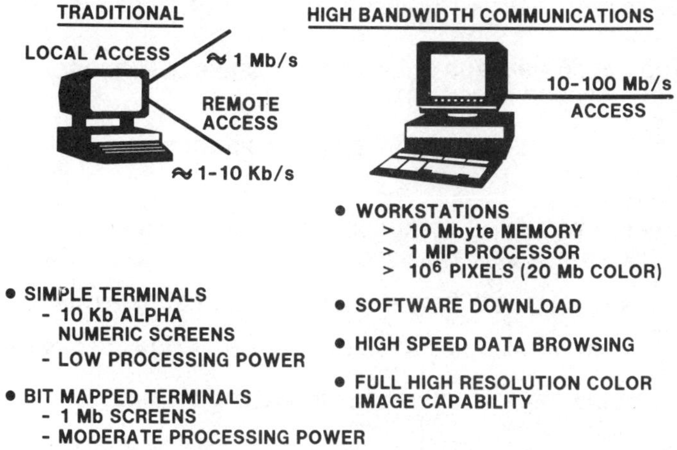

Source: Bellcore

Figure 18. User-Access Communication

become a reality, even to the extent of providing total high-resolution color-image capabilities.

Since I have given considerable attention to multi-media functions in this discussion, a brief examination of the computer power necessary for their proper behavior is in order. Reference to Figure 19 indicates that the transport data rate for voice ranges from 10 to 100 kilobits per second. Low-data rates would require specific encoding at the 10-kilobit per second rate, and the 1-megabit per second rate would suffice for high-fidelity stereo. It is also obvious that we are beginning to do speech processing but the techniques involved suggest reiteration of an important point that was made earlier today: it is unwise to simply count the number MIPS (million instructions per second) required for a process. MIPS can also be very special-purpose functions. As a rule of thumb, perhaps hundreds of MIPS (and a super computer) would be required to achieve some rudimentary form of general-purpose speech recognition, a procedure that comes in many dimensions. Instead of waiting for the super computer, however, these attempts at special-purpose signal processing are being performed on special-purpose machines which may also be adapted for use in real time with the digital bit streams that will be arriving in the near future.

VOICE

10 Kb/s TO 1 Mb/s TRANSPORT (LOW BIT RATE TO STEREO HI FI)

100 - 1000 MIPS FOR RUDIMENTARY GENERAL PURPOSE SPEECH RECOGNITION

⟹ SPECIAL PURPOSE SIGNAL PROCESSING

VIDEO

50 Kb/s TO 250 Mb/s TRANSPORT ("SLOW SCAN TO HI DEF. TV")

- **SIMPLE AXIS TRANSFORMATIONS**
- **SINGLE ITEM STATIONARY PATTERN RECOGNITION**
- **MOVING SCENE PATTERN INTERPRETATION (10,000 MIPS ?)**

⟹ SPECIAL PURPOSE SIGNAL PROCESSING

Source: Bellcore

Figure 19. Multi-Media Digital Information (Real Time Processing of Bit Streams)

Data-transport requirements for video processing may vary from 50 kilobits per second for the highly processed images that are used today in specialized network services such as teleconferencing terminals, to 250 megabits per second for high-definition video. Depending on what we are trying to do with processing, there is an enormous range of processing power necessary, ranging from a still picture with very simple axis transformation in real time, to attempting to obtain some kind of stationary-pattern recognition, to achieving near-perfect interpretation of a moving scene. Again it is not necessary to wait for the supercomputer. Specialized machines will be developed to handle the ubiquitous forms of digital bit streams coming into the terminals. Those workstations will be able to interpret not just text, data, and computation, but also voice and image information.

Finally, in assessing the future of computing in the next communications era, we may examine the set of "new frontiers" listed in Figure 20. I am really passing the baton back to the computer people at this point. Data-base technology will become critical, with storage, searching, and the caching of transfer rates being the dominant features. Distributed software architecture will become a serious factor as one set of problems is traded for another. Design criteria involving fiber optics will be governed by the previously mentioned system-latency constraints. These limitations occur, not only in the

- **DATA BASE TECHNOLOGY**
  - **STORAGE**
  - **SEARCHING**
  - **CACHING**
  - **TRANSFER RATES**
- **DISTRIBUTED SOFTWARE STRUCTURES**
  - **DEVELOPMENT**
  - **ADMINISTRATION**
  - **PERFORMANCE**
- **SYSTEM LATENCY CONSTRAINTS**
  - **HARDWARE AND SOFTWARE INTERFACES**
  - **CHANNEL SETUP**
  - **SECURITY AND PRIVACY TECHNIQUES**
  - **LAWS OF PHYSICS**

Source: Bellcore

Figure 20. The New Frontiers

communication links, but also in the hardware and software interfaces. Channel set-up times and the inevitable restrictions of the laws of physics involving the speed of light, help to put latency into perspective. From a back of the envelope calculation, we find that a gigabit-per-second fiber, running coast-to-coast, has about the same bandwidth as a 747 airliner filled with tapes. That is, a five-hour flight on a 747 full of tapes is equivalent to pumping those tapes over a gigabit-per-second fiber. The difference between the two is about five or six orders of magnitude in latency: it would require about five hours to transport the first bit across the country in one sense, and tens of milliseconds in the other. Clearly, latency becomes an important design criterion, particularly when we are now dealing with latencies in communication in cross-country information transfer that are comparable to those found in disks and access times. In effect, as we go to very-high-speed computers, we are dealing with the latencies of computer-hardware backplanes.

Overall, the way I would like to think of the future is that if today's computer backplanes can become tomorrow's communication network, we will achieve a new degree of freedom in computer structures, so that when we talk about computers and communications they really will be mentioned in the same breath.

COMPUTERS: HARDWARE, SOFTWARE, AND COMMUNICATIONS

Andrew T. Hospodor

Andrew T. Hospodor received the Bachelor of Mechanical Engineering from Cornell in 1960 followed by Masters degrees in Mechanical Engineering and in Business Administration from Lehigh University. After several years with Air Products and Chemicals, Inc. as an engineer and as a product marketing specialist, he joined RCA in 1966 and held a series of marketing posts until 1973 when he became Director of Marketing for RCA's Advanced Technology Laboratories in Camden, New Jersey. In 1976 he completed the Harvard University Program for Management Development. From 1977 to 1979 Mr. Hospodor was Director of Marketing and Advanced Planning for Automated Systems, then served as Manager of Command and Control of Underseas Warfare Programs until 1981 when he was appointed Division Vice President and General Manager of RCA Automated Systems in Burlington, Massachusetts. In 1985 he assumed his present position as President and CEO of RCA American Communications, Inc.

INTRODUCTION

At a Cornell Alumni Engineering Colloquium at Ithaca in May 1984, I raised the question of what Cornell is doing to train engineers to work in the modern world of integrated computers, communications and software. Fortunately, Dean Everhart assigned Professor H. C. Torng to answer my question and we have been communicating over the past year on the University's role in training engineers to work in the information industry. My presentation today, the result of an effort that began with that question, is outlined in Figure 1. I will give a general overview of the industry including its economics, the needs of the marketplace, and the role of systems integrators in this marketplace. My discussion will then focus on examples of recent systems-integration projects that provide a set of typical problems to be faced by electrical engineers in the future of the information industry. Finally, I will recommend a curriculum that will train electrical engineers to solve these problems.

- **INTRODUCTION/PURPOSE**
- **THE ECONOMICS OF COMPUTERS: HARDWARE, SOFTWARE AND COMMUNICATIONS**
- **NEEDS OF THE MARKETPLACE**
- **SYSTEM INTEGRATORS ROLE**
- **EXHIBITS**
- **FUTURE: MARKETPLACE DIRECTIONS AND STRATEGIES**
- **CONCLUSION**

Figure 1. Outline of the Discussion

THE ECONOMICS OF COMPUTERS

On February 18, 1985, *Business Week*,[1] Figure 2, conveniently ran a story on IBM, obviously one of the dominant firms in this business. The cover headline states that IBM's planned future growth . . ." requires new moves in communications and software. It won't be easy." That message in *Business Week* is the same as some of the messages we are seeing in the current industry marketplace. The information industry, as shown in Figure 3, consists of a wide variety of computer hardware manufacturers, communications hardware manufacturers and software companies; a total in excess of 6000 firms. In

---

[1]Harris, Marilyn A., "IBM: More Worlds to Conquer", *Business Week*, No. 2881, February 18, 1985, pp. 84-98.

(© 1985 McGraw Hill Inc.)

Figure 2. Status of IBM in the Information Industry

	# OF FIRMS
COMPUTER HARDWARE MANUFACTURERS	500
COMMUNICATIONS HARDWARE MANUFACTURERS	430
SOFTWARE COMPANIES	5000

Figure 3. Economics of Computers: Number of Firms

1985, the computer and communications industry will ship $141B in equipment - the statistic is informative but clearly understates the real impact.

One of the dominant pervasive trends of 1985 is the consolidation and conglomeration of these companies. Merger, acquisition, and consolidation for growth have been a basic business strategy in the computer service and software industry in 1984. Figure 4 shows that between 1983 and 1984

MERGER, ACQUISITION, CONSOLIDATION, GROWTH

IBM ACQUISITION OF ROLM — A COMMUNICATIONS FOOTHOLD?
IT&T LOOKS HARD AT ACQUIRING SPERRY
AT&T POSITION IN INTERMETRICS
GM ACQUISITION OF EDS

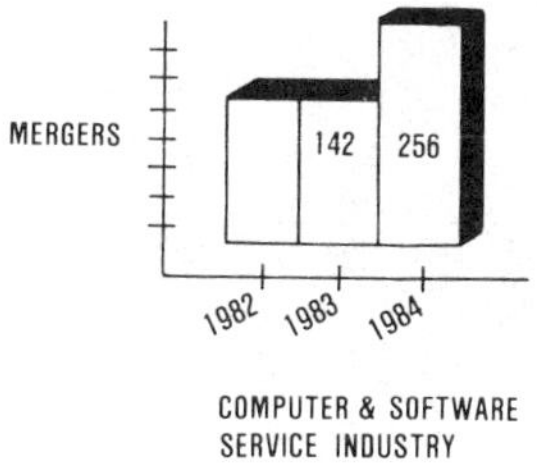

Figure 4. Economics of Computers: Consolidation for Growth

mergers increased by 50 percent, whereas there was no increase between 1982 and 1983. In addition to these macro data, there are some readily recognizable micro phenomena. IBM's acquisition of Rolm Corporation is clearly an attempt to enhance its communications position. The acquisition of EDS by General Motors probably fits somewhere in their strategy of factory automation. AT&T has taken a position in Intermetrics, a dominant firm in the development of the Ada software language system for the US Air Force, and also the developer of significant Space Shuttle software. Finally, IT&T is reported to be looking seriously at the acquisition of Sperry to round out its capability as an integrated supplier of information systems and services.

The statistics of this business are important. Figure 5 presents industry 1983 sales and employment data as published by *Fortune* magazine in April[2] and June,[3] 1984. The total sales in communications, hardware and software are over $230B, and the industry provided employment for over 3.1 million people. These figures include final sales for all domestic computer, communications, semiconductor and software firms having 1983 sales greater than $500M, the industry group usually referred to as "high tech". Not included are

---

[2]Directory of the Largest U.S. Industrial Corporations, "The 500", *Fortune*, Vol. 109, No. 9, April 30, 1984, pp. 224-295.

[3]O'Reilly, Brian, "AT&T: What Was it We Were Trying to Fix?" *Fortune*, Vol. 109, No. 12, June 11, 1984, pp. 30-36.

1983 FISCAL YEAR DATA*

COMPANY		SALES (B)	EMPLOYEES (K)	SALES (K)/EMPLOYEE
• AT&T (BEFORE)		$ 69.4	965	$ 71.9
• IBM		40.2	381	105.5
• IT&T, GTE & AT&T TECH.		38.3	605	63.3
• RCA, XEROX, HONEYWELL, SPERRY, HP, CDC, TI		42.3	593	71.3
• OTHERS (25)		44.0	613	71.8
TOTALS	W/IBM	$234.1	3,167	$ 73.9
	W/O IBM	$193.4	2,786	$ 69.6

*FORTUNE, 30 APRIL AND 11 JUNE 1984

Figure 5. Economics of Computers: Industry Statistics

aerospace firms like General Dynamics, Boeing, Rockwell, TRW, and Lockheed. Estimating that 60 percent of this industry's sales are value added, the GNP contribution is about $180B, or 4 to 5 percent of the national GNP. The table groups companies, with each group being about the same size as IBM. Group 1, which is AT&T before the breakup, includes the Bell Operating Companies as we currently know them, plus AT&T long lines - about $70B of sales. The third group, IT&T, GTE, and AT&T Technologies is about $40B and the subsequent two groups of seven and twenty-five companies, respectively, are also about $40B. The dominant dollars ($147.9B or 63.2 percent) in the business are concentrated in the hands of the top three groups (five companies).

A significant message in Figure 5 is the productivity estimate in the right-hand column. Each of the groups that are about IBM's size, have about an equal level of productivity - this is an old-fashioned way of measuring productivity - but about $70K of sales per employee appears to be an industry norm from top to bottom, from the largest to the smallest firm. IBM is the exception with sales per employee of $105K, a full 50 percent above the industry norm. IBM currently has tremendous productivity advantage in this marketplace, which makes competing with them, head on, a very tough business.

NEEDS OF THE MARKETPLACE

The marketplace will continue to grow and its needs are changing from communications and computers being separate things, to computers, communications and software being integrated entities. Figure 6 contains several statements about the needs of the emerging marketplace. These quotes have

- "PROVIDE CUSTOMERS WITH OFFICE OR FACTORY SYSTEMS THAT CAN 'SEAMLESSLY' LINK A LOADING DOCK IN CALIFORNIA WITH AN ACCOUNTING CLERK IN NEW YORK"
- "A DOCUMENT CREATED ON ONE TYPE OF COMPUTER CAN BE TRANSMITTED TO AND EDITED BY ANY OTHER TYPE . . ."
- . . . WHILE UPDATING DATABASES ON BOTH ENDS . . . WITH RETRIEVALS BY HUMAN OPERATOR QUERY OF SEARCHABLE FIELDS . . .
- ACROSS OFFICE, BUILDING, CITIES, COMPANIES AND NATIONS

Figure 6. Needs of the Marketplace

appeared in the business press over the last several months and state the need eloquently. The computer industry must provide its customers with office or factory systems that can "seamlessly link" its customers to their customers and suppliers . . . . while providing high-quality intra-firm service.

Customers must be provided with the capability to create, on one type of computer, computer terminal, word processor or mainframe computer, a document that can be transmitted and edited by any other type of computer. Further, this process must transfer documents while at the same time updating data bases on all appropriate nodes. The architecture must also support retrievals, by human operators, through query of searchable fields. In other words, the need is for a system in which user-transparent computer capability is provided across office, building, city, company, and national boundaries . . . . "seamlessly".

SYSTEM INTEGRATOR'S ROLE

A critical function in achieving this desired operation is the role of the systems integrator, with some of its attributes shown in Figure 7. A systems integrator is a value-added supplier; a firm that provides turnkey systems capability. Successful systems integrators have a classic, objective view of the heterogeneous world; they take end-to-end system responsibility for guaranteed system performance. The minimum requirement is the capability to provide cradle-to-grave logistics support and an in-place multi-disciplinary team.

EXHIBIT NUMBER 1

With the information industry set in its proper perspective, we may now consider three examples of end-to-end systems (designated as Exhibits 1, 2, and 3) that demonstrate the application of standards, the use of computers,

THE SYSTEMS INTEGRATOR IS —

- VALUE-ADDED SUPPLIER
- TURNKEY SYSTEM PROVIDER

POSSESSING AN —

- OBJECTIVE VIEW OF HETEROGENEOUS WORLD
- END-TO-END RESPONSIBILITY FOR PERFORMANCE
- CRADLE-TO-GRAVE SUPPORT TEAM
- THRU A MULTI-DISCIPLINED ENGINEERING TEAM

Figure 7. Role of the Systems Integrator

communications and software in a single system, and the use of specialized engineering talents to achieve customer satisfaction.

Figure 8 lists the attributes of Exhibit 1, a system with semi-smart workstation, off-the-shelf software and a standardized message format document interchange that provides end-to-end system performance for the customer. Figure 9 is a simplified block diagram of the Exhibit 1 architecture. First of all, at the top right of the figure is a box called "radio protocol". Through this protocol, external users are provided a virtual circuit with end-to-end connectivity, error-correction code, and all the bookkeeping necesssary to provide a 16-kilobit digital service. The radio net has two different 16-kilobit links, one at UHF and the other at VHF. On the other side of the system, internally, the system communicates over an X.25 type protocol at the network layer. Figure 9 shows, in a stylized way, that all seven layers of an International Standards Organization (ISO)-like layered architecture are used in this system. The system did not start out with seven explicit layers, as shown, but the separate functions do exist. Thinking in terms of the physical, datalink, network, transport, session, presentation, and applications layers provides a coherent framework for discussing this system. The system operates by receiving radio messages, entering them into the data base and cueing the operator that messages to his address have arrived. The operator, using his custom workstation, is able to retrieve individual messages, edit them, prepare reports and send messages out of the system - back through either the RF links or through land lines networked to the workstation processor.

Taken alone, this configuration doesn't look like a very complicated system. In order to meet the message and man/machine load of the system, however, it was necessary to build a totally integrated network that connected together three minicomputer hosts, and up to fifteen workstations. In each minicomputer the system integrates about 200,000 lines of structured Fortran.

AN END-TO-END SYSTEM DEMONSTRATING:

- STANDARDIZED COMMUNICATION & INTEROPERABILITY
- SEMI-SMART WORKSTATION
- OFF THE SHELF SOFTWARE
- MESSAGE (DOCUMENT) INTERCHANGE

Figure 8. Exhibit 1 Overview

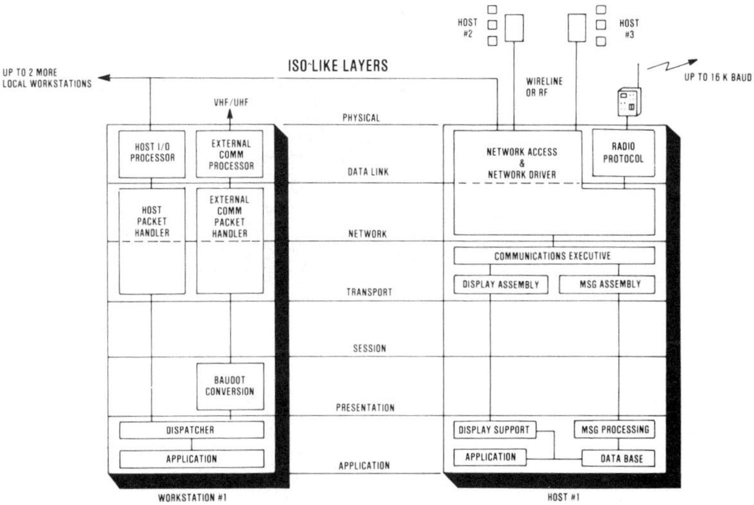

Figure 9. Exhibit 1 System Architecture

(All of the software for this system was developed in about a one-year period.) The system operates through a semi-smart workstation. The operator of the workstation is not a computer operator - he is an end user doing a dedicated user function - he really doesn't care if there is "a computer back there" or not. He has an external role to perform. He is not a computer programmer in any sense of the word. This feature is typical of the kinds of systems that our customers require in the information environment of the future.

The customer saw his problem as having four parts as summarized in Figure 10. First of all, the necessity for the seamless links was clear. Secondly, it was necessary to assign priorities to incoming and outgoing messages so they

- INTEROPERABILITY WITH EXISTING SYSTEMS, I.E. 'SEAMLESS' LINKS
- PRIORITY PROCESSING AND DISPLAY OF CRITICAL INCOMING MESSAGES
- REAL TIME RESPONSE TO DATA BASE QUERIES
- VARIABLE NETWORK TOPOLOGY WITH DYNAMIC RECONFIGURATION TO MEET SURVIVABILITY AND MOBILITY GOALS

Figure 10. The Customer Problem

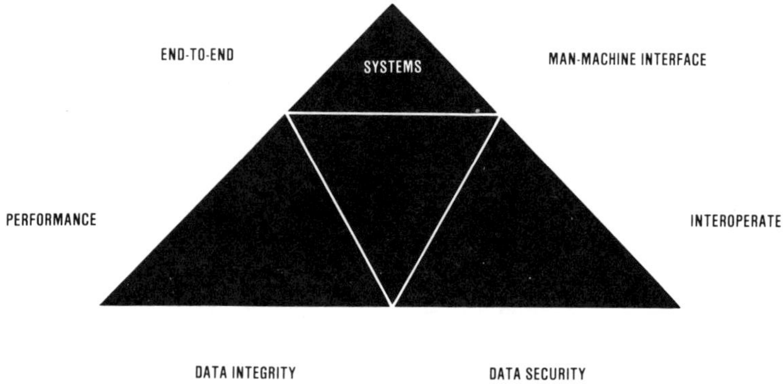

Figure 11. The System Integrator's Solution - System Engineers

were handled in an orderly way. Thirdly, the user response to the data base in real-time was extremely important. Finally, the most unique feature about this system was the ability to break it apart, move the nodes and reconstitute the nodes, using the RF links described earlier to resynchronize the data base without programmer intervention.

The engineering specialists in this business fall into four types: systems, hardware, communications and software. The systems engineers, as shown in Figure 11, have the responsibility to define the full end-to-end system performance, to develop an architecture strategy to provide fast system response, to develop a communications systems architecture and topology to support the break-apart and bring-back-together-operation and, finally, to provide a man/machine interface that a non-computer professional can use to make this system function.

The software engineering team, Figure 12, was required to make maximum use of off-the-shelf systems and applications software. In this case, a

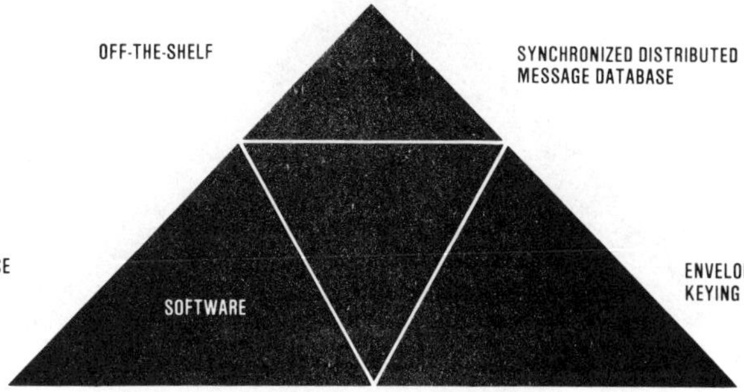

Figure 12. The System Integrator's Solution - Software Engineers

commercial vendor's off-the-shelf operating system was available to be used without modification. All programming was done in structured Fortran, a self-documenting ANSI-77 standard Fortran for source programming. The critical software engineering problem boiled down to building the synchronized data base to operate in a dynamic multi-node environment with distributed lock control at the record level - a state-of-the-art software solution. At the same time, the software engineers had to demonstrate the performance requirements that systems engineering had defined in terms of response to user workstation, message traffic, standard message formats, etc.

The hardware engineers, Figure 13, had the major hardware-integration task. In addition they designed the unique semi-smart workstation that provides the man/machine interface. Under the direction of the system engineers, they considered proper human factors and produced a unique workstation configuration with which the user was very comfortable.

Finally, the communications engineers, Figure 14, defined the land lines, the RF media, and the protocols to be used. They specified, for software engineers, the multiple protocol layers necessary for both internal connectivity and external net performances. They defined termination characteristics for hardware engineering. All the protocols in the system were executed in software or firmware in the processors shown in Figure 9 on distributed architecture.

The result was a system developed through the best efforts of four kinds of people. Figure 15 shows the interrelationships of those four activities with

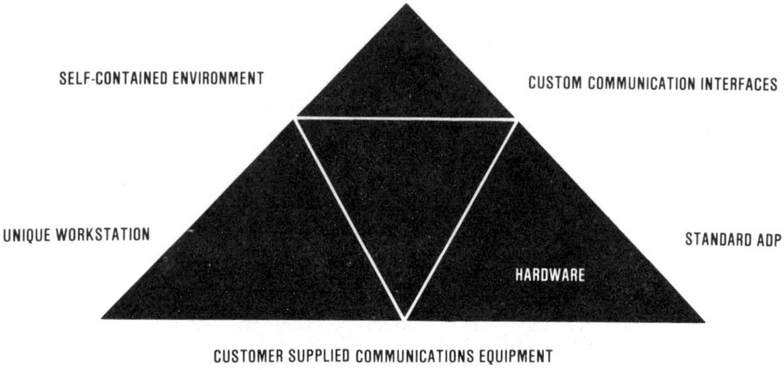

Figure 13. The System Integrator's Solution - Hardware Engineers

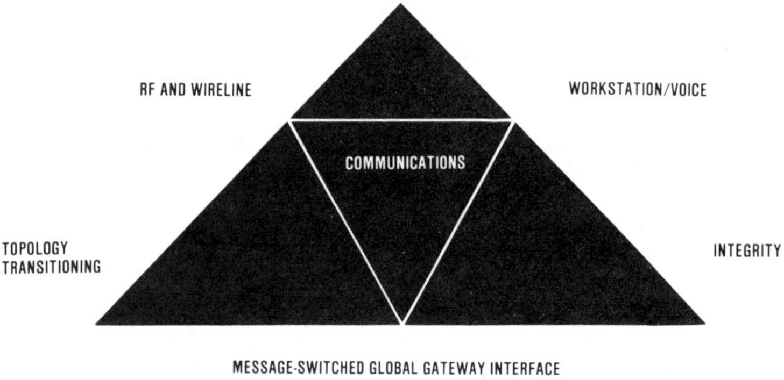

Figure 14. The System Integrator's Solution - Communications Engineers

communications being the central activity. The interfaces between communications and software, hardware, and systems on this chart are very well defined. The real management task was finding engineers who comprehended the technical problems of the rest of the team. Now, where does such a team come from? The team leaders are largely electrical engineers with career specialities, but with broad general experience. In 1985 it is not possible to

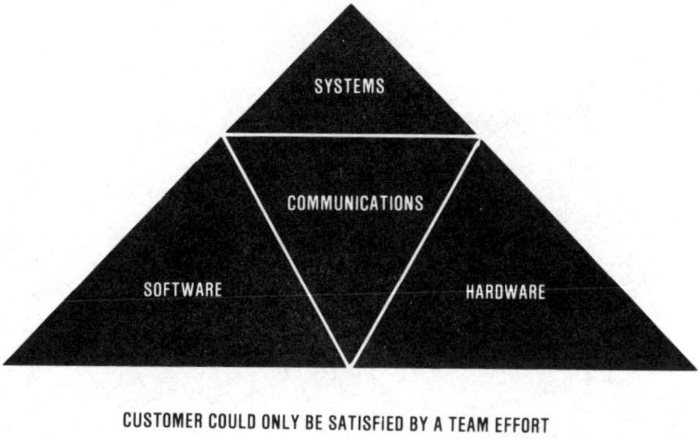

Figure 15. The System Integrator's Solution - Interrelated Engineers

hire people directly out of college to do this work. Universities simply are not providing them. But the skill mix is absolutely critical to system integrators.

Figure 16 shows the two-screen workstation with integrated voice and integrated data capability. From a single keypad, the operator can dial any other user in the net. The workstation's two screens support six simultaneous windows. Two windows are controlled exclusively by the host which provides system-type messages to the operator; four windows are operator controlled. All are fixed-position windows. The operator uses those windows to retrieve information from the data base, work on it, edit it and send it back to the data base; he also prepares and executes messages to the outside world. A system running six windows simultaneously on three workstations connected to a host, and networked with two other hosts with synchronized data bases, is a very complex system. It is a microcosm of the kind of system that will be common in the future.

Figure 17 shows the system in the field. This is a real-world system used currently by our armed forces both in the United States and in Europe. This particular set of pictures was taken in West Germany where the system is online every day, doing its job.

EXHIBIT NUMBER 2

This exhibit, described in Figure 18, is a system that integrates office-automation software, networks, super minicomputers and PC-based workstations into a customer network consisting of 32 primary sites and 70 secondary

Figure 16. Two-Screen Workstation

sites. The basic requirements of the system are listed in Figure 19. The user in this system retrieves, manipulates and sends messages in a general format to any place in the network. Up to 12,000 messages per day are retrieved employing user-defined keyword data, including synonyms. At the start up of this extensive communications system all the tools seemed available for software and communication integration. The major effort would be the man/machine interface. The job should have consisted of designing and integrating the forms managers, text editors, spread sheets, and word processors needed to help the operator do his job at the workstation level. In this case, we used an extensive amount of off-the-shelf hardware and software and by this time the awareness of OSI-type protocols made it clear that a layered approach would be highly desirable. Figure 20 shows that in this system we had data-link, network and transport layers well defined and available off-the-shelf. Further, our selected vendor claimed proprietary integration of

Figure 17. A Field Installation of the System of Exhibit 1

software to the application layer. It appeared that the specific systems-integration function would be to put into the workstation the application programs that were necessary to support the man/machine interface.

It turned out that this integration did not work . . . the vendors' integration was "vaporware" (see Figure 21). Vaporware is not an original phrase, but it describes the phenomena very well. It is a fact that from the same vendor, with the workstation and the host having the same architecture and the same operating system, basic LAN communication and software-to-software interfaces did not exist.

Vaporware is a very dangerous phenomena in this marketplace. There were some things that could have been done about it. First of all, the vendor engineers were not integrated. The hardware engineers controlled access to the protocols, probably understood layers one through six very well, and held them all proprietary. The vendor's software applications engineers, normally

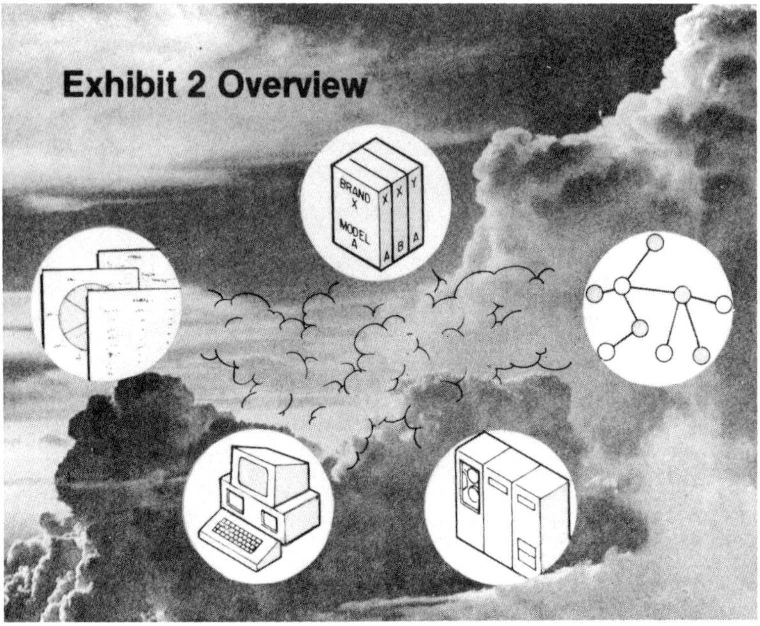

Figure 18. Customer Network of Exhibit 2

those you think of as working at layer seven, had no visibility below that layer. They acted as if the PC in this environment was going to operate in the stand-alone mode - clearly not a winning strategy. In retrospect, it is clear that the systems integrator must be the responsible party for these functions. Unfortunately, this position was not obvious until it was too late and the job was lost. This project did not provide a very satisfying set of experiences but, in fact, did give us some "affordable tuition"; the kind that makes one extremely cautious of commercial marketplace offerings of systems that claim to solve problems at this level.

EXHIBIT NUMBER 3

At about the same time that Exhibit 2 was in progress, we decided to expand our engineering department office automation. There are at least three basic strategies one can use in executing the development of an office

- INTEGRATION OF PC-BASED LOCAL OFFICE AUTOMATION INTO HOST ENVIRONMENT
  - WORD PROCESSING
  - DATA BASE
  - GRAPHICS
  - MENU MANAGER
  - FORMS ENTRY
  - SPREADSHEET
- CONNECTION OF ALL SERVICES TO LAN WITH GLOBAL GATEWAY
  - 32 PRIMARY SITES
  - 70 SECONDARY SITES
- MANIPULATE AND RETRIEVE MESSAGES BASED ON CONTENT
  - USER DEFINED KEYWORDS AND PHRASES WITH SYNONYMS
  - UP TO 3 MONTH INDEXED OFF-LINE; 30 DAY ON-LINE; 12000 PER DAY
- SYSTEM FABRICATED FROM OFF-THE-SHELF, AVAILABLE COMPONENTS

Figure 19. Exhibit 2 Basic Requirements

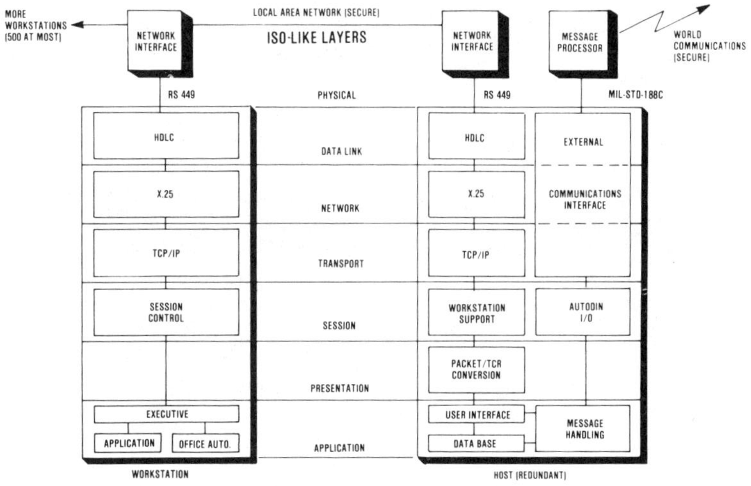

Figure 20. Exhibit 2 System Architecture

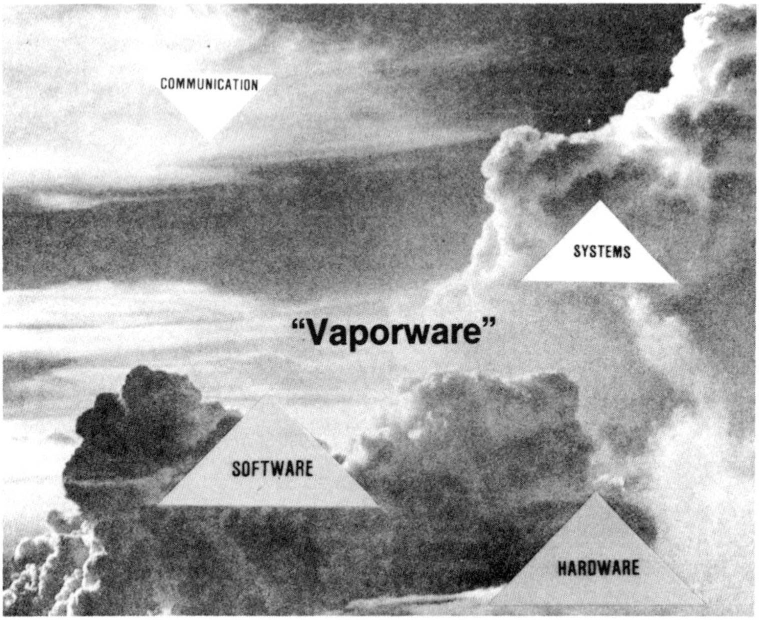

Figure 21. An Illustration of "Vaporware"

automation program. As defined by John Metcalf at Hewlett Packard, the logical top-down strategy seems to be "ready, aim, fire". There are, however, two more prevalent strategies: one is "ready . . . aim . . . aim . . . aim . . . aim . . . aim . . ." and the second is "ready-fire". We executed "ready-fire", with interesting consequences. We bought hardware, we bought a "multiple vendor" LAN, we bought terminal equipment from several vendors, and a number of different word processors (from the same vendor), all of which one would believe could play together in an integrated office. These elements formed Exhibit 3, as outlined in Figure 22.

When we assembled this system we learned that vaporware comes in several different size packages, shapes and colors (see Figure 23). In fact, the vendor software was trivial, particularly in the electronic-mail area. Figure 24 shows the steps that are necessary to produce a word-processor edit of incoming electronic mail. The secretary who performed this feat would hardly define this system as "seamless". Clearly we could have simplified her

A THIRD SYSTEM DEMONSTRATING DISTRIBUTED OFFICE AUTOMATION FOR "IN-HOUSE" USE.

- MULTIPLE VENDOR LAN, TERMINAL EQUIPMENT AND WORD PROCESSORS
- OFF THE SHELF SOFTWARE PACKAGES
- A BASIC 'STARTER SET' FOR OFFICE AUTOMATION

Figure 22. Exhibit 3 Overview

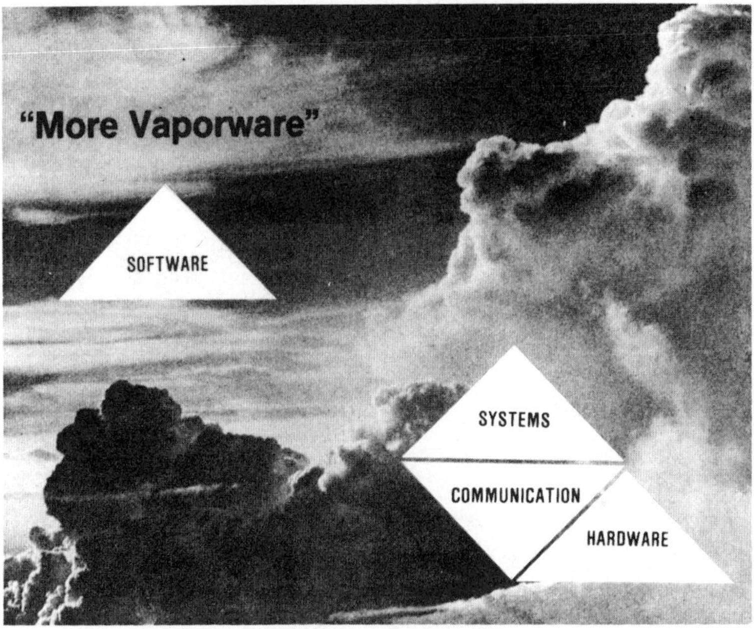

Figure 23. Another Version of Vaporware

task by writing a few dozen simple commands - but this "integrated" software should have been part of the vendor's package. Although the vendor claimed to support open-communications architecture, it was almost impossible to connect third-party hardware or software in this environment. Eventually the vendor-supplied PCs did permit some layer-seven integration - an improvement over the Exhibit 2 system - but it is limited to host-to-PC-to-host file transfer.

PROMPT OR QUESTION	ANSWER
COMMAND:	MAIL
MAIL:	READ
MAIL:	FILE
FILENAME:	(TYPE IN WHATEVER NAME YOU WISH TO GIVE THE FILE)
MAIL:	EX (FOR EXIT OUT OF MAIL)
COMMAND:	XXXX (COMMAND TO TRANSFER OUT OF MAIL INTO WORD PROCESSOR)
CONVERT DOCUMENT TO FILE (DF) OR FILE TO DOCUMENT (FD)?	FD
TYPE NAME OF FILE TO BE CONVERTED	(TYPE IN NAME YOU TYPED IN PREVIOUSLY: "FILENAME:")
SEVERAL QUESTIONS ARE ASKED REGARDING SOURCE CODING, PARAGRAPH FORMATION, ETC.	ANSWER YES OR NO TO EACH QUESTION
THEN GIVES YOU CURRENT RULER SETTINGS AND ASKS YOU WHETHER OR NOT YOU WANT THE CURRENT SETTINGS.	PRESS "RETURN" (NO CHANGE) OR TYPE IN DESIRED MARGINS, TABS
TYPE NAME OF DOCUMENT TO BE CREATED	GIVE IT A DOCUMENT NAME.
PROCESSING	WHEN IT'S THROUGH PROCESSING, IT WILL START THE 'PROMPTING' SEQUENCE ALL OVER AGAIN (IN CASE YOU HAVE MORE FILES TO TRANSFER). PRESS 'CONTROL Z' IF THERE ARE NO OTHERS.
COMMAND:	WPI
WPI:	PRESS "GOLD KEY" FOLLOWED BY "XX"
DOCUMENT TRANSFER MENU	TYPE DESIRED COMMAND FROM MENU (eg., R=RECEIVE DOCUMENT; THEN, TYPE NAME OF DOCUMENT WHICH IS GOING TO RECEIVE THE TRANSFERRED DOCUMENT; TYPE NAME OF DOCUMENT YOU ARE RECEIVING FROM THE SUMERMINI: THE DOCUMENT IS THEN FILED INTO WORD PROCESSOR DISKETTE.)

Figure 24. Downloading from Supermini to Word Processor

FUTURE MARKETPLACE DIRECTIONS

Thus far I have attempted to state the needs of systems integrators from industry and from universities. It is appropriate now to make some observations on future marketplace directions of real importance, as derived from industry and listed in Figure 25. First, the trend toward Open Systems Interconnect (OSI) and UNIX[4] will provide a standardization environment. Second,

[4]UNIX and System V are Trademarks of AT&T.

- TRENDS OSI; UNIX
- IBM OBJECTIVES
- ALTERNATIVE STRATEGIES

UNIX AND SYSTEM V ARE TRADEMARKS OF AT&T

Figure 25. Future Marketplace-Driven Directions

IBM will support its proprietary approach to meet its own objectives and third, alternate strategies are emerging in this information-systems marketplace.

Let's talk first about UNIX. Figure 26 shows that in 1984 there were about 120,000 UNIX installations in the United States running on the machines of some 80 manufacturers in some 150 models of computers. An IEEE/UNIX standardization committee has been formed and there is also an effective users' group for UNIX. In Europe, a six-company consortia has been formed; these same six companies are also involved in an ISO/OSI consortium to assure European inter-operability, across networks, to the operating system level. The European projection is 250,000 UNIX users by 1988.

<u>1984</u>

US — 120,000 'UNIX' INSTALLATIONS
150 MODELS
80 MANUFACTURERS

- FORMATION OF IEEE UNIX STANDARDIZATION COMMITTEE
- FORMATION OF UNIX USERS GROUP, "UNIFORUM"

EUROPE — 8,000; PROJECTED TO BE 250,000 BY 1988

— SIX COMPANY UNIX GROUP

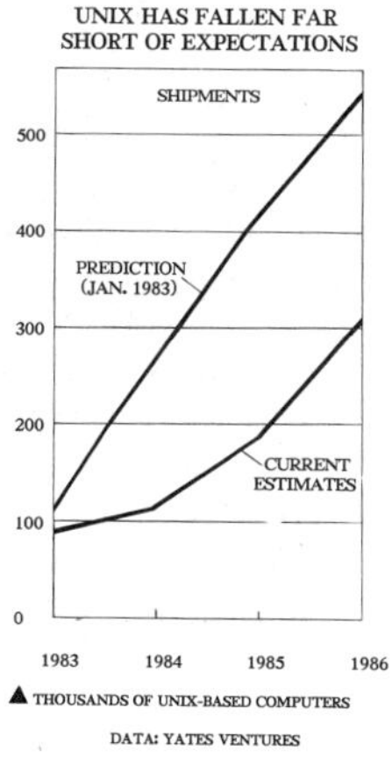

Figure 26. UNIX Installations - *Quo Vadis*

There are many varieties of UNIX that are available. The characteristics of the two dominant ones - UNIX V and those based on Berkeley version 4.2, are shown in Figure 27. Ultrix 32 from DEC is an example of that type UNIX. It should be clear from this chart that the differences between the two exceed the similarities. If a level of standardization does not occur, UNIX will have failed in providing the important underlayment that operating systems must provide for open-system operation.

TODAY	AT&T UNIX SYSTEM V	AS COMPARED TO	BERKELEY 4.2/ ULTRIX 32
UNIQUE SYSTEM CALLS	209 OUT OF 411		212 OUT OF 414
UNIQUE SHELL COMMANDS	208 OUT OF 329		212 OUT OF 333
COMMUNICATIONS NETWORKING PROTOCOLS — SYSTEM	ISO/OSI 1986		TCP/IP
RECORD LOCKING	1986		1985
DEMAND PAGING	IN BETA TEST		YES
DISTRIBUTED RESOURCES SHARING	1986		PARTIAL

Figure 27. UNIX Types - *Quo Vadis*

AT&T recently published a System V interface document that defines the relationship between applications programs and System V operating roles. The document also indicates that there will be extensions to the UNIX System V and that ultimately UNIX System V will pick up the Berkeley 4.2 modifications, since 4.2 already contains commands to implement TCP/IP, the transport layer of DARPANET, and supports its ISO-like protocols. It should be possible to insert the necessary modifications into System V to make it both DoD and ISO/OSI compatible.

On the ISO/OSI side, the interconnect strategy side, Figure 28 describes the 1984 NCC demonstration of GM's manufacturing automation protocols.[5] In *1985 Autofact* some twenty-five companies will get together and show systems operating to ISO/OSI standards, interacting in an environment driven by the General Motors' Manufacturing Automation Protocol (MAP) effort. These demonstrations are indicative of a real commitment on the part of industry in this country to ISO/OSI. At the international level (Figure 29), the ISO committee has already approved the protocols for the transport-layer (4) and session-layer (5), and this process is continuing. At the message level, the

---

[5]Bell, Trudy E., "Technology '85-Communications", IEEE *Spectrum*, Vol. 22, No. 1, January, 1985, pp. 56-59.

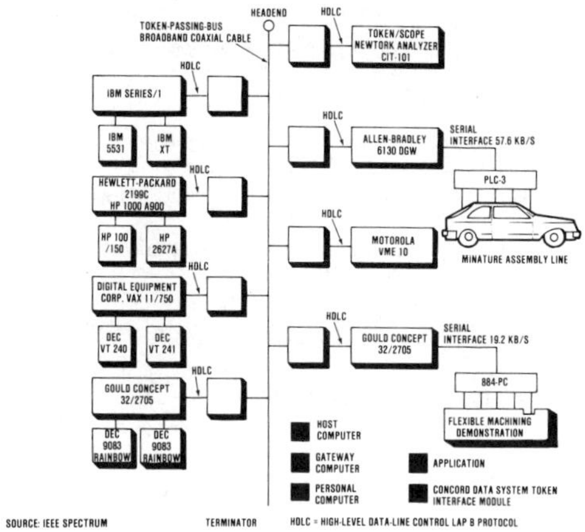

(© 1985 IEEE)

Figure 28. Open Systems Interconnect - Good Signs

CCITT has adopted X.400 as a message standard compatible with ISO/OSI. The ultimate objectives here are standards that will allow the "seamless communication" to be effective.

That is a fairly rosy picture - but what is IBM doing in the meantime? Well, IBM's numbers tell an interesting story. The numbers in Figure 30 come from the February 18, 1985 *Business Week* article mentioned earlier. IBM in 1984 achieved \$46B of sales; their sales are projected to grow to \$100B by 1990 and \$185B by 1994. That is a 15 percent compound growth rate of sales. There are a variety of size and growth estimates in Figure 30 which are provided here without comment. We know that IBM has a commitment to increase their software staff 20 percent per year for the next ten years. Since Figure 2 states IBM's key problem is integrating communications and software, it is safe to assume these new software people will be busy implementing SNA/DIOSS/DIA/DCA across their entire product line. It is not at all clear that the educational system will be able to supply the professionals necessary to meet IBM's growth objectives.

INDUSTRY SUPPORT

1984	— NATIONAL COMPUTER CONFERENCE DEMONSTRATIONS
	— EUROPE — 12 COMPANY GROUP SET UP TO IMPLEMENT COMMON OSI STANDARDS
1985	— AUTOFACT — 25 COMPANY DEMONSTRATION OF COMMUNICATIONS INTERFACING USING A TYPE A CLASS 4 SERVICE

ISO PROGRESS:

1984	— TRANSPORT-LAYER (4) AND SESSION-LAYER (5) PROTOCOLS ACHIEVED INTERNATIONAL STANDARD STATUS
1984	— NATIONAL ACADEMY OF SCIENCE RECOMMENDED DOD REPLACE MIL-STD-1778 (TCP/IP) WITH ISO TRANSPORT
1984	— CCITT ADOPTED OF X.400 AS WORLDWIDE ELECTRONICS MESSAGE HANDLING STANDARD — INCLUDING X.409 WITH ITS TYPE-LINK-VALUE DATA TYPE ENCODING SCHEME — WITH APPARENT NBS COOPERATION
1984	— CCITT X.25 PROTOCOL SPECIFICATION MODIFIED TO INCLUDE OSI CONNECTION MODE NETWORK LAYER SERVICE (VIRTUAL CIRCUIT)
1985	— NETWORK LAYER (3) PROTOCOLS AND IEEE 802.3 (CSMA/CD); IEEE 802.4 (TOKEN BUS) PROTOCOLS ARE DRAFT STANDARDS BEING VOTED UPON
1986	— FILE TRANSFER, ACCESS AND MANAGEMENT (FTAM) WILL BE VOTED

Figure 29. Open Systems Interconnect - More Good Signs

	ESTIMATED MARKET (BILLIONS)			GROWTH RATE	IBM MARKET SHARE	
SOURCE	1983	1984	1990		1983	1990
IBM (BUSINESS WEEK)	$40.2	$46	$ 100	15%	—	—
DATAQUEST (INDUSTRY)	$268	—	$1000	20%	14.9%	10%
FORTUNE (INDUSTRY)	$234	—	$ 582	15%	17.2%	17.2%

IN MILLIONS (1977-81 FIGURES ARE AS OF MARCH; 1982-84 FIGURES ARE AS OF JUNE 30)

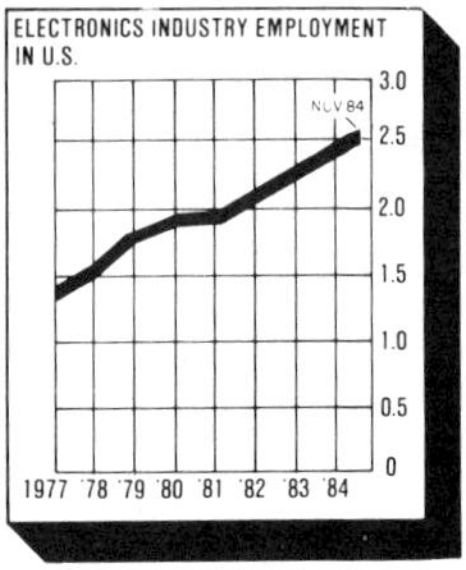

TOTAL EMPLOYMENT (MILLIONS)

	1983	1990	NEW JOBS
INDUSTRY	2.7	6.5	3.8
IBM	0.4	0.9	0.5
	3.1	7.4	4.3

Figure 30. IBM Objectives

- INTERCONNECT STRATEGY
- DECISION CRITERIA
- VALUE CHAIN

(© 1982 Time Inc. All rights reserved)

Figure 31. Alternative Strategies

FUTURE MARKETPLACE STRATEGIES

Figure 31 was provided by the magazine *Fortune* in December of 1982, as the lead-in to a rather critical article[6] on the whole subject of corporate strategic planning. Up until this time, strategists, typically consultants in the Boston area, had suggested that we should all feed the rising stars, shoot the dogs, and milk the cash cows to achieve corporate success. Central to the execution of this concept is a process called differentiation. Even if your product is a boring commodity, differentiated in the eyes of the customer - it can become a rising star. In the information systems business, this differentiation is happening in two areas: operating systems and interconnect strategies. Figure 32 shows IBM, to date, clearly positioned in the proprietary interconnect and operating system sectors along with Apple and Wang. AT&T appears to be

---

[6]Kiechel, Walter III, "Corporate Strategies Under Fire", *Fortune*, Vol. 106, No. 13, December 27, 1982, pp. 34-39.

Figure 32. Interconnect Strategy

taking a different position from that of IBM by means of UNIX (open operating systems) and support of open-network architectures.

By 1990, up to 30 percent of all computer transactions will be *external* network transactions and only 70 percent will be done in the local area. The firm that can connect easily from the local area to the external network is going to have a rising-star strategy. Other than IBM, whose SNA includes gateways to interconnect to the world, firms who cannot or will not connect openly will need to follow a different kind of strategy, perhaps leading to a market sector that grows more slowly than the 15 to 20 percent projected rate. In simple terms, 1990 will consist of an "open" world and an IBM proprietary world with many players gravitating to the open world.

Another way of looking at strategy is the decision-criteria matrix shown in Figure 33. It is not an oversimplification to say that in both the communications and the computer market in the past, buying decisions at the CEO level, have largely been made as separate, safe decisions. In 1985, a significantly different market structure exists now that integrated services are available: one does not buy computers in 1985 without also buying communications and vice versa. Buyers will still attempt to make "safe" decisions but

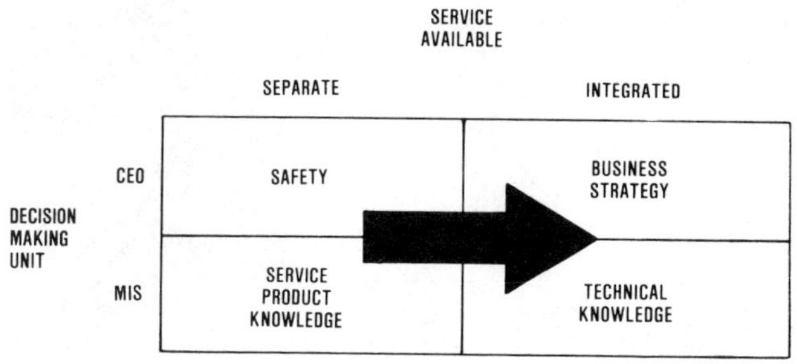

Figure 33. Decision Criteria

the payoff will have to be real. Further, buyers will have to understand the impact of their decisions on an integrated technical basis.

How, then, will a firm differentiate its products in 1990 to achieve rising stars? Another way of looking at strategy is the "value chain" originated by Mike Porter,[7] at the Harvard Business School. The concept defines a firm's value chain in terms of five elements: inbound logistics, operations, outbound logistics, marketing and sales, and service. Figure 34 is a "value chain" for this industry, that combines the value chains of buyers and sellers.

Without question, IBM is unparalleled and unequalled in its marketing and sales, and service, the essential functions of the seller's value chain. Also, they are incredibly efficient in internal operations. That is why they produce $105K of sales per employee, some 50 percent above the industry average. On the other hand, AT&T and other firms pursuing an "open" strategy, are positioning themselves in the buyer's value chain. A fundamental problem in the information society is going to be the handling of inbound and outbound logistics where those logistics are largely the communications transport of data. AT&T is positioned to optimize inbound and outbound logistics; they are also well positioned to support the buyer's internal operations, providing computers that efficiently manipulate and manage inbound and outbound logistics. IBM's positioning in operations is largely a result of their early concepts of industry marketing: outstanding industry marketing people are placed in any pertinent industry. IBM's acquisition of Rolm is a clear attempt to provide the kind of inbound/outbound logistics strategy that is used by AT&T. AT&T does something else here. Note the positions of telemarketing in the buyer's value chain.

---

[7]Porter, Michael E., *Competitive Advantage, Creating and Sustaining Superior Performance,* The Free Press, New York, NY, 1985.

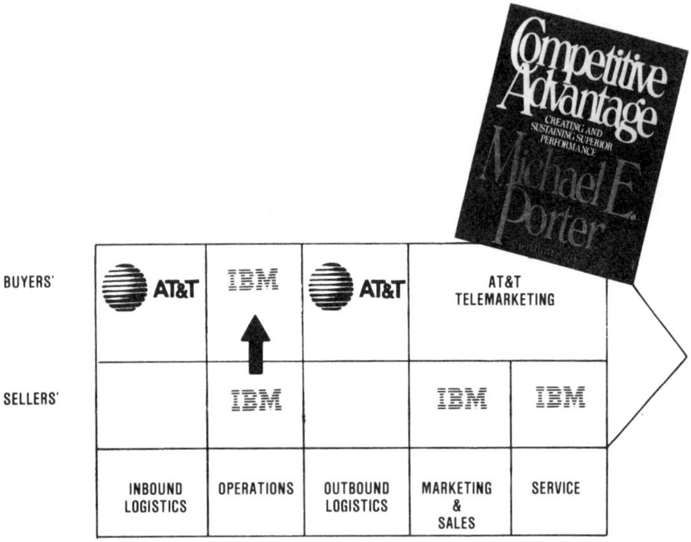

(©1985 Macmillan Inc.)

Figure 34. A Value Chain for the Information Industry

It is clear that AT&T has moved into the *buyer's* marketing sales and service area. To the extent that they can develop novel, unique products in telemarketing, they might provide the buyer with a set of options for his entire value chain from a set of industry standards, open communications, hardware and software.

The value-chain concept also helps us to understand some of the other merger activities currently being discussed. IT&T acquiring Sperry would be a strategic move, vectored at the buyer's value chain. Given this value chain, how does a firm differentiate? By providing products that match the standards of the external world (UNIX and ISO/OSI) with performance, service, and pricing features that enhance the buyer's performance . . . a role not different from the systems-integrator function described earlier. In fact the real opportunity for growth in this industry will occur for people who are trained to see the buyer's end-to-end problem.

The use of computer work stations by other than computer professionals will require custom engineered person-to-machine and person-to-system designs. In a user transparent - "seamless" way these designs will integrate multi-vendor hardware and software, and both local and remote public and private networks.

CONCLUSION

The marketplace is sure to grow and the need for engineers will change accordingly. Electrical engineering has been the basic ingredient that has caused the communication industry to grow from a regulated point-to-point entity to an unregulated multipoint enterprise, and has made computers emerge from a batch-process punched-card orientation to an interactive on-line real-time capability. The world has changed and technology forces us to look at communications and computers as one and the same function. It is essential that electrical engineers be provided with the curriculum and the background to have that same purview. This objective can be achieved by implementing the nine-part undergraduate curriculum (over and above the basics of physics, chemistry, math and electronics) outlined in Figure 35. Note that the curriculum presents a balance of hardware, software and communications - the exact balance called for in the system described as Exhibit 1. No one who wants to work in the information-systems business should graduate from a college of electrical engineering after 1985 who does not have this background. If projections of a $0.5 to $1 trillion market are correct, there could be 4-1/2 million new jobs in this industry in five years. Right now, there are not enough engineers to do all the design and development that must be done to assure this growth. There certainly will not be enough specialists with this background unless schools like Cornell get busy and start training those people today.

HERE IS THE MINIMUM UNDERGRADUATE CURRICULUM REQUIRED TO MEET THIS NEED — OVER AND ABOVE THE BASIC ELECTRONICS, PHYSICS, CHEMISTRY & MATH.

- PROGRAMMING IN A MODERN SYSTEMS IMPLEMENTATION LANGUAGE
- ALGORITHMS AND DATA STRUCTURES
- OPERATING SYSTEM DESIGN
- ARCHITECTURE OF DIGITAL COMPUTER HARDWARE
- SIGNAL PROCESSING AND COMMUNICATIONS SYSTEMS
- COMMUNICATIONS THEORY
- INFORMATION TRANSMISSION SYSTEMS
- COMPUTER NETWORKING AND DISTRIBUTED PROCESSING
- COMPUTER SYSTEM PERFORMANCE EVALUATION

Figure 35. A Suggested Electrical Engineering Curriculum

WHAT IS COMPUTER SCIENCE?

Peter J. Denning

Peter Denning received the B.E.E. degree from Manhattan College in 1964 and the M.S. and Ph.D. degrees in Electrical Engineering from MIT in 1965 and 1968, respectively. After four years as an Assistant Professor at Princeton University, he was appointed as Associate Professor of Electrical Engineering at Purdue University in 1972; in 1975 he became Professor, and in 1979 Head of the Computer Sciences Department. He left Purdue in 1983 to become Director of the Research Institute for Advanced Computer Science (RIACS) at the NASA Ames Research Center in Mountain View, California, his present position. In 1984, he received an honorary Doctor of Laws degree from Concordia University, and in 1985 he received an honorary Doctor of Science degree from Manhattan College.

Dr. Denning's primary research interests are computer-systems architecture, operating systems, and performance modeling. He has published over 150 papers and articles in these areas since 1967, and has coauthored two textbooks. He has been active in the Association for Computing Machinery (ACM) throughout his career and served as President of that organization from 1980 to 1982. He is the recipient of an Engineering Faculty Award from Princeton University, and has won two Best Paper Awards, one from the ACM in 1965, and the other from the American Federation Information Processing Society in 1972. He was elected a Fellow in the IEEE in 1982 and became a Fellow in the American Association for the Advancement of Science (AAAS) in 1984.

INTRODUCTION

At the time that Professor Torng called to invite me to participate in this symposium I was in the middle of thinking about relationships between computer science as a discipline and other scientific and engineering disciplines. For the last several years I have been concerned that computer science has been far too provincial. It is time for our field to grow up and interact with other disciplines as a peer. In fact, my concern with this problem was one of the reasons that I left the academic world and entered a semi-academic regime to pursue interdisciplinary science and engineering. Accepting this invitation, I thought, would offer an interesting challenge: an opportunity to devise a statement to my colleagues about what it is that we all do in our discipline. A casual inquiry addressed to a computer scientist who is met in the street or at a cocktail party, will not result in a well-defined answer as to what the discipline is all about. An attempt to obtain a precise statement by reading a typical university catalog will cause one to conclude that computer science is programming, because that subject is the principal component of the introductory courses in the curriculum. It is impossible to escape questions such as: is computer science related to mathematics? to engineering? both? neither? is it a flash in the pan of history? will it last? does it deal with fundamental questions? will it be reabsorbed into mathematics after a number of years? will it stand on its own? These are all very interesting questions; no one seems to have any answers. Since I am convinced that computer science is a fundamental and enduring discipline, I have set about the task of trying to explain its principles. My efforts are presented here for your consideration.

HISTORICAL DEVELOPMENTS

Computer science, in the simplest possible description of it that I could come up with, is the body of knowledge that deals with the design, analysis, implementation, efficiency and application of processes that transform information. The basic underlying question of computer science is: what can be automated? The discipline seems to be new; it was born in the 1940's with the introduction of electronic computing. In actuality, as we have heard in our previous history lessons today, it has deep roots, not only in mathematics and engineering, but also in logic. When we think about computer science many of us tend to forget about the roots in logic, a very important component that we must comprehend in order to fully understand parts of our discipline. Some additional details are given in Figure 1. The roots in mathematics are basically in calculation. For centuries one of the principal concerns of mathematics has been models of various types of physical processes, resulting in the equations of motion of different systems, and providing us with

Mathematics

- Calculation
  - specific equations
  - general methods

Engineering

- Calculation of properties of mechanical systems
- Calculating machines
  - Surveyors & navigators instruments
  - Pascal, Leibniz (1600s)
  - Babbage (1830s)
  - Bush (1920s)

Logic

- Incompleteness problem (1830s)
- Universal systems (1930s)
  - Goedel (1931) & Turing (1937) theorems
- Symbols both data and operator

Figure 1. Basic Roots of Computer Science

methods of solving these systems. In some cases, the calculation of the equations of a particular system is the problem at hand; we have a specific set of equations and need to solve them. Typical examples are found in weather forecasting or in the calculations of planetary orbits. Mathematics has also given substantial attention to general methods for solving systems of equations and dealing with mathematical functions such as integration, or the solution of linear equations.

Meanwhile, engineering has also had a long interest in calculation, but in this instance, of course, involving techniques that apply to the construction of mechanical or electrical systems. Towards the goal of aiding these calculations and making it possible for engineers to carry out their jobs more reliably, there have been machines for calculating. For example, surveyor's and navigator's instruments have been used for many centuries, if not thousands of years. In more modern times, Pascal and Leibniz had arithmetic calculating devices in the 1600's. Babbage had a mechanical calculator in the 1830's called the analytical engine. In the 1920's Bush had an analog computer for solving general differential equations. At the end of the 1930's we began to see actual electronic computing projects.

Simultaneously, among those who were less concerned with specific methods of computation, there was a completely separate line of development in mathematical logic. For ages, logic had been employed as a means of organizing rigorous argument. All of us are familiar with the ideas that we picked up in geometry: the arguments of Euclid, with his axioms, theorems, and propositions. Although those concepts had been around for many years, by the 1830's a number of inconsistencies and paradoxes had been discovered in the various known reasoning systems, giving rise to the so-called "incompleteness problems". Around that time a search began for a "complete", universal, logical system within which it would be possible to examine a given statement and determine, once and for all, whether or not that statement was a theorem. Logicians had been beating their heads on this problem for a hundred years. Finally around 1930, Goedel published his famous "incompleteness theorem" which said that there is no such thing as a complete, consistent logical system and that every sufficiently powerful system of logic will contain statements that cannot be proved or disproved in that system. One of the foundations of Goedel's proof was the concept of a universal logical system within which it is possible to simulate any other logical system. In his 1938 proof that there exist well-defined functions that cannot be calculated mechanically, Turing also demonstrated a machine capable of simulating any other machine. Both Goedel and Turing exploited the concept that the same abstract symbols could sometimes be interpreted as instructions and at other times as data. The construction of universal machines is based on the use of this key concept.

The convergence of the three lines of thought, as indicated in Figure 2, led to the modern electronic stored-program computer. It is interesting to note that this development is hardly an historical accident. As a student of logic in the 1920's, von Neumann was determined to find the elusive universal system of logic. Although he of course never found it, he carried with him the key idea of universal computation—in this context, the stored-program computer was a natural development.

Who would have thought that esoteric arguments in mathematical logic would have come together with a long line of calculating-machine development to produce a universal electronic calculator that is revolutionizing our world? It is clear that we have a long and venerable history.

(Logic → universal computation)

universal computation
+
electronic calculation

produced

stored-program computer

(ca. 1940)

Figure 2. Birth of Computer Science

EVOLUTION OF THE DISCIPLINE

Since the time that the basic principles merged to produce a recognizable beginning of our discipline, we have experienced an evolutionary trend as more and more fundamental questions have been answered. I believe that computer science consists of eleven identifiable components as given in Figure 3: theory, numerical computation, architecture, programming languages and methodology, algorithms and data structures, operating systems, networks, human interface, database systems, concurrent computation, and artificial intelligence. These items are listed in the order in which they became accepted as parts of the discipline. The term "became accepted" should be considered as my subjective assessment; it is based on a perusal of university catalogs and curricula, in order to determine when a significant number of universities were requiring courses in these sub-areas. Such an inquiry is not the same thing as finding out whether research laboratories around the country are promoting a particular component. I like to think of it as a kind of acid test of the acceptability of a significant part of a discipline: if we are determined to teach it to our students, it must be fundamental. A case in point involves the current status of artificial intelligence (AI) in engineering curricula. Although the area of AI is a direct outgrowth of mathematical logic, it has not been accepted as required subject matter until recently. Now that the

Theory	1940
Numerical Computation	1945
Architecture	1950
Progr. Lang's & Methodology	1960
Algorithms & Data Structures	1968
Operating Systems	1971
Networks	1975
Human Interface	1978
Database Systems	1980
Concurrent Computation	1982
Artificial Intelligence	1986(?)

(©1985 Sigma Xi)

Figure 3. Evolution of Computer Science

trend is in place, it may be that early in 1986 a significant number of universities will require their students to take courses in AI techniques. It is for this reason that the last item in Figure 3 contains a question mark after the 1986 date.

I will now take a brief look at each of the eleven areas of Figure 3. For those who are interested in more detail, I suggest an examination of my article in the January-February, 1985 issue of *American Scientist.*[1] That article was written after I received the invitation to this symposium, when it seemed reasonable to make a formal transcript of my thoughts as I prepared this talk. For each of the areas I will focus on the pertinent fundamental questions and also state some of the major accomplishments. These eleven areas are not independent of each other. Although most of them have their own theoretical components, deal with internally related programming languages and notations, and have specific kinds of algorithms and data structures, some areas, such as software engineering, seem to cut across boundary lines and adopt concepts from all of the others. It is therefore necessary to think of these areas as intellectual components of the discipline but they certainly are not independent.

THEORY

The first area to consider deals with basic mathematics underlying computation. Some of the fundamental questions are outlined in Figure 4. What problems can machines solve–what are the limits of their capabilities? For given classes of problems what algorithms are optimal? What are the intrinsic best- and worst-case performances that we can expect of algorithms in particular classes of problems? What problems are computationally equivalent; that is, what are the equivalent classes of problems in terms of their difficulty? Figure 4 also lists some examples of areas in which there have been significant theoretical accomplishments. We have the theory of computability that was initially concerned with distinguishing between decidable and non-decidable algorithms in order to determine the limits of the various types of automata and formal logical systems. These studies progressed into complexity theory that deals with the problem of measuring the time and space needs of computations, which, in turn, led to problem-class definitions involving deterministic and non-deterministic polynomial questions that are characteristic of finite complexity classes. Also, there is an entire area dealing with automatic theorem proving which has made some progress although not as much as many people would like.

---

[1]Denning, Peter J., "The Science of Computing", *American Scientist*, Vol. 73, No. 1, January-February, 1985, pp. 16-19.

Basic mathematics underlying computation.

What problems can machines solve?
What algorithms are optimal?
Intrinsic best– and worst–case performance?
Computational equivalence?

Computability
Complexity
Problem classes
Automatic theorem proving

Figure 4. Theory

NUMERICAL COMPUTATION

Figure 5 contains a summary of this area that is concerned with general methods of efficient and accurate solution of equations arising from models of systems. The basic questions include: What errors arise when discrete approximations are made of continuous processes? How fast do given methods for solving these approximate equations converge? What symbolic manipulations of mathematical expressions can be automated? How can libraries of mathematical software be proved correct? Some of the principal accomplishments of this area include: a theory of stability and error control; very fast algorithms for certain critical operations, such as the Fast Fourier Transform

General methods of efficient and accurately solving equations arising from models of systems.

Discrete approx's of continuous processes?
Error propagation?
Speed of convergence?
Symbolic computation?
Correct mathematical software?

Stability and error—control theory
Certain fast algorithms (e.g., FFT, Poisson)
Finite-element model
Mathematical software packages
Symbolic manipulators

Figure 5. Numerical Computation

and solution of partial differential equations; the finite element model for structural analysis; the multigrid model for accelerating convergence of equations over grids; symbolic manipulators such as MACSYMA and REDUCE; and high-quality packages of mathematical software.

ARCHITECTURE

The next area on my list considers methods of organizing hardware and software components into efficient and reliable systems. A summary of this topic is given in Figure 6. The following basic questions may be posed. What are the requirements, the best approaches, and the principles that underlie processing memory and communication functions? What are principles for building large systems correctly by combining components? How do we approach the problems of error tolerance and reliability when we put components together into a system? Some of the accomplishments of this area follow in chronological order. In the early days, finite-state machine theory and Boolean algebra were established and gave us relationships between formulas that describe circuits and the structure of those circuits. These developments were followed by the von Neumann machine, the stored-programmed computer that contained an instruction counter. Next came hardware units for fast arithmetic; mechanisms for encoding, storing and transmitting information, especially to correct errors; some steps in the direction of system reliability theory; some methods of synthesizing large systems by the systematic assembly of simple components; prototypes of highly parallel computers; and most recently, the introduction of microelectronic and VLSI circuit design technology.

PROGRAMMING LANGUAGES AND METHODOLOGY

This fourth area deals with notations for algorithms and data, translations among different notations and the effective construction of correct programs. As shown in Figure 7, we may again ask a number of basic questions. What are the basic data types and operations that require algorithms for their execution? What are the basic representations for instructions and data, especially inside machines and also in programming languages? What kind of syntax notations do we need to express the control of the execution of algorithms? How much meaning can we capture in syntax so that it can be analyzed by compiling systems? What methods can we employ usefully in constructing programs that are known to be correct? A list of accomplishments for this area must include procedure-oriented programming languages such as ALGOL, FORTRAN, and COBOL; some recent object-manipulating languages, of which the best example is SMALLTALK; as well as applicative languages such as

APL and LISP that deal with functions that are applied to arguments. Codification of basic programming concepts is an achievement that has developed over the years; for example, basic control structures and basic data structures now seem to appear in many programming contexts. There have been many accomplishments in the field of compiling and code generation. Some steps have been taken in the direction of automatic verification of programs, although most of that effort still remains a manual art despite some recent significant advances. Another technique that has become fairly well known allows a syntax-directed editor to help the user prevent and correct mistakes by taking advantage of programming language knowledge that can be built into the editor.

Methods of organizing hardware and software components into efficient, reliable systems.

Processing, memory, communication functions?
Building large systems correctly?
Error tolerance and reliability?

Finite-state machine theory & Boolean algebra
von Neumann machine
Hardware units for fast arithmetic
Encoding, storing, transmitting information
System reliability theory
Methods of synthesizing large systems
Prototypes of highly parallel computers
Microelectronic and VLSI design technology

Figure 6. Architecture

Notations for algorithms and data, translations, and effective construction of correct programs.

Basic-data types and operations?
Basic representations of instructions & data?
How to control execution?
Capturing meaning in syntax?
Methods of establishing correctness?

Procedure–oriented programming languages
Object–manipulating programming languages
Codification of basic programming concepts
Compiling and code generation
Verification
Syntax–directed editors

Figure 7. Programming Languages and Methodology

ALGORITHMS AND DATA STRUCTURES

Some people think of this area as the whole of computer science. I agree that these elements form a major part of the discipline but they are certainly not the only constituents. In Figure 8 we see that the focus is on specific classes of problems and their efficient solutions, giving rise to another set of basic questions. For given classes of problems what are the best algorithms? How much time and storage do they need? What is the trade-off between space and time? What is the worst-case performance of the best algorithms? What problem classes use similar algorithms so that an algorithm devised for one class of problems can be used elsewhere? One accomplishment in this area that deals with these questions is the identification of good and bad algorithms for various classes of problems, including techniques for searching, sorting, generating random numbers, and matching patterns and text. The ability to categorize the effect of data structures on time and space needs of algorithms is another important achievement. Knuth's series of textbooks on *The Art of Computer Programming*, as well as a number of subsequent texts, all serve as excellent examples of important contributions to this part of the discipline.

OPERATING SYSTEMS

This area is concerned with questions of control mechanisms that allow multiple resources to be efficiently coordinated in executing programs. Figure 9 indicates that some of the questions deal with levels, objects and operations. These questions are important when it is necessary to manage a large amount of software that spans control functions ranging from hardware all the way up to user interfaces. Under such conditions, it is necessary to divide the software or the functions into levels of abstraction that can be identified by means of the following queries: what is their specific nature? what objects can be seen at each level of abstraction? and what can be done to those objects? Other basic questions that are always being addressed include: what are simple interfaces that emphasize abstractions but hide the internal details of the implementation? what are good strategies for controlling resources of various types, such as processor scheduling, memory management, or files? what are the basic construction rules for large systems, especially good hierarchical and data-abstract rules? Figure 9 also contains a list of specific accomplishments in this area involving concepts that may be found in many operating systems such as: time-sharing, interrupts, and schedulers; libraries of utility programs; theories of process coordination, storage management, and job scheduling; memory management; methods of organizing secondary storage into files, directories, and access controls; and high-level interfaces including the recent development of windows and mice.

Specific classes of problems and
their efficient solutions.

What are best algorithms?
How much storage and time do they need?
Tradeoff between space and time?
Worst-case performance of best algorithms?
What problem classes use similar algorithms?

Identification of good & bad algorithms
for important problem classes, e.g., searching,
sorting, random numbers, text pattern matching

Categorizing effect of data structure on time
and space needs of algorithms

Figure 8. Algorithms and Data Structures

Control mechanisms that allow multiple resources to be efficiently coordinated in executing programs.

Levels, objects, and operations?
Simple interfaces emphasizing abstractions?
Resource control strategies?
Basic construction rules for large systems?

Timesharing, interrupts, schedulers, ...
Libraries of utility programs
Theories of process coordination
Theories of storage management
Theories of job scheduling
Files, directories, access controls
High-level interfaces (e.g., windows, mice)

Figure 9. Operating Systems

NETWORKS

This area is involved with the organization of systems that consist of interconnected computers, as outlined in Figure 10. Basic questions include: what are efficient methods of error detection and correction? what are efficient ways of transmitting information through various media? what are efficient ways of mediating contention, especially when the media are shared? what are protocols for short and long-haul communications? what are techniques for hiding the network and providing transparency? Some of the accomplishments in this area include prototypes for computer networks: long-haul interconnections such as the ARPANET, or local networks such as the ETHERNET; reliable protocols for use in unreliable networks; protocols for shared channels; cryptographic protocols that allow secret communication, if desired, or at least good secure authentication of the sender; and finally, distributed computation: taking components of a computation, placing them on different machines that intercommunicate, and still getting the computation to work.

HUMAN INTERFACE

A somewhat different aspect of computer science is displayed in Figure 11 where the focus is on the efficient and the accurate transfer of information between humans and machines. Some pertinent questions in this domain are concerned with the efficient representation of objects: how does one take an object and represent it internally so that answers to questions about the object can be obtained? how are pictures generated from object representations? Other questions relate to human/machine interactions: how is input received and output transmitted between a man and a machine? how are misperceptions minimized? for example, is it possible to determine if the communication process is taking place between a human and a machine? alternatively, how do we maximize the probability of correct perception of the information? Major accomplishments in this area include: core-graphic systems that provide interfaces for dealing with standard graphical operations such as drawing objects, doing perspectives, presenting rotations, making translations, displaying zooms, and suppressing hidden lines; interactive computer-aided methods for designing CAD and manufacturing systems; advanced forms of input and output, including some recent marketplace developments such as bitmap displays containing icons, as well as mice and related devices for pointing at things on the screen. Finally in this area, developments in psychological studies of human/computer interactions are of direct importance to the concerns raised in the previously stated question on the minimization of chances of error in man/machine communications.

Organization of systems of
interconnected computers.

Efficient error (det)(corr)ection?
Efficient transmission in various media?
Mediating contention?
Protocols for short and long haul?
Hiding the network?

Prototype computer networks (e.g., ARPANET)
Prototype local networks (e.g., Ethernet)
Reliable protocols for unreliable networks
Protocols for shared channels
Cryptographic protocols
Distributed computation

Figure 10. Networks

Efficient and accurate transfer of information between humans and machines.

Efficient representations of objects?
Generating pictures from representations?
Receiving input or transmitting output?
Minimizing misperceptions?

Core-graphics systems
Interactive computer aided design systems
Advanced forms of input and output
Psychological studies of interaction

Figure 11. Human Interface

DATABASE AND INFORMATION-RETRIEVAL SYSTEMS

In this area we are concerned with organizing large sets of data for efficient queries. A series of relevant questions as outlined in Figure 12 include: what are the models for data and the relationships among data elements? what are the basic operations that need to be performed on data? What kind of notation should be used for making queries? what are fast methods for selecting and joining elements from these very large sets of data? what methods should be used to protect the data against unauthorized access? and what methods should be used to ensure consistency when different parts of the database can be updated separately and independently from one another? Important accomplishments include database systems that incorporate each of the various major models of data, protocols for concurrent record access, protocols for ensuring consistency among copies of data bases independently maintained, protection against disclosure and tampering, and the development of special-purpose machines that are designed for high-performance database operations.

CONCURRENT COMPUTATION

This area has been around for a long time and is receiving considerable attention these days because of the advent of the super computer. It is centered on the organization of computations that require many processing elements working concurrently. Some basic questions for this mode of calculation are given in Figure 13: what are basic models for concurrent computation? what are the best matches between models and problem domains; for example, in the field of computational chemistry, what kinds of architectures are best suited for computations in that discipline? what machine properties are implied by each model? what are efficient methods of managing the many resources that will be called into play during these types of computations? Some related accomplishments include: specific machines such as arrays of machines organized as meshes, data-flow machines, a recently released product from Intel--the hypercube, cooperating sequential processes as a model for computation, and other types of parallel algorithms. The construction of computations for these machines has benefited considerably from the development of suitable interactive programming and debugging aids.

ARTIFICIAL INTELLIGENCE

This last area on my list has the simplest definition--as measured by the number of words in the definition--the simulation of intelligence. The first question in the display of Figure 14 is the most difficult to answer: what is

Organizing large sets of data
for efficient queries.

Models for data and relations among them?
Operations on data?
Notations for queries?
Fast methods for selecting and joining?
Protection against unauthorized access?
Consistency control?

Systems incorporating major models of data
Concurrent record access protocols
Consistency protocols
Protection against disclosure or tampering
High-performance database machines

Figure 12. Database and Information-Retrieval Systems

intelligence? In dealing with that question, of course, the emphasis must be on answers that are potentially realizable in computer processible form so that it becomes possible to simulate the particular intelligence that has been defined. Another important question is concerned with what models of intelligence should be considered. In recent times there has been a particular interest in rule-base models that can lead to the construction of expert systems. Still another question concerns the relationship between intelligence and rule

Organization of computations requiring many processing elements working concurrently.

Basic models of concurrent computation?
Best match of models to problem domains?
What machine properties implied by each model?
Efficient methods of programming?
Efficient management of many resources?

Mesh–array machines
Dataflow machines
Hypercube
Cooperating sequential processes
Parallel algorithms
Interactive programming & debugging aids

Figure 13. Concurrent Computation

evaluation. Perhaps the most troublesome question of all in this area is how to deal with the unpredictable aspect of intelligence: the creative impulse, the unexpected turn. Accomplishments in AI have included computational theories of cognition, knowledge representations and searches through knowledge data bases (otherwise just called knowledge bases), logic programming, theorem proving, some specific applications in robotics, image processing, vision, speech, and finally, rule-based expert systems.

It is clear from this quick tour through computer-science country that much has been done. It is also clear that quite a number of interesting but

Simulation of intelligence.

What is intelligence?
Models of intelligence?
Relation between intelligence and rule evaluation?
Unpredictability of intelligence?

Computational theories of cognition
Knowledge representation and searching
Logic programming
Theorem proving
Robotics, image processing, vision, speech
Rule-based expert systems

Figure 14. Artificial Intelligence

fundamental questions are being addressed throughout the discipline. Computer science embraces its own theory, experimental method, and engineering. Some of the areas deal explicitly with theory, other areas deal explicitly with design, and some of them emphasize experimental techniques in both understanding theoretical objects and in understanding designed products. Historically, the physical sciences have tended to separate engineering or the application of the science from the discipline itself, so one typically finds physics separated from mechanical and electrical engineering and chemistry separated from chemical engineering. In contrast, we see that computer science is a discipline that is trying to embrace its own theory, experiment, and engineering. A look around the colleges in this country reveals that computer science departments of every kind of flavor are found in engineering schools as well as in math or science schools. Yet all of them have one common feature: none has been successful in separating the theory and the engineering of the discipline. I think that a specific characteristic of the discipline is that computer science and computer engineering cannot be separated because of the fundamental emphasis on efficiency that exists in the discipline. Engineers are interested in efficient implementations, the theoreticians are interested in efficient algorithms. Because efficiency is so important, the science and the engineering are inextricably intertwined with one another. In many cases, computer science and computer engineering are situated in different departments but are offering identical curricula. I cannot predict the outcome of processes that universities may use to solve this problem, but I am sure that any solution must take account of the fact that computer science and engineering cannot be separated.

The discipline has been addressing many fundamental questions for years and many questions remain. The supply of fundamental questions is seemingly inexhaustible! Much flows from the basic question of the discipline: *what can be automated?* That question is not going to be answered soon. It is for this reason that I think computer science will be with us for a long time.

MILESTONES AND MILLSTONES

David H. Ahl

David H. Ahl received the Bachelor of Electrical Engineering degree from Cornell University in 1961 and the MBA degree from Carnegie-Mellon University. Following positions as Manager of Marketing Communications with AT&T, Education Product-Line Manager at Digital Equipment Corporation, and as a consultant with Management Science Associates, he founded *Creative Computing* in 1974, the first magazine in the world devoted to personal computing. He is the editor-in-chief of that publication. Mr. Ahl has been involved with computers since 1956 in a wide variety of applications such as acoustics, space-craft cockpit simulations, grocery-product sales research, early-childhood education, investment analysis, and computer/videodisk games.

It is an honor to be here. When I first volunteered to attend a planning meeting for this symposium, I never dreamed that I would end up actually on it.

BRIEF BACKGROUND

As some of you know, I have been involved in the use of computers for some 25 years. I cut my teeth on such systems as the IBM 650, Bendix G-15 and 20, and some specialized avionics computers at Grumman Aircraft. From the very beginning, my involvement has been as a user, not a designer. Furthermore, most of the applications with which I have been involved have been simulations and analysis, not data processing. For example, I worked on an Apollo cockpit simulation at Grumman and the very first management game at Carnegie-Mellon University in 1962.

As a consultant with Management Science Associates in the late 60's, I devised several models (or simulations) of consumer purchasing behavior. Some of these same models are still used today by companies such as Scott Paper, General Foods, and AT&T.

In the 50's and 60's and even most of the 70's, it took a fair amount of special training and experience to devise anything but the most trivial computer program. With the possible exception of DEC and HP, no one was claiming their computers were user friendly. And indeed they weren't. You might almost call them "human hostile".

Thus, it wasn't surprising that most executives gave the computer room wide berth and placed an insulating layer of middle managers between themselves and the computer gurus. However, as a consultant with Management Science Associates, I found myself regarded by executives at our clients as one who could interpret all those reams of computer output, yet back at home base in Pittsburgh, I was the one writing the programs and producing the reams of output.

I mention this because I feel I had the somewhat unique opportunity to observe both sides of the computer user spectrum in the early days and can contrast it with the situation today—which, surprisingly enough, is not all that different than the situation 25 years ago.

That is not to say that there has been no progress. There has—both on the hardware and software fronts. I've always liked the comparison with the automotive industry. If, since the invention of the transistor in 1948, the automotive industry had made the same progress as the electronics industry, today you would be able to buy a Rolls Royce for $2.80 and drive it one million miles on a gallon of gasoline.

Progress in the electronics industry has been accelerating—I was astounded to read of the packing densities achieved at the Sub Micron lab at Cornell—and more recently the progress has extended to the user interface as well. Today it is a rare manufacturer that does not use the term "User friendly" at least five times in their ad or brochure. Incidentally, some of these systems are still just as human hostile as the ones of the 50's and 60's. Here in the Bell Labs heartland, I dare not say anything about UNIX, the new darling of the microcomputer industry.

In any event, just consider some of the microcomputer related milestones of the past ten years: the first microcomputer, the Altair 8800 in 1975; the first microcomputer-based word processing package, Electric Pencil in 1977; self-contained computers–the Sol 20, Apple, TRS-80, and PET–in 1978; the first spreadsheet, Visicalc, in 1979; a truly portable computer, the Epson HX-20, in 1982; integrated software in 1983; and software that deals with human interactions in 1984.

But despite these advances, most executives and managers today continue to avoid computers–and probably will continue to so do. And for much the same reasons as an earlier crop of executives avoided computers in the 60's.

EXECUTIVES AVOID COMPUTERS

Why do executives avoid computers? The easy answer is technophobia. The fear of typing. The fear of loading a disk wrong, the fear of losing something–which is reinforced by the stories told by data processing managers: "We had a glitch and lost a day's worth of data entry." Or, "there's a bug in the program and everything we've given you for the past three months is wrong." The fear of not being able to understand or use the information that comes out. The fear of losing face in front of younger employees. Everybody knows these young hotshots all have a computer at home; the boss doesn't want to be humiliated by having one of them tell him that he just loaded his disk upside down.

There's the feeling that typing is for secretaries and not for executives. That somehow the use of computers is not really for decision-makers. Tied in with this is the fear that the computer will somehow diminish the need for the skills of the executive. Shades of the Luddites–except this time the executive doesn't have to smash the machines–he can just avoid them.

Another popular reason for executives not using computers was expressed by one executive, "After all, my job has to do with people. I make decisions about those people. How can a computer help me in that? Frankly," he continued, "the name of the game is power–and computers don't equal power."

Put another way, the nature of the typical executive's job is unstructured and, as such, it is not particularly susceptible to computerization.

Couple all these fears and feelings with reports such as a recent one from McKinsey which said that to use a personal computer effectively would take at least 100 hours of learning time, and you can start to see the reasons for the slowdown in the growth of the personal-computer market.

Although it sometimes seems that personal computers have penetrated just about every level of every business, it just isn't so. A recent study by the Diebold Group indicates that only 13% of the top executives—and, indeed, only 17% of all managers—have access to a personal computer. And another study among executives in New York City indicated that 89% still use pencil and paper to write out memos and letters longhand, 10% dictate them, and only 1% use a computer.

Moreover, most executives and managers who use computers tend to be in industries that rely upon numbers such as banking, insurance, engineering, and consulting. Even computer users in other industries tend to be those involved with financial data. Most of the rest are taking a "wait and see" attitude.

CONVINCING THE HOLDOUTS

However, there are several recent developments that might just convince the holdouts that computers may have some real benefits for them after all.

FIRST is an increasing number of dramatic success stories which indicate that companies that make good use of computers are gaining an enormous advantage—sometimes one which can never be made up—over non-users.

SECOND are some entirely new types of software. One, for example, focuses on human interactions such as negotiating, selling, and managing. Another are the expert systems of which we are just beginning to see a smattering.

And the THIRD development to convince the holdouts is some significant new hardware, such as low-cost, easy-to-use local area nets as well as the notebook portable (or, if you prefer, the laptop or kneetop computer).

Let me cite some examples of success stories from the September 1984 issue of *Fortune*. A regional business manager for a bank draws up budgets and designs compensation programs for the 1600 people under him. He writes his memos and letters at home on a second PC and hands his secretary a disk in the morning. In all, he spends about two hours a day with his two computers and is getting answers in a fraction of the time it took just two years ago. As a result, he is able to explore alternatives and ask new questions that no one had the time to examine before.

Art Hanke is a sales planning manager at Maxwell House. These days Hanke is busy supplying the field sales force with personal computers with which to make presentations. He soon found that the same hardware can transmit information on competitors prices over a modem back to General Foods headquarters, thus dramatically speeding up crucial pricing decisions.

Or take the case of Norman James, manager of analysis and controls at Mobil. Using a PC, his budget reports now take ten days or so to prepare compared with the usual four to six weeks. As he said in an interview, "No more mucking in draft after draft and waiting for people to bring me calculations. When I shave in the morning, I find myself contemplating all the ways I can solve a problem using the computer instead of worrying if there's a solution at all."

One interesting case was that of Robert Weissman, executive vice president of Dun & Bradstreet. He decided to put the billings of D&B's time-sharing computer services onto a PC--which may seem a bit odd. However, it paid off. He said, "We had been operating on the common wisdom that small customers paying small fees turn into big customers paying big fees. I quickly found out that big customers get bigger, but small ones don't. Worse, small companies weren't even covering the cost of our service. We revised prices and stopped losing millions." He says it never would have occurred to him to go through the hassle of getting the analysis on the corporate mainframe.

One attraction of the PC, as opposed to the corporate megamachine, is the security of the information contained in it. Jon Kropper, executive vice president of Wang Laboratories uses his home computer to calculate the pay bonuses of his staff. As he says, "It's all on one disk. No assistants. No leaks."

At E. F. Hutton, about 100 senior managers now have personal computers that are linked in a local area net. Vincent Pica, a senior vice president, reports that about half of his time at the keyboard is spent sending messages. "When I finally get to see people, I know what we're going to talk about. Half-hour meetings turn into ten-minute meetings."

Perhaps most telling is the comment by Samuel Fleming, a senior vice president at Arthur D. Little. He warns that managers who delegate computer tasks may find they have unwittingly given away policy-making functions. He should know. In 1982, he assigned two separate groups the task of designing data bases for use by ADL clients. The health care one turned out to be incredibly simple to use, but not very powerful while one for the telecommunications industry was very sophisticated but almost impossible to use.

The second development that I believe will convince many executives to use computers is a new wave of software that is not only easy to use, but is truly useful to an executive. For example, the Negotiating Edge, a package from Human Edge Software practically leads you by the hand. On a personality inventory, it asks you for an honest appraisal–in multiple choice format–of your own traits. It then asks you, to the extent you know them, for the traits of the person with whom you will be negotiating.

With this information–which takes all of ten minutes to enter–it then produces a six-page negotiation strategy report. For example, in a live run of the program, I was advised to avoid deadlocks. "Stubbornness is a quality you both share which makes arguing pointless. If you reach an impasse, the simple tactic of taking turns can be used to please all parties." I was also advised to stop stalls with deadlines, avoid threats, give minimal answers, and use Mr. Mellon's shortcomings to my advantage.

After one negotiating session, you can run the program again and feed it up-to-date and more detailed information. Each time, you get a revised game plan of 12 to 16 specific points.

This is a low-level version of an expert system, but much more sophisticated ones are in the works. With such packages, all of a sudden the argument that "my job has to do with people and computers can't help" doesn't hold much water anymore.

The third significant factor in increasing executive use of computers is the notebook portable. Why?

First, it is small enough to take home or use on a plane or train. Thus, it encourages a manager to use it where he is at ease and comfortable. No one is looking over his shoulder. And as learning takes place, technophobia evaporates. The user can prove to himself that his fears–of typing, of doing things incorrectly, of privacy, of not being able to recover a file–are groundless once he is familiar with the machine.

Moreover, the likelihood is high that the notebook-computer user will discover the very thing that has propelled the computer into such a position of importance today: that computer applications are limited only by the imagination of the user.

MAKING THINGS HAPPEN

Are there some lessons in all this for you people at this conference? Yes, I think so.

FIRST, I think it is clear that we are in the information age. Valuation is no longer just a matter of buildings, machines, and furniture. What's important is the people inside the buildings: the return on the people. And I think there is good evidence that personal computers increase this return, sometimes in unexpected and unpredicted ways. Lesson 1: Give managers and executives computers even if you don't know what they are going to do with them. You must break away from the conventional thinking that office automation starts with word processing and expands to local area networks; that it starts at the bottom and spreads upward. Often, that has happened, but it is not necessarily the best way.

SECOND, it is quite clear that younger managers take to computers more readily than older managers. Nevertheless, the nonthreatening notebook computer makes the adjustment easier for an older, more traditional manager. Lesson 2: Give every executive over 40 a notebook computer.

Lesson 3: Don't look over people's shoulders. Run your personal computing program like a college—have a counselor available for those who wish to consult one, but don't force anything on anybody. Believe me, they'll soon find out for themselves what they've been missing.

Remember, there are three kinds of people in the world: those who make things happen, those who watch things happen, and those who wonder what happened. And these days, it's the people with personal computers who are making things happen.

FUTURE DIRECTIONS IN ELECTRICAL ENGINEERING

COMMUNICATIONS

PROCEEDINGS of the

Cornell Electrical Engineering Centennial Symposium

at Newton, Massachusetts—April 10, 1985

Simpson Linke, editor

Promethean Press
Ithaca, New York

PROCEEDINGS OF THE
CORNELL ELECTRICAL ENGINEERING CENTENNIAL SYMPOSIUM:
FUTURE DIRECTIONS IN ELECTRICAL ENGINEERING-
COMMUNICATIONS

Copyright 1987
by the School of Electrical Engineering
Cornell University, Ithaca, New York
All rights reserved

Published by Promethean Press
Ithaca, New York

Printed in the United States of America

December, 1987

CONTENTS

PREFACE

The celebration of the Centennial Year of the School of Electrical Engineering at Cornell University featured a series of symposia based on the central theme of "Future Directions in Electrical Engineering". Each seminar concentrated on a major facet of the discipline, with individual topics presented by outstanding specialists in each field. The symposia were held at sites around the country that are widely known as centers of activity for each particular subject area, thereby providing significant updates on pertinent electrical engineering technologies at appropriate locations, and allowing a large number of Cornell Electrical Engineering alumni to participate in the celebration in a meaningful way.

The ultimate success of the Symposium Series must be credited to the dedicated efforts of the faculty, symposia, and local alumni chairmen who were responsible for planning the programs, engaging the participants, and attending to local arrangements within specific time constraints. The following schedule of events attests to the overall magnitude of the task and also displays the broad range of topics that were presented:

March 15, 1985—Microelectronics——Stanford, Calif.
April 1, 1985—Computers——E. Windsor, N. J.
April 10, 1985—Communications——Boston, Mass.
April 17, 1985—Atmospheric and Space Sciences——St. Louis, Mo.
April 25, 1985—Energy for the Future——Washington, D. C.
June 12, 1985—Societal Impacts and Future Directions——Ithaca, N. Y.

Each session was videotaped to facilitate preparation of the proceedings of each meeting and for possible use of the individual presentations in classroom seminars and elsewhere. The audio portions of the tapes were transcribed into preliminary text and coordinated with figures obtained from the video images and from hard copy provided by some of the authors. Each presentation was edited with a view toward achieving continuity and coherence without

compromising the natural style of the author. In several instances it was necessary to introduce substantial changes in format and phrasing in order to achieve this goal; in others, the hard copy supplied by the authors made the task easier. In all cases, the final versions were sent to the authors for their criticisms, corrections, and final approval. I am grateful to all of the contributors for their excellent cooperation and assistance in bringing their presentations to this present form.

I wish to thank Professor Edward D. Wolf for his enthusiastic interest in this project and for authorizing financial support from the National Research and Resource Facility for Submicron Structures to assist in the publication of the Symposia Series *Proceedings*. Additional funding from the College of Engineering and the School of Electrical Engineering is also gratefully acknowledged.

It would not have been possible to produce this work without the skillful secretarial assistance of Joan Hammond, Catherine Kuhl, Deborah Overslaugh, and Linda Struzinsky who performed the difficult tasks of transcribing the tapes and retyping the edited manuscripts with meticulous attention to detail. The valuable assistance of John Ralston in the preliminary editing of several of the manuscripts is also acknowledged with thanks. The expert guidance provided by Mrs. Diana Nathanielsz on format and production of this volume is very much appreciated.

I am particularly grateful to Catherine Kuhl and Deborah Overslaugh for their diligence and care in forming the camera-ready copies of the manuscripts on that versatile new instrument that is so illustrative of "future directions in electrical engineering" — the laser printer.

Simpson Linke

Ithaca, New York

INTRODUCTION

John A. Nation

Director, School of Electrical Engineering

The celebration of our Centennial Year affords a unique opportunity to review past accomplishments, to evaluate our present circumstances, and to make some reasonable estimates of prospects for the coming years. A glance at the history of the University reveals that the Cornell *Register* for 1883 contained a "Special Notice" from President Andrew Dickson White announcing the establishment of a "Course in Electrical Engineering." President White had convinced the University's Trustee Executive Committee to create the new program "provided such course occasions no additional expense."

The early curriculum was rigorous and in addition to the traditional engineering topics of mathematics, physics, mechanics, mechanical drawing, and shop work, required rhetoric, French and German. The senior year also included courses in several Cornell specialties such as photometry and electric lamps, telegraph instruments, lines and cables, the dynamo, and electric motors. The first baccalaureate degrees in electrical engineering in the United States were awarded by Cornell (as well as by M.I.T.) in 1885, and the first Ph.D. in the new discipline was conferred by President White at Cornell in the same year. Thus we are truly celebrating 100 years of electrical engineering education.

One of the major figures in the E.E. School of those days was Professor Harris J. Ryan who ran the Electrical Engineering Department essentially as a one-man faculty for 17 years and is credited with having trained almost 1000 electrical engineers during that period. Ryan left Cornell in 1905 to become head of the Electrical Engineering Department at Stanford. It is interesting to note that 12 of the first 13 members of the Stanford faculty were educated at Cornell.

This brief summary of the beginning of electrical engineering instruction at Cornell would not be complete without mention of the internationally known Professor Vladimir Karapetoff who joined the faculty following Professor Ryan's departure and made many contributions to early electrical engineering theory and analysis in several textbooks and many technical articles. Something of the flavor of electrical engineering studies of the time is suggested by the following excerpt from Karapetoff's *Electrical Laboratory Notes*, published in 1906:

> "In coming to the laboratory, bring with you a slide rule, an inch rule or tape, a speed counter, a screw

driver and a pair of plyers [sic]. This will save you time and trouble of looking for them or borrowing them. Do not forget to have a pocket knife for skinning off wire; a bicycle wrench is also sometimes very handy to have."

In these days of hand calculators and personal computers, it is rare indeed to find a student who has even seen a slide rule, much less to have used one.

Classical electrical engineering education at Cornell was devoted principally to electric machinery, power systems, telecommunications, and early electronics. As the technology developed from the 20's through the 40's, the curriculum was, of course, continually upgraded as new and advanced concepts were introduced, but the basic format remained essentially unchanged until the mid-50's when research began to play a central role in the E.E. School, coincident with the move from Franklin Hall to the newly constructed Phillips Hall. The following figure among the highlights of our research development over the years:

An early solar telescope project initiated an era of Radio Astronomy research that culminated in the establishment of the world's largest radio-radar telescope at Arecibo, Puerto Rico, now known as the National Astronomy and Ionosphere Center.

A research program in the new field of Physical Electronics gradually expanded into the present National Research and Resource Facility for Submicron Structures housed in the Knight Laboratory adjacent to Phillips Hall.

The Old High Voltage Laboratory on Mitchell Street, converted to an Energy Laboratory in 1960, served as the staging area for the Cornell Extra-High-Voltage Cable Test Program and housed E.E. research on high-voltage vacuum switches and arc physics until the formation of the interdisciplinary Laboratory of Plasma Studies in 1966. This facility, together with adjacent installations, now contains the largest particle-beam program in any university in the world and has made many important contributions to the understanding of intense electron and ion beams, particularly as they relate to inertial confinement thermonuclear fusion experiments.

The national energy crisis in the early 70's created renewed interest in power and energy research within the School leading to the combination of power-system elements, control theory, and new computer facilities, that now constitutes the unique Kettering Power Systems

Laboratory for the analysis of modern complex power-generation and transmission networks.

Two recent events attest to the high regard that the University and the E.E. School enjoy in the national research community:

1. The Program in Microscience and Technology was established in 1982 and receives about $2 million annually through the Center of Excellence Program of the Semiconductor Research Corporation.

2. The NSF-sponsored Cornell Center for Theory and Simulation in Science and Engineering, with its associated supercomputer production facility, will augment and expand research capability in the School and will encourage greater interdisciplinary activities with other scientific units on campus.

One of our strengths in recent years, as the Electrical Engineering School has evolved from a purely teaching organization into a teaching and research institution, has been our ability to attract good new faculty. A positive measure of this quality is in our receipt of Presidential Young Investigator Awards from the National Science Foundation in the first year of the establishment of these awards. The E.E. School received 5 of the total of 26 awards made available to electrical engineering departments in the nation. Our success in attracting these high-quality young people bodes well for the continued future growth of the School.

Remarkable changes in electrical engineering education have occurred in the School within the past 25 years as the influence and impact of new faculty members and their research have brought modern concepts to curriculum and laboratories that had been essentially identical to those of World War II days. The advent of computers has been particularly significant. We have been very fortunate in the last decade in the support that we have received from industrial corporations. Their generous donations of equipment including computers, work stations, and state-of-the-art measurement devices have been of great value in upgrading our laboratories to present-day standards.

On the other hand, the current huge interest in computer engineering and electrical engineering has introduced major enrollment problems in the School. Using 1965 as a year of reference, the size of the faculty has remained fixed at about 40 members whereas the number of undergraduate students has tripled. Our senior class in 1985-86 numbered 186 students. We have 265 graduates students in the School with about 145-150 MS/PhD students and the remainder in the Master of Engineering Program, the latter being the second largest professional degree program in the University. This year we graduated

30 MS/PhD students and 88 Master of Engineering degree candidates. An encouraging factor is that 12 of these advanced-degree students were women. Today about 15 percent of the students in the E.E. School are women and 25 percent are minority students.

The increase in student enrollment is coupled with a research effort that is growing at a compound rate in excess of 10 percent per year in constant dollars. Clearly we are approaching the limits of our present capability. It is hard to realize that within 30 years of our initial occupancy of Phillips Hall we are now bursting at the seams. Further, the building which was then state-of-the-art, is now an antiquated facility in terms of present-day technology, with little in the way of the clean-air environments, air conditioning, and other amenities that are necessary for the proper functioning of modern equipment.

Plans are in progress to build a new engineering complex on campus in which electrical engineering will have substantially more space than in the past. When these new facilities are completed, our faculty will be able to work at its full potential. We expect to see a relatively modest growth in the number of faculty members. Since a faculty lifetime post is forty years, and student-interest time cycles are rather shorter than that, the response to large student increases in enrollments and interest in electrical engineering cannot be followed instantly by new faculty positions. Attempts will be made to maintain our undergraduate size at about its present level with some growth allowed in the number of our graduate students. It is certain that there will be many more interactions with industry in the near future, and these will take on a much broader form as the years go by. For example, an exchange program may be arranged with industrial laboratories so that their scientists will come to work in the School for a period while a faculty member will have the opportunity to spend corresponding time in industrial environments.

This brief perspective of past, present, and future encourages us to begin our second century with a true sense of excitement. We may be justly proud of the important contributions that our School of Electrical Engineering has made to many of the developments that are reported in these *Proceedings*. As we celebrate these past achievements, we can look forward with confidence to producing equally significant additions to our challenging discipline.

PART 3

COMMUNICATIONS

PROCEEDINGS of the

Cornell Electrical Engineering Centennial Symposium

at Newton, Massachusetts—April 10, 1985

"It is my pleasure to welcome you today to this third meeting in our series of cross-country symposia dealing with frontiers of electrical engineering. This program is not just a random event. We are celebrating the national centennial of electrical engineering education that began both at Cornell and at M.I.T. in 1885. I am especially pleased, therefore, to also welcome Professor Richard Adler from M.I.T. who has graciously consented to make a few remarks from the viewpoint of his institution before we begin the formal part of this program.

"On behalf of the School of Electrical Engineering, I wish to thank James Broadhead, our Symposium Chairman, and Glenn Thoren, our Local Arrangements Chairman, for the dedicated efforts that have made this meeting possible. Thanks also to the members of our distinguished panel of speakers who have taken time from their busy schedules to give us the benefit of their experience and insight in the communications field. The School is also grateful to our many sponsors for their contributions to this meeting and indeed for their support of the entire symposium series.

"Our General Chairman for this meeting, James L. Broadhead, received the B.M.E. degree from Cornell University in 1958 and the L.L.B. degree from Columbia University in 1963. Following several formative years in industry, he joined St. Joe Minerals Corporation in 1968, held a variety of positions in that organization, and became President of the Corporation in 1981. In 1982 he was appointed as Executive Vice President and Director of U. S. Industries, and became Senior Vice President of the G.T.E. Corporation in 1984, his present position. This broad executive background and experience has made Mr. Broadhead an authority on the complex problems related to policy issues in the communications field."

– Professor Toby Berger, Symposium Faculty Chairman
From Welcoming Remarks, April 10, 1985
Newton, Massachusetts

"It is a pleasure to join our Cornell colleagues in celebration of approximately 100 years of electrical engineering history and education, a field, which in that time, has gone from direct-current electric power generators and primitive telegraph and telephone facilities to the immensely varied and complex electrical systems of today. Indeed, the diversity of the field is so great that in the context of our Department of Electrical Engineering and Computer Science at M.I.T., we sometimes think of 'E.E.' as meaning adapt continually to the inevitable proliferation of new ideas that has characterized our provocative discipline.

"The future directions in communications that will be described today reflect the strong sense of excitement and enthusiasm that is prevalent throughout the communications industry. Unfortunately, the same degree of excitement does not exist within the university environment, including our own E.E. & C.S. Department at M.I.T. The main reason for this unhappy state of affairs lies in the industry's dependence upon the extensive hardware and software systems that have always been so difficult for universities to handle. I believe that the educational community should attempt to match the zeal and the magic, if you will, that have inspired the remarkable industrial developments in the communications field. I must confess my inability yet to accomplish this worthy objective, and I offer it as a challenge to my university colleagues and to the communications industry."

– Professor Richard B. Adler, Assoc. Dept. Head, M.I.T.
From Opening Remarks, April 10, 1985
Newton, Massachusetts

"Milestones such as this Centennial Celebration are frequently occasions for mellow reflections on the distant past; however, today's observance is somewhat different. It is dedicated to the future, to the exciting and dramatic advances in electrical engineering, and to the remarkable effect that these advances have had – and will continue to have—on our lives.

"It is particularly fitting that this Symposium Series includes a broad look at the field of communications, because Cornell University has its roots in the first of the modern communications technologies – the telegraph. In 1844, Ezra Cornell directed the construction of the first telegraph link between Washington and Baltimore. Over these very lines, Samuel F. B. Morse would eventually transmit his famous words, 'What hath God wrought.' Mr. Cornell's subsequent financial success with Western Union and other enterprises became the means for establishing the great University that bears his name. The technological success of his early enterprise helped us emerge into the information age – an era that is largely the product of wide-ranging research and innovation in microcircuitry, electronics, computers, and telecommunications.

"There is probably no area in which technological advances have occurred more swiftly or more dramatically than in the latter. The marriage of telecommunications and computer technologies has made it possible for people to carry out many tasks by telephone. Businesses are able to send vital information electronically to widely separated locations. Meetings can be held at considerable savings in time and money through audio and video teleconferencing.

"There seems little doubt that all of society has felt the impact of modern communications. For example, the immediate and generous world response to the famine in Africa was made possible to a large degree by telecommunications. The misery and despair of those poor people was brought quickly and clearly into our living rooms *via* satellite images transmitted directly from Ethiopia.

"Considering the importance of communications in so many facets of our lives, it is appropriate that this topic has been selected for discussion. We are fortunate today to have six men who have all made significant contributions to our progress in communications, and who can speak from experience and with authority about the prospects for this dynamic field in the coming years."

– *James L. Broadhead, Symposium Chairman*
From Introductory Remarks, April 10, 1985
Newton, Massachusetts

THE PROMISE OF OPTICAL COMMUNICATION

Robert W. Lucky

Robert W. Lucky received the B.S., M.S., and Ph.D. degrees in Electrical Engineering from Purdue University in 1957, 1959, and 1961, respectively. He joined AT&T Bell Laboratories in 1961 where his initial assignments involved theoretical studies in the transmission of digital information over telephone lines, leading to his appointment as Superviser of the Signal Theory Group in 1964. From 1965 to 1976 he was Head of the Department of Advanced Data Communications, and in 1977 became Director of the Electronic and Computer System Research Laboratory. In 1982 he was named Executive Director of the Communication Sciences Research Division, his present position. Dr. Lucky has served as Vice President and Executive Vice President of the IEEE, and as President of the Communications Society of the IEEE. He has been editor of three technical journals, including the *Proceedings of the IEEE*, and is a consulting editor for Plenum Press. He is a Fellow of the IEEE, a member of the National Academy of Engineering, and Chairman of the USAF Scientific Advisory Board.

Editor's Note: In 1987, Dr. Lucky was awarded the 13th annual Marconi International Fellowship.

When I completed my studies at Purdue so many years ago, I could not have believed that one day I would be presenting a talk on optical communication–the subject did not exist at that time. Now, however, optical communication is the harbinger of what has come to be known as the information age. On the other hand, it is my perception that we are not really in an information age; this is really the "data age." We in the communication business are busy bringing bits to people rather than transferring information, and I think there is an important difference. In my discussion today I will show that technology will favor us with many more bits, but that we are still uncertain about exactly how clever the world will be in using these bits, a subject that is certain to provoke considerable fascinating controversy.

I was at a brainstorming session on a very general subject last week as the only technologist among eight people representing a diverse collection of fields. During their opening statements, all of the other speakers attacked technology. The last thing we want, they said, is an increase in the number of boring inventions that lead to more technology. Indeed, an architect in attendance claimed that there is an unlimited growth of information that must be stopped before we are overwhelmed. To this end he had a marvelous suggestion that I found very appealing: perhaps in the near future all elevators should have a coin slot in the wall that would provide three minutes of silence for a quarter – a "negative juke-box" of sorts to furnish us with a brief respite from the incessant "piped-in" Muzak that is so hard to escape these days. After all, the goal of communication is to bring more information to people and hence bring people closer to each other. On the other hand, we must consider the need for privacy along with the need for communication. I think that there is a fundamental dichotomy between these two requirements which must be addressed while we charge ahead and put telephones on airplanes, in cars, and everywhere else.

EARLY CAPACITY CONSTRAINTS

For the past decade, the information transmission business has focussed on the search for more capacity. Construction of the American telegraph network began in the 1840's so that in 1875, when the telephone was invented, this country was already spanned by a digital network owned by the then largest corporation in the world–Western Union. Although installed capacity had reached substantial proportions in the 1880's, it is hard to believe that the telephone poles were already as congested as illustrated in Figure 1. Since we had not started running cables underground at that time, the search for more capacity led to such daring innovations in the industry as the forty-arm pole of Figure 2 – a somewhat primitive space division multiplexer.

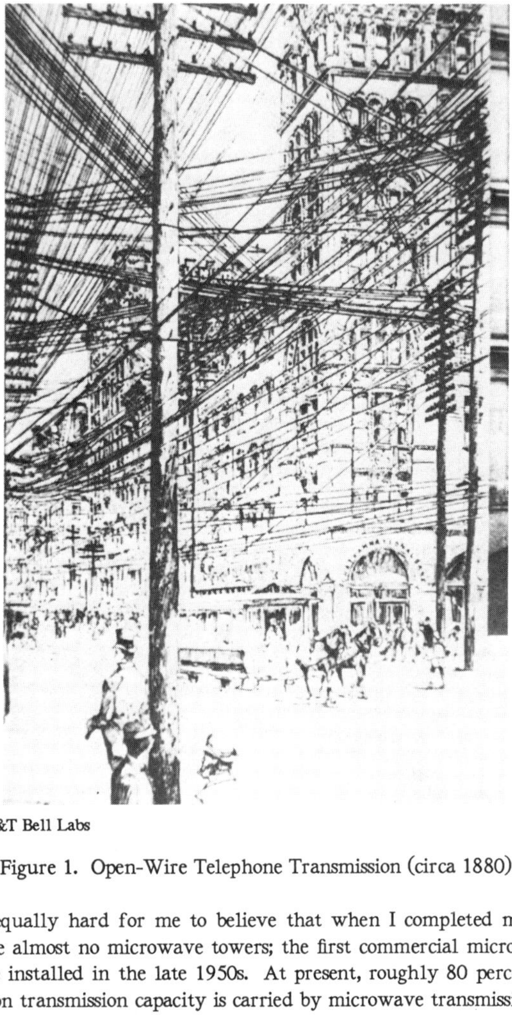

Source: AT&T Bell Labs

Figure 1. Open-Wire Telephone Transmission (circa 1880)

It is equally hard for me to believe that when I completed my studies, there were almost no microwave towers; the first commercial microwave systems were installed in the late 1950s. At present, roughly 80 percent of our information transmission capacity is carried by microwave transmission, due to the build-up of microwave radio capability in the 1960's. Satellites came into their own in the 1970's, helping international communications to become so cheap and ubiquitous that we have coined the cliche "global village". Having

Source: AT&T Bell Labs

Figure 2. Forty-arm Telephone Poles (circa 1920)

brought the world closer together in many ways, there is no question that satellites will play an important role in communications for a long time to come, but I think that the spotlight in the 1980's undoubtedly will shine on fiber communication.

We have been interested in optical communication for several decades, although the first systems utilized free-space communications after the invention of the laser in the early 60's. Figure 3 shows a laser beam as seen at our Holmdel laboratory after propagating from our Murray Hill laboratory 35

Source: AT&T Bell Labs

Figure 3. 35-Mile Propagation of a Narrow Laser Beam

Source: AT&T Bell Labs

Figure 4. Experimental Free-Space Laser Link

miles to the north. The beam lights up the night sky to an observer whose head is in the same position as that of the camera. Beam divergence is so small, however, that movement of the head one foot to the right or left causes the beam to disappear. Transmitting only a few milliwatts of power, this system provided data rates on the order of a hundred megabits per second. The presence of fog, of course, creates serious problems! With the invention of optical fibers in the late 1960's, and subsequent improvements in their transparency, people lost interest in the free-space optical link.

Occasionally, however, we revive our efforts in free-space links. Figure 4 shows an experimental link in operation right now in Washington, D.C., between an office building, and an AT&T communications building. By simply strapping a cheap transmitter on the side of a building, we are able to provide transmission at rates of at least 90 megabits per second using conventional low-cost equipment. Such a system is perfectly adequate as long as inclement weather does not obscure the transmitted laser beam; on a foggy day, the free-space link is cheap but not available. Communications are always that way: since no link is continuously available, it is necessary to trade off costs and other parameters against such contingencies. I was at a meeting last month where such matters were being discussed; and one fellow in the room, having fallen half asleep, woke up to see this picture of a free-space link on the screen. Shaking himself like a dog coming out of the rain, he wondered if he had been asleep for lo, these many years; had we not worried with these things 20 years ago, he asked, and had they not been replaced by better gadgets? This question was resolved immediately by a manager who stopped all discussion on the subject with just two words: "It's unregulated." That's very important--it's funny, but it is important. One could simply strap these devices up anywhere right now, and not have to contend with the FCC, the spectrum-allocations people, and all other regulatory bodies until someone stopped the transmission with a bulletin board or some equally effective obstruction.

ADVENT OF FIBER TECHNOLOGY

As shown in Figure 5, in just the past 10 or 12 years, optical-fiber attenuation has evolved from the previous range of 20 to 100 db per kilometer down to about 0.16 db per kilometer, thereby allowing transmission through about 60 miles of fiber before a repeater is required. The quality of present-day silica-based fibers has reached a plateau due to the Rayleigh scattering limit, so we do not expect to see much improvement in their transparency or their transmission capabilities. There are other glasses, however, composed of fluoride and halide-based materials, that are known theoretically to have lower attenuations of as much as two to three orders of magnitude.

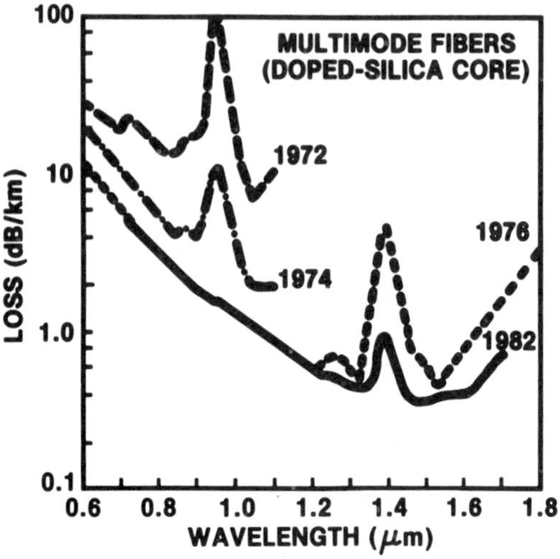

Source: AT&T Bell Labs

Figure 5. Evolution of Optical-Fiber Attenuation

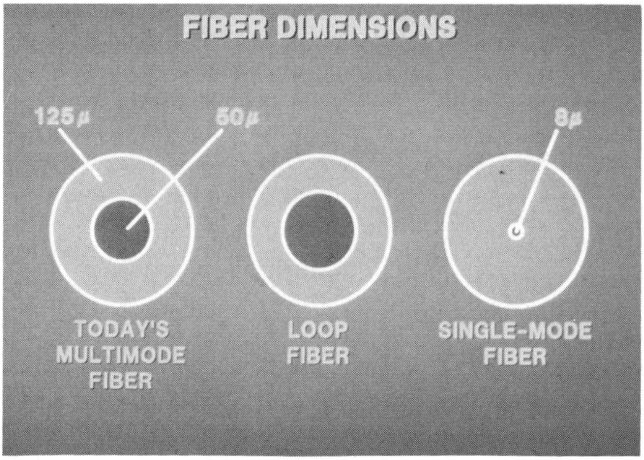

Source: AT&T Bell Labs

Figure 6. Physical Dimensions of Long-Distance, Single-Mode Fibers

Researchers at the Naval Research Laboratory have decreased the attenuation in such glasses to a few db per kilometer, but this is still far from the theoretical limit of about 0.001 db per kilometer. Such attenuation would allow repeaterless transmission half way around the world through a fiber, making possible transatlantic or transpacific transmission channels without any repeaters whatsoever, and at the same time providing enormous data-transmission capacities.

In today's world all newly-installed long-distance transmission uses single-mode fiber with a very small diameter core–about 8 microns–as shown in Figure 6. The wavelength of light that is being transmitted is about 1 1/2 microns, so the fiber core is on the order of a wavelength (in silica) in diameter. For this reason such fibers propagate only one mode and hence suffer much less dispersion than do earlier fibers. One of the surprises to many of us in the field is how the whole world has gone to single-mode fiber, including local-distribution applications which do not really need the capacity at present. The driving force appears to be the desire on the part of the buyer for a sophisticated high-tech product. Engineers are often in the position of telling people that something so elaborate is not necessary, with the customer replying, "I don't care, I want it." We sometimes get caught in that dilemma. We are often wrong, simply because we have not yet learned that the customer is often perceptive about investing in the future. There is a fantastic high-tech market out there; people want the best, particularly when investing in rapidly evolving technologies.

We are seeing a tremendous build-up in long-haul fiber transmission capacity. Not just AT&T, but MCI, GTE, and many others are installing fiber links; for example, the railroads are installing fiber along their right-of-ways. In the next two years data rates will increase from the current 90 megabits and 180 megabits per second to 1.7 gigabits per second of capacity per fiber. It is possible that fiber-transmission networks will become overbuilt in the near future. Later in this talk I will briefly discuss what the world might do with all this capacity.

Placing fiber capability in the hands of potential customers is yet another issue, reminding me of the time I appeared on a TV talk show in Cleveland. As I ran into the studio, the program staff told me I had to have a prop, and proceeded to hand me a piece of fiber cable. The intention was for me to display the fiber on the TV and tell the viewing audience what it did for them. I looked at the thing in my hand, and I couldn't think of anything that it did for the people; the fact is that fiber doesn't yet reach into the home. At present, fiber is being installed in some business offices, but in the residential area it only reaches an interface unit (the vertical structure in the center of Figure 7) that is still several thousand feet from the average home. Final

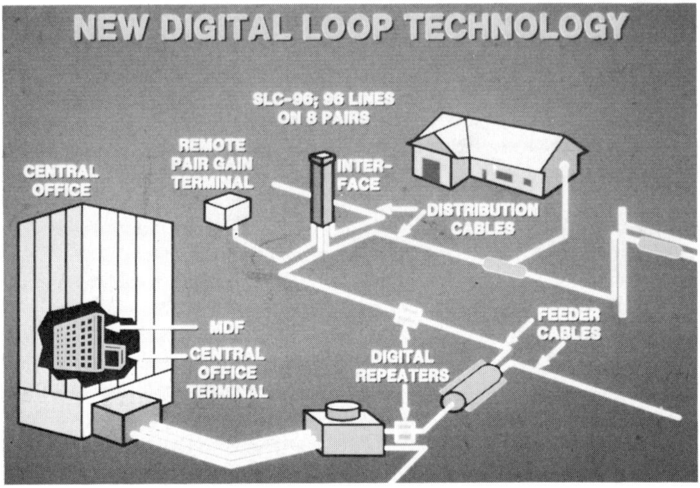

Source: AT&T Bell Labs

Figure 7. New Digital Loop Technology

distribution to the household still utilizes copper wire. In effect, we are piping information through a fire hose to a central terminal and tapping into it with a drinking straw . . . we are building communications super highways without entrance and exit ramps.

The pathway to the home is blocked by economics. The cost of optical systems is much more than that of Cable TV, which already supplies most of the services desired by its present clients. There are a few experimental locations, however, where we have installed fibers all the way to the house. Such experiments evoke in the minds of many people an image of a broadband integrated services digital network (ISDN) that would be capable of bringing a high-capacity bit stream into the home and the office, thereby providing extensive audio, video, and data services.

Let us turn now to a brief examination of current research in the fiber area. We can say that, simply speaking, we have a flashlight that turns on and off very rapidly (a laser), a piece of glass fiber, and a photodiode detector at the other end to receive the pulses. At present we are able to transmit at rates of several billion pulses per second. One of the earliest problems we faced was chromatic dispersion. Various spectral components of the pulse arrive at the far end of the fiber at different times due to their different phase velocities, and the pulse gets spread out. Beyond that we ran into something called mode-partition noise that virtually halted progress in fiber communication for a period of about one year. The output spectrum of a laser contains a dominant longitudinal mode at a particular wavelength, and several side modes with much smaller intensities. (A laser cavity typically supports many possible longitudinal modes of oscillation.) Basically, the side modes should have powers that are 30 to 40 db below that of the primary mode. Every now and then, however, a side mode will leap in intensity, as shown by the dotted line in Figure 8, and all the transmitted power is contained in that side mode for a microsecond or so. It is not yet fully understood exactly why this happens, but in a microsecond, at a data rate of a billion bits per second, a half-million errors can occur, representing a significant degradation in information integrity. Consequently, the control of mode-partition noise was an important problem.

In order to place these remarks in their proper perspective it is necessary to recognize that my background is similar to that of many people in this room–I come from the electronic world and I am very comfortable while working at audio frequencies, but I have never been comfortable with anything faster than audio. In audio transmission, mode-partition noise would be no problem; one would simply insert a narrow-band filter in the system. Since filters, amplifiers, equalizers, and all the things that we are used to working with, do not yet have functioning analogs in optical systems, we bypass our problems by depending upon new devices that have been built by our colleagues in physics. This mode of operation has been frustrating to communication types like myself because all the responsibility for progress has been handed to these physicists. The art of optical systems design is so primitive that we do not understand the intrinsic limits or the important parameters in our models of such communication systems. A typical communications theorist will isolate a problem and go off and study it for a month. When he returns with his results, the physicists will inevitably tell him that his solutions are no longer applicable because "we have invented something different". In all fairness, however, they have indeed invented a succession of lasers with improved spectral characteristics.

Moving from technology to economics, for the moment, one of the great mysteries to me now is why semiconductor lasers cost $2,000. Thousands of

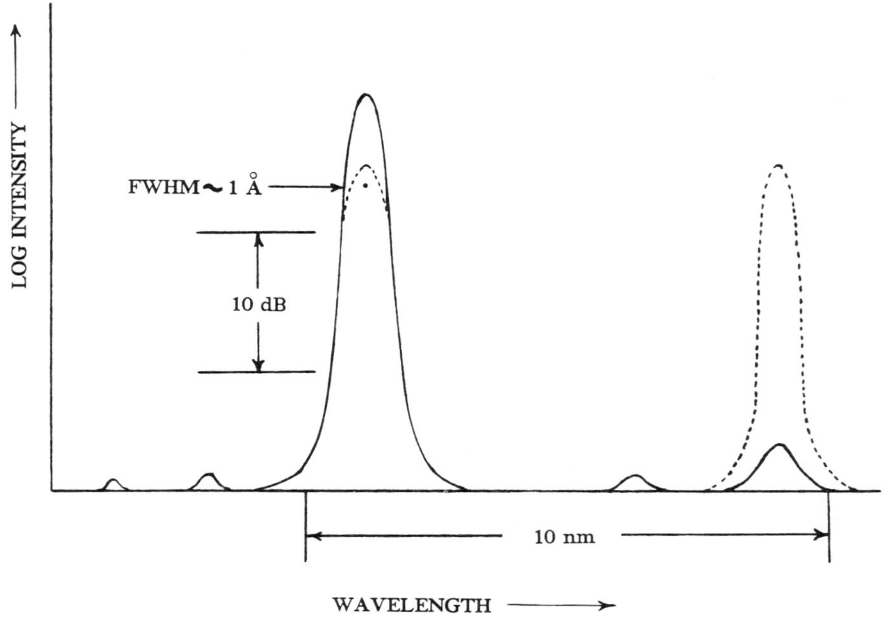

Source: AT&T Bell Labs

Figure 8. An Illustration of Mode-Partition Noise

lasers can be fabricated from a single good wafer, and there is no reason why the laser itself should cost so much in the future. Of course it is very expensive to package them so as to ensure temperature stability, mechanical stability, and efficient coupling to the fiber. Perhaps all of these functions soon will be incorporated into a very inexpensive package, as has been achieved in the lasers used in compact-disc players.

The next problem we encountered was "chirp". When a laser is turned on and off, the injected carrier density increases sharply and then decreases rapidly, as shown at the bottom of Figure 9. Both the output power and wavelength of the laser are directly related to the carrier density. As a result, when the laser is turned on it shifts toward the shorter wavelength—toward the blue—and then shifts toward the red when it is turned off. Due to dispersion in the fiber, these different wavelengths have different propagation velocities. Hence, the various pulse components arrive at the detector at different times, leading to broadening of the pulse. For a while this factor limited transmission rates, but then the physicists went to work and devised systems that either did not chirp or were not seriously affected by the phenomenon.

Still another problem concerns small reflections that come back to the laser from the end face of the fiber and jar the laser into the wrong modes. The judicious use of optical isolators helps ease this vexing phenomenon.

In the future we would like to have coherent optical transmission systems that incorporate heterodyne detection (Figure 10). Heterodyne detection would give us both increased sensitivity and flexibility. The optical output signal of a local oscillator that could be mixed with an incoming optical signal would, in the course of demodulation, give us as much as a 20 db improvement in signal-to-noise ratio. Moreover, the ability to extract various sub bands by tuning the oscillator would allow us to have frequency-multiplexed signals, just as we do in the regular communication world. Such systems, however, are only now being demonstrated in the optical domain. The key component we lack is a tunable laser with a very narrow spectral output; it is sure to become available in the near future. The upswing of progress in the optics field is dominated by the device people who are generating new inventions at a remarkable rate. Unfortunately, this rapid advancement has meant that the communications people really have had little opportunity to gain control since the problem changes every month.

OPTICAL SYSTEM PERFORMANCE CHARACTERISTICS

Figure 11 shows the progress we have made in fiber system performance: these are research results, not what you could go out and buy at the moment. New commercial systems being installed operate at 1.7 gigabits per second. An appropriate figure of merit for these systems is the product of the number of

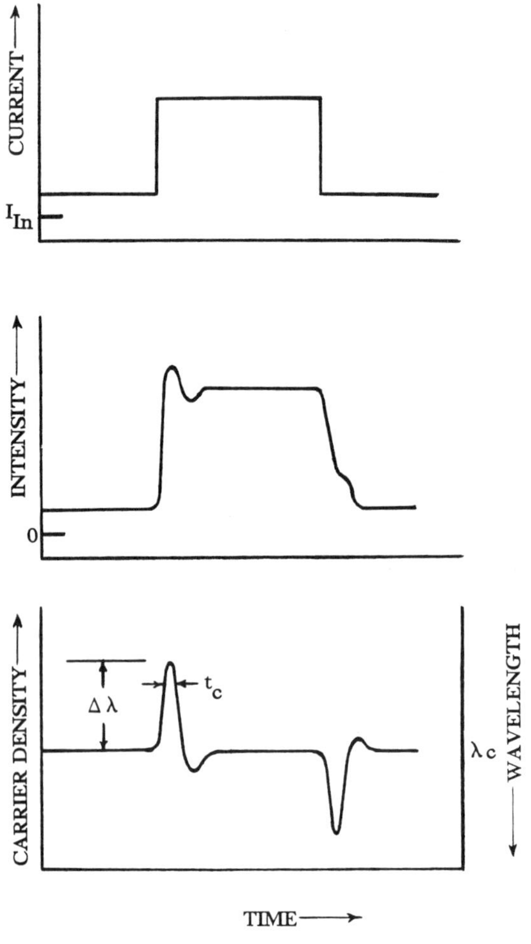

Source: AT&T Bell Labs

Figure 9. Chirp Phenomena in a Laser

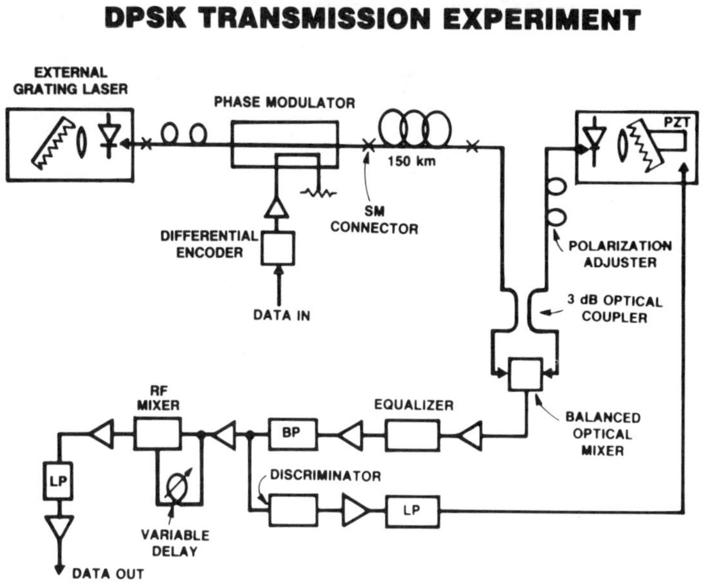

Source: AT&T Bell Labs

Figure 10. Heterodyne Detection in a Coherent Optical-Transmission System

gigabits per second and the distance one can transmit without a repeater, in other words, the speed-distance product. Examination of Figure 11 reveals that the research advances, plotted as a function of time, yield a straight line on log paper; the speed-distance product has doubled every year for the last decade or so. Similar progress should continue for perhaps another decade or two because the end is not yet in sight. The current data-transmission record is 8 gigabits per second for 70 kilometers, but such records are broken every month. At these data rates, the electronic circuitry barely functions—the bits come spilling out of the end of the fiber onto the floor and down the drain because present-day electronics cannot process bits that rapidly. One of our major problems is that the pace of progress in fiber communications has outstripped the pace of progress in the speed of electronic circuitry. The two must function hand-in-hand unless we can devise suitable optical logic or an optical demultiplexer to sift such high-speed data streams into a number of

LIGHTWAVE PROGRESS

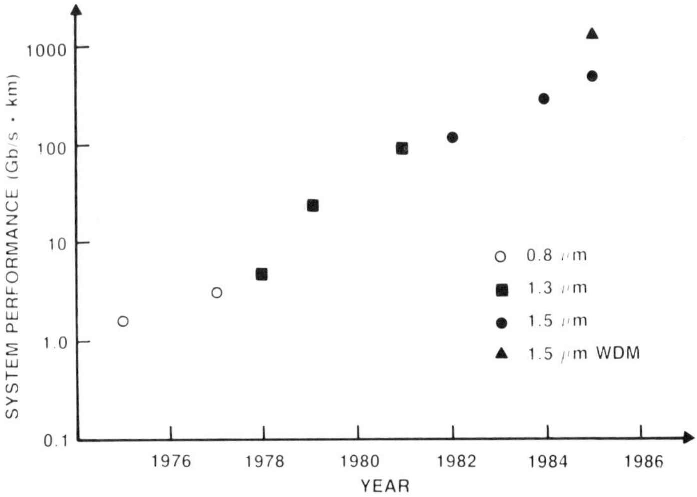

Source: AT&T Bell Labs

Figure 11. Progress in Fiber-System Performance

low-speed parallel streams suitable for electronic manipulation. Some degree of optical processing may ultimately be required.

Considering the possibilities for the future, we have hardly begun to utilize the bandwidth that is available in a fiber system. The attenuation characteristics of silica glass fibers show a pronounced minimum between the wavelengths of 1.2 and 1.6 microns (see Figure 12). Taking into account typical laser bandwidths, there is room for about 5,000 channels in this spectral range, and at several gigabits per channel, the total available capacity is on the order of tens of terrabits per second in a single fiber, enough to provide everybody in the country with 64 kilobits per second on one fiber. Such capacity represents a tremendous resource. On the other hand, the non-linear response of the fiber and interaction between different channels become important considerations when such high-density transmission is attempted. These problems will be fun to solve in the decades to come. I do want you to appreciate the fact that we are still dealing with almost unlimited potential in terms of fiber capacity, especially with the promise of new fibers which are a thousand times

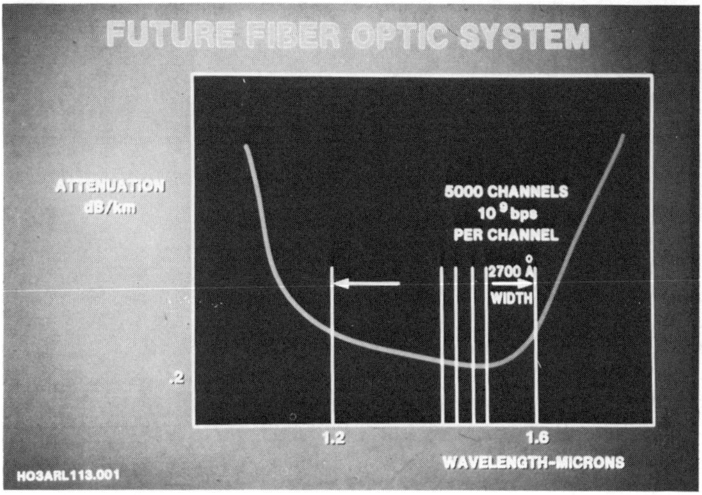

Source: AT&T Bell Labs

Figure 12. Attenuation Characteristics of Future Silica-Glass Fiber-Optic System

more transparent. Putting all of the above together, I just don't know what should be done with all these bits.

Perhaps we will have to find ways to squander bits, just as we have become very successful at squandering transistors in this age of integrated circuits. I think that a plot of all the curves that might relate to the squandering of bits would show that each one climbs exponentially as a function of time. Look at the size of computer programs, for example. The trouble with us as communication workers is that our thoughts are often earthbound. Many of us were trained on 1,200 bits per second and consequently would have no idea of what to do if several gigabits per second were brought into the terminals in our homes and offices.

One of the problems that we have to face in our transition to the new technology is the lack of available optical plumbing compared to what we have in other communications domains. Many necessary optical components do not yet exist. For example, as ridiculous as it may seem, we do not have an optical amplifier, although there are some people who are working on the development of such a device. Whenever I complain about the lack of a particular component, I get arguments from physicists who say they have made one. The patent office tells me that I should not complain so much because many of these items are already patented and are coming along quite well. As mentioned before, the physicists are hard at work: every day they are making new components, putting them out on the shelves, and waiting for someone to come and get them.

APPLICATION OF OPTICAL BUILDING BLOCKS

Development and application of optical techniques have generated considerable excitement in communications circles. In particular, optical switching is a hot topic at present. We have had optical transmission for a decade now, perhaps because it is so much easier to work in the transmission than in the switching of data. Typically, transmission problems are easier and more amenable to analysis. In the past year, however, switching has come of age with two devices currently paramount in our efforts. The first utilizes an electro-optic material called lithium niobate that has been studied for more than a decade. It is useful because its index of refraction can be changed with the application of an external voltage. By photolithographically fabricating coupled wave guides on a substrate of this material and applying a voltage across the coupling region, the propagating mode can be switched from one wave guide to another. This basic element can serve as both an optical switch and a modulator. Fabrication of lithium niobate is now a relatively mature technology, with 4 x 1, 4 x 4, and 16 x 1 cross-point switching elements already in existence. The major disadvantage is the very large interaction length required—a typical device is about a centimeter long. In our communications world we prefer to work with components of considerably smaller dimensions. Several years ago some very competent physicists informed me that practical devices could not be made smaller than this length, basically because the interaction between light and matter is very weak, requiring many thousands of wavelengths of propagation before significant switching can be obtained. Figure 13 is a photograph of a lithium niobate waveguide coupler/switch which has two fibers attached to either end.

The second device is based on a newly discovered phenomenon that has radically altered perceptions of what can be done in the optics world. This discovery relies on the unique optical properties of artificially created layered semiconductor structures known as "superlattices", or multiple quantum-well structures. Using the techniques of molecular-beam epitaxy, alternating layers of different semiconductors can be deposited in a sandwich-like fashion with precisely controlled layer thickness as small as tens or hundreds of angstroms. The transport properties and optical response of holes and electrons in such

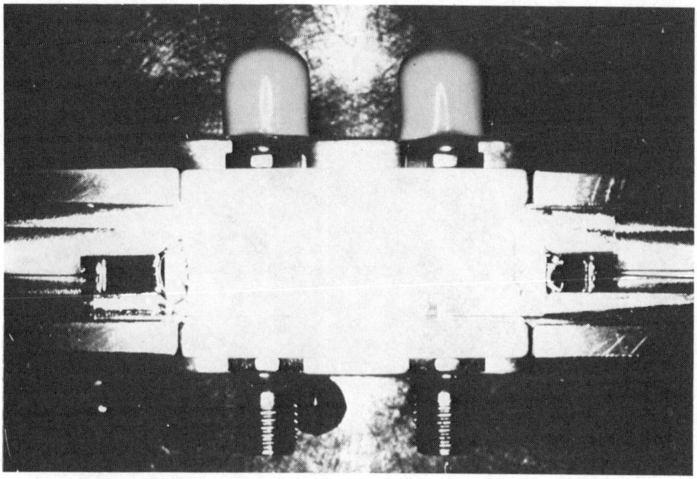

Source: AT&T Bell Labs

Figure 13. A Lithium-Niobate Waveguide Coupler/Switch

artificial semiconductor structures are drastically altered from those of bulk materials due to quantum-mechanical confinement in the direction perpendicular to the plane of the layering. A novel optical modulation structure, as shown in Figure 14, utilizes the unique properties of a superlattice consisting of alternating layers of the compound semiconductors GaAs and AlGaAs. Very strong optical absorption peaks are apparent in the vicinity of the fundamental absorption edge of the structure, and the positions of these peaks can be shifted through a substantial energy range by applying an electric field in the direction perpendicular to the layers. The peaks are due to the formation of "excitons", or bound electron-hole pairs, that are created when an incident photon excites an electron into the conduction band of the crystal and leaves behind a hole in the valence band. These two oppositely charged carriers can then be attracted to each other to form an exciton. Exciton absorption peaks are much more pronounced in a superlattice than in a bulk semiconductor due to confinement of the electron and hole in the GaAs layers of the superlattice. The interaction is an order of magnitude stronger than in any previously known material. Many people have become very excited about these new structures that could represent the long-awaited ideal optical switching

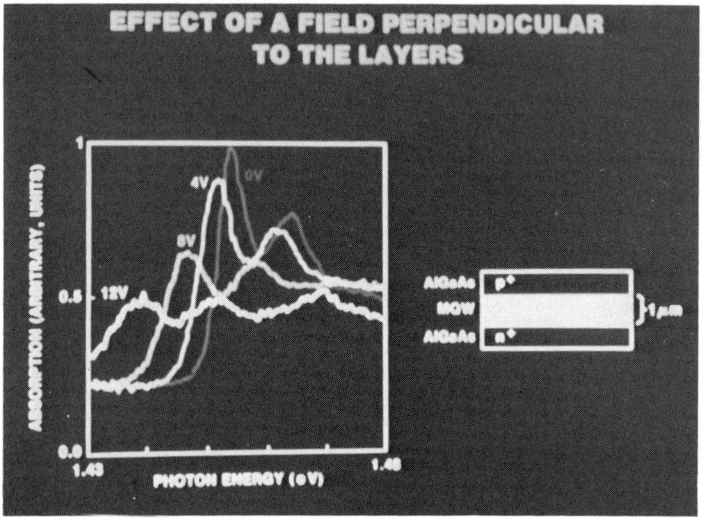

Source: AT&T Bell Labs

Figure 14. Absorption Characteristics of a GaAs/AlGaAs Superlattice Structure

elements that are necessary for development of the full potential of optical communications.

I like the philosophy that George Heilmeier at Texas Instruments once shared with me. He said that we invented the transistor and then spent the last several decades shrinking it and shrinking it until it is almost so small that it won't work. What we need is a new viewpoint whereby we start with a device that depends on being small in order to work. The superlattice modulator shown in Figure 14 is such a device. There are daring plans afoot concerning the building of optical arrays from the same material. Optical computing also has received a major boost from this research, and we have launched some exciting optical computing projects which, I am happy to say, do not simply replicate electronic computational structures using optical logic gates instead of transistors. Instead, new architectures are being developed that take advantage of the inherent benefits of optical hardware.

Another key area of application is in optical sorting. At present we are most interested in fast-sorting networks to perform the sifting of high-speed optical data streams. High-speed packet sorting can be done today with fast-sorting algorithms on a computer using VLSI circuitry. It is not possible, however, to switch high-speed optical communication paths at the corresponding high rates (gigabits per second) using the kind of digital electronic switches that are available today. We have traded bandwidth for bit speed in today's digital switch, and now we are stuck with that trade and will never be able to force a billion bits per channel through similar switches and expect them to function. We need something that does not rely for control and coordination on a centralized intelligence. Optical sorting networks that do not require a central processor-type architecture may become feasible.

Fiber optics local-area networks comprise another rapidly evolving technology, but people always decry the lack of an optical tap for such networks. For years I joined those who complained about the lack of an optical tap, but I now think that I was wrong. What we really ought to do is make networks that do not need taps and trade off the bandwidth properties of fiber against ease of access—perhaps develop a spread-spectrum network, or something similar, without trying again to duplicate the world of coaxial cables, this time with fibers. A fiber tap has fundamental limitations, regardless of implementation, in that power must be removed from the fiber at each tap. Placing a succession of access nodes on the fiber will result in severe attenuation of the available signal.

FUTURE PROJECTIONS

Following this brief survey of the status of optical communication technology, it is appropriate to give some attention to how it might all be applied. It is clear that we in the communication field are committed to bringing this capacity to the world, but I think that no one has a good idea as to what the world might do with it. It is rather like building a highway: first the road is built and then the cars show up. We have a famous example in New Jersey, that I am sure is not unique. Route 287 was built 15 to 20 years ago, at which time people thought it was a boondoggle for the road-construction lobby, since it went from nowhere to nowhere. But today, Route 287 is unbelievably jammed. Traffic appears to generate more traffic, which then attracts new businesses along its routes. I think that fiber capacity will evoke a similar scenario, creating new traffic and new business, although it will also draw "Sunday drivers", junk traffic, and, perhaps, even traffic jams. We will probably also suffer from information pollution that could be worse than ever.

One way to use optical transmission is to change the world from audio to video, a transfer that would provide a factor of a thousand in squandered bits right off the top. We all know that a picture is worth a thousand words, and picture transmission basically requires a thousand times the bandwidth of voice transmission. If we were to change all of our present voice channels into video channels, we would go a long way towards using up our excess capacity. Of course, we have tried this before with the videotelephone. In 1971, we offered such a product for about $100 a month. We made a large number of the devices and waited for people to come and buy them but nobody showed up. I think I had the last videophone in the world on my desk for a while until there was no one left to call, and that was the end of the videophone. Many people through the years have had dreams of video conferencing, but it has never yet been a real commercial success. We still ponder why we come in person to meetings like this; why we get up at 5:00 A.M. and fly hundreds or thousands of miles instead of using a service such as video conferencing. In his book, *Megatrends*, John Naisbitt states "video conferencing is so rational it will never succeed." I think that sociological behavior cannot be changed easily. Social revolutions occur at a much slower rate than technological revolutions. Technology grows exponentially and will continue to do so, and there will be an exponentially increasing communications capacity. Again, however, exactly how this capacity will be used remains to be seen. I have the feeling that much of the expanded usage will be devoted to image transmission. We live in a world so information-rich and so image-rich that by comparison, communications links are very poor. In the office I believe that imagery in the form of document storage and retrieval systems will become increasingly prevalent. In the home perhaps we can use the huge bit rates for entertainment services. One of the services I would like to develop is holographic transmission of television, allowing the viewer to move around a 3-D image. The bits will be available for this service, but since prior possibilities have been so limited by the much slower bit rates in today's narrow data channels, it will be difficult to think freely in terms of our newly relaxed constraints. Certainly, many people did not foresee the full potential for television when it came along, and I think the same thing will happen in the world of optics. With a promising future ahead, the data age is sailing along smoothly . . . It remains to be seen if the information age will follow readily in its wake.

SATELLITE COMMUNICATIONS

John V. Evans

John V. Evans received the B.Sc. and the Ph.D. degrees in Physics from the University of Manchester, England in 1954 and 1957, respectively. Following several years as a Research Fellow with the Jodrell Bank Experimental Station at the University of Manchester, he joined the staff of Lincoln Laboratory at MIT in 1960. In 1966-67 he took a leave of absence to serve as a Visiting Professor at the University of Illinois-Urbana. From 1972-74 he was Associate Leader of the Surveillance-Techniques Group, became Associate Head of the Aerospace Division in 1974, and was appointed Assistant Director of Lincoln Laboratory in 1977. From 1980 to 1983, in addition to his directorial duties, he was a Professor of Meteorology and Director of the Haystack Observatory at MIT. In 1983 he became Director of Research of COMSAT Laboratories, his present position. During his professional career, Dr. Evans has been responsible for programs in advanced electronic research, radar studies of the upper atmosphere and ionosphere, and atmospheric physics. He has served on national and international committees concerned with radio and radar science and solar terrestrial research, and has contributed many articles to various scientific journals on these topics. In 1968, he was made a Fellow of the IEEE for "numerous contributions to radar astronomy", and in 1975 he was awarded the Appleton Prize in Ionospheric Physics of the Royal Society of London.

Bob Lucky is a hard act to follow but I will do my best.

Since the telegraph and the telephone were inventions of the last century, satellite communication is truly an invention of this century. Figure 1 displays a chronological listing of satellite communications developments. Until about 1929 it was not recognized that a satellite could be placed in a 24-hour "geosynchronous" orbit around the earth and would appear to remain stationary above a point on the equator. In 1946, Arthur C. Clarke, the author of *2001 - A Space Odyssey* (the science-fiction novel followed by the movie) first proposed that placing an object in such an orbit would make it possible to obtain widespread communication over the earth's surface. Since Clarke did not foresee the invention of the transistor, his satellite model was, in fact, a manned space station in which the occupants were required to maintain and service the equipment by replacing the vacuum tubes in the receivers and transmitters carried on board. The use of space for any purpose was greatly stimulated, of course, by the launching of Sputnik in 1957, thereby allowing investigation of this new medium for information transfer.

The very first attempts at satellite communication were literally experiments performed by bouncing signals off passive reflectors placed in earth orbit. Signals were reflected either by large metalized balloons or from little copper dipoles, as in the WESTFORD experiment. If developments had been limited to these exercises, it is highly unlikely that satellite communication would have grown to its present state. The decisive breakthrough occurred in 1962 when the Bell Telephone Laboratories launched TELSTAR I. This was the first active repeater satellite placed in orbit that received signals from the ground, changed their frequency and retransmitted them back to earth. That development came much sooner than most of us had anticipated, and precipitated some concern in the United States that the large AT&T organization would move into international telecommunications to a degree that was not considered to be desirable at the time. Consequently, the Communication Satellite Act was passed by the Congress in 1962 and signed into law by President Kennedy at the end of that year. The measure was intended by the United States to take this technology and give it to the world for peaceful purposes to bring about global telecommunications. The Act established the COMSAT organization as the U.S. agent for that process.

Two Air Force satellites were the first to be placed in geosynchronous orbit in the following year. While SYNCOM I failed it was soon followed by SYNCOM II which achieved a nearly perfect orbit. During the next few years a debate raged in the technical community of this country over the choice between high-altitude *versus* low-altitude satellites for telecommunications purposes. The TELSTAR satellite (and later the RCA RELAY satellite) had been launched into elliptical orbits and therefore could only be seen from two fairly close points on the ground. Moreover, they had to be tracked by the pair

1929	PROPERTIES OF 24-HOUR SATELLITE ORBIT DESCRIBED BY HERMAN NOORDUNG
1946	COMMUNICATIONS RELAYS USING GEOSTATIONARY SATELLITES PROPOSED BY ARTHUR C. CLARKE
1957	SPUTNIK LAUNCHED BY USSR
1960-64	COMMUNICATIONS EXPERIMENTS PERFORMED USING PASSIVE REFLECTORS IN ORBIT
1962	BELL LABORATORIES TELSTAR I LAUNCHED
1962	COMMUNICATIONS SATELLITE ACT SIGNED INTO LAW BY PRESIDENT KENNEDY
1963	COMSAT INCORPORATED IN WASHINGTON, D.C.
1963	SYNCOM I LAUNCHED IN FEBRUARY, BUT EXPLODED SYNCOM II LAUNCHED IN JULY --- A SUCCESS
1964	INTELSAT FORMED WITH 11 PARTICIPANTS
1965	INTELSAT I (EARLY BIRD) LAUNCHED IN APRIL
1967	COMSAT LABORATORIES ORGANIZED IN WASHINGTON, D.C.
1969	COMSAT LABORATORIES MOVED TO NEW QUARTERS IN CLARKSBURG, MARYLAND

Figure 1. Early History of Satellite Communications

of antennas for the duration that the link was to be established, and they remained above the horizon for only a short time. A high-altitude synchronous satellite, on the other hand, does not pose a tracking problem; it is, moreover, visible from about one-third of the earth's surface so that only three of them are necessary for a global communication system. A low-altitude orbiting system, by contrast, would require a fleet of satellites that would be tracked across the sky by pairs of antennas, as each individual one rose and set. Although this system is clearly more complex, it has an advantage over the geosynchronous satellites in that the signal delay times are much shorter than the quarter-second period required for a signal to travel to the high-altitude satellite and then return to the receiving point. Many people, including John R. Pierce of Bell Labs, were convinced that the overall half-second response time would cause satellites in geosynchronous orbit to be unacceptable for telephone conversations.

The successful establishment of a satellite in geosynchronous orbit, however, inspired COMSAT to opt for this approach. Working with the State Department COMSAT helped found an international communications satellite organization, known as INTELSAT, with the express purpose of creating a global satellite system. Initially, there were eleven participants in partnership with the United States, consisting principally of friends of this country. The first INTELSAT satellite, the so-called "Early Bird", was placed into orbit over the Atlantic Ocean in 1965. It was the first satellite to bring television from Europe to the United States. COMSAT Laboratories, the group that I represent, was organized in 1967.

I will concentrate on the growth of INTELSAT, as charted in Figure 2, because it was the first communications satellite system to be organized. Several present-day domestic systems have experienced similar growth. People who work with satellites count their sales in terms of "half-circuits", since the forward path for a telephone circuit, through a telephone system, is separate from the return path once the signal leaves the two-wire copper line at the local switchboard. The load on the satellite is thus the sum of all the half circuits and grew very rapidly initially. Extrapolating the solid curve in Figure 2 indicates that the INTELSAT system is carrying about 100,000 half-circuits at present; this comprises about two-thirds of the world's overseas traffic. The dashed curve shows the total growth across the Atlantic, and the dotted curve shows the growth in the Indian Ocean region. The least growth has occurred in the Pacific region, as shown by the fourth curve in the chart. The number of paths in the system, as displayed in the bar graph of Figure 3, has also grown enormously, from essentially one to something on the order of 1200 or more at the present time. The beauty of a satellite is that it can be seen from many places on the ground, thereby allowing extra paths to be established rapidly just by putting up one more antenna. Consequently, a satellite is

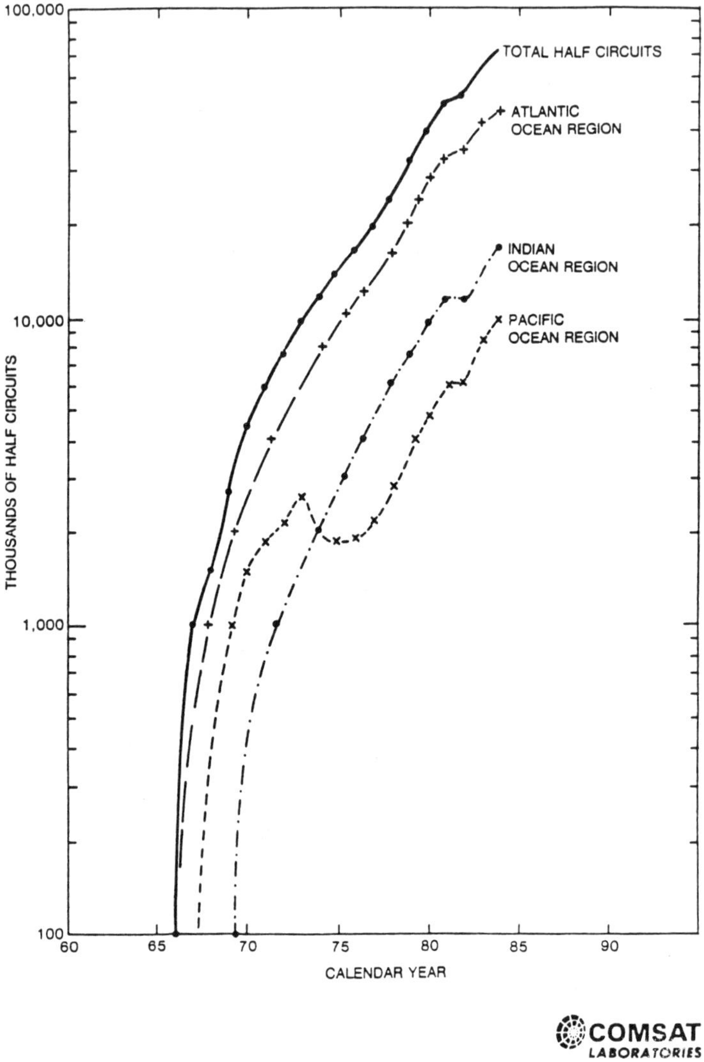

Figure 2. Growth of Traffic in Half-Circuits

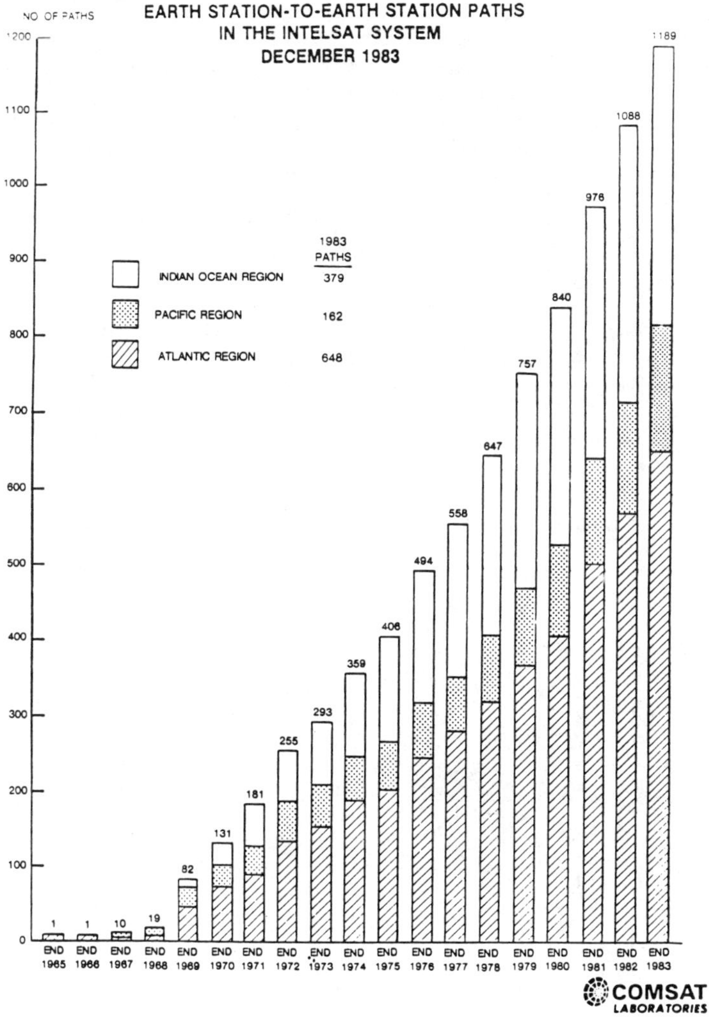

Figure 3. Number of Paths in the INTELSAT System

almost like a switching station in the sky that can "connect" services to all antennas on the ground that are already within its view. An examination of a world-wide installation map would show that INTELSAT serves approximately 150 nations and territories and is operated by about 108 member countries. It has been one of the most successful international endeavors in that it has managed to run its affairs without getting embroiled in membership debates. Of the three oceanic services, a typical installation consists of four or five satellites, of which one is kept as a spare. In actual practice there is a tendency to use one satellite for the major trunk traffic (*e.g.* the United States to Europe traffic) and to use the other satellites to carry the thinner routes such as communications between the United States and South America.

This total amount of traffic growth has required enormous improvements in the capabilities of the satellites. The microwave repeaters carried by the satellites (Figure 4) consist of an antenna that picks up the signals from the ground, low-noise amplifiers that increase the signal strength and convert the signals to an intermediate-frequency band, and a transmitter that returns the amplified signals to earth. The bulk of the traffic is carried in C band, with

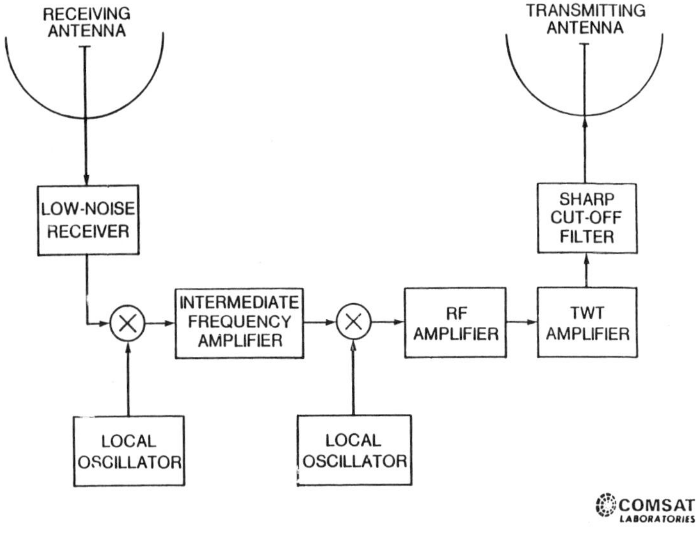

Figure 4. Typical Transponder Layout

the uplink at six gigahertz and the downlink at four gigahertz. The final amplifier is typically a traveling-wave tube (TWT) that provides from 10 to 20 watts of output power. Traveling-wave tubes are nonlinear devices that always require filters in their output circuits to insure that the signals from the satellite are confined to the allocated band and do not interfere with other radio signals assigned to adjacent frequency bands.

The overall growth of communication channels is well illustrated by the fact that during World War II there were no voice channels across the Atlantic. Churchill and Roosevelt talked to each other on HF radio, and each morning Hitler generally had a report on what was said. Scramblers were not introduced until late in the War. By the time the INTELSAT I (Early Bird) was launched in 1965, there were still only about 300 cable circuits across the Atlantic and this one satellite doubled that number. In order to carry the many thousands of half circuits now in existence it has been necessary to design the satellite systems to make maximum use of the limited frequency bands assigned.

One of the ways to accommodate the traffic growth has been to reuse the frequency bands several times over. The INTELSAT I (Early Bird) Satellite, for example, had a simple whip antenna that radiated in all directions. Since the power and complexity of these early satellites was kept to a minimum, very large earth stations were needed in order to allow satisfactory communications. One of the U.S. facilities (at ETAM in West Virginia) that operates with the Atlantic satellites has a 105-foot antenna, a 95-foot antenna, and a 45-foot antenna.

INTELSAT I was a simple, solar-powered, spinning satellite that was equipped with a single transponder. Subsequent INTELSAT satellites were designed to include more directive antennas to beam the signals toward the earth. Since the satellites were stabilized by spinning the entire structure, the necessary earth-oriented antenna was mounted on a despun platform. The required despinning was accomplished by mechanically coupling the platform to the satellite through a bearing and turning it in the reverse direction with an electric motor. These designs allowed reuse of the frequency band by means of multiple antenna beams. Two provide hemispheric coverage, as sketched in Figure 5. One of these beams would look at, say, North and South America, and on the same frequency band another beam would look at Europe and Africa. Using the opposite sense of circuit polarization, two new so-called "zone beams" would be formed to look at North America and Europe, respectively. Thus the same frequency band was used four times by one satellite. By going to Ku band it was possible to add the spot beams shown as cross-hatched areas in Figure 5.

The need for frequency reuse spurred developments in antenna technology. The essential problem is to provide the correct shape for the beams and

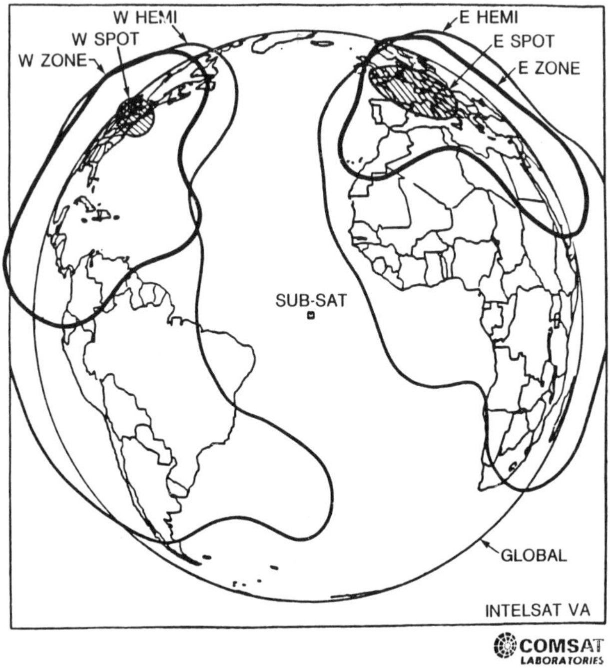

Figure 5. "Hemi" Beams in the Eastern and Western Hemispheres

provide the requisite isolation between beams. Arrays of feed horns are employed to illuminate long-focal-length parabolic reflectors. On INTELSAT V, an array of about 105 feed horns is used. This is driven by a microwave-distribution network, that carries the energy to the feed horns and excites each one with the correct amplitude and phase. Since the same satellite design is used for each of the three ocean regions, it is necessary to install three distribution networks on each satellite and arrange to switch the feed-horn array to the appropriate one, depending upon the particular region that is to be covered. In the INTELSAT VI satellites being built at the present time by Hughes, four distribution networks can be switched alternately to an array of 150 feed horns contained in a structure that is the size of a king-size bed. I think that this technology has reached an absolute limit, however, and will be replaced by arrays that are computer controlled. This topic will be discussed later in this presentation.

A procedure known as "channelization" provides another method of gaining capacity, and its use is illustrated in the thumbnail sketch of the INTELSAT series of satellites shown in Figure 6. The first satellites employed only one or two transponders and attempted to gain capacity by operating them over wider and wider bands (INTELSATs I, II, and III). The problem with that approach was that the traveling-wave tubes are so nonlinear that they will mix carriers coming from different stations and produce beat products that limit the number of carriers that can be placed in the satellite. It became apparent that a much better technique was to partition the spectral window into narrow frequency channels and operate a separate transponder on each channel. As detailed in Figure 6, the number of transponders increased in succeeding models, with 12 in INTELSAT IV, and 20 in INTELSAT IV-A. INTELSAT V, now being launched, contains 27 transponders, and INTELSAT VI, now under construction, will have 48. The corresponding number of circuits that can be handled by each model is shown in the fifth row, rising from an initial 240, to 33,000 on an INTELSAT VI.

In order to accomplish these results it was necessary to improve the microwave cavity filters employed to define the channels. Investigators discovered that simply cascading one microwave cavity after another resulted in a Chebyshev response characteristic that limited how sharply it was possible to define the edge of the band. The necessary step to remove this difficulty involved exciting each of the cavities with two orthogonal modes and coupling the first cavity to the last cavity, independently of the intermediate cavities, by separately adjusting the coupling of the orthogonal polarizations through the filter. This process, invented at COMSAT Laboratories, made it possible to produce the extremely sharp microwave filters that are used in all present-day satellites.

This same technology was also necessary to make it possible to connect all the transponders to the same antenna. On the INTELSAT IV-A, for example, two transmit antennas, were employed with the output of the traveling-wave tubes arranged so the odd-numbered channels were fed to one antenna and the even-numbered channels were fed to the other.

Subsequently, when more advanced filters had been designed this solution to the coupling problem was abandoned. Had we not been able to make these improvements, the INTELSAT V satellites would have been so heavy that a full *Titan* launch would have been required instead of the *Atlas-Centaur* rocket that was actually used. The difference in launching price is about 30 million dollars. Since there are 15 of those satellites in orbit, these savings far exceed the cost of the research that has been done on communications satellites.

	INTELSAT I	INTELSAT II	INTELSAT III	INTELSAT IV	INTELSAT IVA	INTELSAT V	INTELSAT VI
TRANSPONDERS	2	1	2	12	20	27	48
BW/TRANSPONDER (MHz)	25	130	225	36	32.36	40. 80. 240	40. 80. 160
COVERAGE	N. HEMISPHERE	GLOBAL	GLOBAL	GLOBAL & SPOT BEAMS	GLOBAL & SPOT BEAMS	GLOBAL. REGIONAL & SPOT BEAMS	GLOBAL. REGIONAL & SPOT BEAMS
e.i.r.p. (dBW)	11.5	15.5	23	22.5 (GLOBAL) 33.7 (SPOT BEAM)	22 (GLOBAL) 29 (SPOT BEAM)	22-29 (4 GHz) 44 (11 GHz)	23-31 (4 GHz) 41-44.4 (11 GHz)
NO. OF TEL.CIRCUITS	240 (NO MULT. ACCESS)	240	1200	4000(AVG.)	6000 (AVG.)	12000	30000
LIFETIME (YR)	1.5	3	5	7	7	7+	10
SATELLITE COST/CIRCUIT YR ($K)	30	10	2	1	1	0.9	0.5

COMSAT LABORATORIES

Figure 6. Channelization Chart of the INTELSAT Satellite Series

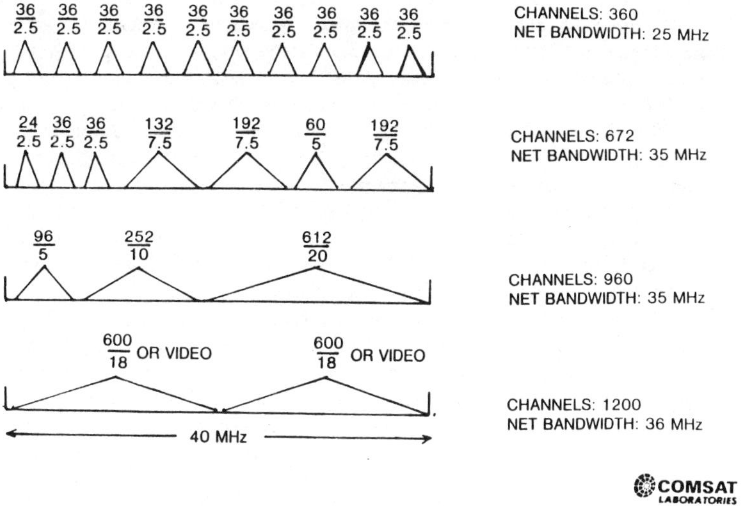

Figure 7. Frequency-Division-Multiple-Access (FDMA) Scheme

Another new development is concerned with changes in the way that satellites are being used. At present, most of the satellites in the INTELSAT program are part of a frequency-division multiple-access (FDMA) system in which a number of countries are allowed to put up a carrier into a given transponder. Figure 7 shows a transponder with a bandwidth of about 40 Mhz. Each of the 10 triangles in the top row of the figure represents a country that is putting up a carrier bearing 36 voice channels. Since the nonlinear characteristics of the traveling-wave tubes mandate the use of frequency modulation, each carrier is spread over about 2.5 Mhz. When we allow for guard bands between the carriers, we end up using about 25 Mhz of the 40 Mhz bandwidth, for a total of about 360 channels. Several other combinations are shown in Figure 7, with the bottom row being of particular interest from the standpoint of more efficient use of the system. In that situation there are just two carriers, each carrying 600 voice channels that have been multiplexed together, providing a total of 1200 voice channels through that transponder. Obviously this is more efficient use of the transponder but only a few countries can generate that much traffic. Efficient use of all of the transponders can be achieved by going to time-division-multiple-access (TDMA) operation. In this mode each station is allowed use of the entire transponder for a very short space of time. During a pre-set time interval, a pulse is sent up to the satellite that contains all the bits of all the voice channels to be transmitted by a specific station. Full utilization of the complete bandwidth of the transponder is accomplished by an appropriately chosen bit rate and maximum throughput is achieved by adjusting the length of time assigned to each station.

The TDMA approach is becoming the standard for future satellite communications. INTELSAT, for example, is installing equipment that will operate in this mode at 120 megabits per second. An actual traffic terminal in a TDMA system will contain a digital speech interpolation (DSI) unit that detects and seizes empty channels, thereby allowing an approximate two-fold increase in system capacity. A TDMA system also requires a master control station that tells all the traffic terminals when to transmit.

The current rapid introduction of speech encoding, based on bit-sampling techniques, will have a strong influence on communication practices in the next few years. Since the generation and transmission of bits will certainly be discussed later today, I will touch only briefly on the subject. The simplified encoders of Figure 8 indicate a typical approach to the process. Present digital circuits carry voice signals at 64 kilobits per second. Essentially, the analog signal from the microphone is sampled eight thousand times a second and assigned an eight-bit word that is the measure of the output. Those eight bits, sent at eight thousand times a second, give us the 64-kilobit digital bit stream. Taking the same analog signal and subtracting from it an *estimate* of its value based on prior samples, it is possible to reduce the size of the digital word down to four bits since only a difference signal is being sampled. If the sampling rate is maintained at eight thousand times per second, the output is a 32-kilobit stream. It is necessary, however, to use this output to drive a predictor circuit that will provide an estimate of the expected signal. This predictor is basically an electronic model of the voice cavity that the speaker is exercising at that particular moment. International standards for 32-kilobit voice channels of this type are now being adopted and are being rapidly introduced. Similar advances have been made in the actual modulation methods that are being used on the satellites. A typical INTELSAT VI satellite transponder has a bandwidth of about 70 megahertz. By going to, say, eight-phase modulation, it is possible to obtain approximately two bits per hertz of bandwidth, resulting in about a 140-megabit signal through the transponder. This capability is desirable because the first transatlantic fiber-optic cable (TAT-8) will also transmit at that same frequency, thereby permitting the switched-digital network, now being built in Europe, to be brought to the United States *via* satellite as well as by the cable.

Combination of the several techniques outlined in Figure 9 makes it possible for satellites to match the enormous signal capacity of fiber-optic cables.

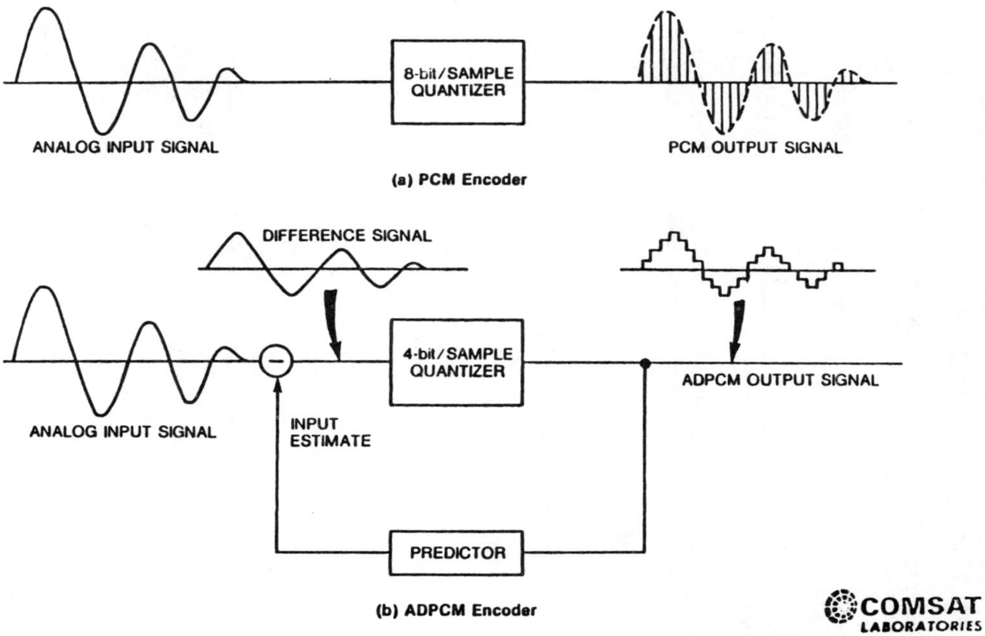

Figure 8. Simplified PCM and ADPCM Encoders

- **32-kbit/s Voice Encoding Is on the Verge of Implementation, and 16-kbit/s Voice Coding Is Being Developed**
- **Digital Speech Interpolation (DSI) Offers More Than a Two-fold Increase in Capacity**
- **Coded 8-Phase Modulation Should Permit 140 Mbit/s Rate in a Single 80-MHz Transponder**
- **For 32-kbit/s Digital Voice, These Lead to 10,000 Equivalent Voice Channels/Transponder**
- **With Six-Fold Frequency Reuse at C-Band, the Total Number of Voice Channels Is 450,000 Per Satellite in 600 MHz of C-Band Alone**
- **For 16-kbit/s Digital Voice, These All Double**

Figure 9. Digital Voice Capacities of Satellites

The previously mentioned eight-phase modulation technique can pass about 140 megabits through one transponder. When this capability is joined with DSI and 32-bit voice coding, a capacity for handling 10,000 talkers in each transponder can be achieved. Consequently, a single INTELSAT VI will be able to carry about half a million channels. Since there are only about 50,000 channels in operation now between the United States and Europe, it is not clear at the moment how this potential capacity will be put to use. When the need arises, however, the channels will be ready.

Turning now to the future, it is very clear that the inherent time delay in communicating by satellite will cause large amounts of traffic to be transmitted by fiber-optic cables. Consequently, we can expect that the dominant need for satellites will cease to be for trunking capacity that simply replicates the cable connections from the shore of one country to the shore of another. The specific advantage of a satellite is that it can connect a single point to many other points on the globe simultaneously. The fact that signals may be carried everywhere and broadcast over wide areas has led to many applications, other than providing voice circuits, that benefit from these flexible capabilities. For example, in the United States the greatest domestic use of satellites is for distributing TV to cable networks, and in Europe and Japan there will soon be direct broadcast of TV from satellites to people's homes.

At the present time the INTELSAT satellites are not designed to operate with very small stations because of insufficient transmitter power, but the difficulty can be bypassed by taking advantage of a "double-hop" technique.

As an example, consider a situation where a satellite has insufficient gain to allow a very small low-cost earth station to communicate with another small earth station. It is possible for that station to transmit to a large central station which, in turn, retransmits the information through the satellite down to the other earth station. This approach is being used for voice communications in some remote parts of the world, such as in Alaska, and in the Pacific where there are many small islands without enough traffic to justify the installation of a very large antenna. The same strategy can be exploited in Africa and South America where it is difficult to lay cable. In such areas the double-hop scheme will bring modern communication facilities to villages where all those bits mentioned by Bob Lucky are not accessible to the people at the present time.

An unanticipated application for the double-hop approach is evolving in the United States for data transmission and electronic mail. Many companies are considering the possibility of by-passing the entire telephone system and establishing their own private networks for these purposes. Federal Express, for example, would like to set up a facsimile system involving every major office building in the country. A user would place a document in a xerox-like machine and dial an address. The contents of the document would be encoded and sent, *via* a small roof-top antenna, to a satellite that would transmit the information to one of, perhaps, four regional hubstations. A large computer at the hubstation would store the data until it senses that a proper transmission channel is free. Following perhaps a one-second delay, the message would be routed to its ultimate destination. Clearly, a voice-communication channel would find such a delay to be intolerable. The beauty of double-hop is that delays are of no consequence in the bulk transfer of data. Once a channel is open, data may be moved at any appropriate speed.

For telephone traffic, where double-hop is not appropriate, it is necessary to develop satellites that are more sophisticated than current models. In particular, there is a requirement for advanced satellite-mounted antennas capable of forming narrow pencil beams that can put the energy where it is needed rather than spreading it all over the surface of the earth. One of the presently active areas of satellite-antenna research is devoted to the design of such antennas. Given these narrow beams, it becomes feasible to replace the large earth-based antennas, now used to carry the bulk of the traffic, with small antennas that may be of the order of six to ten meters in diameter. One of the problems with the narrow beams, however, is that in any one beam only a small number of stations would be able to hear each other, thereby causing the loss of connectivity that was formerly available with simple satellites and wide beams. Since satisfactory service depends upon maintaining the connectivity, the satellite must be fitted with some kind of switch that will allow the traffic to be transferred between appropriate beams. Hence the satellite will become a more complicated device, not only because of its physical size or the design of the antenna, but also by reason of requiring this on-board switching.

The generic block diagram of Figure 10 illustrates an advanced satellite with several pencil beams and a number of receiver amplifiers. One way to perform the desired switching function is to demodulate all of the upcoming carriers, recover the bits, route the traffic through an on-board digital switch, remodulate, and retransmit. This process has a number of attractive features. In present satellites it is necessary to ensure that each upcoming carrier has about the right intensity in order to avoid having any one carrier seize too much of the transmitted power in the downlink. The problem is basically one of power sharing that makes it necessary to orchestrate the network of users carefully so that all the carriers arrive at the satellite with the correct power level. In the satellite of Figure 10, however, the power-sharing problem has been eliminated. The design of the uplink is essentially divorced from the design of the downlink because of the demodulation feature. This type of satellite can work with a wide range of antenna sizes and carrier powers.

A number of technologies, as listed in Figure 11, are needed for this advanced satellite. Work is in progress on all of these items, including antennas that provide multiple pencil beams formed with arrays containing lightweight solid-state amplifiers, on-board demodulators, forward-error-correction (FEC) routing, and components of a compatible earth segment. Rather than going through this list in great detail, I will concentrate briefly on the antenna problem.

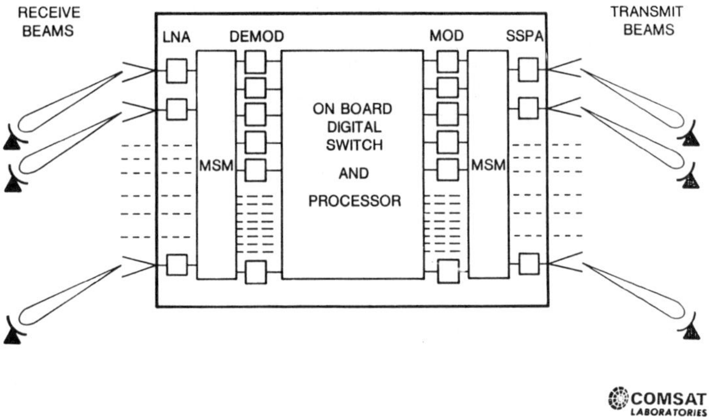

Figure 10. Generic Satellite Block Diagram

ANTENNAS PROVIDING MULTIPLE PENCIL BEAMS

- LIGHTWEIGHT ANTENNA ARRAYS
- LIGHTWEIGHT SOLID-STATE AMPLIFIERS AND PHASE SHIFTERS (MMIC)
- RELIABLE, PROGRAMMABLE CONTROL SYSTEMS (LSIC)

ON-BOARD DEMODULATION, FEC, ROUTING AND REMODULATION

- LIGHTWEIGHT SOLID-STATE DEMODULATORS (MMIC)
- RELIABLE DIGITAL VITERBI DECODERS (LSIC)
- RELIABLE DIGITAL SWITCH (LSIC)
- RATECONVERTERS AND REMODULATION

COMPATIBLE EARTH SEGMENT

- LOW SIDELOBE ANTENNAS
- NETWORK CONTROL
- LOW-COST EARTH TERMINALS

COMSAT LABORATORIES

Figure 11. Needed Technologies for Satellite Development

Figure 12 is a sketch of an array design, performed at COMSAT Laboratories, that contemplated the use of 576 feed horns. The array was imaged by two reflectors in order to expand its size. Depending on how the system was to be used, this array would form either a large number of pencil beams, or a number of scanning-spot beams. The system was arranged so that it could be designed for either approach. The feed array contained the amplification units internally instead of having a beam-forming network behind it driven by many heavy, power-consuming traveling-wave tubes. In this case the feed horns were arranged to transmit opposite polarizations, with a polarizing unit mounted behind each single-array element. The amplifiers and phase shifters that are needed to generate power control were embedded in the two wave guides that feed into the polarizing unit.

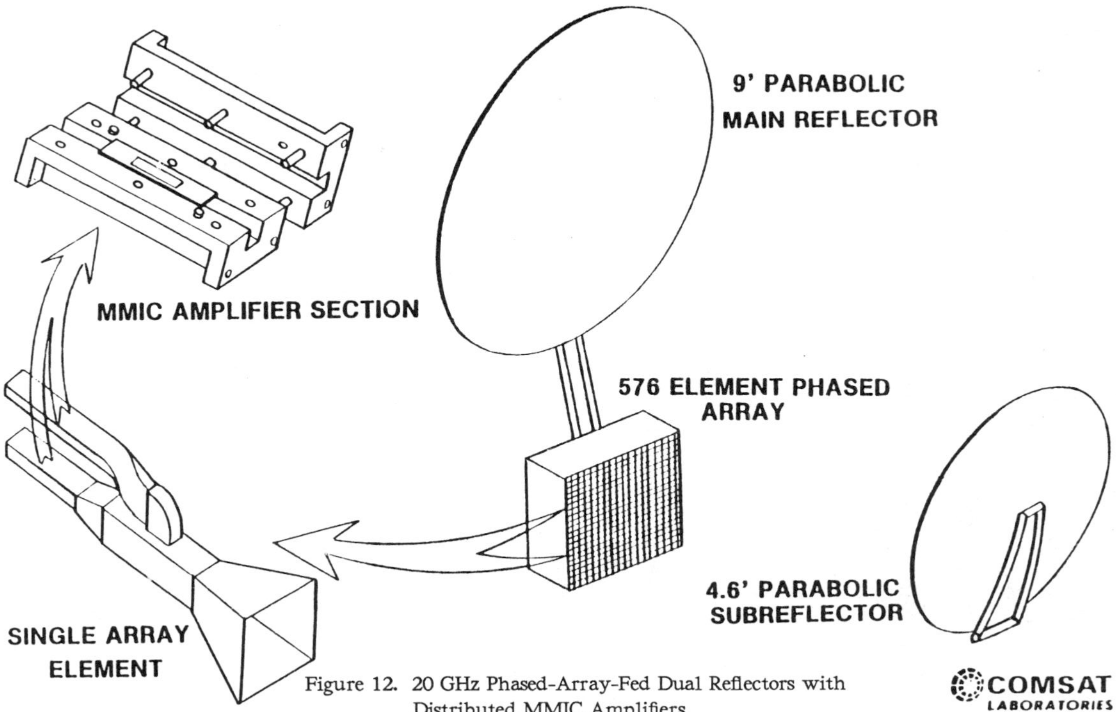

Figure 12. 20 GHz Phased-Array-Fed Dual Reflectors with Distributed MMIC Amplifiers

Actual models of the feed horns and the polarizer have been constructed based on the preliminary designs. An amplifier carrier, inserted in the middle of the wave guide, coupled all of the energy coming down the waveguide onto a chip for amplification and retransmittal. The isolation between input and output circuits was found to be satisfactory. At present, a typical microwave amplifier at these frequencies consists of a strip-line circuit composed of transistors bonded to an alumina substrate mounted on a ground plane. The much smaller chip circuit that is needed for amplification in this system is provided by a monolithic microwave integrated circuit (MMIC) in which the transistors and all other components are built into the device at the same time via printed circuit fabrication techniques. The resulting chip would fit easily on a dime. This kind of technology is progressing well, with good prospects for general use by the mid-1990's.

At present, the U.S. NASA Advanced Communications Technology Satellite (ACTS) program is conducting an experiment that is concerned with the development of these technologies. A number of simultaneous approaches are being taken, as outlined in Figure 13. Since C band is becoming full, communication satellites, especially the domestic ones, are using Ku band at 12 and 14 gigahertz. Additional higher frequencies of 20 and 30 gigahertz have been assigned for commercial use. These bands have serious fading problems caused by rain, but the ACTS satellite will be built to operate at those frequencies in order to explore the extent of this problem. On-board control

DEMONSTRATE ADVANCED MULTIBEAM SATELLITE TECHNOLOGY VIA AN EXPERIMENTAL FLIGHT TEST PROGRAM

- 30/20 GHz, BOTH FIXED AND SCANNING SPOT BEAMS
- SATELLITE-SWITCHED TDMA (SS/TDMA) WITH ON-BOARD RF SWITCH AT HIGH BURST RATES: 220 Msps (HBR)
- ON-BOARD BASEBAND PROCESSING AND SWITCHING AT LOW TDMA BURST RATES: 27.5 AND 110 Msps (LBR)
- DEMAND ASSIGNMENT OF 64-kbps CHANNELS IN LBR SUBSYSTEM
- AUTOMATIC UPLINK AND DOWNLINK POWER CONTROL TO COMBAT RAIN FADING (HBR)
- AUTOMATIC FORWARD ERROR CORRECTION CODING TO COMBAT RAIN FADING (LBR)
- ADVANCED NETWORK AND PAYLOAD CONTROL

Figure 13. ACTS Program Objectives

functions will cause the transmitter power to be switched from 8 watts to 40 watts when the satellite is looking in the direction of heavy rain. Signals from the ground will also use forward-error-correction to overcome the effects of the rain. Fixed-spot and scanning-spot beams will be included in the design, with time-division-multiple-access transmission through the fixed-spot beams at the rate of 220 megabits per second, and at the rate of either 110 or 27.5 megabits per second in the scanning-spot beam system. One of the more interesting aspects of arranging to control this type of satellite is illustrated by the projected use of a 64-kilobit-channel network demand-assignment control that will allow anyone who picks up a telephone to be assigned a circuit at any point in the entire system.

Inside the satellite, data can be moved through two paths. Since there is no need to recover the bits in the fixed-spot-beam system, it is necessary only to arrange for the signals to be sent to the right path by switching the IF into the correct down beam. In the scanning-spot-beam system, however, the procedure is not so simple. The destination path may not exist at the particular instant of time when a signal is coming up because the beam is not pointing in the correct direction. It is therefore necessary to demodulate, store the bits, and retransmit them at a suitable time using a base-band processor of the type illustrated in Figure 10. The NASA Advanced Communication Technology Satellite, shown in Figure 14, is scheduled for launching in 1989 and will provide a first test of these concepts.

In summary, it is clear that satellites are going to change. As much of their current role of "bent pipes in the sky" is taken over by fiber-optic cables they will increasingly be applied in the areas for which they are eminently qualified, such as: improving the current poor communications facilities in the third world, establishing private communications and data networks for large corporations, and for broadcasting of television and radio signals. Unfortunately, direct-broadcast television has experienced a very rocky start in this country. The U.S. financial community does not believe that there is money to be made in that business. Consequently, most of the players who had ambitious plans for the field have withdrawn their entries. The technology will surely succeed, however, and it is just a question of time until the market will follow. Finally, we will see increasing use of sophisticated satellites that will carry telephone traffic to regions where it is uneconomical for an AT&T or an MCI to install fiber-optic cables.

Update:

Following this presentation at the April 10, 1985 Symposium, Federal Express appears to have abandoned its plan for a satellite-connected, document-distributed network, and the launch of the NASA ACTS satellite has been postponed to late 1990.

Figure 14. Model of Advanced Communication Technology Satellite (ACTS)

DEVELOPMENTS IN VIDEO BROADCASTING DIRECT FROM SATELLITE TO HOME

Irwin M. Jacobs

Irwin M. Jacobs received the B.E.E. degree from Cornell University in 1956 and the M.S. and ScD. degrees in Electrical Engineering from MIT in 1957 and 1959, respectively. He was appointed Assistant Professor of Electrical Engineering at MIT in 1959 and was promoted to Associate Professor in 1966. He served in that capacity until 1969 when he became Professor of Information and Computer Science at the University of California at San Diego. In 1968, he cofounded the LINKABIT Corporation. In 1972 he resigned his professorship in order to devote full time to LINKABIT. After the 1980 merger of LINKABIT with M/A-COM Inc., Dr. Jacobs became a member of the M/A-COM Board of Directors, and in February 1983 assumed the position of Executive Vice President. Dr. Jacobs has specialized in the fields of information theory, coding theory and applications to digital communication, satellite multiple access, and microprogrammed communications systems and packet switching. He has received a number of awards for his achievements in these areas, and has consulted with several research and industrial laboratories on related problems. He is a Fellow of the IEEE and a member of the National Academy of Engineering.

Editor's Note: In April 1985, Dr. Jacobs left M/A-COM and in July 1985 he founded QUALCOMM, Inc. in San Diego, California. His present position is Chairman, CEO, and President of that organization.

It is a pleasure to be here to help celebrate the hundredth anniversary of the School of Electrical Engineering at Cornell and that of MIT, since I attended both institutions. It is also good to see many of my old friends and colleagues again. I believe that much of the enthusiasm and excitement that exists today in the world of industrial communications comes from the inspiration that EE students have received from schools such as Cornell and MIT. I received a fine undergraduate education–most of my courses turned out to be very, very useful–I even took a course from Les Eastman on how to build vacuum tubes that I have never quite forgotten for a variety of reasons. Although I was kept pretty busy at Cornell, I did manage to meet my wife there, a fellow Cornellian who is here with me today.

My year of graduation may be cause for a little confusion. I actually graduated from Electrical Engineering in '56 although some records may show that it happened in '54. The problem is due to the fact–I think I am relatively unique in Cornell history–that I started out in the School of Hotel Administration and spent a year and a half there. I had a roommate who insisted that it was not possible to make the kind of grades in engineering that you could get in Hotel Administration. You can listen to that kind of talk for only so long. Luckily, I asked for a transfer, the EE School accepted me, and that is how I came into this particular career.

In my talk today on direct satellite broadcasting, I will discuss the latest developments in the medium and its future promise.

Perhaps it seems a little odd to be talking about television and television transmission in a colloquium on the future of communications. No doubt, we all think of television as being very old hat; it has been with us for such a long time that we probably do not expect to see any major changes. In actual fact, however, new ideas, such as direct broadcasting and signal scrambling, have appeared on the scene only recently. Personally, I have found these particular two developments to be very interesting and exciting–some of my activities for the last few years have been devoted to these areas. As much as anything else that I have encountered in my professional life, this experience has caused me to appreciate the meaning of the term "the real world".

Video technology, itself, is undergoing a number of modifications. We are all aware of the recent changes in TV viewing equipment, ranging from the wide-screen units to the very tiny sets, that demonstrate the potential of the newly-developed picture tubes. A very interesting new technique, known as "digital TV", performs digital rather than analog processing of the video signal within the television set. Of course our familiar National Television Standards Committee (NTSC) transmission will still be used–but once the signal enters the TV set and receives a bit of analog processing, it experiences A/D conversion, and thereafter is processed digitally. It is clear that we cannot begin to change the transmission format at this time because of the many tens of millions of TV sets out there that would become obsolete. Those sets will be around until the year 2000 or longer, but in the meantime we can certainly upgrade the capabilities of new units as well as prepare for advanced transmission facilities. Some of the implications of digital TV, including clearer outlines, sharper detail, and truer color, are so impressive that even the Sunday *New York Times* saw fit to run an article on the subject in mid March of this year.

The ability to place a second channel in one corner of the picture tube is an immediate but not too exciting application for digital TV. Going beyond that first thought, there are some very interesting things that can be done with this new technique. In particular, it will be possible to make significant improvements in filtering and signal processing that will result in a better picture. Both horizontal and vertical resolution will be increased without having to change present-day picture tubes and transmission facilities. Some of the problems that occur with color effects mixing with luminance will be eliminated. Even the audio channels will be improved considerably. Consequently, I believe that initially, digital processing will allow some upgrading of the performance of an otherwise standard TV set.

Looking ahead a bit, we can expect to have pictures of very high quality. As large memories become cheaper, insertion of a frame buffer inside the TV set will become economical. When the manufacturers start to use some of the new color tubes with very high resolution capabilities, we probably will see sets with twice the present scan rates and twice the number of lines on a screen. If that is done, however, without changing the transmission format, it will be necessary to find the information for those additional lines in some way. The solution to that problem will emerge from current research on algorithms that ensure the desired picture quality by allowing interpolation between lines and providing the necessary filler lines. Similar procedures will make it possible to have frame-to-frame interpolations in time. Overall, we can look forward to improved television without ever having to change the transmitted signal, a fortunate circumstance in view of the previously mentioned difficulties that accompany transmission modifications. Still, some parallel development in high-definition television is in progress in this country, in Japan, and probably in Europe, that depends upon additional transmission channel capacity. This work is being conducted in the analog mode, a step in the wrong direction in my opinion. I suspect, however, that these efforts will be considered as somewhat specialized, at least for quite a number of years to come, because of the expanded channel capacity requirements.

Turning now to the area of satellite-to-home television transmission, I will begin with a brief discussion of the background of this utility, called "backyard television" by some people and "direct broadcast service" (DBS) by others. Several years ago, a number of local TV cable systems came into being in areas where hills and valleys made ordinary video transmission either difficult or impossible. Placement of a central antenna in a line-of-sight position with respect to a broadcast station allowed signals to be picked off the air and brought into a subscriber's home by means of a distribution cable. It happens that one of the early systems of this kind was installed in San Diego, my present home. Those systems were not very exciting, they grew very slowly, but they did provide a service. Soon afterwards, someone realized that a video signal could be transmitted from an uplink to a satellite (as sketched in Figure 1) which would then broadcast it down to all parts of the United States. The signals could be detected by a satellite-oriented antenna at the cable-head end, thereby permitting, for example, premium broadcasting by distribution cable. HBO was one of the first broadcasting services to take advantage of this technique. Cable television (CATV) thus became a rather special, fast-growing, and very successful new industry, since it could now offer to sell much better programming than had been available previously.

As the satellite-delivered channels proliferated, CATV subscribers began to say to themselves: "Perhaps we should have satellite antennas in our backyards and pick up all of the channels ourselves instead of paying the cable company to receive them". This attitude inevitably resulted in a major increase in the number of these DBS-type stations. At present, much of this transmission occurs in the 4 to 6 GHz C-band, although the term DBS also is used to refer to a higher-power-satellite service involving the 11.7 to 14.5 GHz KU-band, a subject that I will touch on later.

Figure 2 indicates the domestic satellites that are in orbit around the United States. A recent FCC monitoring survey of satellite transmissions disclosed that there are about 408 C-band transponders available in those satellites. The survey was performed one day in the month of January for both day and night transmission activity. It was found that about 175 of the transponders were not in use at any time of the day or night when the monitoring occurred. (It is possible, of course, that many of these inactive units were being held as spares or being used for some other purpose). Of the remainder, some 115 were transmitting either television or FM signals during the day or the night, with the predominant use being for television. While much of the transmission by this relatively large number of active transponders is confined to only two of the satellites, a significant amount of traffic is carried by a number of other satellites including SATCOM, GALAXY I, and SATCOM F3, with roughly all of their transponders being used for television.

Examination of the listing in Figure 3 indicates that quite a potpourri of programming services are available from the satellites. The "premium" services, such as HBO, Showtime, and Cinemax, are usually offered to the public at extra charges above the basic cable service fee. A number of the basic services are advertiser-supported, to some extent, and are provided without extra

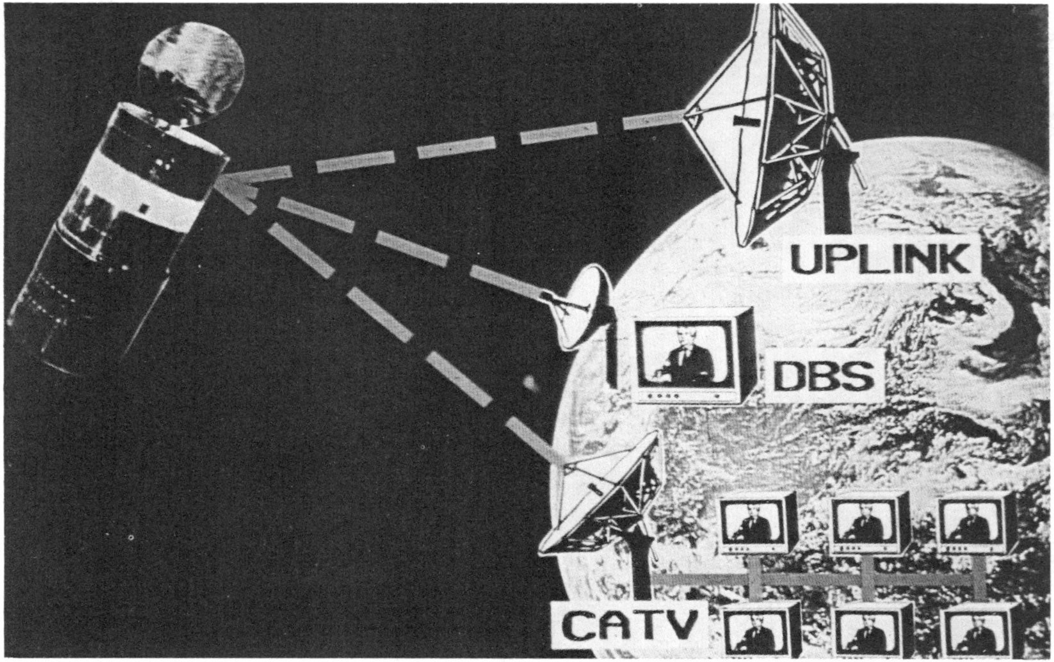

Figure 1. Satellite Broadcast Links

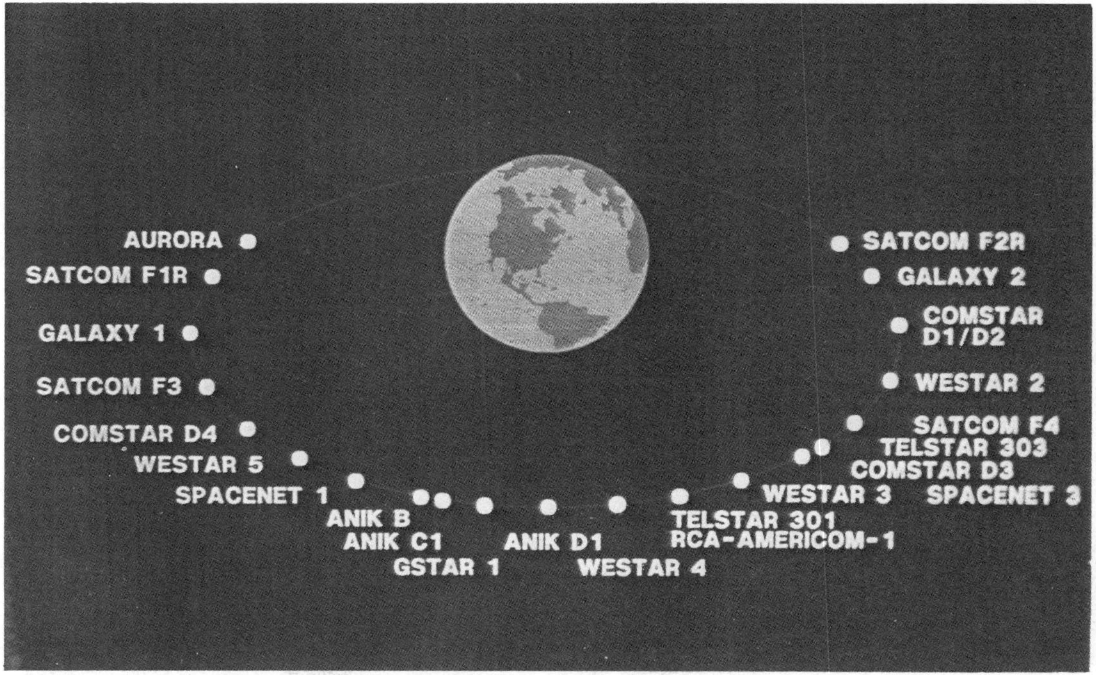

Figure 2. Domestic Satellite Orbits

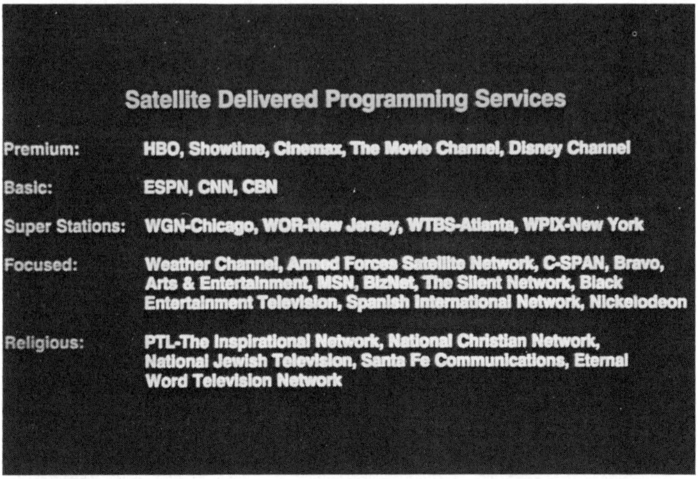

Figure 3. Satellite-Delivered Programming Services

charge. There are some so-called "super stations", originated by Ted Turner and other entrepreneurs, with the ability to cover the entire United States, thereby allowing additional advertising revenues. Many programming varieties are available, as detailed in Figure 4, including some adult kinds, lots of national and local sports, and widespread news and data services. Some of the channels use transponders to accommodate their programs to the time shifts across the country. HBO and Showtime, for example, use two transponders for their transmissions, one for the East Coast time feed and the other for the West Coast time feed.

The availability of all these services has caused the backyard business to expand, initially in rural areas where the cables were not installed. Various kinds of satellite antennas are now making their appearance in suburban areas. Recent legislation concerned with cablevision-company operations holds that it is not illegal to receive signals from a satellite if they are used for one's own personal enjoyment. If the satellite operators scramble the signals, however, and sell them as a service, unauthorized descramblers are considered to be illegal. Once general satellite reception became legal, a number of newspapers started carrying ads for the programs and for the antenna systems such as: "Heavenly Pictures at Down-to-Earth Prices" and "Everything You Always

Figure 4. Satellite-Delivered Program Channels

Wanted to Know About Satellite Television (but were afraid to ask)". The resulting rapid proliferation of dish antennas began to cause some problems. Those of you who either have a dish antenna, or can see one from your premises, may agree that the larger ones can be somewhat of a blot on the landscape if they are not mounted discretely. A number of communities are starting to bring pressure on their residents by passing ordinances that restrict the location of the dishes. The cable operators have also become quite sensitive to the growing presence of the antennas and are beginning to take action either to find a way to live with them or to fight their proliferation. Even in well-established cable areas, such as San Diego, antennas are popping up as people take advantage of the free signal reception.

Figure 5 gives us an idea of how fast the home-satellite-antenna market is growing. Perhaps the first ad for a backyard terminal appeared in the Neimann-Marcus Christmas Catalog in 1979. I think it was for a Scientific-Atlanta System that was offered for about $36,000. At that price they did not move very rapidly. As the years passed, however, the sales rate rose and is projected to continue to rise. It is possible that public confusion over the scrambling issue may slow down the growth rate, but this is difficult to predict. At present, there are over a million backyard terminals in service. A

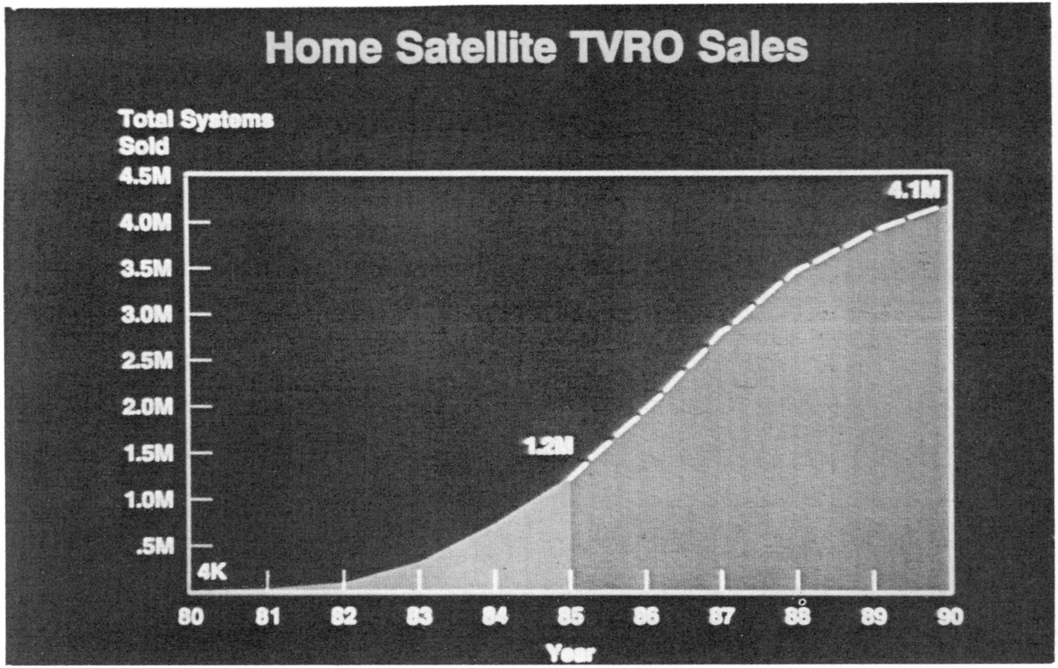

Figure 5. Home-Satellite-Antenna Sales

reasonable estimate for the near future lies somewhere between four and ten million terminals. One of the driving forces is that the terminal price has dropped dramatically. Both American and Japanese manufacturers have been improving the electronics and the low-noise block-down-converters (LNB's) mounted at the antenna, so that an entire terminal at C-band, without too many bells and whistles, is now approaching the $1000 cost level. At that price, many people are tempted to go ahead and install them.

The rapid growth of the home-satellite television market poses a major question: "Will TV viewers be able to continue receiving signals for free or will all satellite transmission be scrambled?" A few years ago my company became involved with trying to "unscramble the future" when HBO began to think about the need for scrambling and investigated what would be available in the market place. The ultimate business potential of scrambler/descrambler devices depends heavily on the video industry's rationale for scrambling its signals. Early in the game, the premium services felt that perhaps they were losing some of their signals to non-paying commercial operators, such as hotel/motel chains, whom they were willing to service but could not identify. In fact, after we began to manufacture and ship the scramblers, we were told that orders increased by well over 10 percent just due to people who called the pay services and said, "Gee whiz, we forgot to tell you that we are affiliates of yours and would you please send us a descrambler." Now this sector of the industry uses the scramblers to ensure future collection of a larger portion of its revenue stream.

Providers of premium movies and similar entertainment services are interested in the use of scramblers for protection of their copyrights. These groups are paid on the basis of the number of viewers. The free signals received by the million or so backyard terminals in service obviously constitute a significant revenue drain. Consequently, the entertainment studios have put pressure on the video system to provide scrambling equipment that will make it impossible to receive the signals unless a user fee has been paid. As mentioned before, the cable operators are concerned about the loss of some of their business to the backyard area, but the large number of units and homes connected to cablevision systems opens the possibility of creating a profitable market for scrambled satellite services that the cable companies can distribute to their customers. Improved audio, pay-per-view, and other specialized services that I will discuss later, could become very popular. There seems to be every reason for going ahead with experimentation and placement of scramblers into the satellite system.

At this point it is reasonable to ask why scrambling did not occur earlier in the development of the satellite systems. The answer lies in a combination of the operating philosophy of the premium services and the costly technology development required for the scrambling equipment. Originally, the premium services were anxious to buy the scrambling devices only because they wanted to protect their cable feeds. Upon further reflection, they realized that there was potential new business to be gained if the price of a descrambler could be made reasonable enough to allow installation in everybody's home while maintaining high-quality cable service. Such a development would also tend to offset some of the protest from the backyard owners who were receiving free signals. The key point here is that the signals are already being paid for by their delivery to cable-head ends. A DBS service that wishes to establish an entirely new high-quality system aimed specifically at the home will face large start-up and overhead costs that need not be incurred by the providers of service to cable companies. Thus far, three or more DBS services have tried to establish new home delivery systems at KU-band without success. The scrambled-TV service industry has to experience a period of technology growth so it is too early to call it a sure thing. It is certainly possible, however, for this kind of service to become well-established in the United States and to spread to other countries.

A block diagram of a scrambler/descrambler satellite-delivery system is given in Figure 6. The intent here is to apply special treatment to input video and audio signals before they are introduced into the uplink. When these signals are received on the downlink, they cannot be seen or heard clearly unless a descrambler is available for decoding. A number of devices that perform these functions have been developed for use by the cable services. Unfortunately, the scrambled signals can be pirated easily. A pertinent question was whether a high-quality scrambler system could be designed that would make pirating very difficult and economically unfeasible. The development of equipment with that capability made it necessary to look at a new generation of scrambling and descrambling devices that would also upgrade the quality of the pictures as well as that of the audio. Adequate security of audio services is a difficult problem that has been around since World War II. The solution is to employ high-quality digitization of the audio, similar to the procedures used in the production of digitally mastered computer discs. When the resultant signals are fully encrypted, it is essentially impossible to break the code. Application of this technique adds exceptionally good stereo audio to the television signal. Still another enhanced service can be obtained by designing the second audio channel for use in alternate-language broadcasting if the high-quality stereo service is not desired. Depending on the kind of programming, one might have a Spanish and English broadcast (as in San Diego, for example) or a full stereo broadcast. In terms of distribution of signals to the home, it must be recalled that the major advantage of satellite broadcasting is its ability to cover large sections of the country and thereby serve a huge number of customers. We must therefore design a system that can individually address and authorize service for any one out of ten, twenty, or one

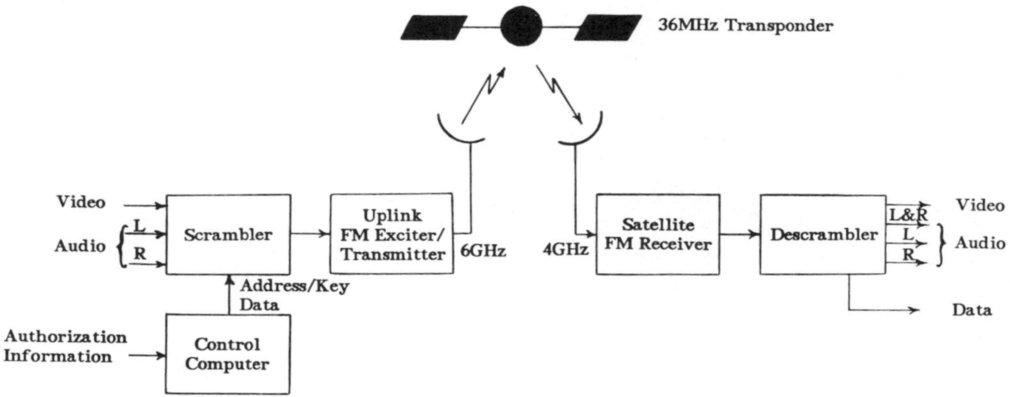

Figure 6. Satellite TV Delivery System

hundred million different subscribers. It is obvious that successful scrambling can be achieved only with the aid of carefully engineered communications control devices.

Photographs of a scrambler and a cable-head-end descrambler with these characteristics are shown in Figures 7 and 8, respectively. Some video services are now placing these units in service at quite a rapid rate. Therefore, we can expect to have a quick answer to the question of whether or not the scrambling business will succeed.

Whenever a new technology is introduced for public use, it is important to subject it to thorough testing, particularly when complicated communications issues are involved. The new scrambling devices were being applied to a

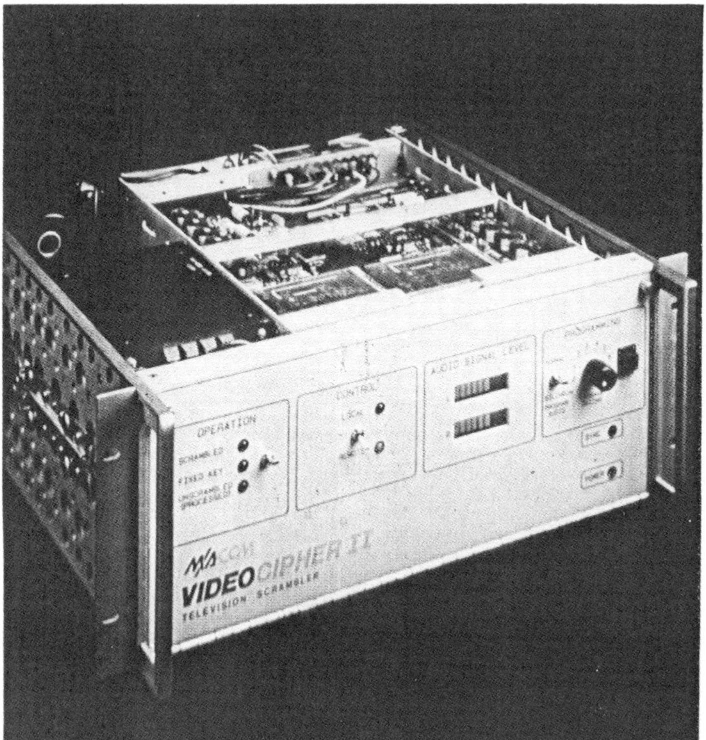

Figure 7. Typical Scrambler Unit

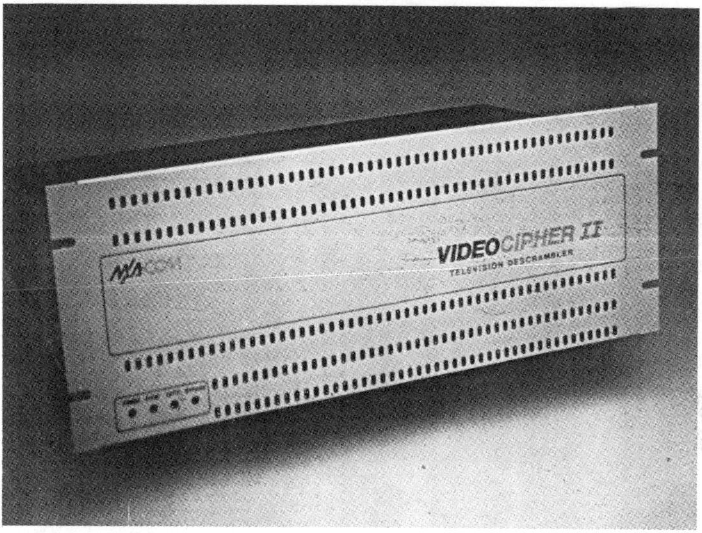

Figure 8. Typical Descrambler Unit

system that originally was not designed for high-quality communications. In addition, many new techniques were being introduced in an attempt to bring down the cost. Consequently, we monitored the signals from the satellites (see Figure 9) for over a year as the equipment was tested to ensure that it would function properly in all kinds of weather and under adverse electrical situations. The scrambler must be capable of distorting the picture (see Figure 10) if the subscriber does not have a descrambler, and, of course, the picture must be absolutely clear if a descrambler is in service.

Once a digital audio system has been established, it also becomes possible to include digital data and other types of videotext services that can be displayed on the TV screen. For example, some of the devices that have been developed can provide a temporary caption when a new channel is selected. As illustrated in Figure 11, the caption can supply channel and program identification, or other kinds of textual information. On the other hand, if your subscriber fee has not been paid, your descrambler can send you a succinct message, as in Figure 12. A caption of this kind, however, raises the issue of consumer sensitivity. We found that the "Use not Authorized" phrase was not really acceptable. The customer apparently wants both the computer and the TV to be user-friendly. A more desirable caption, involving a bit of quick hype

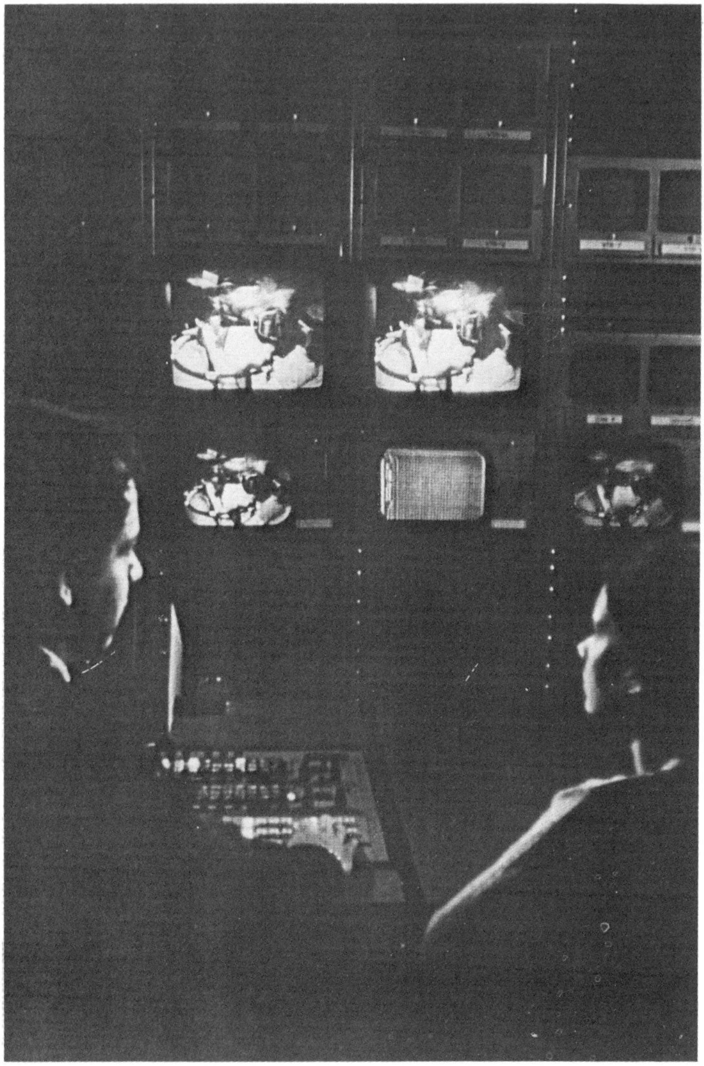

Figure 9. Monitoring of Satellite Signals

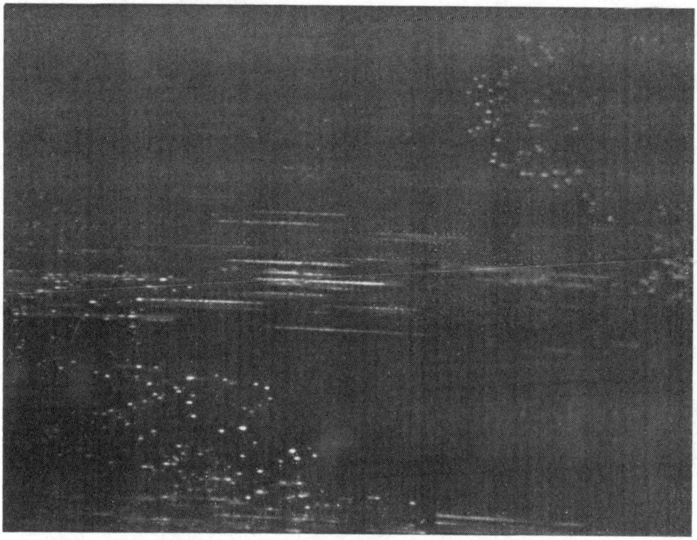

Figure 10. Scrambler TV-Signal Distortion

perhaps, would be: "If you had paid your bill you could be watching such-and-such at this time". It makes good business sense not to ignore such problems.

When a system has acquired the ability to scramble and descramble at an individual address, a number of other services can be added, as suggested in Figure 13. Program guides are clearly of interest and can always be updated to include channel content, special events, and different headline services. Perhaps advertisements can be used effectively. The general-broadcast, global-downlink nature of a satellite system, however, has introduced some special problems. For example, local news broadcasts are somewhat more difficult to execute. But the ability to address information directly to a group of receivers in a certain area makes it possible to provide some level of local service, such as farm and weather data.

The occasional necessity to create communication "blackouts" is another problem that was not anticipated. Sporting-event fans realize that in some situations there are league rules against having a program visible in some area. Even though a game or an event is being broadcast down from a satellite, and can be picked up everywhere, it is necessary to have equipment with the ability to black out individual regions.

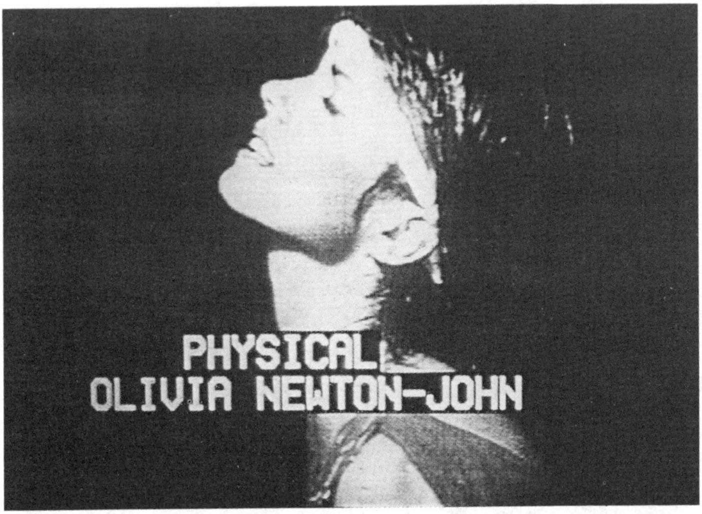

Figure 11. Program Identification Caption

Figure 12. "Out of Funds" Message

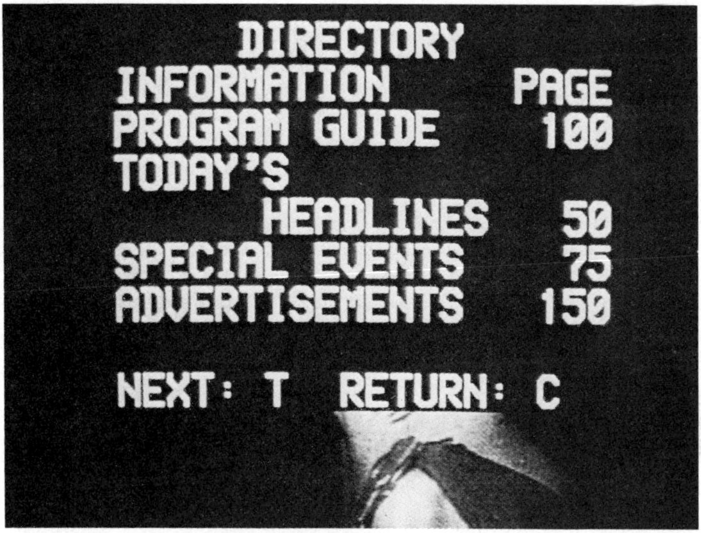

Figure 13. Program Directory

The procedure for accomplishing area-restricted broadcasting is relatively straightforward if the location of each receiver is known. A subscriber's zip code, obtained from bill payments or other correspondence, is sent to the appropriate descrambler box. When an event occurs that must be blacked out in one or more areas, pertinent center points and radii are broadcast. Each descrambler then calculates whether it falls inside or outside of the corresponding area. Receivers located within the area are automatically switched to another channel; those falling outside the circle experience no change. This additional communication capability is certainly an essential part of these new systems.

The entertainment field offers another possibility for the large-scale use of video-text service combined with controlled communications. Most cable systems, such as HBO and Showtime, offer entire month-by-month service for monthly fees, payable in advance. Some experiments have also been tried with broadcasting very special sporting events, or other major attractions, with the viewer paying only for a specifically chosen event. One of the problems with this so-called "pay per view" or "impulse pay per view" service is that people generally do not like to make up their minds very much in advance. Consequently, the design of scrambling and descrambling equipment must include the necessary provisions to allow a subscriber to buy an event on impulse. One of the detriments to establishing this type of service in the past has been the possibility that nearly everybody would call up at about the same time to schedule a particular event, resulting in complete swamping of the telephone system. A service of this sort will not be successful unless it is very well organized.

Although the devices that I have been describing are one-way systems, they may be programmed to handle the scheduling and financing of impulse-pay-per-view service in an efficient manner, as illustrated in Figure 14. Each subscriber establishes a credit balance that is monitored in the descrambler. When the system is queried about a particular upcoming event, the screen displays the credit balance, the cost of the desired program, and other pertinent information. If a decision is then made to watch the program, the fee is automatically subtracted from the credit balance. If the balance goes negative, the screen would indicate that fact and remind the viewer that money must be placed in the account before the program can be seen.

The typical satellite receiver for home use, as shown in Figure 15, includes the descrambler, the satellite signal demodulation equipment, and a hand-held remote-control unit. Such equipment must be designed so that it is

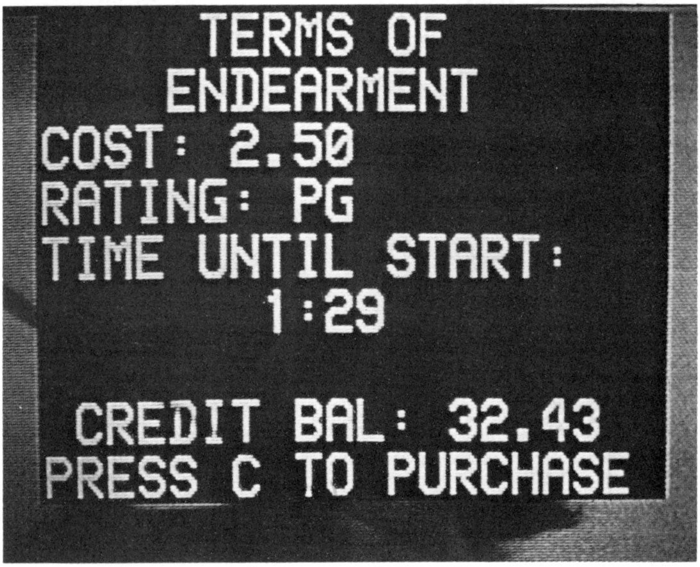

Figure 14. "Pay per View" Service Information

Figure 15. Typical Satellite-to-Home Receiver Unit

entirely acceptable in the home, and must be relatively convenient for people to use. Once again, a significant part of the engineering involves issues related to human factors; the various services must be presented and controlled in ways that people will be willing to tolerate. Since most of us are perhaps just beyond our capabilities in dealing with video cassette recorders, the necessity of having to cope with still another complicated control device may give us considerable pause. It is therefore essential that a great deal of careful thought be given to the development of simple control procedures accompanied by clearly stated "help" screens, just as in the home computer.

I will turn now to a brief discussion of the technical issues that relate to the actual performance of the various areas that have been mentioned. The upper sketch of Figure 16 is essentially an indication of the electrical waveform for one line, using NTSC transmission. During the time that the line is being retraced, a synchronizing (sync) pulse is sent along, followed by a color-reference burst, which, in turn, is followed by the video luminance signal. The color normally is sent on the NTSC subcarrier, and the FM-modulated audio is sent on another subcarrier, all resulting in the standard waveform of the sketch. If it is now desired to include digital information for audio, as well as for control and data purposes, it is necessary to find a suitable

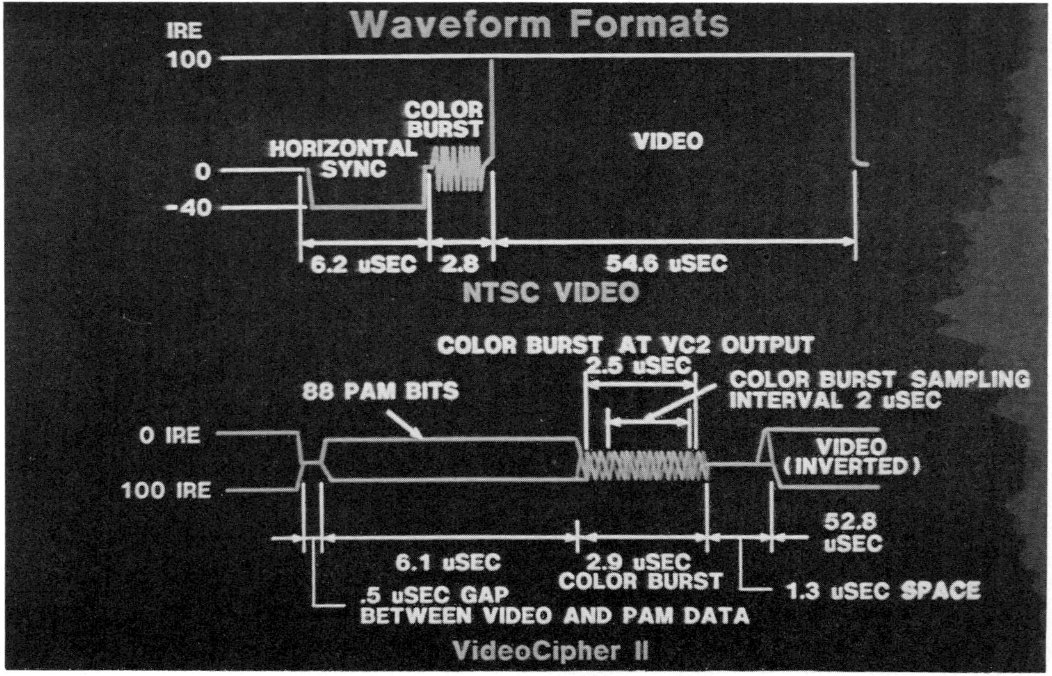

Figure 16. Transmission Waveform Formats

location. The easiest solution is to remove the horizontal sync entirely and replace it with frequency-shift-keying (FSK) modulated digital data transmission. It turns out that modulated signals of this type are able to pass through the nonlinear transponders, and associated gear, in reasonable fashion. In a typical system, therefore, the audio is removed from the subcarrier and digitized, after which it is inserted into the horizontal retrace interval, as shown in the lower sketch of Figure 16. In addition to providing a place for the audio, the removal of a subcarrier also typically enhances the signal-to-noise ratio in the transmission of information to a satellite by about 2db, giving us the opportunity to use a somewhat smaller antenna. Since the horizontal synchronization has been eliminated, television receivers that do not have descramblers will experience synchronizing problems with these signals. On the other hand, the presence of a descrambler, with its ability to synchronize very reliably to a specific word, will result in TV reception of synchronizing waveforms with very low signal-to-noise ratios. Overall, this procedure provides some scrambling of the picture, adds needed space for control of audio and data, and, at the same time, improves the signal-to-noise ratio.

Other techniques have been proposed for television transmission, including a procedure known as multiplexed analog components (MAC), a technique that transfers the color information from a subcarrier, as in NTSC, to the horizontal interval. This approach has been adopted in some systems, particularly in Europe. Unfortunately, it is not possible to squeeze the luminance and all of the other information into the horizontal interval without going to signals of greater bandwidth. Existing cable systems, however, would not be compatible with such signals, so it is necessary to arrive at some kind of acceptable tradeoff. While it is true that some of the color-luminance problems in current transmission systems could be removed by the suggested procedure, those problems can also be solved by the processing that occurs in digital TV, a technique that must work on standard NTSC signals in any case. Determination of the future directions of television-transmission signaling is certain to generate considerable controversy in the industry.

Figure 17 is a photograph of a circuit board that does all the scrambling, descrambling, video displays, and the other functions that I have described. With modern electronics and microcircuits, these units can be designed to be fairly small.

The question of how to circumvent piracy of the signals poses an interesting problem on encryption issues. The initial approach that is taken is to bury a unique key within the memory of the descrambler, with each key being different for every single card that is manufactured. The usual procedure is to insert the key in the very last manufacturing stage. Encrypted descrambling information can then be transmitted only to that card, thereby allowing each TV receiver to be individually addressed in a very secure fashion. Normally

Figure 17. Scrambler/Descrambler Circuit Board

the key is part of a standard data-encryption algorithm that has been devised for these systems. Since in the real world of communications it is always possible for lists of information to become compromised, a typical counter move is to insert a number of keys and then use the control channel to select the preferred key. A list that has been compromised is simply written off and a new list is generated, using a different key, with the hope that security will not be lost too quickly.

Once the key is in hand, it is necessary to determine how to encode the video and audio signals and to arrange to transmit them with optimum security. The problem here is that when several million subscribers must be served, it is difficult to send frequent key information to individual stations without having a very high data-rate channel.

A typical solution is to provide another level of encryption that is illustrated by the following example. A consumer or cable operator would buy a particular level of service with an associated service key that is sent at leisure to each of the individual descrambler cards around the country. That key would be maintained in service for a period of time up to perhaps a month. When a subscriber wishes to receive a given program that is included within certain service groups, the working key would be sent out after having been scrambled to conform with the service group constraints. If, for instance, there were ten different service groups that were able to receive that program, only ten versions of the working key would be transmitted. The pertinent scrambled key would then be received by the customer, and the descrambler would activate the desired service program or service, always assuming, of course, that reliable reception of the transmitted signal has occurred. Clearly, we need complicated systems in order to maintain a proper level of security and still provide service to a large number of users.

I have an additional comment on the issue of piracy and its possible impact on the industry. Since we realize that any information that is buried in a card can be "reverse engineered," the essential problem is to determine how to deny unauthorized copying and subsequent widespread sale of that material. This problem is similar to that of providing protection against duplication of computer software, but it is not as difficult to perform. The first step is to go to great lengths to bury the key within the integrated circuit in such a way as to make it very hard to carry out reverse engineering. It is essentially impossible, however, to have complete security. Some individual would always be able to find which unique key is buried in a particular card, and could then proceed to manufacture any number of duplicates. If a system operator determines, however, that the pirates have sold their copies to a number of users, he can make all of the duplication effort useless by simply transferring his authorization information to another key. Since it is relatively easy to take such corrective action, it appears that guarding against reverse engineering is not a matter that will cause major economic concern.

An attempt to predict the operating future of satellite broadcasting, and its associated services, probably should start with prospects for the use of smaller dish antennas. At least four broadcasting services have announced some initial activity in the KU-band. I believe that KU-band service will be introduced at some point, and may indeed appear fairly quickly, perhaps using 100-watt transponders instead of the 200 or 250-watt units now contemplated. The corresponding dish antenna diameters may shrink to 1.2 meters, or even down to 0.8 meters, resulting in fairly reasonable roof-top installations, as illustrated in Figure 18. As I mentioned earlier, however, present prospects for expansion appear to be limited to the C-band because of pressures that are building in the satellite broadcasting industry. The growth of scrambling activity has caused groups of users with backyard dishes to say, in effect, "You can't deny us service now; you have to permit us to be able to buy that service". The threat of legal action by these people has created considerable incentive to provide such a service, and has posed the problem of how to make it economical to operate.

Several issues must be resolved before a satisfactory solution to the financial problems can be solved. For example, all of the broadcast services have

Figure 18. Small Roof-Top Antenna

been unsuccessful in assembling a proper collection of entertainment software that can be sold effectively. Referring once again to conditions in the real world, the basic obstacle is that many of the software suppliers are very competitive with one another. It is therefore very difficult to achieve the kind of cooperation that is necessary to produce a viable service. A somewhat less obvious issue is concerned with the fact that of the large number of available television channels, only a few are of real interest to many viewers. It should therefore be possible to scramble these premium services and generate some level of demand. Some of these services, however, carry sporting events and other interesting programs that are supported by advertising. Unless those services are also scrambled, the demand for scrambling equipment will be small. Consequently, the scrambling business will not be successful without the development of a strong incentive for the use of the devices. Since the service operators obtain substantial revenue from advertising, they are anxious to increase the number of viewers. They are not under any pressure to scramble, however, and will not do so without strong economic justification.

I am reasonably optimistic that these issues will be sorted out to everyone's satisfaction for several reasons. The additional capabilities that scrambled services can provide will obviously be attractive to suppliers of DBS customers. The large amount of essentially unused data capacity that is available in a video signal is another encouraging factor. It is feasible, for example, to insert 0.5 megabits in the entire vertical retrace interval. This potential data rate could carry from ten to twenty million 300 to 400-character messages each day over a single channel. Multiple channels, of course, would allow transfer of even greater quantities of information. Various new services could therefore be provided to the home, such as: downloading of software programs to personal computers, video text facilities, and a limited form of electronic mail. It is possible, therefore, to supply a composite service where the price and maintenance of the equipment will be carried by the entertainment component of the service, and all of the ancillary capabilities will come as a bonus. I am certain that this type of service will become available. Once a service with such capabilities is announced, it will be sure to spread rapidly as viewers begin to appreciate the possibilities for the additional TV features that can be brought into their homes by this encrypted, addressable, broad-band pipeline.

Update:

Following the presentation of this material at the April 10, 1985 Symposium, the scrambling of satellite-delivered television signals at C-band, using the technology discussed here, has become widespread. Many tens-of-thousands of descramblers for home use are now sold each month. Some services are beginning to transmit pay-per-view programming to homes via satellite. KU-band-based services to the home are still in the future. The product, which was developed at the LINKABIT facility of M/A-COM in San Diego, has since been sold to General Instruments.

THE EVOLUTION OF COMPUTER NETWORK SPECIES

Paul E. Green, Jr.

Paul E. Green, Jr. received the B.A. degree from the University of North Carolina in 1944, the M.S. degree from North Carolina State University in 1948, and the Sc.D. degree from MIT in 1953. From 1951 to 1969 he was at the MIT Lincoln Laboratory where he pioneered several innovative applications of communication theory. In 1969 he joined the IBM Research Division in Yorktown Heights, New York, where his interests have centered on speech and signal processing, network performance modeling, and computer network architectures and protocols. Since 1975 he has divided his time between IBM's Corporate Technical Committee and the Research Division. Throughout his career Dr. Green has been active in IEEE Professional Society affairs with particular attention to editorial functions. He has been a member of the Editorial Boards of the *IEEE Press* and of the *IEEE Proceedings.* At present he is Director of Publications of the IEEE Communications Society. In 1980 he was one of four recipients of the Pioneer Award of the IEEE Aerospace and Electronic Systems Society. He is a Fellow of the IEEE, and a member of the National Academy of Engineering. In 1982 he was named a Distinguished Alumnus of North Carolina State University.

It is a great pleasure to be here to meet with old friends, and also to help Cornell celebrate the Electrical Engineering Centennial. The previous announcement of the coming of the supercomputer to the campus is good news indeed and I offer my congratulations on Cornell's success in obtaining the funding for that exciting new facility. Speaking of funding, some of my friends at NSF told me of an amusing incident that occurred recently. It seems that the Deity Himself applied to NSF for a grant to do some research on the origins of life. The proposal was given a peer review but was finally rejected for three reasons. First of all, the Principal Investigator had not worked in the field for over 4000 years; secondly, nobody has been able to reproduce His results; and thirdly, His publications are limited to one chapter in one book. I'm glad to see that Cornell has been more successful than that.

My talk today will cover the current state of the computer-network art and speculate on which forms will survive, which will adapt, and which will disappear. In the twenty years that computer networks have been around, they have evolved from simple tree topologies for time sharing to the point where they now encompass a large variety of both form and function. We have seen the emergence of large commercial multi-system networks, comprehensive layered structures, local-area networks, videotex, Open System Interconnect, protocol converting gateways, and other structures. Such forms emerge, flourish, mutate, and are selected out by technical innovation and evolving user needs.

INTRODUCTION

Three main species of computer networks can be identified and will be discussed in this presentation: communicating mainframes, communicating minicomputers, and communicating microcomputers. Networks of all three types are prevalent today, but they had their respective origins in the 1960s, 1970s, and 1980s.

It will be seen that there are several identifiable themes in the evolution to be described, *variegation* in types of networks and of attached systems within a network, *convergence* in a conscious effort to reduce this variegation, *attrition* of the less viable species, *decentralization* of intelligence in each network, and *system hiding, i.e.* making the communication aspects of the system less visible to the user.

MAINFRAME-BASED COMPUTER NETWORKS

Figure 1 shows a typical network of the 1960s, for example one based on IBM's Basic Telecommunication Access Method (BTAM) host software. A typical application for such a network was time-sharing of the host's instruction

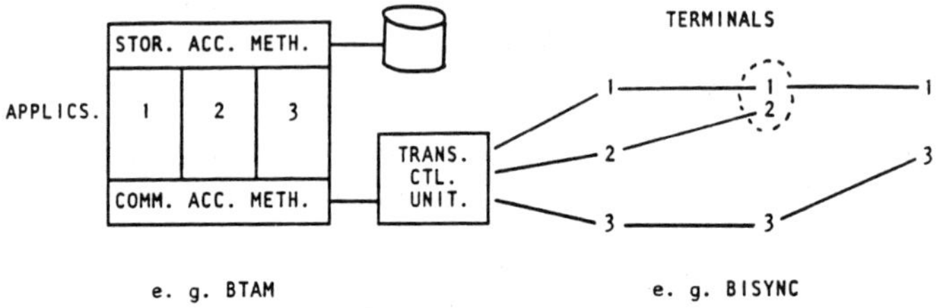

- All intelligence in host mainframe
- All terminals on a line talk to the same application
    - Result: duplicated lines, terminals
- Typical application: time sharing
- Communication details show through to end user and application
- 34 different DLCs, 5 access methods in IBM alone

Figure 1. Typical 1960s Computer Network

execution, memory, and secondary storage across many users in, for example, an airlines-reservation application. All the intelligence resided in the host mainframe; only the most rudimentary communication function (*e.g.* check-sum checking) was carried out in the communication controller, which was non-programmable. In this figure, and those to follow, the number given to each terminal shows which application it is talking to. Each multidrop line, together with all its attached terminals talked to the same application program, so that if a location needed two types of terminals talking to two applications, separate terminal devices and lines were required, as shown by the dashed circle. Communication details like polling and packet sequencing showed through to the user's application. One of the most troublesome problems was the proliferation of incompatible implementations, *e.g.* 34 different incompatible data link controls and 5 important different access methods, just in the IBM product line alone.

System Network Architecture (SNA), introduced by IBM in 1974, addressed many of these problems. (See Figure 2.) SNA standardized on one data-link control and two communication-access methods. To allow the network owner to keep up with the latest network technology simply by program reloads, the communication controller was made programmable. The host and controller software allowed any terminal on a line to talk to any application, independently of what the other terminals on the line were doing. As a reflection of its centralized past, SNA involved (and still involves) a high degree of centralization of both "control" (creating and managing a path between users) and "management" (administering the network).

The arrival of the ARPANET technology (and related European packet network experiments) coincided with the needs of large corporations to allow a given terminal to talk not just to any application in the owning host, but to a choice of applications anywhere in a multi-host network. Figure 3 depicts an SNA-3 or 4 network showing such paths, and also showing the proliferation of intelligent minis doing significant application processing of their own at the periphery of the network. Such networks are used for both terminal-to-application and application-to-application purposes. By now there are some 20,000 SNA licenses, probably 10-15 percent of which involve multi-host configurations. There are four levels of hierarchy visible in SNA networks, namely hosts, communication controllers, "peripheral nodes" - *e.g.* terminal controllers, and the dumb-terminal devices (not considered part of SNA). Both control and management are centralized in host processors, each presiding over a "domain" of the network; three domains are indicated in the diagram. ARPANET has three levels of hierarchy: hosts, intermessage processors (IMPs), and terminals. Both SNA and ARPA (and also DECNET, to be discussed later) supply "real-time" connection paths; other large networks that evolved during

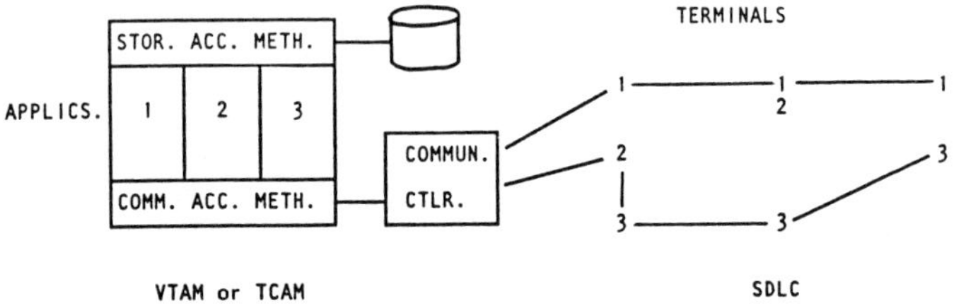

- E.g. System Network Arch.-1,-2 (one DLC, 2 AMs)
- Programmable communication controllers
- Different terminals can talk to different APs at same time (one host)
- Both control and management highly centralized

Figure 2. Typical Network of the Mid- and Late-1970s

the 1970s, for example IBM's internal VNET (over 1000 hosts today) supplied "store and forward" non-real-time paths.

Perhaps one of the most important achievements of the computer network architects of the 1970s was to successfully divide the involved myriad network functions into cleanly defined layers (Figure 4), thus allowing separate layers to evolve independently, and also shielding a given layer from having

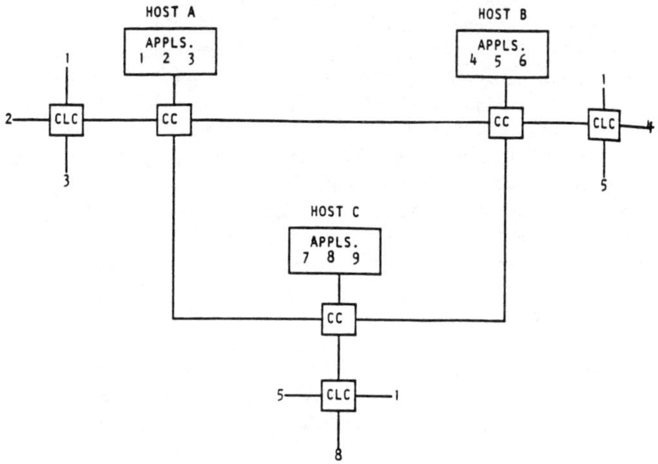

- E.g. ARPANET, SNA-3 (real-time paths), VNET (store/forward paths)
- Different terminals can talk to different APs in different hosts; hosts talk to each other
- Outboard intelligent minis
- For SNA, both control and management centralized in large pieces (domains)
- Clean layer separation

Figure 3. 1970s Resource-Sharing Network

to deal with the details of a layer below. Figure 4 shows several paths of data flow threading their way from node to nodes. A node is indicated as a stack of layers.

The LSI revolution has led to the growth of more and more computer power in the smaller more peripheral nodes of the network, to the point where even the simpler terminals contain a microprocessor to handle such things as driving hammers and painting screens. This trend has led to defining the higher levels of SNA as a peer (*i.e.* symmetrical non-master-slave) interaction between two programs to replace interaction between a program and a device. Advanced Program to Program Communication (APPC) is the name given to this peer extension of the higher levels of SNA, as depicted in the top layers in Figure 4. Meanwhile, the lower layers have continued to operate in a master-slave (primary-secondary) fashion.

Today there are several technical challenges that characterize the leading edge of mainframe network technology. The first is the sheer size being attained by some individual networks, in terms of number of hosts (already over 1000 in some cases) and number of terminals per host (also over 1000 in some cases). This, in turn, imposes the need for better "dynamics" (ability to lose or gain a node or line nondisruptively), better partitioning into subnetworks connected by gateway nodes, and improved network management software. Another challenge is the demand for "nonstop" operation, since many businesses are becoming so computerized that more than a few minutes' network down time is totally unacceptable.

MINICOMPUTER NETWORKS

The leading example of computer networking in the minicomputer context is DECNET, illustrated in Figure 5. It supports a "bottom-up" style of system evolution and the desire of many users to make their own modifications, as has been characteristic of much of the community of minicomputer users. By bottom-up is meant that individual systems are procured at fairly low management levels in the organization, with interconnection to other lower levels and to higher levels coming only later.

In DECNET, full topological generality is allowed (as with ARPANET) by the decentralized nature of the network control, so that each node takes its own responsibility for membership in the network. Unlike SNA, which was much influenced by the need for terminal-to-application requirements, DECNET has emphasized program-to-program support from the beginning. Network management requirements have recently been strongly felt and are being supported by later software releases. Recently, ETHERNET local-area networks have been introduced as one option for physical connection with DECNET.

- **Definitions: peer (coequals) vs. primary-secondary (master-slave)**

- **Peer program-to-program upper protocol layers (e.g. SNA Advanced Program to Program Communication)**

  - **Driven by decentralization of compute power**

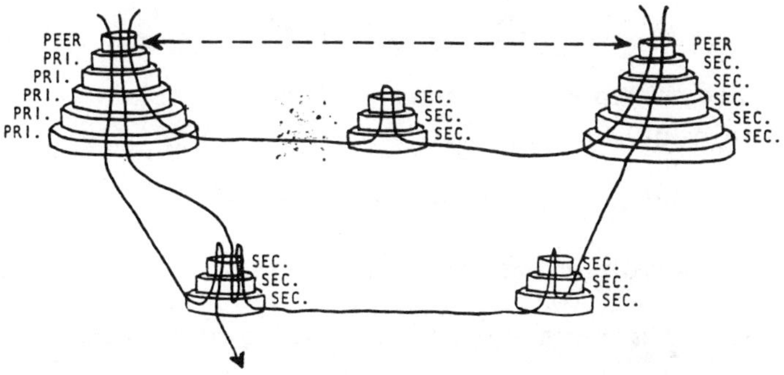

Figure 4. Peer Interaction in Mainframe Networks Showing Layering of Network Protocols and Peer Upper Layer

COMMUNICATION FOR MICROCOMPUTERS

We are seeing the three different types of terminals – data-processing terminals, word-processing terminals, communicating personal-computer terminals – all converge to the last of these, personal computers. The dominant mode of connection for PCs today is by voice-grade Telco lines or in-plant twisted pair using intelligent modems at 300 or 1200 bits per second. This

DECNET the leading example

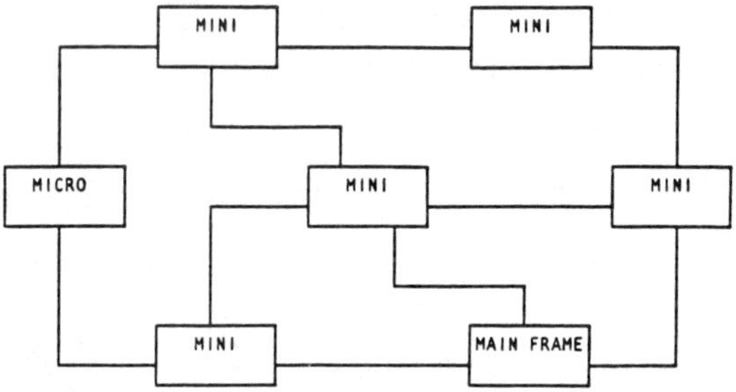

- Basically program-to-program, not terminal-to-program
- Topological generality - no hierarchy
- Decentralized network control
- Network management being added - centralized
- Supports bottom-up growth pattern
- Open - much user modification
- Local area network (Ethernet) support

Figure 5. Typical Minicomputer Network

statement is true whether connecting to another PC or to a host in the role of an ASCII-code terminal (*e.g.* to information providers such as SOURCE or the emerging Videotex services). Installations that are also growing rapidly are based on in-house connections to mainframes via twisted pair or coax using PC software and feature cards by which the PC emulates such a terminal as a 3270 or VT-100.

The local-area network (LAN) phenomenon is most prevalent with PCs as the inheritors of the role of an intelligent office workstation. LAN technology provides a high-speed unified way of interconnecting a collection of machines of wide size range (PCs to mainframes) within one building or campus, and allowing selected larger ones to act as file or printer servers to the remainder. Whenever traffic attempts to leave the LAN, however, and go to another station or LAN, the low bit rates and long installation leadtimes of non-voice-grade common-carrier facilities form a considerable bottleneck.

The problems and solutions that characterize the leading edge of microcomputer communication are dealt with in the current (May, 1985) *IEEE Journal on Selected Areas in Communication* (*JSAC*). That issue contains a description of one promising solution to the inter-LAN interconnection problem: "personal earthstations", small (four-foot aperture) send/receive customer-premise satellite stations at about the cost level of one to two PCs. As for the LANs themselves, today's users are not sensitive to the difference in architecture (baseband vs. broadband, star vs. ring), or the bit rate on the medium. As applications become more demanding, however, and utilization grows, there is apt to be a shakeout of the earlier LAN species that have poor growth, flexibility, and problem-determination capabilities.

The very rapid growth of the use of PCs as intelligent partners in communication with mainframes, has led to rethinking the nature of the operating systems internal to both the PC and the host. The host environment essentially lacks the PC-environment capabilities of close coupling to the user, the ease of accessibility and control, and the ability to rapidly incorporate innovative new products. On the other hand, the PC environment suffers from lack of large file capability, high-quality printers, large highly-evolved transaction processing software, and easy access to shared databases. This need for a better symbiosis between PCs and hosts than that provided by simple terminal emulation in the PC, has led to the development of "cooperative processing", which is described by Parr, *et al* in the May, 1985 *IEEE JSAC*. By an addition to the operating systems within such disparate computers as PCs (running, *e.g.* DOS) and mainframes (running, *e.g.* MVS), service requests to the operating system are intercepted and handled as "remote procedure calls". That is, they are "function shipped" to the other machine where they are fulfilled and shipped back, to reappear to the requesting operating system as though they had been satisfied locally.

LOW-ENTRY NETWORKING

It has been noted that mainframe networks have evolved from totally centralized structures to ones in which the upper layers, at least, have become peer connected. Meanwhile, the management and administration functions of minicomputer networks have become more centralized. At the same time, more and more networks are being built with a wide range of processor sizes in the same net.

It is a natural next step to extend the peer-to-peer style of coordination to the lower layer levels of Figure 4. A number of us at both the Raleigh Development Laboratory and the Yorktown Research Laboratory of IBM have been working on an experimental extension to SNA that provides a unified architecture for peer networking of mainframes, minis, and micros. This extension is called Low-Entry Networking (SNA/LEN). The network control is decentralized and the network management is centralized. The current status of LEN is described by Baratz, *et al* in the May 1985 *IEEE JSAC*.

The specific needs driving the LEN effort were extreme ease of use, simplicity and openness (well documented interfaces), support of arbitrary topologies, reduction of the number of hierarchical levels, availability, and ability to interwork with existing SNA networks. LEN can be implemented as an extension to existing SNA "peripheral nodes".

The nature of LEN may be of interest and can best be described in terms of the five stages of operation: (1) A new node that dials into the net learns the entire existing topology. (2) An end user seeking to connect (*e.g.* logon) to another logical resource, causes an automatic distributed directory search to locate the node containing that resource. (3) The best route leading to the target node for the particular class of service specified by the user is computed at the source node. (4) The session is set up and frozen onto the chosen route. (5) Message flow is deadlock-free and is controlled to make good use of the line and buffer resources. At Yorktown, all of these stages have been accomplished by a code in a prototype of less than 20 thousand lines of code.

WHAT TO DO ABOUT VARIEGATION?

Even as new forms of network, such as LANs or LEN, arrive on the scene, the old ones that have been mentioned continue to persist for years. There are still thousands of active BTAM licenses that have been in service, since the 1960s. There is ever wider use of one of the earliest link controls (namely start-stop ASCII). New innovations in technology are increasing the variegation daily — cooperative processing, decentralized control as with LEN, distributed database technology, paging across LANs, and on and on. Equally strong pressures come from innovative new applications such as videotex, short response-time CAD/CAM, voice-data integration, and so forth.

Still other pressures exacerbate the tendency for an individual network to be "inhomogeneous", *i.e.* to include components not originally intended to work together. These include the fact that reorganizations bring together groups of people who used different networks in the past, the fact that deregulation now encourages connections across enterprises, the "bottoms up" evolution of systems where interconnection is an afterthought, widespread purchasing practices emphasizing multivendor solutions, the incompleteness of most standards, and the addition of new carrier offerings, *e.g.* X.25 networks.

There are two basic approaches to this "zoo" or variegation problem: protocol convergence or protocol conversion. Convergence refers to the enunciation of a standard followed by attrition of the nonstandard implementations. Conversion refers to the splicing of paths at gateways between disparate network portions. Considering its importance, there has been little more than *ad hoc* implementation of the latter and until very recently essentially no attempts at a deeper methodology.

The constant battle between variegation and convergence is evident from the present capsule history of computer networks. Examples of convergence covered in this presentation have included SNA's replacement of 34 DLCs by one, DECNET's coverage of 8, 12, 16, and 32-bit machine-based operating systems with one network design, replacement of seven pre-existing types of sessions ("LU Types") with APPC, and the idea that LEN could suffice for mainframes, minis, and micros.

These are all within single product lines. There have been equally strong pressures for more global convergence across product lines. This has led to such successful standards as RS232 for machine-to-modem interfacing, X.21 for fast circuit switching, and X.25 for packet switching, and last, but not least, Open-System Interconnection (OSI).

THE ROLES OF OSI

Open-System Interconnection has been cast in four roles, depicted in Figures 6-9. The "CP" shown at each node is its control point. The first is as a replacement layered architecture for SNA, DECNET, etc. The second has the lower OSI layers providing a facility upon which someone else's upper layers can be put intact. In the third, some non-OSI lower layers can be used intact to support some new OSI services, *e.g.* videotex or file transfer. The fourth is as an intermediary in protocol conversion from, say, architecture A to architecture B.

There is a great deal of controversy surrounding the first option. In my view, this role for OSI has fallen a victim to its own lateness. With the huge

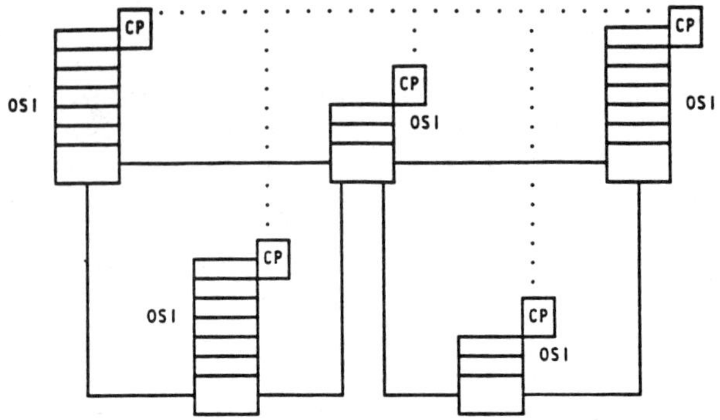

- Control (currently missing) must be identical across vendors
- Late (Compare 20,00 SNA licenses, 3000 DECNETs, 100s of IP/TCP, . . .)

Figure 6. Option 1: Open System Interconnection as a Total Architecture

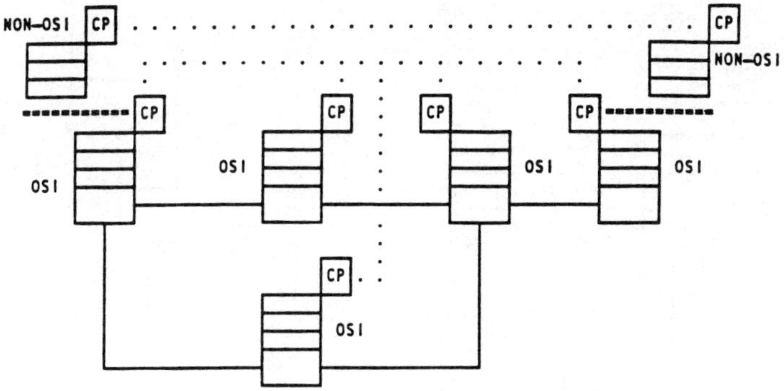

- Control need not be standardized - two weakly coupled pieces
- Exists (DECNET/X.25 and SNA/X.25 products)

Figure 7. Option 2: OSI for Interconnection

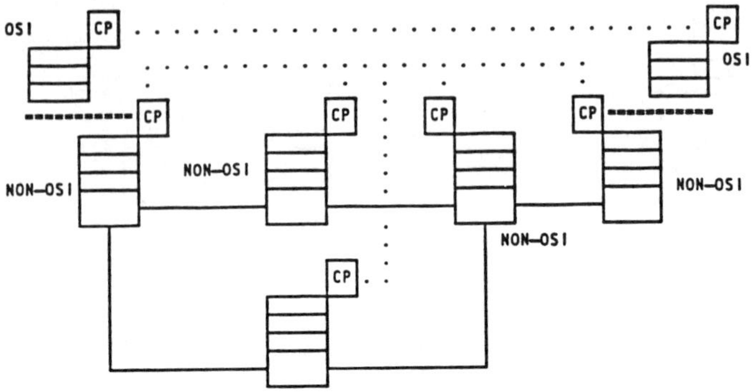

- Again control is in two weakly coupled pieces
- Will happen with availability of OSI applications (videotex, file transfer, . . .)

Figure 8. Option 3: OSI Upper Layers with Interconnection Provided by non-OSI Protocols

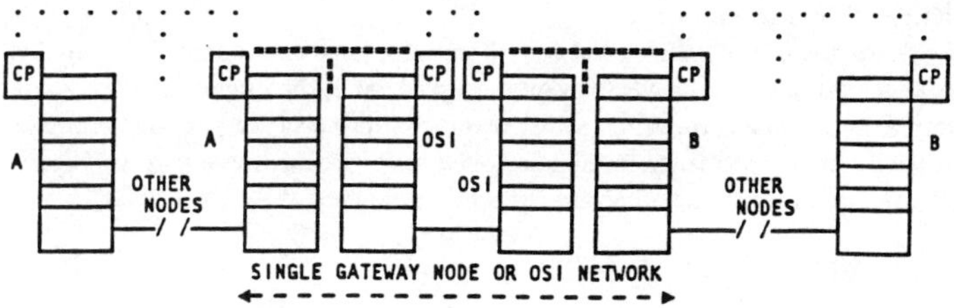

- Control need not be standardized - three weakly coupled pieces
- K conversions, not K-squared
- Active prototyping projects exist (e.g. IBM and NTT)

Figure 9. Option 4: OSI as a Protocol Conversion Intermediary

installed base of SNA, DECNET, etc. in place and growing, and agreement on an OSI architecture that includes the necessary network control (dotted lines in Figure 6) years away, it seems naive to hope that nodes from different vendors will soon be built to a common standard architecture and will provide highly efficient networking.

With the second, third, and fourth options, network control between the various portions can be very loosely coupled, and these should work well. Option four is a promising solution to the protocol conversion problem, although the performance implications of going through three protocol sets to get from one user to the other are not well understood.

PROGNOSIS

It is a safe bet that many different varieties of networks will persist and therefore coexist. Evolution and innovation seems to be particularly rapid at the smaller machine sizes, as part of the role inversion of "hosts plus dumb terminals" into "servers plus intelligent workstations". Protocol conversion may or may not be a growth industry, but it will be a fact of life for many years. While standardization is an important enabling factor in all of telecommunications, a true convergence on a worldwide network architecture is for the far distant future after computer communications has become a tired, stable field with only second- and third-order improvements being made. Computer networking technology is anything but that today.

CODES AND MODULATION FOR MODERN COMMUNICATION SYSTEMS

Richard E. Blahut

Richard E. Blahut received the B.S. degree in Electrical Engineering from MIT in 1960, the M.S. degree in Physics from Stevens Institute of Technology in 1964, and the Ph.D. degree in Electrical Engineering from Cornell University in 1972. He has been with the IBM Federal Systems Division in Owego, New York since 1964 where he has general responsibility for the analysis and design of coherent signal processing systems, of digital communications systems, and of statistical information processing systems. In 1980 Dr. Blahut was named a Fellow of the IBM Corporation, and became a Fellow of the IEEE in 1981. He was a member of the IEEE Activities Board of Governors from 1978 to 1984 and served as President of the IEEE Information Theory Group in 1982. He was the recipient of the IEEE Information Theory Group Outstanding Paper Award in 1974. He is the author of two textbooks, *Theory and Practice of Error Control Codes*, and *Fast Algorithms for Digital Signal Processing*. Dr. Blahut was a Distinguished Visiting Professor of Electrical and Computer Engineering at the University of Illinois in 1984-1985. At present he is a Courtesy Professor of Electrical Engineering at Cornell University.

Someone has just asked me what a Courtesy Professor of Electrical Engineering actually does and my response is: I teach students to say "thank you".

Seriously, I would also like to congratulate Cornell on a hundred years of Electrical Engineering, and I also thank the organizing committee for inviting me to speak today since I consider it a personal honor to be included. As I listened to the beginning of some of the earlier talks I realized that I have something in common with Irwin Jacobs. It seems that we both went to the same schools, MIT and Cornell, although we received our degrees in the opposite order. But I must say that although Irwin is unique in very many ways, in at least one matter he is not as unique as he thinks he is. I have a son who not only transferred from the Hotel School at Cornell into the Electrical Engineering School but he went one better than Irwin because he started out as an Electrical Engineer, transferred into the Hotel School, and then moved back into the Electrical Engineering School. Breaking that record will require at least three transitions.

This afternoon I will talk about coding, give a little time to modulation, and then move on to digital communications, all subjects that are near and dear to my heart. The message that I would like to convey is that in the last few years, digital communication has become very different from what it was five or ten years ago when I was in the earlier part of my career. The principal reason for the change is the introduction of large-scale integration. It is possible to do completely new things with our present capability of putting a hundred thousand transistors on one chip for the same cost once required to put one transistor on a chip. In the last few years, applied mathematicians, and electrical engineers with an interest in mathematics, have been applying advanced mathematical techniques to communications theory and practice. Although many of the new approaches are becoming well established, they are not widely known as yet, so I will take this opportunity to try to spread some of these ideas around the community.

When we think about digital communications most of us visualize something like a horizontal line with a square pulse above it to mean a 1, and another square pulse below it to mean a 0. Present communication systems already look very different from this elementary picture, and future systems will change even more so, in spite of the fact that waveforms and digital techniques will remain basic to their operation. I brought along a compact disc as an example of what I consider to be very innovative engineering in communication. Probably most people by now have seen a compact disc. I was astounded when I first read of some of the things that have been incorporated in the compact disc, features that a few years ago I imagined would be found in some of the more esoteric and advanced digital communications systems of the future. Here the Dutch and the Japanese have gone on to do all of these things well ahead of the time when I thought that they would become available. Not only have they done them, but they have prepared a product for the mass consumer market that sells for just over ten dollars, together with developing a player that is cheap enough to be accepted by the market. As I go along, I will use the compact disc as an example of some of the things I am talking about as well as of some of the things that they have done that are pertinent. I will also use it as an example of some things that I think they should have done and did not do.

In order to make my point that coding of every kind is a very timely subject that is penetrating all elements of digital communications, I will have to give a somewhat technical talk as outlined in Figure 1. The elements in the boxes labeled "encoder" represent some of the areas that have undergone many very advanced, very complicated, and very mathematical investigations, and have experienced the application of the results of those studies.

I would like to give most of my attention to data compression, data transmission, and data compaction, but I will comment on the contents of all of the boxes in Figure 1 and define the terms as I use them. By the term "data compression," I mean something that plays the role of an A/D converter, an element that quantizes analog information regardless of whether it is picture

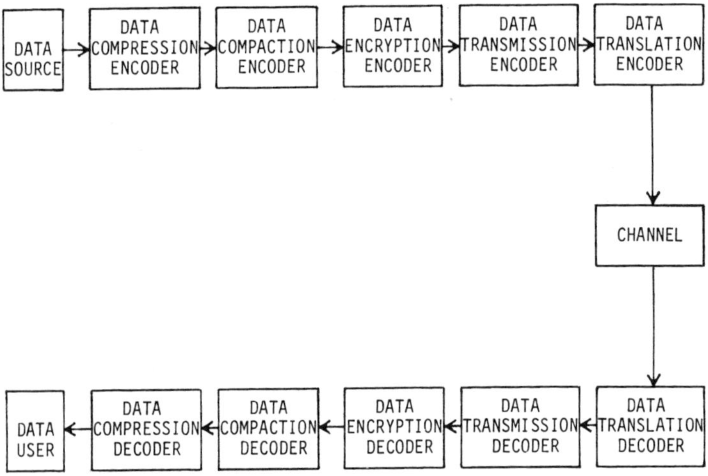

Figure 1. A Digital Communications System

information for TV, voice information, or whatever. By the term "data compaction," I mean a kind of code sometimes called a noiseless source code. I define it as an operation that takes a string of previously digitized data in some alphabet and represents this string more efficiently by a string of bits. The difference between the two functions of data compression and data compaction is that the data compression operation involves some loss of fidelity. Any kind of quantization or digitalization incurs some loss of fidelity, whereas the data-compaction encoder is a completely lossless operation, one that is perfectly invertible with absolutely no loss of even a single bit.

I will not say very much about data encryption except to make the point that the proper place to put it in a string of these functions is just where the data is compacted the most, so that the cryptanalyst does not have the opportunity to exploit redundancy in the data to break the code. We all know that English cryptograms are broken by using redundancy such as "the" or "qu...". In highly compacted data, such redundancy is suppressed.

I do want to give some attention to the two types of channel codes known as data transmission codes and data translation codes. In the terminology that I have adopted I use the term "data transmission code" to include channel codes such as an error-correcting code. I prefer the general term "data transmission code" to "error-correcting code" in the general case because many of the systems using these codes, as used today, never get around to making a hard error; neither the demodulator nor the decoder ever work at a level where an error is made. Consequently, it is best to employ "data-transmission encoder" as the more general term. In the second kind of channel code, the data-translation code, the encoder prepares a string of bits or symbols for a channel that has constraints on the kinds of sequences of symbols that can be recorded. This function will become clear when I discuss the input to this encoder.

The bottom of the chart in Figure 1 is the inverse of the top. As data passes through, layer by layer, the data coming out of the channel becomes decoded, layer by layer, until it is received by the data user.

An interesting point to be made here is that the data user may not necessarily perceive that all of these things have occurred within his communication system. In fact, in many systems being designed today different functions are the responsibility of different individuals who are not even aware of what some other designers are doing in their functions or how their operations affect the system. The end result is that the total system might be made up of layers of hardware and software that are physically distinct from one another.

I will start with data-translation codes, as indicated in Figure 2, and discuss some of the pertinent elements. This is an area where some of the neatest things have happened in the last few years. There does not seem to be any general term that ties them all together so I call them "data-translation codes".

PURPOSE

PROHIBIT CERTAIN SUBSEQUENCES OF SYMBOLS THAT ARE TROUBLESOME

EXAMPLES

MANCHESTER CODES AND MILLER CODES

TERNARY LINE CODES

FREIMAN-WYNER CODES

FRANASZEK CODES

ADLER-HASSNER CODES

Figure 2. Data-Translation Codes

Their purpose is to permit a channel to admit certain subsequences of symbols that may cause difficulties unless corrective action is taken. As an example, suppose a channel is available that accepts discrete data that may be binary, ternary, alphanumeric, or of whatever nature, but that may include troublesome sequences of symbols. Let us say that for a binary communications system certain patterns of 0's and 1's are present that cause interference data in the receiver. Ten years ago, the solution was to choose the clock rate low enough to avoid the problem. Simply increase the clock of the system until an operating problem appears, then back off the clock a few percent until the problem disappears. A claim could then be made that such a clock rate was normal for the communication system. The more modern approach now is to forbid the troublesome subsequences from existing in the data stream as it is sent to the channel, and then setting the clock rate just below the onset of some other problem. Of course if certain subsequences of data are removed from the data stream, there is a reduction in the effective information that is transmitted. It is natural to inquire whether this technique allows anything to be gained with an increase in the clock rate. The question one should ask is: "Do you pay more than you get or do you get more than you pay?" The answer is that you get more than you pay in most instances. Some of the procedures that you may have heard about have been around a long time and were designed in an *ad hoc* way without any formal approach to the problem. The technical papers of the late 1950's, for example, inform us that ternary line codes, especially, were used by the telephone systems of the time. Franaszek codes have played a very important role in the development of magnetic tapes and magnetic-disc storage systems, allowing a 50%

improvement in storage density for a trivial cost in the encoder and decoder. Among the more recent developments now in use, the Adler-Hassner codes are my favorites both from a practical point of view and because they are the most elegant from the theoretical point of view.

Figure 3 displays a simple example of a data-translation encoder that is similar to ones that have actually been designed. This particular binary channel uses four 1's to mean "end of message" as soon as they are received. Consequently, four 1's can never appear in the data stream because their receipt would cause the receiver to truncate the message immediately at that point. It is necessary, therefore, to remove all sequences of four 1's from the data. A naive approach to the solution of this problem was to insert a false 0 after every three data digits so that the transmission consisted of three bits of actual data in four bits of code in the channel. We say this code has a "rate" of 0.75. If we now go back and look at the so-called capacity of this channel, we find that it has a capacity of about 0.9. Thus it appears that much of the capacity is wasted by the unsophisticated code. A modern solution is to design a data-translation code that operates at or just below the capacity of 0.9 bits/bit, as illustrated in some of the following techniques that have been used successfully.

The Franaszek data-translation codes shown in Figure 4 are variable length codes of self-punctuation that are known as "prefix codes" to students

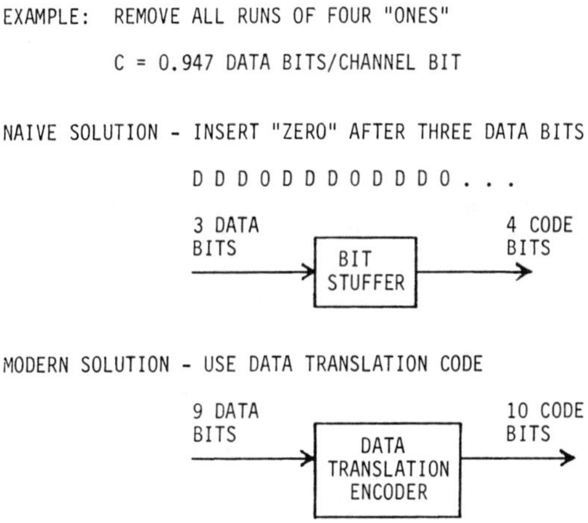

Figure 3. Data-Translation Encoder

EXAMPLE - FRANASZEK CODES
(MAGNETIC RECORDING)

- AT LEAST TWO "ZEROS" AFTER EVERY ONE
- AT MOST SEVEN "ZEROS" IN A ROW

INFORMATION WORD	CODE WORD
1 1	0 1 0 0
1 0	1 0 0 0
0 0 0	0 0 0 1 0 0
0 1 0	0 0 1 0 0 0
0 1 1	1 0 0 1 0 0
0 0 1 0	0 0 0 0 1 0 0 0
0 0 1 1	0 0 1 0 0 1 0 0

- SELF-PUNCTUATING VARIABLE LENGTH CODE
(PREFIX CODE)

1 1, 0 1 0, . . .

Figure 4. Franaszek Data-Translation Code

of information theory. These codes, however, are different from the prefix codes that are used for source coding since they expand rather than reduce the number of symbols. The columns in Figure 4 list the binary information and the corresponding encoded words. It is necessary to mention here that the magnetic recording channel for which this code is designed has two constraints: every data 1 must be followed by two 0's, and no string can contain more than seven 0's in a row. I will discuss the reasons for imposing such constraints in a moment, but first let us suppose that we have a binary channel where it is necessary to have these two constraints. A calculation of the capacity of the channel gives us a result that is somewhat larger than 0.5, so we endeavor to design a code whose rate will be at least 0.5. Franaszek designed this code some time ago. It turns out that in order to obtain a good code he had to go to a variable-length code, a prefix code, but the ratio between the number of information bits and number of code bits is always one to two.

Now suppose that we want to encode data 1 1 0 1 0. Since the code is self-punctuating, it has to be punctuated as 1 1 , 0 1 0.... The encoder can place the commas uniquely just by knowing that it has to encode the data stream starting at the beginning. The nature of the data stream does not matter since any arbitrary stream of data can be uniquely punctuated in the codewords that are used to map the data. Thus, 1 1 gets mapped into 0 1 0 0; 0 1 0 gets mapped into 0 0 1 0 0 , and so on. The encoded strings, in turn, are also uniquely punctuated so that the decoder can do its job.

Figure 5 contains a ground rule of magnetic recording that helps to explain why we need a translation code. The rule states that transitions must be at least T seconds apart. The magnetic recording people say transitions must be λ microns apart, but in communication terminology that is equivalent to

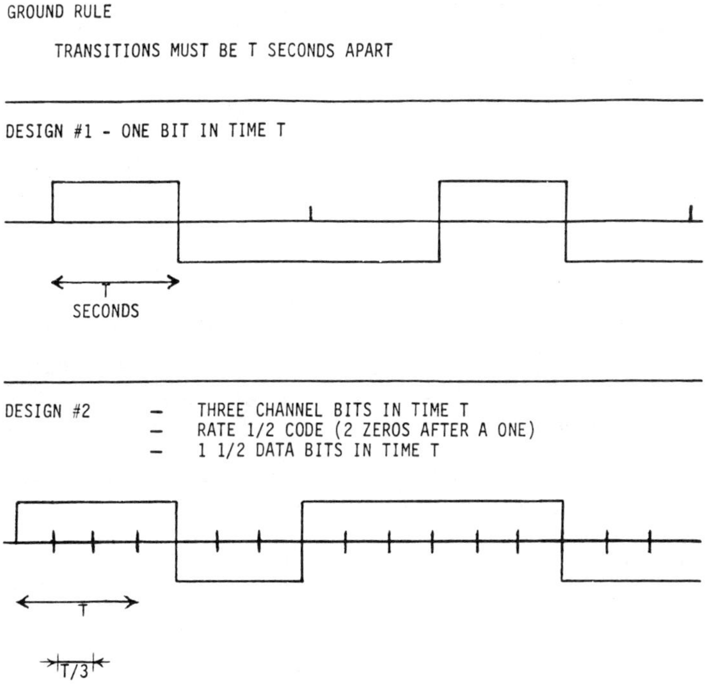

Figure 5. Ground Rules for Magnetic Recording

saying T seconds apart. The naive approach that was used a decade ago was to take a clock with transitions T seconds apart, and to fit one bit to every clock transition. The magnetic system is designed to transmit one bit in time T. The alternative scheme, shown in the lower part of Figure 5, is a second design approach that is much more subtle. The second design approach takes advantage of the fact that the constraint only says that the transitions must be *at least* T seconds apart, it does not say that they must be integer multiples of T seconds apart, so why not just introduce more granularity on the time axis. If this tactic is followed every T/3 seconds for example, there is a possible transition at any one of those times which seems to result in three channel bits in time T. In order to accommodate the constraint, however, we will use a rate one-half code that requires two 0's after every 1, an approach that suits a kind of differential encoding that causes 1's to become transitions while preventing 0's from doing so. But if we use the constraint that at least two 0's must follow after every 1, then it is trivial to do the differential encoding to put it into that format. What is the trade-off here? Although we now have three times as many channel bits per second, we had to use the rate one-half code, so we have to pay half of the bits back as code overhead. The overall result is that, although half of the channel bits are returned as a tax, we are still 50 percent ahead. There needs to be absolutely no change in the magnetic disc, the magnetic head, the servomechanisms or any of the mechanical components – just the addition of a small amount of silicon that suddenly introduces a 50 percent increase in density. That increase is the payoff for using the data-translation code. At one time this technique would have been turned down in the magnetic recording industry because the encoding of the information would require discrete silicon transistors. Now, with integrated circuits the circuit complexity is of little consequence.

Adler-Hassner codes do the same thing, and in a way that I consider to be more appealing, and not only from a practical point of view. Figure 6 shows an encoder and decoder for a code of rate one-half. The Adler-Hassner code satisfies a set of particular constraints: constraints on certain subsequences of 1's and 0's. The point is, however, that the Adler-Hassner codes are more akin in spirit to the convolutional codes that are familiar to those people who are involved with error-correcting codes. It is better to call them tree codes since they are typically nonlinear and, in fact, they can have potentially infinite encoding memory. They have the flavor of a convolutional code in that everything is sliding; there is no beginning nor ending. In the encoder circuit, bits are shifted into the input end, and at the output end the bits will emerge at twice the entry rate; two bits out for every bit that goes in. The output of the channel then goes through the decoder and the original data is reconstructed. What happens in between is that the prohibited data subsequences do not appear in the serial bit stream that passes through to the channel. I like

NOTES:

- EVERY 1 FOLLOWED BY A 0
- AT MOST THREE CONSECUTIVE 0'S
- CODE RATE EQUAL TO ONE-HALF
- CHANNEL SYSTEM RATE TWICE SOURCE SYMBOL RATE

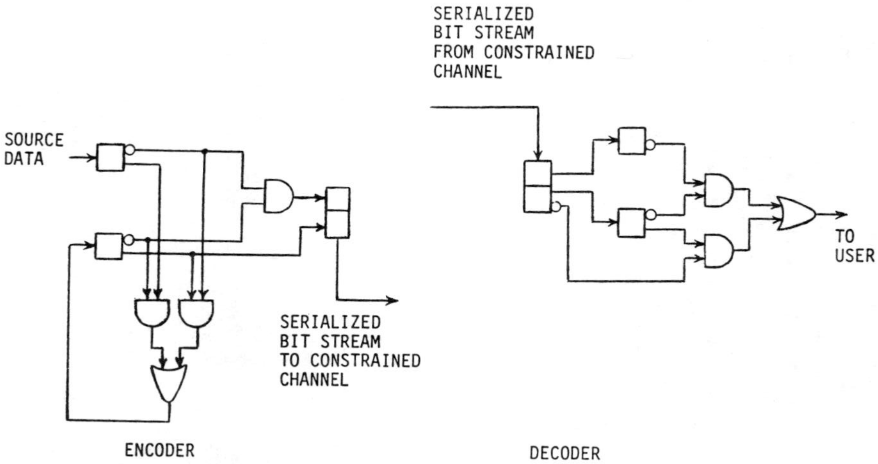

Figure 6. Encoder and Decoder for Rate 1/2 Adler-Hassner Code

Adler-Hassner codes because of the interesting history and theoretical content of their construction. Information theory is one of my loves and I am always happy when it has a spectacular pay off. Here we have an example of how theory has made it possible to show how to design these codes. The history of the development of these procedures is based on many years of theoretical work that evolved from background studies concerned with the isomorphism theorem and the concept of entropy as an invariant. In summary, we can now say that in principle any set of prohibited subsequences can be specified, and the Adler-Hassner procedure will allow the design of the necessary encoders and decoders. In fact, although Figure 6 only shows a simple logic drawing, the whole thing can be implemented with read-only memories. Source data is shifted in and the next code bits are looked up in read-only memory. The decoder also can be implemented as a read-only memory.

Figure 7 is a picture of the Adler-Hassner code for a ternary channel that I worked out as a simple example. The upper sketch is a practical channel that exists, as far as I know, in some kinds of telephone circuit or in a submarine cable. This is a ternary channel in which every positive pulse is a 1 and every negative pulse is a 0. For reasons of circuit design, transmitter cooling, or whatever, let us assume that the channel requires a 0 to come after every positive or negative pulse. The lower sketch is a sliding-block code or tree code that achieves that purpose. An example of binary data is shown entering the circuit and the resulting waveform is an example of the output – positive and negative-going pulses with at least one 0 in between. A shift-register type of decoder with a sliding structure is shown at the bottom of the figure. It is a ternary shift register. Two symbols are provided to the circuit, the decoder tables supply one bit of output data, and the circuit then slides the data one symbol further.

Additional information on the Adler-Hassner codes is given in a very nice article by Gina Kolata in the February 22, 1985 issue[1] of *Science*. Although I have not described how to do these things I want to leave the message that the theory is now known. If you study the structure of these techniques you can make your own examples for any desired application.

Before leaving the subject of data-translation codes, I will use the compact-disc player as an example. The compact disc has a built-in data-translation code but it is not an Adler-Hassner code; it is a block code that was designed before the Adler-Hassner scheme was invented. As you all know, the compact disc uses a laser read-out technique based on the presence of little dots that are pulse-position modulated. Since the dots are on the order of a micron in size and are spaced on the order of a micron apart, proper pulse-position

---

[1] Gina Kolata, "Changing Bits to Magnetic Blips", *Science* Vol. 227, No. 4689, 22 February 1985, pp. 932-33.

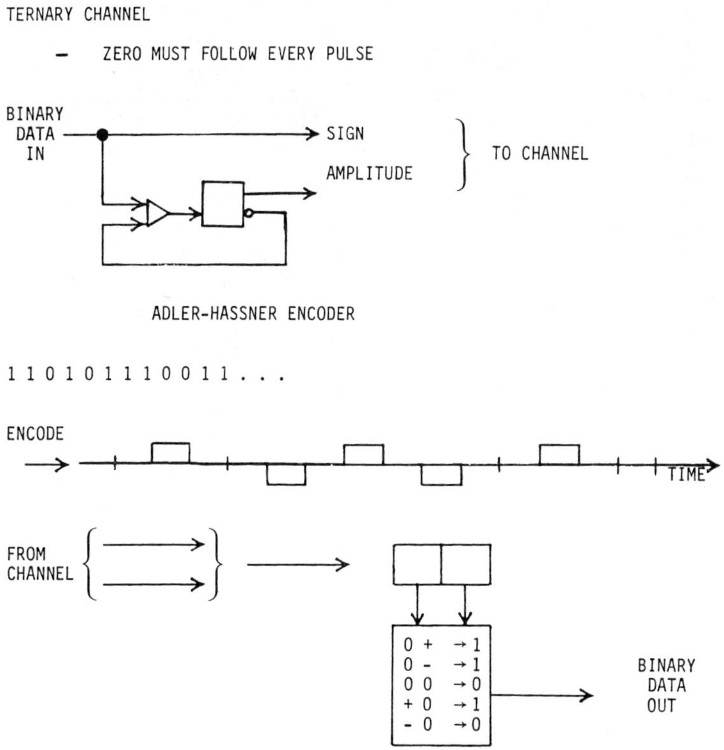

Figure 7. Adler-Hassner Code for a Ternary Channel

modulation requires that the modulated dots should not be allowed to move back and forth too much or they will overlap. One can achieve high density, however, without a possibility of overlap. It is only necessary to introduce the constraint that every 1 in the compact disc must be followed by three 0's. Further, in order not to lose the bit timing, there can never be more than ten 0's in a row because the bit timing is obtained directly from the disc. In order to accommodate these diverse constraints, a data-translation block code is used. In fact, it is a rate 8/17 block code, which means that 8 data bits are mapped into 17 code bits so as to satisfy the constraints. Thus for every 8 bits of data there are really 17 code bits stored on a compact disc.

I will turn now to data compaction, as outlined in Figure 8. Many things have happened in this area in the last two or three years that are of great interest, so I will try to bring you up to date about them. I will start the subject in this way: While I was drinking my glass of orange juice this morning (the high point of my day), I was thinking that the orange-juice industry, in a certain sense, is really ahead of the electronics industry, and the satellite communications people, and all of the speakers that we have heard today. I do not know how much research is done by the orange-juice industry, but they found out a long time ago that it is very expensive to ship truck loads of water from Florida to Massachusetts. Since orange juice is composed of about 75 percent water and 25 percent orange-juice concentrate, they figured out

- ENGLISH TEXT
- FACSIMILE
- TELECONFERENCING
- AUDIO
- VIDEO
- COMPUTER PROGRAMS
- RAW RADAR/SONAR DATA

Figure 8. Data-Compaction Applications and Process

that a lot of money would be saved by first removing the water, shipping the concentrate, alone, and then putting the water back in at the receiving end. Now I believe that we are starting to learn that the same thing is true in the engineering business – just ship the data concentrate (the information) around, leave the "water" bits out and put them back in at the receiving end wherever you wish.

Let us look again at the compact disc. If I remember all the numbers, the audio signal is sampled at a rate of 40,000 samples per second, with 16 bits to a sample. It is a stereo system so there is a right track and a left track. All of this information is converted into a 75-minute recording in this case. I have not done the work (perhaps someone here is interested in studying such things and going into the business of making supercompact discs) but I believe that in spite of all of these bits that are assembled here, a recording of Luciano Pavorotti's voice does not change very much in 25 microseconds within a dynamic range of 16 bits, and does not change very much from the right channel to the left channel. I think that there are 7 billion data bits on this disc, and of all of those, let's guess that only 10 percent, or 700 million bits are really required to make the recording. So let us take a lesson from the orange juice industry and concentrate the bits to put 10 hours of Luciano Pavorotti on the same disc (if you like him). The compact disc player can put the "water" back in and pop it back up to the normal 7 billion bits. That is what I mean by data compaction.

I want to make it very clear that I am not talking about anything that throws away data, or about any kind of approximation. I am talking about 100 percent exact, perfectly invertible mappings. Information is placed in the data-compaction computer, the data that goes through the channel has a rate that is lower than that of the source data, and then at the receiving end the encoded bits are mixed with "water" and restored to their original form. Since I realize that it is now possible to do these things so well, I am understandably irritated with some of the relatively poor telephone service that I have encountered recently (although I am not going to name any specific names). In particular, when I am making overseas calls, and the system is digital, there are, let us suppose, 8,000 samples per second each with 8 bits into the digital channel, thereby providing a high-fidelity communication channel at 64,000 bits per second. But everybody says that a 64,000 bit per second rate is too high to send through the channel, so we will solve the problem by cutting the speaker's audio off whenever the speaker is in a pause. So every time that I am talking to somebody, this thing thinks that I am taking a pause and I lose half of his voice, or I get someone else's voice. I do not know exactly what goes on in there, but the point that I want to make is that my brain is not putting out 64,000 bits per second by any means whatsoever. It is not possible for me to do so. I suggest that the 64,000 bits per second could be processed by some excellent data-compaction devices that use LSI chips that are already available. I am not talking about things that are imaginary, these things exist. If it is necessary to sample 64,000 bits, just run those bits through an array of boxes, as shown in Figure 8; the data compaction will reduce those 64,000 bits by a factor of say 10 to 1, or whatever is appropriate. Then the "water" is inserted at the receiving end and the information pops back up to the identical 64,000 bits in the original message. The customer never knows that all this is happening. There is no longer a need to cut off a customer's voice abruptly when he is pausing. This process that I have described does something like that, in effect, but it is done in a very deep mathematical way that is not so unsatisfactory.

In Figure 9 I have listed the names of some additional data-compaction codes. The Huffman codes have been known for a long time and are the oldest ones. My reading of the industry is that the Huffman codes are out of favor perhaps because they have been around for so long and are no longer glamorous. Elias proposed another kind of data compaction code some years ago. The Elias codes have a strange property (for a code that is manipulating bits) in that the encoder performs actual real-number multiplication, rather than Boolean operations. If the multipliers do not have infinite precision, however, the codes break down. Since the Elias code has a precision problem it was viewed as an impractical code for many years until Pasco came along and suggested that the precision problem could be eliminated by inventing some

- HUFFMAN CODES
- ELIAS CODES
  - PASCO-JONES
  - LANGDON-RISSANEN
- ZIV-LEMPEL CODES

Figure 9. Data-Compaction Codes

very clever scaling and rounding strategies, thereby avoiding the necessity for the multipliers to have an infinite number of bits. When this approach was used, the Elias codes became workable data-compaction codes. The practical versions are the Pasco-Jones code and the Langdon-Rissenen code.

Now I will discuss the Ziv-Lempel data-compaction codes. Elias codes and Ziv-Lempel codes both came out of information theory (much to the credit of that field). The Ziv-Lempel codes are quite different from the Elias codes; there is no reason to conjecture that the two codes are the same thing disguised differently.

Ziv-Lempel codes scan previously transmitted data attempting to find a data pattern waiting to be sent that is appearing in data that has already been sent. It then transmits the bit addresses of any found patterns to the decoder in the user terminal. Suppose the input data is a Dickens novel that is to be transcribed onto a tape or a disc. It happens that single pages or a series of pages in a Dickens novel tend to have several repetitions of the same words, the same person's name, or the same phrases. At any point in the process, let us say that the encoder stores the previous two pages of running text into its working memory. It then looks at the next passage to be transcribed. For the new passage, the encoder does not send the passage to the decoder as it is written. Instead, it sends a bit address that indicates where a similar passage can be found in the already transmitted material. The decoder now reconstructs the new passage by looking back at this bit address, pulling out the correct phrases, and tagging them onto the portion that has been reconstructed previously. The computer works its way through the entire Dickens novel in this manner. This kind of procedure can compact data down to somewhere in the range of two to three bits per English character. Let us imagine another application, a computer operator sitting at a remote terminal. Perhaps he feels that the computer's reaction time is too slow. But perhaps the fault is in the communications channel. The large number of symbols on his monitor must go through a standard conversion to binary data, known as the ASCII code, which then passes down a cable to the host computer. The digital host may be quite slow in receiving the data and transmitting the reply only because of communication limitations. If a data compaction code is used instead, there is less data to send and the user perceives a much faster reaction time.

Facsimile service is another example of the application of data-compaction codes. The information to be transmitted is converted into a string of bits that may then be compacted. The string of compacted symbols is reconstructed at the receiving end.

The next function, in Figure 10, is data compression. I will not dwell very long on data compression because I do not have any personal experience in this area, but the point should be made that many algorithms are now available. I have been told that in order to maintain reasonable voice quality

VECTOR QUANTIZATION

VOICE AT 300-400 BITS/SECOND

Figure 10. Data-Compression Codes for Voice Quality

it is sufficient to transmit at a rate of from 300 to 400 bits per second. The use of vector quantization schemes rather than scalar quantization, allows reasonable fidelity at such low data rates by quantizing the entire wavelength instead of a single sample.

At this point it is appropriate to summarize what I have said so far by means of the quotations given in Figure 11. For many years, I have felt that there has been a dichotomy between information theory and engineering practice. Figure 11 contains a gross paraphrase of Shannon's actual words in order to use terminology that makes my point. Since 1948, Shannon has said essentially, that digital communication reduces the channel bandwidth required in order to be able to attain a specified fidelity; regardless of whether we are concerned with digital TV, or any form of digital communication, the required channel bandwidth is reduced. On the other hand, examination of the literature of working engineers reveals a firm belief that digital communication increases the bandwidth required to obtain the specified fidelity. Basically, the two statements come to opposite conclusions because of different premises. Shannon recommends digital communication systems with optimal coding at

SHANNON SAYS -

"DIGITAL COMMUNICATION REDUCES CHANNEL BANDWIDTH NEEDED TO ATTAIN A SPECIFIED FIDELITY"

EXPERIENCE SAYS -

"DIGITAL COMMUNICATION INCREASES CHANNEL BANDWIDTH NEEDED TO ATTAIN A SPECIFIED FIDELITY"

??

THE DIFFERENCE IS CODING

Figure 11. Dichotomy Between Information Theory and Engineering Practice

every level. The literature of the working engineer largely ignores the possibility of coding.

I believe that the information-theory viewpoint will prevail for two reasons. (1) We are able to digitize the source efficiently by data compression and data compaction; source coding, vector quantization, and other quantization schemes allow us to minimize the number of bits required to represent the source. Although the techniques I have spoken about are sophisticated techniques in place of simple quantization or differential PCM, the degree of sophistication does not matter because our ground rule is based on the fact that today a hundred thousand transistors can be purchased for the same price that we had to pay for only one transistor many years ago. (2) We are able to improve our waveforms for communication to get more bits per hertz with essentially zero error rates. If we use channels that have the high signal-to-noise ratio required for high-fidelity TV pictures, or any other form of high fidelity system, we can transmit multibits per hertz. Consequently, not only do we have a smaller number of bits at the source end, but at the channel end we also receive more bits per hertz. The combination of all of these factors results in a decrease in bandwidth.

Data-transmission codes, or in better-known terms, error-correcting codes, as illustrated in Figure 12, are something with which I have had some personal experience. Here we have an encoder, followed by a channel that has distortion because of noise or other interference. That channel might contain the data-translation code that I mentioned before. For the present discussion, however, I will just look at it as a channel that makes errors or includes some sources of noise. Many channels make errors. Indeed, channels should make

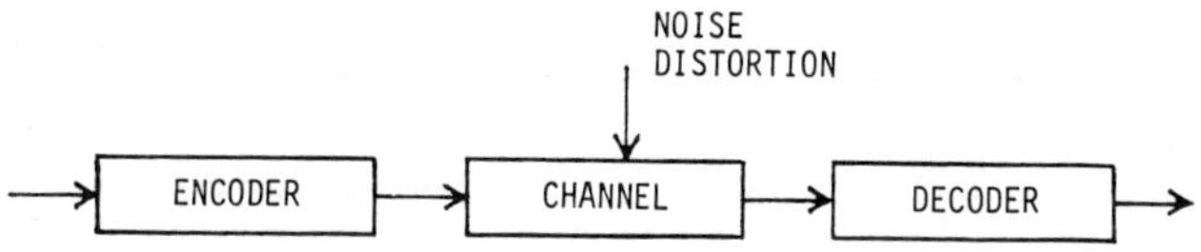

BLOCK CODES

- REED/SOLOMON CODES

CONVOLUTIONAL CODES

- UNGERBOECK CODES

Figure 12. Data-Transmission Codes (Error-Correcting Codes)

errors. I sometimes talk to people who will make what I consider to be an outlandish statement. They will say, "My channel is so good that I don't make any errors, so I don't need an error-correcting code". I always tell them that an error-free channel is probably a poorly designed channel. This may sound like a facetious response but I am very serious. I believe that when such a channel is built it is necessary to pay a high price for its error-free characteristic, perhaps by paying a price in power-transmitting capability or by paying for the error-free characteristic in some other way. For example, any attempt to make an error-free magnetic recording requires that defect-free material is available. I am always glad to see people make errors because it proves to me that they have designed their systems correctly. We now have available encoders and decoders that can be placed at the input and output of a communication system, such as the one in Figure 12. The decoder will correct all of the errors that occur and make the channel virtually error free. A compact disc, for example, contains a Reed-Solomon decoder so that scratches and dust will not degrade the reproduced sound. My view is that future error-correcting codes will be primarily of two kinds: Reed-Solomon block codes for heavy errors, and convolutional codes or Ungerboeck trellis codes for cleaning up isolated errors and reducing required signal-to-noise ratios. I will discuss both kinds of codes because I believe that they are the two most important contenders in this area. Before doing so, however, I will give some attention to the Hamming codes.

The Hamming code, as shown in Figure 13, is an example with 16 codewords, exactly the correct number of codewords to represent four bits. We can use the four bits of data to match the first four bits of these 16 patterns. When that has been done, we can designate the last three bits as parity bits. With these 16 codewords we can transmit any of 16 four-bit messages. The codewords are designed in such a way that every codeword differs from every other codeword in at least three places, so if the channel makes one bit error, we can correct that bit with the Hamming code. These codes are easy to understand and may be useful for certain kinds of operations, but they are much too weak for our desired applications in the communications field.

Reed-Solomon codes, much more powerful generalizations of the Hamming codes, are now important or will become important in all the areas listed in Figure 14. They are used in laser discs, as in the compact disc, in mobile radio, and in military communications. Even the *Voyager* spacecraft used a Reed-Solomon code to down-link pictures from Uranus.

I have a little story that illustrates the importance of the Reed-Solomon codes. I happened to be in Europe during the time of the war in the Falklands, and my knowledge of the war came from the London newspapers that were available to me at the time. One of the items that I read, but have never confirmed independently, was that the *Sheffield* had a priority message from

0 0 0 0 0 0 0
0 0 0 1 0 1 1
0 0 1 0 1 1 0
0 0 1 1 1 0 1
0 1 0 0 1 1 1
0 1 0 1 1 0 0
0 1 1 0 0 0 1
0 1 1 1 0 1 0
1 0 0 0 1 0 1
1 0 0 1 1 1 0
1 0 1 0 0 1 1
1 0 1 1 0 0 0
1 1 0 0 0 1 0
1 1 0 1 0 0 1
1 1 1 0 1 0 0
1 1 1 1 1 1 1

Figure 13. Hamming (7,4) Code

London via satellite, but could not receive the message because of interference from the Air Defense radars aboard the ship. The *Sheffield* commander decided to shut down the radars while the message was coming in. Unfortunately during that interval, the ship was struck by a missile and sank with great loss of life. I have a lot of personal experience with using Reed-Solomon codes in military applications in order to counteract radio interference from a ship's own radar. Codes that will read through any kind of interference from radar pulses are fully achievable today. I speculate that the *Sheffield* might be afloat today had there been a Reed-Solomon code in the satellite communication link.

For those who are unfamiliar with the Reed-Solomon codes, I have prepared the information shown in Figure 15. A Reed-Solomon code is a byte-organized code with the symbols represented as bytes rather than bits. Assuming that there are k bytes of data followed by 2t bytes of parity, and

- OPTICAL DISKS
- MAGNETIC DISKS AND TAPES
- MOBILE RADIO
- DEEP-SPACE PROBES
- MILITARY COMMUNICATIONS
- FREQUENCY HOPPING
- TELEPHONE LINE MODEMS
- SATELLITE COMMUNICATIONS

Figure 14. Reed-Solomon Code Applications

that the total number of bytes to be transmitted is k + 2t, the decoder will then correct any pattern of t byte errors. (A byte error is a byte that is wrong in any possible way whatsoever). Several examples are given in Figure 15, but I will concentrate on the compact disc, a device that works on an 8 bit byte. The blocklength is 32 bytes and there are 4 parity bytes, which means that the code will correct two byte errors. In the compact disc, this Reed-Solomon code is concatenated or nested with another Reed-Solomon code of blocklength

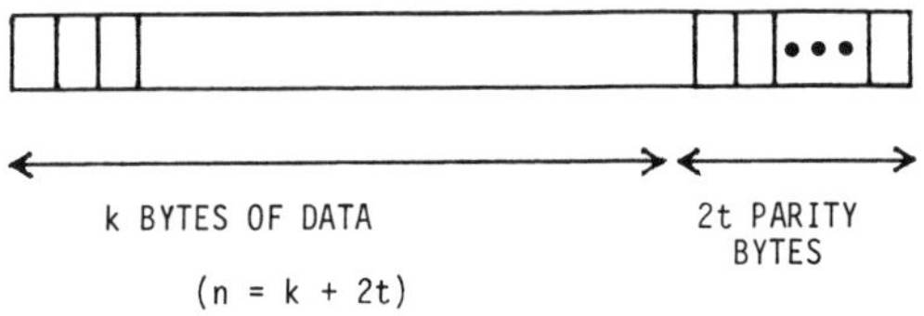

WILL CORRECT ANY PATTERN OF t BYTE ERRORS

EXAMPLES

JTIDS -	5 BIT BYTE	n = 31	k = 15
GALILEO -	8 BIT BYTE	n = 255	k = 223
COMPACT DISK -	8 BIT BYTE	n = 32	k = 28
		n = 28	k = 24

Figure 15. Reed-Solomon Code Examples

28 bytes and again 4 parity bytes. The reason for cascading two Reed-Solomon codes is to allow the second decoder to correct those errors that the first decoder might have missed.

As for the state-of-the-art in Reed-Solomon decoders, I feel that it is entirely possible today to build Reed-Solomon codes, working up to block lengths on the order of thousands, and with byte sizes, let us say up to 12 bits. Such codes could correct hundreds of byte errors and byte erasures. I believe that the kind of things I am talking about are entirely possible today, with data rates in the range of tens of megabits. The conclusion is that really powerful and massive error correction is possible with the Reed-Solomon encoder and decoder and all to be done on a little silicon chip. The user need not know that those decoder chips are designed into the receiver, just as the consumer does not need to know that there is a little silicon chip inside the compact disc player.

I also want to talk about Ungerboeck codes. Ungerboeck codes represent a certain kind of trellis code that I think is very important today. Figure 16 is a chart that portrays Ungerboeck codes used as a way of reducing signal-to-noise ratios, in contrast with the capability of the uncoded waveform, known as QPSK, that is so popular for satellite communications. Suppose that a typical satellite communication system, using QPSK, sends a certain number of bits per second, and has a certain signal-to-noise ratio, and a certain E_b/N_o required in order to make the system work. On the other hand, let us contrast it with an Ungerboeck code that replaces the signal constellation in the QPSK waveform with an 8-ary PSK signal constellation so as to introduce eight points in the complex plane that represent the amplitude and phase modulation (or in this example, phase modulation, only). The modulation into eight points is very simple, but is done in such a way that the data rate is not increased. Instead of using the eight points to increase the data rate by modulating three bits at a time into a channel symbol, the trellis code modulates only two bits of data at a time into a channel symbol. The encoder maps two bits of data into three code bits every single clock time. The three code bits define one of eight possibilities that are labeled in the phase diagram on the right side of Figure 16. Hence, every pair of incoming data bits is mapped into three code bits by the Ungerboeck trellis encoder in order to create a waveform. This action is performed by shift-register circuits that are encoding data based on the memory of the two previous pairs of bits into the encoder, rather than on the current two bits alone.

There are some complicated structures built into this process that are hard to understand. The result that gets transmitted, however, now has some very remarkable characteristics, as indicated in the summary statements at the bottom of Figure 16. The transmitted waveform has the same data rate as a QPSK waveform, and the code is transparent to the user. The QPSK

1) UNCODED QPSK

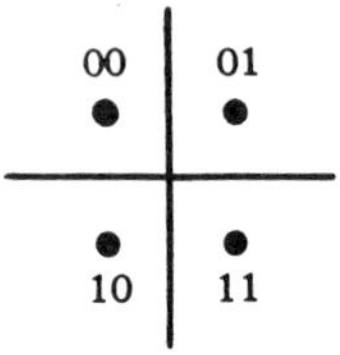

VERSUS

2) UNGERBOECK CODED 8-ARY PSK

- PARITY IN SIGNALING CONSTELLATION NOT IN BANDWIDTH

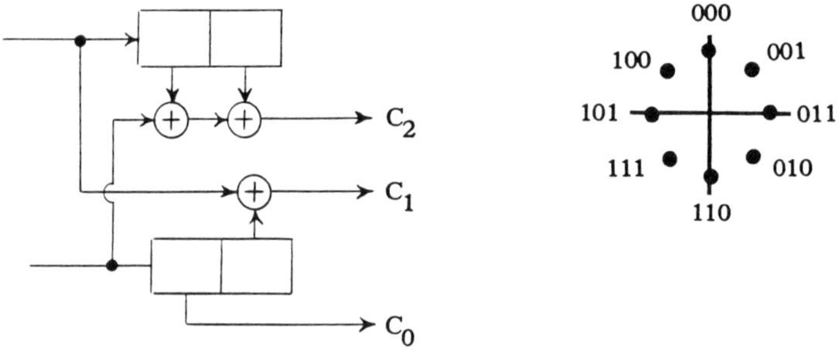

- SAME BANDWIDTH
- SAME DATA RATE
- 4.1 dB LESS POWER

Figure 16. Ungerboeck Codes for Reduction of Signal-to-Noise Ratios

modulator/demodulator can be snipped out of that receiver, the new modulator/demodulator can be installed in its place, and the user does not know that anything has changed. The circuit has the same bandwidth and the same data rate as before, but suddenly the power requirement is reduced by 4.1 db. We are free to spend that 4.1 db of now unneeded power in any desired fashion. If we use a trellis code that is a little more complicated we will have a power reduction of a nice even 6 db. In a roof-top antenna dish, 6 db less power corresponds to a 2 to 1 difference in the diameter of the dish. So if you are transmitting digital data through a satellite and you are worried about the esthetics of putting the dish antenna on the roof, use the Ungerboeck code instead of the QPSK code, it only takes a speck of silicon. Nobody will even know that the code is there except the guy who checks the roof to determine the size of the dish, and the neighbors who are enjoying the nicer view.

In addition to reducing power requirements and improving signal-to-noise ratios, Ungerboeck codes can be used instead, in a different dimension, as a means of increasing the transmitted data rate. Much more complicated signal constellations are required for this purpose, as illustrated in Figure 17. Examination of such complex signal constellations may lead one to believe that they are impractical or only of theoretical interest, but I assure you that there are modems in existence that work in signal constellations of this and even higher complexities.

The sketch in Figure 18, a typical example of the output of the first stage of a receiver, may provide an intuitive feeling for these complex signal constellations. Here is a case with a 4 × 4 array that results in 16-ary transmission. Suppose we are using a code to transmit three bits at a time, through 16 levels, so there is some kind of code structure that ties the history together in a kind of trellis code. A three-bit out of four, 16-ary code symbol goes through the channel as the amplitude and phase of a transmitted pulse. The complex amplitudes of the received pulses will be noisy at the receiver. The complex amplitudes can be measured and plotted in (x,y) coordinates. This plot will look like the sketched pattern when it is displayed on a 'scope in (x,y) coordinates. Examination of this noisy pattern may suggest that the bit-error rate is going to be horrible because it is very hard to be sure which transmitted point corresponds to many of the intermediate dots that have been received. When the system is required to have very small bit error rates, of the order of 10^{-6} or 10^{-10}, it might be distressing to see a display on the 'scope like the one shown in Figure 18. The Ungerboeck decoder, however, does not look at that picture on the 'scope; it uses the history of the received signal in order to determine from the context where the data points should be so it will demodulate with a low error rate.

My main purpose in this talk has been to make this audience aware of the wide variety of communication codes, such as error-correcting codes, that are

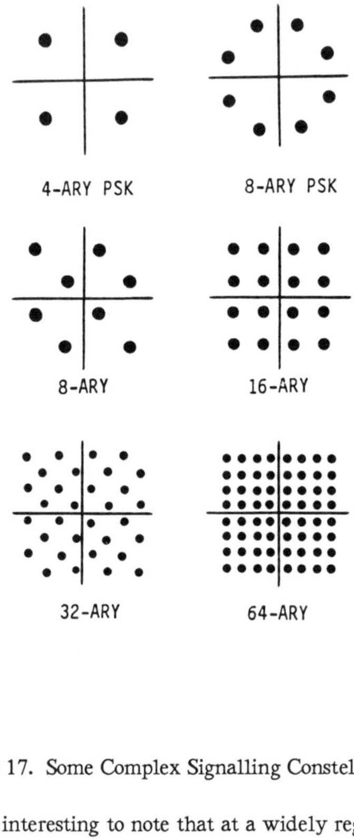

Figure 17. Some Complex Signalling Constellations

now available. It is interesting to note that at a widely regarded conference in Florida, more than 10 years ago, there was a general conclusion that coding was a dead issue because no one could afford it. Today it is clear that coding is not a dead issue because we *can* afford it. VLSI makes us able to afford it, and the massive needs of communications today makes us want to afford it. I will close with the following two quotations that I believe will help me to make this final point:

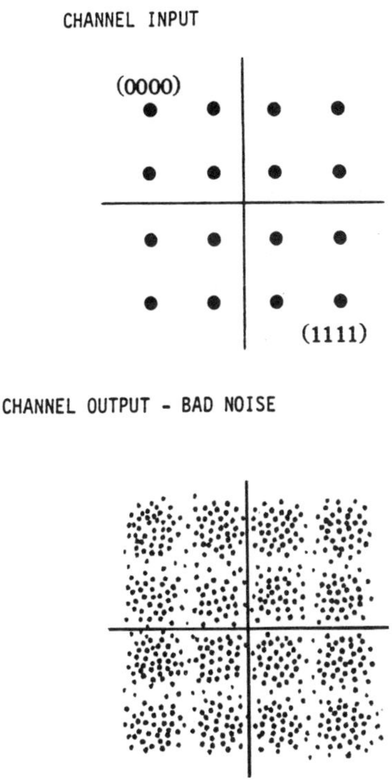

Figure 18. Typical Input/Output Signals of the First Stage of a Receiver

"There are two theories of light–the wave theory and the corpuscular theory–We used to believe in the corpuscular theory of light–Now we believe in the wave theory of light because all those who believed in the corpuscular theory of light have died".

—James Clerk Maxwell

"There are two kinds of digital-communication systems–those with coding and those without–We used to believe in communication systems without coding–Now we believe in communication systems with coding because all those who believed in communication systems without coding have died.

—Author Unknown

THE FUTURE OF DATA COMMUNICATIONS

G. David Forney, Jr.

G. David Forney, Jr. received the B.S.E. degree from Princeton University in 1961 and the M.S. and Sc.D. degrees in Electrical Engineering from MIT in 1963 and 1965, respectively. He joined Codex Corporation in 1965, became Vice President of Research in 1970, and Vice President of Research and Development in 1975. In 1982 he was appointed as Vice President of Motorola Inc. and Director of Technology and Planning of their Information Systems Group. He is currently Vice President of the Technical Staff and Chairman of the Business Development Committee at Codex. In 1971-72 he was a Visiting Scientist at Stanford University. Dr. Forney has specialized in the fields of communications, data communications, and information theory. He was a member of the IEEE Information Theory Group Board of Governors from 1970 through 1976 and began a new term on the Board in 1987. He served as Editor of the IEEE *Transactions on Information Theory* in the 1971-74 period. In recent years he has been active in policy matters in his fields of interest. He served on the IEEE Committee on Communications and Information Policy from 1978-80, and was appointed to a three-year term on the IEEE Communications Society Policy Board in 1983. In 1970 he received the Information Theory Group Prize Paper Award and won the Browder J. Thompson Memorial Prize Paper Award in 1972. Dr. Forney is a Fellow of the IEEE and a member of the National Academy of Engineering.

Like the other speakers, I want to thank the organizers of this symposium, and of the entire symposium series, for what I think is a brilliant way of celebrating a 100th anniversary. If the presentations at the other symposia are of the same high quality as the ones that have been made today, Cornell will have done an outstanding service to the whole electrical engineering community. I am privileged to be a part of this program and I have been fascinated by what I have heard so far.

When Toby Berger asked me to give this talk as part of a program that was generally concerned with future directions in electrical engineering, he suggested that it might be interesting to talk about the future of data communications. This topic is one that I had not attempted to address in quite such breadth before, but it seemed reasonable to give it a try, since I have been in the communications business for quite some time, and since I am paid in part to have a well-informed view on where data communication is going. I thought that drafting a few simple propositions on the subject would be a useful exercise, both for myself and for my listeners.

It did not take me very long to find out why most sensible people avoid giving talks on the future of data communications. For every trial proposition that I wrote down, I quickly came up with a counterexample, or at least a strong qualification. In fact, it appeared that there was very little that could be said without fear of contradiction, or even in which I could generate much conviction myself.

Consequently, the main theme of my discussion will be the term that Paul Green used earlier today: *variegation*. In the 15 to 20 years that I have been a part of this rapidly growing and fascinating industry, it has been my experience that the data communications world is generally unpredictable and full of variety, diversity, complexity, and messiness. So the series of propositions that I have formulated are deliberately paradoxical. Perhaps this approach is chicken-hearted, but I hope that at least it will stimulate some thought and a few questions.

The first proposition (Figure 1) states in part the well-known fact that within what is now known as the "information industry", data communications has emerged as a distinct industry in its own right, indeed one of the most dynamic and fastest-growing in this sector. When I entered the commercial communications field in about 1970, there was no such thing as a "data communications industry". At that time there were two separate worlds that were only marginally connected: the traditional telecommunications world, largely voice-oriented, and the computer industry, described for us so well by Paul Green. To send data over voice-grade telephone lines from computer to computer, the telephone company used what they called "data sets" (now known as "modems"), with a top rate of about 2400 bits per second. As for the computer industry of 1970, perhaps its most salient characteristic in retrospect

DATA COMMUNICATIONS

IS BOTH AN EMERGING AND

DISAPPEARING INDUSTRY

Source: MOTOROLA INC.
Information Systems Group

Figure 1. Proposition Number One

was the relative lack of extensive communication. Most computer systems existed as isolated islands that performed local functions. Of course, there were a few specialized applications, such as airline reservations and time-sharing, that depended intimately on data communications. Of the mainframes that existed at the time, however, probably only 10 or 20 percent were equipped with any kind of communication capability. As Paul mentioned, when communications were required, an *ad hoc* methodology was developed for each application, unlike the integrated architectural approach that is used today.

Between these two worlds, there was hardly anything that could be called an industry, although a few companies were attempting to make "black boxes" for commercial data communications applications. As I recall, our largest competitor had about $10 million in annual revenues in 1970, far greater than our own.

Today, in 1985, Wall Street classifies data communications as a $2 billion industry that is growing at about 30 percent per year, depending on how one counts, with projections for continued growth at about the same rate. In that sense, the industry has indeed emerged. Within the past 15 years, there has been an explosion of trade journals in this business, and a cadre of stock-market analysts are following the fortunes of the various participants. Thus we see all of the indicators of an established industrial segment, in the same sense as the forest-products business, say.

At the same time, however, it is common knowledge that a so-called "information industry" is being formed by a merging of all information processing and transmission activities, and that data communications is becoming part of this colossus. As expressed in Figure 1, just as we are emerging into the limelight, we are ironically being enveloped in a larger movement that foretells our ultimate disappearance as a distinguishable segment.

If we examine the current status of communications and computing, we still find surprisingly little overlap. Within the public telecommunications networks there have been tremendous advances in the application of digital techniques. Stored-program-control switches are essentially special kinds of computers. Digital-transmission facilities are available, together with their analog counterparts, for both long and short distances. Large numbers of bits are being transferred within the network. However, if we ask how much of this capacity is being used for end-to-end digital transmission, we find that the amount is quite small; I would guess only a few percent of all current traffic.

On the computer side, as Paul has described so well, there has been substantial rationalization of the way communication is done. As a result of much careful thought, data communications architectures have been developed and widely implemented. At the same time, however, these orderly structures are being tested by the advent of an extraordinary variety of new technologies, a trend that I will discuss further later in this presentation.

Within the telecommunications world, there is still a tendency to take the traditional telephony viewpoint that the important thing is the data network, a sort of all-pervasive cloud to which one can attach little boxes called "data-terminal equipment", be they host computers, personal computers, banking terminals, or whatever. Within the computer world, on the other hand, there remains the traditional tendency to try to handle all communications functions within the computer, down to the point where there has to be a connection to a physical link, at which point, as a kind of necessary evil, the task is turned over to the telephone company.

The communications and computer industries have had many dialogues and some convergence, but enough room has remained in between for my industry to do about two billion dollars worth of business in the interconnection of the two worlds, including the provision of modems, multiplexers, protocol converters, digital-data switches, and similar equipment.

At this point it is reasonable to ask what may happen over the next 15 years. Will the data communications industry, as such, disappear as its components are absorbed in the convergence of these two worlds?

On the computer side, there is no doubt that more and more communications intelligence will become embedded into computers. The computer people will naturally try to provide as much communications functionality as possible and leave the communications people as little to do as possible.

On the telecommunications side, there is the relatively new concept of an integrated services digital network (ISDN), which has evolved from the traditional communications world. I hope I am not being unfair if I characterize ISDN as growing out of the notion that once voice traffic has been digitized and sent everywhere as bits, wouldn't it be the most natural thing in the world, as well, to carry everyone's data traffic in an intergrated common public utility facility; and having done that, wouldn't it then be natural also to provide all kinds of shared services?

As far as I am aware, the ISDN idea had its genesis in Europe where the telephone companies are generally part of the government and enjoy a much more monopolistic position than in this country. Indeed, early reaction to ISDN was somewhat skeptical even by telephone people in the United States. Within the past year, however, the concept has been embraced heartily. Standards have been adopted, and extensive implementation activity is in progress.

It is clear that the world views of the computer and the telephone industries contain abundant potential for conflict. There will be an ongoing turf battle over the relative amount of communications functionality that will be performed within common-use networks *versus* within computer systems.

My prediction is that every one of these forecasts will come true, in some degree. There will be not only an ISDN, but multiple ISDNs. Value-added networks of all kinds will be available. Private corporate networks will continue to be operated for economic, functional, and strategic reasons. The U.S. Department of Defense will have its own separate networks. An incredible variety of local networks will be established. Within the computer industry, many different kinds of networks will arise for different purposes.

In other words, no one approach will become universal. Consequently, the communications world will become very complex as the various networks interact with each other. The enormous tasks of interweaving these networks and managing and controlling the subsequent immensely complicated systems will provide many opportunities for the current enterprises that call themselves data communications companies. The larger players in this game are already joining the general movement to embody sophisticated processing techniques into network equipment for integrated transmission of voice, video, and data. Smaller-niche players will continue to proliferate as they develop boxes that incorporate the most advanced technologies.

The industry has already had a rather spectacular financial record. It is difficult to think of anyone who has been completely unsuccessful, and it is easy to think of people who have made admirable progress. While most small start-ups have a rather limited life expectancy, the relative failures in the data communications industry have occurred among those who have not moved with the times; those who have kept up have done very well. One may conclude that it is not that difficult to succeed in this industry.

My second proposition (Figure 2) has to do with the common perception that the way to deal with a proliferation of complexity is through standards.

THERE WILL BE BOTH

GREATER STANDARDIZATION

AND GREATER PROLIFERATION

Source: MOTOROLA INC.
Information Systems Group

Figure 2. Proposition Number Two

There has been an explosion of work on standards, as mentioned by Paul Green. There is an increasing desire for "promiscuous" communications: the ability for a local terminal to talk not only to a similar terminal at some other location but to anybody, anywhere. Achievement of this objective requires a universal standard language, an Esperanto of some kind. In the 1970's it was not very important if communication between IBM stations was restricted to IBM equipment, or if DEC terminals could only talk to other DEC terminals. Now, in contrast, computer people realize that the sustained growth of the entire industry is highly dependent upon openness in communications, thereby creating one of the chief pressures for standardization.

Another pressure arises from the desire of the user not to be locked in to the equipment of any one particular vendor. Since he would like to be able to "mix and match" his components, he must be certain that they will all play together when they are assembled. Important users, such as government agencies, tend to exert great pressure for standards, partly to prevent any one group from dominating the field, and also to avoid the establishment of *de facto* standards that merely reflect how one company happens to do things.

The final pressure comes from VLSI designers, for whom the adoption of a standard is the signal to begin fabrication of the 100,000-transistor VLSI circuits that will implement the latest protocols, or whatever. Here we have a case of a virtuous circle: an agreement on a standard sets in motion forces that bring down costs, which, in turn, make the new standard economically viable.

The various pressures that I have mentioned have produced extensive standardization in the industry. Another recent trend, however, is the development of multiple standards for what would seem to be the same function. For example, in the local-area network arena, six or more standards have already been developed, and there will no doubt be more to come. In my opinion, this trend should not be considered as tragic. Rather, we are developing one standard for each technical category, and then letting "survival of the fittest" decide between categories.

Another change is that whereas in the past, standards tended to follow the acceptance of a technique in practice, nowadays standards are often being written well before the development of the equipment that is to use them. Within the last year, for example, we have seen the development of a standard for dial-line transmission at 9600 bits per second before anyone had a product that was capable of such performance. In my experience, this anticipatory process can work quite well when the membership of a standards committee contains the appropriate experts, although inevitably, some weeding out will occur.

My third proposition (Figure 3) is concerned with the persistence of old technologies. For example, Paul Green mentioned the many "Bi-Sync" communications links that are still in use today. To me, the most dramatic illustration of this tendency is given by my firm's annual internal forecasts of the future of modems. For the past 15 years or more, we have forecasted that we

NEW TECHNOLOGY INTRODUCTION

WILL ACCELERATE, WHILE OLDER

TECHNOLOGIES WILL PERSIST

LONGER THAN PREDICTED

Source: MOTOROLA INC.
Information Systems Group

Figure 3. Proposition Number Three

can expect an increase of from 30 to 40 percent in modem production in the coming year, followed by only about 20 or 25 percent for the following year. We can then expect to remain on a plateau for about five years, after which modems will begin to disappear as the general transition to digital transmission systems comes into effect. That was the forecasted curve in our 1972 five-year plan, and it has been our forecasted curve every year since then. Every year we simply changed the axes to reflect increased volume and to move the curve over by one year. To me, there is no more telling instance of the difficulty of prediction and the fallacy of believing that the world is going to go in only one way.

My central message, therefore, is that the communications world will go in many ways. Although there will be some Darwinian fallout along the way, we may be sure that everything that is being discussed today will be given a trial. It will all happen; none of it, however, will be adopted universally.

We can draw several conclusions from these overall themes of an increasingly diverse and complex world. In the area of public policy, I think that it is urgent to resist simple views of the world as well as simple solutions. While it is easy to make this statement, it is not so easy to convince non-technical people of its validity. Just yesterday I had a conversation with a distinguished professor who is an authority in the fields of economics, public policy, and organizational behavior. Following some recent talks with representatives of the communications industry, he had a very exciting idea. He felt that a strong consensus could develop for the notion that the world is in great need of a cheap universal digital service. Since such a service would obviously require some standardization, it would only be necessary to decide on the proper standards, the chip makers could then go into high gear, and economic nirvana would arrive. Although my reading of this man's work reveals him to be a person of great subtlety, I had considerable difficulty in convincing him that the communications world is just not that simple. In fact, the one thing that the growth of communications is clearly outstripping is the competence of anyone to regulate or set policy for it.

Every year, the ability of the structures we have in this country, and in other countries, to keep up with what is possible, what is feasible, what is desirable, and what is economic, is gradually deteriorating, simply because of human limitations. For example, there are many views on the wisdom of Judge Greene's modified final judgement that came out of the antitrust trial against AT&T. Speaking for myself, I think that of all the people in policy positions who have tried to master the complexities of this industry, Judge Greene has done, and continues to do, by far the most impressive job. Of course that was a very expensive education that we had to give him and that he, in turn, is giving us. Although I am not a fan of settling things in the courts, I think that it is only necessary to compare Judge Greene's exceptional grasp of the industry with what can be expected from harassed congressmen, FCC commissioners, and their staff members, in order to realize how remote is the possibility of sensible central planning and policy making for the communications industry.

We have a strong bias in this country toward letting "a hundred flowers bloom": that is, toward letting the marketplace decide upon the most effective approach. Consequently, I believe that there is a high probability that in the year 2000, everyone's data communications needs will be served more effectively in this country than in countries where central planning is more in fashion. I am not necessarily referring to the Soviet Union; in Germany, for example, what communications capabilities are provided is decided by the *Bundepost*. It will be interesting to check out this prediction 15 years from now.

As for technology, it seems to me that our industry is moving away from being driven by the kinds of mathematical theory and analysis that I have worked on myself, or that Dick Blahut has just talked about. In the future, more attention will be given to the management of complexity. In that sense the communications industry will become more similar to the computer industry.

After listening to Bob Lucky's talk. one might conclude that bandwidth will eventually become free. Unfortunately, that happy event will never occur. In fact, the cost of bandwidth seems to be coming down more slowly than anything else in the information industry, such as the cost of computation or memory. Nonetheless, the importance of using transmission channels efficiently will moderate.

The central concerns in the communications industry will be with systems issues such as network management and control, and giving the user a way to deal with these complex networks. Within the networks, software issues will take precedence over hardware. In the past seven or eight years, for example, our company has employed more software than hardware personnel. Again the communications industry will appear to be more like a branch of the computer industry than it was 15 years ago.

Finally, it seems appropriate to discuss these changes in data communications in terms of their possible implications in a university environment, particularly in the development of the electrical engineering curriculum. I have very strong feelings about the role of the university. Even in the face of the trends I have described, I am firmly opposed to the introduction of special courses in the management of complex systems, or in attempts to have students build and operate complicated networks. As Professor Adler suggested earlier today, academic efforts of this kind have always proved to be difficult to bring up to the standards of industry, and disappointing in their impact. In general, I cannot become very excited about having students build toy versions of what is going on in the outside world just to provide them with some practice, expose them to the issues, or perhaps to have something to show to distinguished visitors.

The special mission of the university is to advance the discipline of each field of study. I am sure that all of us on this panel today were exposed to the most challenging intellectual problems during our university experience. None of us suffered notably from trying to find the best solutions to those problems. It's somewhat like the traditional justification for learning Latin: regardless of whether it has practical use, it is an excellent exercise for training students to use their minds.

On the other hand, I am also not enthusiastic about theoretical university research that is pursuing sterile, secondary, tertiary themes – in effect, attempting to find the next digit in the decimal expansion of pi. How is that kind of theoretical research to be distinguished from the truly vital, interesting research that is characteristic of the best that I know is being done in the universities? The ability to make this distinction is dependent upon having a faculty with a well-developed sense of taste. In my experience the best engineering research has been done by people, whether faculty members or engineers in industry, who have managed somehow to become sufficiently immersed in "real-world" problems to understand the real issues, including complexity. Then, having achieved some meaningful insights about these matters, they step back and pose illustrative thought problems that illuminate those real issues. At that point, theoretical investigations can really pay off.

In that sense, the superb work that was discussed by Dick Blahut perfectly epitomizes the kind of work that is the special mission of the university. For this to take place, fruitful connections with industry must be established by the university, and an equal burden falls upon industry to establish those connections. Although tribute to this sort of cross-fertilization has become a cliché, it is the only way that the taste of which I spoke can be developed in the individuals who are guiding university research.

FUTURE DIRECTIONS IN ELECTRICAL ENGINEERING

ATMOSPHERIC AND SPACE SCIENCES

PROCEEDINGS of the

Cornell Electrical Engineering Centennial Symposium

at St. Louis, Missouri—April 17,1985

Simpson Linke, editor

Promethean Press
Ithaca, New York

PROCEEDINGS OF THE
CORNELL ELECTRICAL ENGINEERING CENTENNIAL SYMPOSIUM:
FUTURE DIRECTIONS IN ELECTRICAL ENGINEERING-
ATMOSPHERIC AND SPACE SCIENCES

Copyright 1988
by the School of Electrical Engineering
Cornell University, Ithaca, New York
All rights reserved

Published by Promethean Press
Ithaca, New York

Printed in the United States of America

September, 1988

CONTENTS

PREFACE

The celebration of the Centennial Year of the School of Electrical Engineering at Cornell University featured a series of symposia based on the central theme of "Future Directions in Electrical Engineering". Each seminar concentrated on a major facet of the discipline, with individual topics presented by outstanding specialists in each field. The symposia were held at sites around the country that are widely known as centers of activity for each particular subject area, thereby providing significant updates on pertinent electrical engineering technologies at appropriate locations, and allowing a large number of Cornell Electrical Engineering alumni to participate in the celebration in a meaningful way.

The ultimate success of the Symposium Series must be credited to the dedicated efforts of the faculty, symposia, and local alumni chairmen who were responsible for planning the programs, engaging the participants, and attending to local arrangements within specific time constraints. The following schedule of events attests to the overall magnitude of the task and also displays the broad range of topics that were presented:

March 15, 1985—Microelectronics—Stanford, Calif.
April 1, 1985—Computers—E. Windsor, N. J.
April 10, 1985—Communications—Boston, Mass.
April 17, 1985—Atmospheric and Space Sciences—St. Louis, Mo.
April 25, 1985—Energy for the Future—Washington, D. C.
June 12, 1985—Societal Impacts and Future Directions—Ithaca, N. Y.

Each session was videotaped to facilitate preparation of the proceedings of each meeting and for possible use of the individual presentations in classroom seminars and elsewhere. The audio portions of the tapes were transcribed into preliminary text and coordinated with figures obtained from the video images and from hard copy provided by some of the authors. Each presentation was edited with a view toward achieving continuity and coherence without

compromising the natural style of the author. In several instances it was necessary to introduce substantial changes in format and phrasing in order to achieve this goal; in others, the hard copy supplied by the authors made the task easier. In all cases, the final versions were sent to the authors for their criticisms, corrections, and final approval. I am grateful to all of the contributors for their excellent cooperation and assistance in bringing their presentations to this present form.

I wish to thank Professor Edward D. Wolf for his enthusiastic interest in this project and for authorizing financial support from the National Research and Resource Facility for Submicron Structures to assist in the publication of the Symposia Series *Proceedings.* Additional funding from the College of Engineering and the School of Electrical Engineering is also gratefully acknowledged.

It would not have been possible to produce this work without the skillful secretarial assistance of Joan Hammond, Catherine Kuhl, Mary Root, and Linda Struzinsky who performed the difficult task of transcribing the tapes into typed versions of the unedited manuscripts. The valuable photographic assistance of John Koumjian in the reproduction of the video images is also acknowledged with thanks. The expert guidance provided by Mrs. Diana Nathanielsz on format and production of this volume is very much appreciated.

I am particularly grateful to Catherine Kuhl for her diligence and care in forming the camera-ready copies of several of the manuscripts on that versatile new instrument that is so illustrative of "future directions in electrical engineering" — the laser printer.

Simpson Linke

Ithaca, New York

INTRODUCTION

John A. Nation

Director, School of Electrical Engineering

The celebration of our Centennial Year affords a unique opportunity to review past accomplishments, to evaluate our present circumstances, and to make some reasonable estimates of prospects for the coming years. A glance at the history of the University reveals that the Cornell *Register* for 1883 contained a "Special Notice" from President Andrew Dickson White announcing the establishment of a "Course in Electrical Engineering." President White had convinced the University's Trustee Executive Committee to create the new program "provided such course occasions no additional expense."

The early curriculum was rigorous and in addition to the traditional engineering topics of mathematics, physics, mechanics, mechanical drawing, and shop work, required rhetoric, French and German. The senior year also included courses in several Cornell specialties such as photometry and electric lamps, telegraph instruments, lines and cables, the dynamo, and electric motors. The first baccalaureate degrees in electrical engineering in the United States were awarded by Cornell (as well as by M.I.T.) in 1885, and the first Ph.D. in the new discipline was conferred by President White at Cornell in the same year. Thus we are truly celebrating 100 years of electrical engineering education.

One of the major figures in the E.E. School of those days was Professor Harris J. Ryan who ran the Electrical Engineering Department essentially as a one-man faculty for 17 years and is credited with having trained almost 1000 electrical engineers during that period. Ryan left Cornell in 1905 to become head of the Electrical Engineering Department at Stanford. It is interesting to note that 12 of the first 13 members of the Stanford faculty were educated at Cornell.

This brief summary of the beginning of electrical engineering instruction at Cornell would not be complete without mention of the internationally known Professor Vladimir Karapetoff who joined the faculty following Professor Ryan's departure and made many contributions to early electrical engineering theory and analysis in several textbooks and many technical articles. Something of the flavor of electrical engineering studies of the time is suggested by the following excerpt from Karapetoff's *Electrical Laboratory Notes*, published in 1906:

> "In coming to the laboratory, bring with you a slide rule, an inch rule or tape, a speed counter, a screw

> driver and a pair of plyers [sic]. This will save you time and trouble of looking for them or borrowing them. Do not forget to have a pocket knife for skinning off wire; a bicycle wrench is also sometimes very handy to have."

In these days of hand calculators and personal computers, it is rare indeed to find a student who has even seen a slide rule, much less to have used one.

Classical electrical engineering education at Cornell was devoted principally to electric machinery, power systems, telecommunications, and early electronics. As the technology developed from the 20's through the 40's, the curriculum was, of course, continually upgraded as new and advanced concepts were introduced, but the basic format remained essentially unchanged until the mid-50's when research began to play a central role in the E.E. School, coincident with the move from Franklin Hall to the newly constructed Phillips Hall. The following figure among the highlights of our research development over the years:

> An early solar telescope project initiated an era of Radio Astronomy research that culminated in the establishment of the world's largest radio-radar telescope at Arecibo, Puerto Rico, now known as the National Astronomy and Ionosphere Center.
>
> A research program in the new field of Physical Electronics gradually expanded into the present National Research and Resource Facility for Submicron Structures housed in the Knight Laboratory adjacent to Phillips Hall.
>
> The Old High Voltage Laboratory on Mitchell Street, converted to an Energy Laboratory in 1960, served as the staging area for the Cornell Extra-High-Voltage Cable Test Program and housed E.E. research on high-voltage vacuum switches and arc physics until the formation of the interdisciplinary Laboratory of Plasma Studies in 1966. This facility, together with adjacent installations, now contains the largest particle-beam program in any university in the world and has made many important contributions to the understanding of intense electron and ion beams, particularly as they relate to inertial confinement thermonuclear fusion experiments.
>
> The national energy crisis in the early 70's created renewed interest in power and energy research within the School leading to the combination of power-system elements, control theory, and new computer facilities, that now constitutes the unique Kettering Power Systems

Laboratory for the analysis of modern complex power-generation and transmission networks.

Two recent events attest to the high regard that the University and the E.E. School enjoy in the national research community:

1. The Program in Microscience and Technology was established in 1982 and receives about $2 million annually through the Center of Excellence Program of the Semiconductor Research Corporation.

2. The NSF-sponsored Cornell Center for Theory and Simulation in Science and Engineering, with its associated supercomputer production facility, will augment and expand research capability in the School and will encourage greater interdisciplinary activities with other scientific units on campus.

One of our strengths in recent years, as the Electrical Engineering School has evolved from a purely teaching organization into a teaching and research institution, has been our ability to attract good new faculty. A positive measure of this quality is in our receipt of Presidential Young Investigator Awards from the National Science Foundation in the first year of the establishment of these awards. The E.E. School received 5 of the total of 26 awards made available to electrical engineering departments in the nation. Our success in attracting these high-quality young people bodes well for the continued future growth of the School.

Remarkable changes in electrical engineering education have occurred in the School within the past 25 years as the influence and impact of new faculty members and their research have brought modern concepts to curriculum and laboratories that had been essentially identical to those of World War II days. The advent of computers has been particularly significant. We have been very fortunate in the last decade in the support that we have received from industrial corporations. Their generous donations of equipment including computers, work stations, and state-of-the-art measurement devices have been of great value in upgrading our laboratories to present-day standards.

On the other hand, the current huge interest in computer engineering and electrical engineering has introduced major enrollment problems in the School. Using 1965 as a year of reference, the size of the faculty has remained fixed at about 40 members whereas the number of undergraduate students has tripled. Our senior class in 1985-86 numbered 186 students. We have 265 graduates students in the School with about 145-150 MS/PhD students and the remainder in the Master of Engineering Program, the latter being the second largest professional degree program in the University. This year we graduated

30 MS/PhD students and 88 Master of Engineering degree candidates. An encouraging factor is that 12 of these advanced-degree students were women. Today about 15 percent of the students in the E.E. School are women and 25 percent are minority students.

The increase in student enrollment is coupled with a research effort that is growing at a compound rate in excess of 10 percent per year in constant dollars. Clearly we are approaching the limits of our present capability. It is hard to realize that within 30 years of our initial occupancy of Phillips Hall we are now bursting at the seams. Further, the building which was then state-of-the-art, is now an antiquated facility in terms of present-day technology, with little in the way of the clean-air environments, air conditioning, and other amenities that are necessary for the proper functioning of modern equipment.

Plans are in progress to build a new engineering complex on campus in which electrical engineering will have substantially more space than in the past. When these new facilities are completed, our faculty will be able to work at its full potential. We expect to see a relatively modest growth in the number of faculty members. Since a faculty lifetime post is forty years, and student-interest time cycles are rather shorter than that, the response to large student increases in enrollments and interest in electrical engineering cannot be followed instantly by new faculty positions. Attempts will be made to maintain our undergraduate size at about its present level with some growth allowed in the number of our graduate students. It is certain that there will be many more interactions with industry in the near future, and these will take on a much broader form as the years go by. For example, an exchange program may be arranged with industrial laboratories so that their scientists will come to work in the School for a period while a faculty member will have the opportunity to spend corresponding time in industrial environments.

This brief perspective of past, present, and future encourages us to begin our second century with a true sense of excitement. We may be justly proud of the important contributions that our School of Electrical Engineering has made to many of the developments that are reported in these *Proceedings*. As we celebrate these past achievements, we can look forward with confidence to producing equally significant additions to our challenging discipline.

PART 4

ATMOSPHERIC AND SPACE SCIENCES

PROCEEDINGS of the

Cornell Electrical Engineering Centennial Symposium

at St. Louis, Missouri—April 17, 1985

"It is a pleasure to welcome you to this symposium on 'Future Directions in Electrical Engineering—Atmospheric and Space Sciences,' the fourth in a series of cross-country meetings on major electrical engineering topics that are being held this Spring in celebration of 100 years of electrical engineering education at Cornell and in the United States. We have assembled what promises to be an exciting program that is organized, roughly speaking, from the ground up. The presentations this morning will deal with conventional weather and the very interesting recent advances in the measurement and prediction of weather phenomena. This afternoon we will move farther out in space, so to speak, as we listen to discussions of activities at the Arecibo Observatory and learn how satellite probes are making discoveries in the outer solar system.

"On behalf of the School of Electrical Engineering, I wish to thank Chuck Knight, our Symposium Chairman, as well as Anatole Browde and L. Keever Stringham, our Local Arrangements Committee, for the expert guidance, assistance, and attention to detail that have made this meeting possible. Our thanks also to our distinguished panel of speakers who have consented to give us the benefit of their knowledge and experience in these provocative disciplines.

"Our General Chairman for this meeting, Charles F. (Chuck) Knight, received the B.S.M.E. and M.B.A. degrees from Cornell University in 1958 and 1959, respectively. Following a two-year period of management training in West Germany, he entered into an extended career in corporate management with Lester B. Knight and Associates and with Emerson Electric Company. In 1973 he became Chairman of the Board and Chief Executive Officer of Emerson Electric, his present position. His broad executive and corporate background have made Mr. Knight an authority on policy issues in a number of fields that are related to the subjects of this symposium."

—Professor Donald T. Farley, Symposium Faculty Chairman
From Welcoming Remarks, April 17, 1985
St. Louis, Missouri

"On this beautiful spring day in St. Louis I am pleased to welcome you to Emerson Electric and to this Cornell Electrical Engineering Symposium on Atmospheric and Space Sciences. The organization of these symposia, of which this meeting is the fourth in the series, is a broad undertaking by the Cornell School of Electrical Engineering. I understand that all of the presentations and lectures in the series will be published ultimately and will be made available to the technical community. I believe that those proceedings will be well received and will serve as valuable reference works for years to come. As the day progresses, our speakers will be introduced in turn, so at this point I would like to introduce two distinguished members of the Cornell faculty who are present today: Professor Joseph Ballantyne, newly elected as the Vice President for Research at the University, and Professor William Streett, Acting Dean of the Cornell College of Engineering. I come now to what must be the easiest part of this program: the pleasant task of introducing the people who will make this meeting happen."

—Charles F. (Chuck) Knight, Symposium Chairman
From Introductory Remarks, April 17, 1985
St. Louis, Missouri

Editor's Note: On June 1, 1985, Professor William B. Streett became Dean of the Cornell College of Engineering, his present position.

WEATHER AND CLIMATE PREDICTION: A VIEW OF THE FUTURE

Wilmot N. Hess

Wilmot N. Hess received a B.S. in Electrical Enginering from Columbia University in 1946, an M.A. in Physics from Oberlin College in 1949, and the Ph.D in Physics from the University of California at Berkeley in 1954. He was awarded an honorary D.Sc. degree by Oberlin College in 1970. He joined the staff of the Lawrence Radiation Laboratory in 1954 and became Head of the Plowshare Division of Livermore Laboratory in 1959, a position he held until 1961 when he was appointed Director of the Theoretical Division of the NASA Goddard Spaceflight Center. In 1967 he became Director of Science and Applications of the NASA Manned Spacecraft Center. In 1969 he left NASA to become Director of the Environmental Research Laboratories of the National Oceanic and Atmospheric Administration (NOAA). In 1980 he was appointed Director of the National Center for Atmospheric Research (NCAR), his present position. Dr. Hess has served as an associate editor of the *Journal of Geophysical Research*, the *Journal of Atmospheric Science*, and the *Journal of the American Institute of Aerospace and Astronautics (AIAA)*. He is a recipient of the Arthur S. Fleming Award, and the AIAA G. Edward Pendray Award. He is a member of the National Academy of Engineering, and a Fellow of the American Geophysical Union and the American Physical Society.

Editor's Note: Since mid-1985, Dr. Hess has been the Associate Director for High Energy and Nuclear Physics of the Office of Energy Research in the U. S. Department of Energy.

I know that this audience consists principally of electrical engineers so my discussion this morning could perhaps be classified as "cultural enrichment." I will talk mainly about general atmospheric physics and present some ideas related to the problems of predicting the weather. Later, Ben Balsley and Gordon Little will probably concentrate on subjects that are more appropriate to the program of this seminar. I am involved in an organization that is concerned both with atmospheric research and with attempts to tie this research into the real problems of prediction. Of course we are aware that the usual public response to weather prediction is an automatic snicker and comments to the effect that "those crazy guys couldn't fight their way out of a paper bag; they miss more often than they hit; they really are inept and inert when they should be ept and ert." I intend to show you today that such opinions are unjustified and unfair.

Quantitative weather forecasting is a subject that is only 40 years old. During that period it has been severely limited to the so-called "synoptic scale of weather," based on observations made over a wide region. For example, typical Weather Service stations may be 300 to 400 miles apart. It is obvious that if you're looking for elephants, you don't use a microscope; if you're looking for tornadoes, you don't use a network with stations separated by 400 miles. The scheme of things must be appropriate to the problem at hand. In the past the Weather Service has attempted to forecast tomorrow's weather in general: the probable temperatures, the possibility of rain, major changes in frontal passage, and other climatic phenomena of that order. Smaller-scale activities, such as individual severe storms and other modest-sized events, pose a different set of meteorological problems that must be considered.

In this discussion today I will outline the advances that we can expect to see in weather forecasting over a sequence of time. Some techniques are available now; they are ready to be made to work in the field. In particular, Gordon Little will describe NEXRAD, a procedure that should remove any surprises about the presence of tornadoes. With that scheme it should be possible to predict the occurrence of a funnel, perhaps 10 or 20 minutes before it actually arrives, with information on its location and probable direction of movement. The technology for this very significant step forward has been in hand for six or eight years but has not been put into operation because of lack of funds for the necessary field engineering that still must be performed. Other developments are not so close to fruition, certainly none as dramatic as the promise of 100 percent accurate, real-time, short-time forecasting of tornadoes.

My first major topic is concerned with the phenomenon that is called a "downburst." This city, the home of McDonnell Douglas, is especially interested in the problem of aviation safety. Starting about four years ago, a study was made of one particular facet of small-scale weather that has very significant impact on this problem. As illustrated in Figure 1, ordinary cumulus thunderstorms (they don't have to be big ugly looking things) frequently have a weather pattern that contains a downpouring of cold air from the base of a moist-air mass, either from emerging evaporation in combination with rain, or for other dynamic reasons. It is not too unusual to find rain and cold air coming out the bottom at velocities of 30 to 40 miles per hour, going down toward the ground and spreading out into a gust front. In Figure 2, an aircraft is shown encountering one of these events as it is trying to come in for a landing. The first indication of trouble is a headwind that tends to lift the plane, so the pilot either throttles back a bit or perhaps noses down somewhat. Then the plane moves through the null point and experiences a tailwind. Now the pilot is in trouble, particularly in modern jets that are unable to execute rapid power recovery. Air-speed losses of from 50 to 100 knots during such transitions can cause serious landing difficulties. These effects are large enough to have been clearly responsible for a number of accidents.

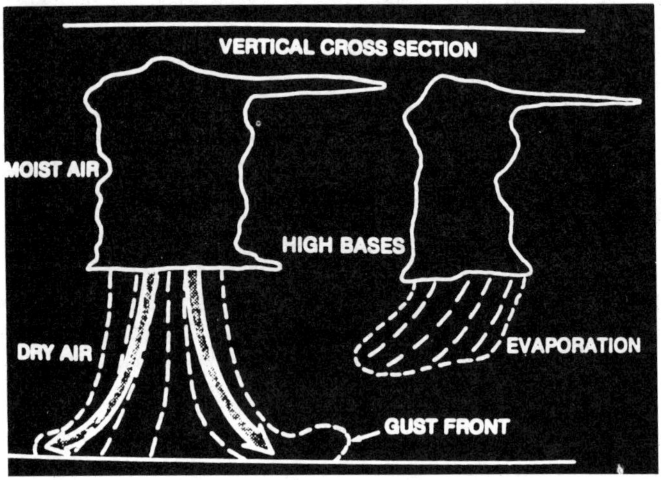

Source: NCAR

Figure 1. Anatomy of a "Downburst"

Three summers ago, when we were doing our first serious measurements on the downburst phenomenon, a Pan American flight crashed in New Orleans. At the time there was positive evidence that the accident occurred because the aircraft encountered a downburst. Another accident that was clearly related to this same phenomenon took place three years ago when a Continental flight crashed at Stapleton Denver Airport. Although there are insufficient

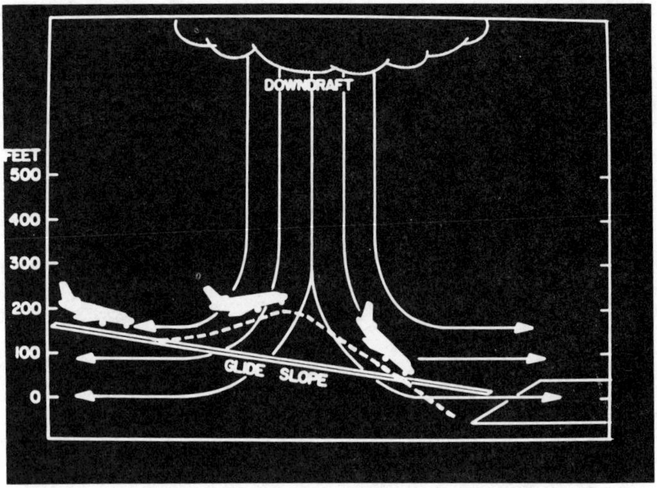

Source: NCAR

Figure 2. An Aircraft Encounter with a Downburst

data in many cases to identify the exact cause of a crash, recent studies indicate that as many as three dozen serious accidents, over the last one or two decades, may have been caused by encounters between major jet aircraft and downbursts. At least six of these events are well enough documented for us to conclude that they were due to the downburst effect. I am being somewhat cautious in these conjectures because it is necessary to have really good instrumentation in order to make positive measurements of these phenomena. At present the instrumentation available at airports is generally inadequate for this purpose.

In an attempt to improve the measurement of downburst events, we began some experiments at Stapleton three summers ago where we placed Doppler radars, ground stations, and upper-air sounding stations in the area around the Airport. For the benefit of those present who are not electrical engineers, a Doppler radar can do more than merely sending out a signal that scatters off some particles, thereby returning some information about the presence of rain. It also can measure the motion of those scattering particles by means of the Doppler shift (the well known fire-engine-siren effect). The velocity of the particles can then be measured, and it is possible to determine whether the particles are moving toward or away from the observer. If two such radars are looking at the same cloud from different directions, the separately measured components can be combined to produce a complete picture of the activity that is occurring inside the cloud. Three Doppler radars were placed at the vertices of a triangle, with one unit at Stapleton Airport and the other two northeast and northwest of the Airport, respectively, with a spacing between units of about 20 kilometers. Since we were looking for the details of quite small-scale events, it was necessary to use this close spacing, unlike the greater distances that are typical in normal practice.

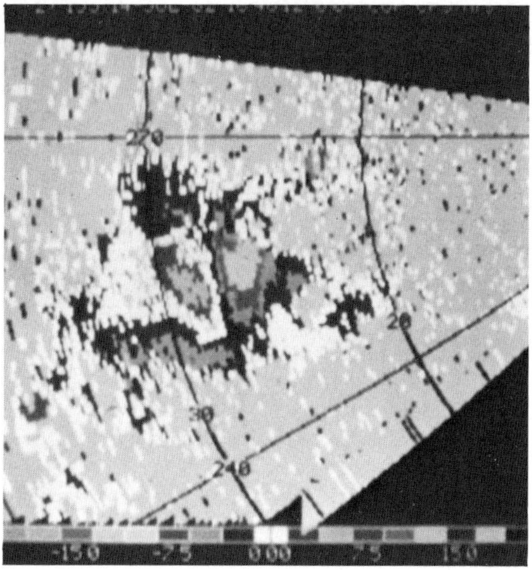

Source: NCAR

Figure 3. A Doppler Radar Signature of a Downburst Event

With a configuration of this kind in place, and with sufficiently powerful radars that have the proper characteristics, it is possible to detect these small events in the clear air. Figure 3 displays a Doppler signature of one of these episodes. In this image of the storm, the light oblong area that is partially surrounded by the dark region in the center of the figure represents winds that are moving toward the Doppler radar. The elliptical area just to the left of the central dark region depicts winds that are moving away from the radar. This rather unique signature shows the outflow on the ground from this particular storm. The 10-kilometer intervals on the chart indicate that the disturbance is from two to four kilometers across—two to three kilometers is a typical dimension for these relatively small-scale events.

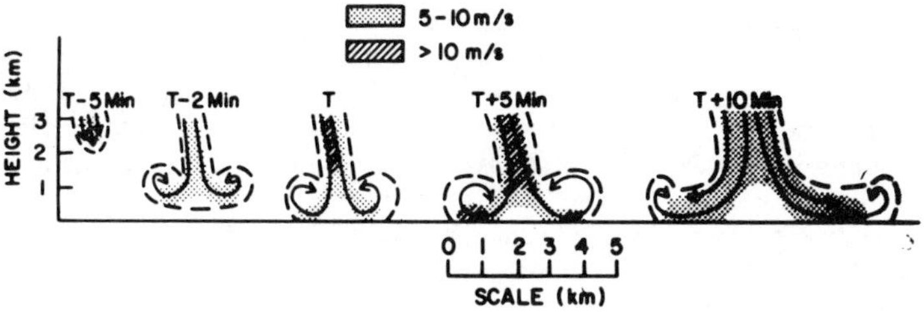

Source: NCAR

Figure 4. Typical Life History of a Microburst

The microbursts are also relatively transient in nature. The typical life history drawn in Figure 4 shows that they may last from eight to ten minutes: at T minus 5 minutes, the storm is still descending; at time T, it has made contact with the ground; at T plus 5, it has attained maximum strength; and at T plus 10 it is starting to decay.

One might very well ask why we should bother with these rather small and short-lived storms. The obvious answer is based on concern for the welfare of an aircraft that is in the unfortunate circumstance of encountering one. These disturbances occur quite frequently. Last summer at Denver Stapleton there were eight or ten occasions when it was clearly appropriate to divert aircraft from microbursts on the runway. Another dozen or so times the decision was marginal— the choice between diverting the aircraft or taking a chance was a close call. Although these decisions are generally made by the captain in the plane, the Stapleton control tower now has the authority to temporarily close a runway, or the entire airport, because of the presence of downbursts.

A downburst may come out of a great benign-looking cloud that contains "virga" (rain that is evaporating and not reaching the ground) with an attached shaft of cold down-flowing air. One particular downburst that was observed at Stapleton had a downward velocity of over 40 miles per hour spread out at the ground. Assuming vertical and horizontal continuity, an aircraft coming into that disturbance would have experienced an 80-mile-an-hour gust, creating a serious problem.

My organization has been moving forward on an examination of the nature and frequency of occurrence of these phenomena. Although we are concerned principally with basic research, we consider this problem to be important enough to justify starting a program of technology transfer. We have made films for use by United, Continental, and Boeing aircraft companies, and have introduced programs into flight simulators that are used for testing Continental, American, and United pilots. The first time a senior pilot encounters a downburst in the flight simulator without warning, he almost invariably crashes. In the simulator, as the pilot is coming in for a landing he suddenly encounters a big headwind. What should he do? Should he move the control forward or try to back off on the power? The correct procedure is to fly at constant ground speed. Alternatively, if a downburst occurs when the plane is taking off, the pilot has to do his best to gain speed and to gain altitude. Examination of aircraft characteristics has revealed that the proper solution is to put the aircraft into a semi-stall configuration.

One flight that took off from Stapleton this past summer encountered a downburst halfway down the runway after it had reached a sensitive point of no return. Rather than trying to abort the flight, the crew decided to continue the liftoff. At the end of the runway they had achieved an altitude of 10 feet when they should have been at 1500 feet. The plane got out of the downburst at about one thousand feet off the runway, but not before it had crashed into one of the light-fixture support towers at the end of the runway and hit the roof of a shed. Miraculously, the plane landed safely with the only damage being some ripped skin on the side of the aircraft. The second officer said that the incident was not more serious because he took proper evasive action as he had been taught to do in one of our courses in Denver. At Stapleton, downbursts are now well understood and have been well documented, but they undoubtedly occur at many other airports, including the one here in St. Louis. The local situation is reasonably bad, especially during summer afternoons, because measurements have not been made for pertinent weather conditions.

Although we are pushing them, the FAA has not taken any action on this serious problem. We think there should be a program that calls for the installation of Doppler radars at major airports so that wet measurements of these phenomena can be made in order to determine if pilots should be issued warnings to stay away. This past summer we had a meteorologist in the tower at Stapleton who informed the controllers about conditions in the neighborhood based on the availability of Doppler radar information. On 30 different occasions, we put out microburst advisories. (An advisory warns that a disturbance is in the vicinity and provides a chart of its approximate location. The pilot is also told that he will have to estimate its exact location and probable seriousness, and must decide on the proper evasive action.) This trial program turned out to be quite valuable. During the test period, 19 percent of the runway approaches were changed, seven approaches were missed, and six departures were delayed.

Another valuable use for Doppler radar is in the detection of windshifts. These phenomena are of interest because of their impact on a plane-stacking schedule for a particular runway. Since the occurrence of a windshift necessitates changing the stacking to another runway, it is useful to have advance information. It turned out that the Stapleton program paid for itself in fuel savings last year because of early warnings about the presence of windshifts.

The Stapleton study has demonstrated that the necessary technology is available for the widespread installation of Doppler radar systems. Unfortunately, similar studies have not been done at other major airports. Since it is now possible to predict where potentially dangerous disturbances are likely to occur, it is clear that the FAA should encourage action by appropriate authorities to bring these modern instruments to major airports. Downburst information will then become common knowledge to pilots who are trying to land or to take off.

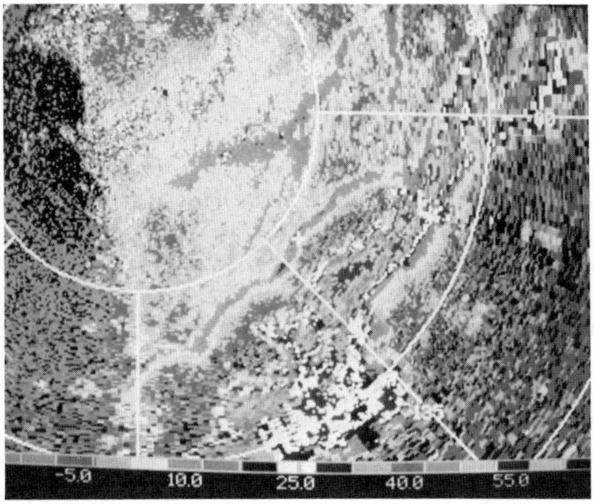

Source: NCAR

Figure 5. Doppler Radar Picture of Thunderstorm Activity

In conjunction with the Stapleton downburst experiment, we came across an interesting bit of serendipity related to thunderstorm forecasting. Figure 5 displays a Doppler radar picture with the location of Stapleton indicated by the cross mark (+) in the center of the chart. The concentric grid lines are at 30-kilometer intervals. The original figure is color-coded with the numbered bars at the bottom of the chart representing relative signal-to-noise ratios of reflectivity. A line of storms is indicated by the bright reflections just east of the airport. The thin dark line that passes through the cross mark is a feature of interest that shows what a good modern radar can do when it is looking at clear air. This line indicates a "gust front," the outflow of cold air that leads a storm and moves in a direction away from the storm. Further to the west is another dark line of unknown origin. Additional photographs gave evidence of the motion of the two lines. Twenty minutes after the situation shown in Figure 5, the two lines had formed into a strong line of convergence somewhat to the west of Stapleton. At that point the airflows came together to give a lifting motion to the air, thereby introducing the proper initial conditions for producing the conductive activity that will result in the formation of a storm. Figure 6, taken forty minutes after Figure 5, shows that a line of storms had indeed formed along the previous line of convergence west of Stapleton.

Since individual thunderstorms are more or less casual events based on small disturbances that cannot be identified easily, prediction of where and when they will appear is generally considered to be very difficult. Topographical factors also have an impact upon their occurrence. South of Denver, for example, there is a natural feature, known as Palmer Ridge, that runs east and west and is about 1000 feet high. Despite its small size it is large enough to initiate thunderstorm activity due to the presence of lifting motion and convergence lines. Last summer, when the measurements of Figures 5 and 6 were made, 90 percent of the thunderstorms near Denver formed over convergence lines, and 68 percent of observed convergence lines produced thunderstorms over an 80 by 80-kilometer inspection area. If it becomes possible to identify situations that produce lifting motions and convergence lines, or quite similar behavior, perhaps future thunderstorm prediction will become more reliable.

In addition to good current radars, especially Doppler radars, modern reliable modeling of the weather is another important tool that has emerged in the past 20 years. Numerical weather models based on a synoptic scale have been studied by weather forecasters for 40 or 50 years. Development of a good model is rather like growing a lawn in England. Once, at a college garden party at Oxford, I asked a gardener, "How do you get a lovely lawn like this one?" "It's very easy," he replied, "You sow seed and weed for 300 years." The same procedure is characteristic of model development. Early attempts are checked, improved, and manipulated for a long period of time until a reasonably trustworthy version is finally achieved. For their predictions of tomorrow's weather, the Weather Service now uses synoptic-scale models that were obtained in this fashion.

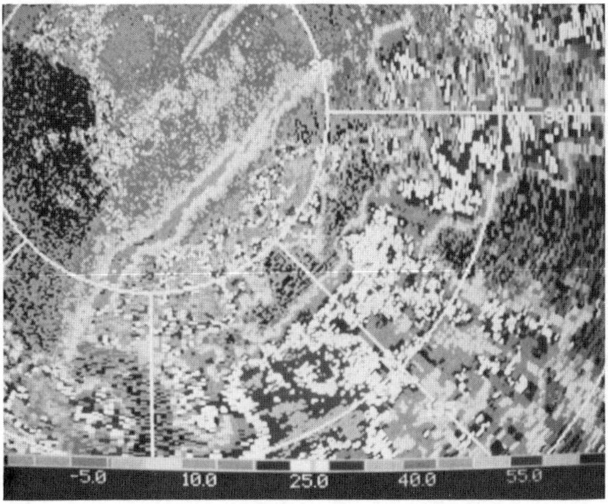

Source: NCAR

Figure 6. Thunderstorm Formation along a Gust Front

Mesoscale models are smaller-scale versions that are used to examine features that slip through the Weather Service net. A reasonable amount of progress in their development has also occurred in the past 10 years. A group at Pennsylvania State University has made major contributions. One of their models, with the characteristics shown in Figure 7, is in use in my organization. For our purposes here, the important item in the list is that the spacing of

- **Horizontal grid 46 × 61 ΔS = 80 km**
- **15 layers**
- **P_T = 100 mb**
- **Blackadar PBL**
- **Prediction of ground temperature**
- **Variable surface parameters based on land use**
- **Surface fluxes of heat and water vapor**
- **Longwave and shortwave radiation at ground interactive with clouds**
- **Modified Anthes (1977) cumulus parameterization**
- **NMC analysis + radiosonde data**
- **Removal of vertical mean divergence**
- **Observed lateral boundary conditions**

Source: NCAR

Figure 7. Characteristics of Meteorological Model

the grid points in the model is only 80 kilometers, compared to the typical spacing of 300 or 400 kilometers that is used in normal Weather Service models. At this spacing we are getting down to a level where it is possible to look at small features. This model, however, cannot detect an individual cumulus cloud. It is still necessary to apply a more complicated averaging process called parameterization, and to feed the results back into the mesoscale model.

The capabilities of these models, and the results that can be obtained, are illustrated by the following example. Figure 8 is a picture of an initial condition that was fed into our model. The original is a color-coded print that shows measurements of temperatures and winds over most of the United States at a given time. The bright area at lower center represents a mass of warm moist air coming up from the Gulf of Mexico. Along the east coast the figure shows the presence of a cold high-pressure cell with the arrows representing the winds flowing around it. The model operated on this information for three days and produced a number of useful outputs. One particularly interesting result is demonstrated by a comparison of Figures 9 and 10. The infrared satellite picture of Figure 9 covers the eastern portion of the United States and depicts approximately the same conditions as shown in Figure 8. A cold front is going offshore with a tongue of dry air being pulled in

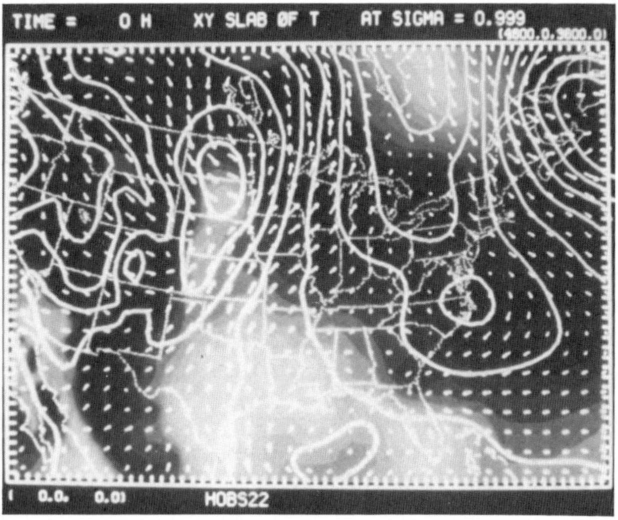

Source: NCAR

Figure 8. Cross-Country Temperature and Wind Measurements at a Given Time

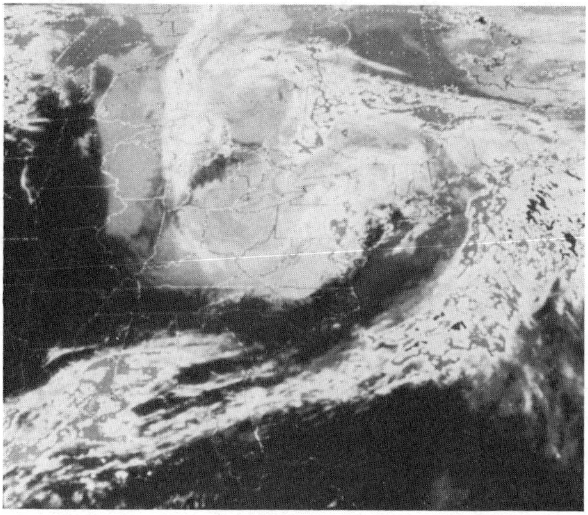

Source: NCAR

Figure 9. Infrared Satellite Photograph of Conditions Approximating Those of Figure 8

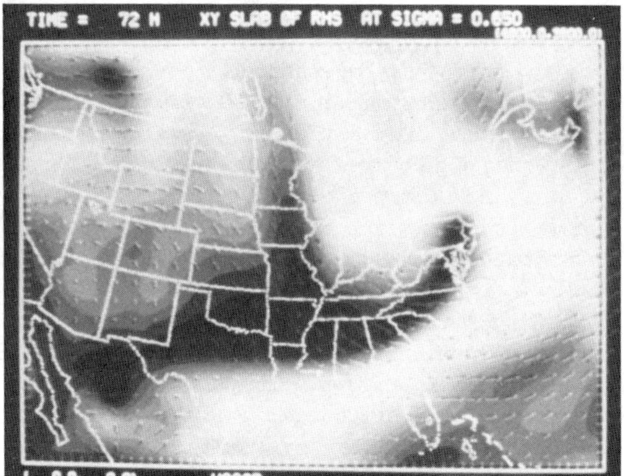

Source: NCAR

Figure 10. Meteorological-Model Prediction for the Conditions of Figures 8 and 9

behind the front. The low-pressure region over the northeastern states takes the typical form of a comma-shaped cloud. Figure 10 is a picture of the relative humidity in the region as predicted by the model. All of the features of the satellite photograph, including the comma-shaped cloud and the channel of dry air, are shown on the output chart. It is not possible to achieve this level of detail with a synoptic-scale model.

The development of these advanced models opens up the possibility of doing a better job on quantitative precipitation forecasting (QPF), the most difficult prediction task when a synoptic scale is used. Those of us who live in Colorado are very interested in knowing where it is going to rain or snow, and how much precipitation to expect. We get annoyed when the forecast for the mountains is say, two to four inches of snow, and the next morning we learn of a 15-inch snowfall in Denver. We can complain on two counts: one, they said it was going to fall in the mountains, not in Denver, and two, there is much more snow than was predicted. In all fairness, the Denver region is a tough place for weather forecasting, and the toughest assignment is to predict how much precipitation will fall. The advent of these improved models allows me to make a "fearless forecast" for the future: in a decade from now we ought to have improved our "skill score" in QPF by a factor of two. Skill score is an objective scheme for evaluating our performance.

We are developing another new model that may be important in the future forecasting of air pollution. The specific format is an acid-rain forecast model that couples the previously described meteorological model with a reasonable amount of chemistry based on the presence of some 40 separate molecules in air, and 40 separate molecules or ions in cloud droplets or raindrops. Both air and water must be considered since the chemistry is quite different in these two environments. If these 80 species are combined to form all of the sensible reactions that are needed for the study, the current total is about 150. Introduction of all of these reactions into our Cray computer caused it to overflow, so it was necessary to keep the model down to a size that could be handled inside memory. Within these constraints, the scheme then is to make predictions based on good mesoscale meteorology and good chemistry. The basic problem is to determine how much sulfuric acid will come down in a certain location for a given set of weather conditions. This particular calculation, of course, also requires input data on the distribution of SO_2 gas emissions from power-plant stacks and from other sources.

A few years ago the EPA conducted a measurement program, called OSCAR, that obtained data on sulfate deposition from a number of stations in the Great Lakes area. Last month some of these data were applied to our air-pollution model in order to try a semiquantitative prediction of sulfate deposition. When the measured rainfall in the area was compared with the amount predicted by the model, calculations for specific points were found to be off by about 23 percent, a rather good performance for QPF. This result indicates that there is a reasonable amount of skill on the meteorological side of this problem. For the chemical side of the model, when measured sulfate deposition per hectare was compared with an equivalent prediction from the model, the individual points were found to be off by a factor of two. Several of the points were quite close, however, and we have some ideas about the reasons for the low readings in some cases.

This model is the first attempt at combining good meteorology with real chemistry to obtain realistic air-pollution predictions. All previous models have used crude approximations to chemistry. A typical approach has been to make an arbitrary assumption that one percent of the SO_2 in the air is converted to sulfate every hour. Instead, this model starts with actual SO_2 data, tracks the chemistry through the clear air and through the clouds, and finally deposits the sulfur compounds on the ground to give us a reasonably good prediction. Further development will probably lead to models that will be used for regulatory purposes and also in forecasting. Perhaps in a decade from now, their main purpose will be to determine the amount of sulfate deposition from a big acid-rain event so that action can be taken to reduce the potential damage. These possibilities are certainly something to think about.

My next topic is climate forecasting, the process of trying to predict next year's events. The procedures are obviously different from those of weather forecasting where the principal concern is with what will happen this afternoon or tomorrow. I will give particular attention to the El Niño, the occurrence of a body of very warm water in the equatorial Pacific up against the west coast of South America, as shown in Figure 11. This phenomenon has been known for over 100 years. The nutrient-poor warm water of the El Niño upsets anchoveta fishing since normal aquatic growth in this region depends upon the nutrient-rich cold water supplied by a southerly flow of the Humboldt Current. In addition to its effect upon the fishing industry, we are quite certain that an El Niño may have direct influence upon weather patterns in other latitudes, as listed in Figure 12. It is clear that these events are not limited to regions near the equator.

The El Niño phenomena apparently increases the north-south circulation of wind velocity on the planet (the Halley Circulation). The presence of this warm puddle of water causes more evaporation, more precipitation, and stronger air flows upward and outward, both to the north and to the south. These enhanced air flows are carried forth from the mid-latitudes and change storm tracks in Central and North America. A typical example of this condition occurred in the winter of 1976-77. Storm tracks entering Alaska came down in the eastern United States and caused Alaska to be unusually warm, California to be dry and warm, and the eastern states to be generally miserable with cold wet weather and heavy snowfalls.

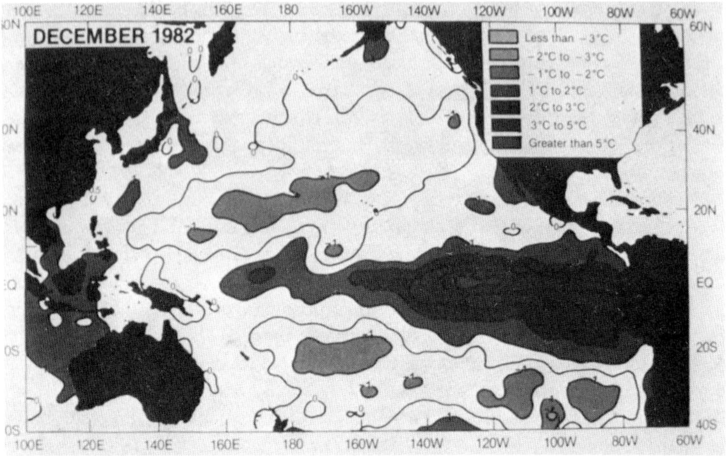

Source: NCAR

Figure 11. Occurrence of an El Niño in the Equatorial Pacific

- **Excess rainfall in equatorial Pacific**
- **Floods in Peru**
- **Drought in Australia and South Africa**
- **Weakening of summer monsoon in India**
- **Strong PNA pressure pattern over North America**
- **Storms in southern California and Gulf Coast**

Source: NCAR

Figure 12. Probable Effects of 1982 El Niño

The existence of these relationships between widely separated events has caused our people to play a whole new set of games called "teleconnections." For example, the well-established empirical teleconnection between the El Niño and mid-latitude storm-track variations has been modeled on our climate model. We are quite sure that the models are valid enough to allow some predictability: if we expect that an El Niño will occur next year, we can say something about odd storm tracks, predict their probable effect on normal weather patterns, and pinpoint the problems that will arise from these potential changes.

Well before these long-term events actually occur, we expect to be able to forecast them by using oceanographic measurements that were taken a year in advance. Since a typical El Niño begins around the month of June in the southern winter, it is necessary to examine the conditions that existed within pertinent areas in June of the previous year. In the southern hemisphere, between Australia and South America, a growing low-pressure region forms over the ocean during the summer months. For example, a measurement of the pressure difference between Hobart in Tasmania (due south of eastern Australia) and Chatham Island (to the east of New Zealand) might produce the characteristic curves plotted in Figure 13, with a peak showing up during the southern winter months. If an El Niño occurred the year before, the corresponding chronological peak would be much smaller. Pressure-difference measurements over a 12-year period have shown a consistent pattern of

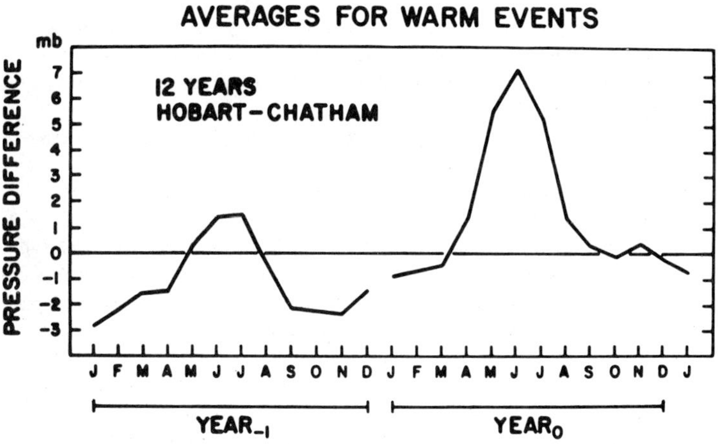

Source: NCAR

Figure 13. Characteristic Pressure Differences in the Southern Hemisphere

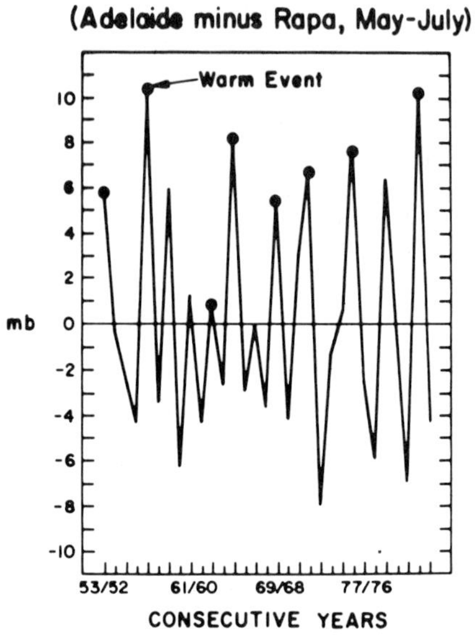

Source: NCAR

Figure 14. Interannual Changes in Pressure Differences

reduced peaks in the years when El Niños were present. These below-normal peaks are accompanied by anomalous southerly flows of unusually warm water off New Zealand. It is possible that a series of such related events could serve as a promising indicator or predictor.

An examination of the 25-year record, as drawn in Figure 14, reveals the potential value of this information. When the pressure differences are compared from year to year, we find that a high number preceded by a low number corresponds to an actual occurrence of a warm event, an El Niño. Unfortunately, the discrepancies in the record limit us to an overall fifty-fifty chance of making a correct call. Considering that we are dealing with a first look at this technique, however, the potential for eventual success appears to be reasonably good. A climate-prediction scheme that enables us to call these events a year in advance would be a major accomplishment. Information on their general characteristics, combined with what we already know, would allow us to predict the probable major effects of an expected El Niño.

The ability to make predictions about the size of a coming hurricane season is a somewhat more controversial aspect of climate forecasting that may now be possible. Professor Bill Gray, of Colorado State University, has studied this problem over the last 12 years and suggests three predictors that can be used to call the number of hurricanes that are likely to occur in the tropical Atlantic each year:

- Presence of an El Niño

- Strength of quasi-biennial oscillation (QBO) of stratospheric winds

- Pressure anomalies in the Caribbean and the Gulf of Mexico

QBO is a well-known and predictable event that augments the other two factors that I have already discussed.

Hurricane tracks for the last few years have followed paths from the Caribbean into the Gulf of Mexico, or northeastward into the Atlantic, with almost all of the Atlantic storms occurring on shore at moderate levels during this period. The Weather Service is very concerned that the general public in Florida may not be sufficiently alert to the problems that would accompany an on-shore storm during a big hurricane year. Neil Frank, the head of the Hurricane Center in Miami, often tells a horror story about what would happen if we had a repeat of the disastrous 1928 hurricane that came up through the Florida Keys and caused widespread damage. Unlike present conditions, the population at that time consisted of only a few hundred people. If a storm of that magnitude came up through that area again, it would be impossible to evacuate the Keys through the single highway in the region. There is just no way to guarantee the safety of the people living in the Keys. In Miami Beach the answer is "vertical evacuation," meaning that people climb to the upper floors of a building for safety. This solution is not practical in the Keys. Since removal of all of the residents could not be accomplished, it is likely that one hundred thousand people would be killed.

This uneasy scenario emphasizes that it would be interesting to have the capability to predict the size of a coming storm season. The quality of Bill Gray's predictors for the last several years can be observed in Figure 15. Distribution of these forecasts is limited to the Weather Service, which uses them for their own purposes. Bill says that making these predictions is a new venture for him, and that he is beginning to understand why the actual numbers for the listed years are low compared to the long-term average climatology figures. In his forecast for 1985 he believes that an El Niño will not occur but that the stratospheric wind conditions are favorable for the production of hurricanes. There is no indication, however, that pressure differences will be unusual, so that factor will have little effect. His overall conclusion is that 1985 should be above-average in the number of hurricanes and tropical storms that will take place. Since these predictions are completely empirical, it is

TROPICAL CYCLONE ACTIVITY
DURING 1982-84 (Gray)

	1982	1983	1984	Expected from Climatology
Number of hurricanes	**2**	**3**	**5**	**6**
Hurricanes and tropical storms	**5**	**4**	**12**	**9**
Hurricane days	**6**	**5**	**21**	**25**
Hurricane and tropical storm days	**16**	**15**	**61**	**38**

Source: NCAR

Figure 15. Comparison of Actual and Predicted Storm Activity

necessary to live with them for a number of years to see how well they turn out. We simply do not know the physics behind these storm formations. Over the last decade we have learned something about the physics of the El Niño, and about the teleconnections between the El Niño and the mid-latitude storm tracks. We may become more knowledgeable about hurricanes in the next decade, but in the meantime, even an empirical scheme is of value.

Although a number of the prediction techniques that I have mentioned so far look promising, there are still many things that we want to do but lack the knowledge to perfect at present. I will point out some of these ideas and describe steps we are taking to gain the necessary information that will lead to improved forecasts.

Perhaps the biggest problem area in forecasting is the one that I alluded to earlier, the so-called "mesoscale gap." The synoptic scale, involves distances from 1000 to 20,000 miles. Weather Service forecasts in this range are reasonably good. Looking below those dimensions, the mesoscale can be defined in terms of three decades: alpha, from 200 to 2000 kilometers; beta, from 20 to 200 kilometers; and gamma, from 2 to 20 kilometers. These three decades of the mesoscale represent the major uncertainties in knowledge and lack of predictability that produce most of the annoyance, and worse, with Weather Service forecasts. Most of the things that they do not do well involve these small distances. On one occasion, for example, they failed to forecast a severe storm of the type shown in Figure 16. Atmospheric disturbances of this nature move through the Midwest creating hail and damaging winds, but they also bring considerable beneficial rainfall.

Source: NCAR

Figure 16. An Example of a Mesoscale Convective Complex (MCC)

The storm shown in Figure 16, called a "mesoscale convective complex" (MCC), is a major disturbance of a type that has not been studied in the past. In the last decade we have developed the necessary instruments so we are now ready to begin investigations of such complicated events. These objects are much larger than single thunderstorms. In this infrared satellite picture, the dark areas within the overcast regions represent the tops of clouds, and the large black tops are active storms that are very big and very cold. The MCC characteristics of individual thunderstorm cells, such as the one above the Oklahoma Panhandle in the photo, can be summarized as follows:

- Shape: roughly circular
- Area: > 100,000 km^2
- Lifetime: from one hour up to three days
- Frequency: 50 each year in the U. S.
- Summer rain produced: 60 percent of total rainfall
- Severe weather sources: 80 percent of summer storms

The life history of an MCC can be traced in a series of consecutive satellite photographs. Figure 17 shows the appearance of the same object of Figure 16, taken five hours later. The storm position has not moved very much but the total system has become larger, while the size of the cold higher portions of the cloud have gotten smaller. Figure 18, taken six hours later, shows that the top of the original cell is starting to fade and that another structure has taken over. It is clear that the new cell is the result of a merger of two or more individual cells operating as a system that is now in place over southern Illinois and Indiana. Underneath one of these large circular objects there may be 12 to 15 individual active cells that somehow or other cooperate to produce these big circular systems. An individual cumulus storm has a lifetime of one or two hours; big ones may survive for several hours. On the other hand, the circular systems may last for an entire day.

At present, we do not fully understand how these systems operate, but we now have the tools that are required for comprehensive studies. An array of good Doppler radars is a necessity, surrounded by geostationary satellites that take soundings to produce the kind of pictures shown in Figures 16, 17, and 18. Another requirement is to fly aircraft through the storms in order to make wind measurements. We are now in the process of installing Doppler radars on

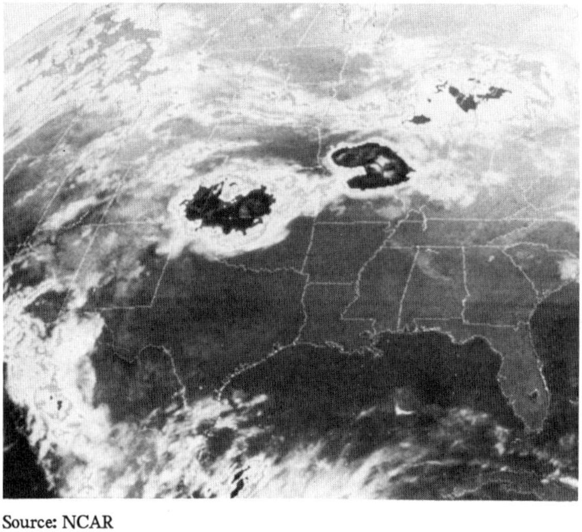

Source: NCAR

Figure 17. Progressive Behavior of the Storm of Figure 16

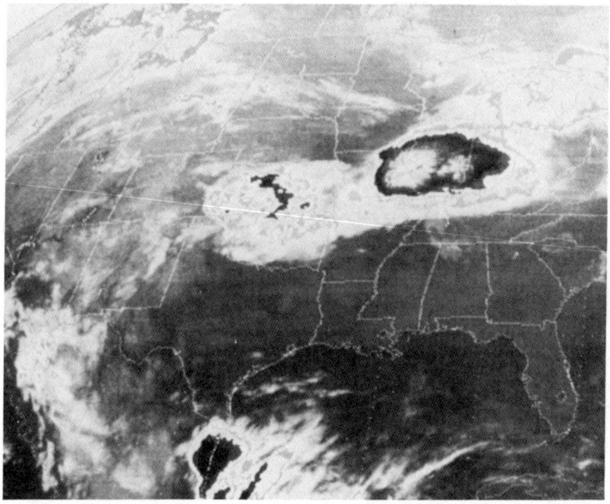

Source: NCAR

Figure 18. Further Development of the Storm of Figure 16

the aircraft so that the radars can be taken directly into the storm: a much better procedure than to wait for the storm to be detected by ground-based radar. Gordon Little will describe devices, known as "profilers," that can make continuous measurements of upper-air winds. Rapid communication facilities between automatic surface stations are also essential for obtaining pertinent data. Combining all of these instruments with good computer facilities will provide the necessary capability for the accurate study of these complex phenomena.

Since most of these MCC storms occur in the Midwest during the summer months, our initial studies are planned for an experiment that we have designed for that region. The current Midwest instrumentation of the National Weather Service is shown in Figure 19. The balloons indicate the points where regular upper-air wind measurements are made. The other symbols show the location of ordinary (non-Doppler) radars known as WSR-57s. In order to conduct the investigations that we have in mind, it will be necessary to instrument the entire area in a much tighter configuration than exists at present, as proposed in Figure 20. At each of the open boxes in the figure, a new upper-air sounding system will be installed. A ground station will be placed at each of the boxes inside the zone defined as the "meso-beta network." Inside the region bounded by the two concentric circles, we plan to install 14 Doppler radars to

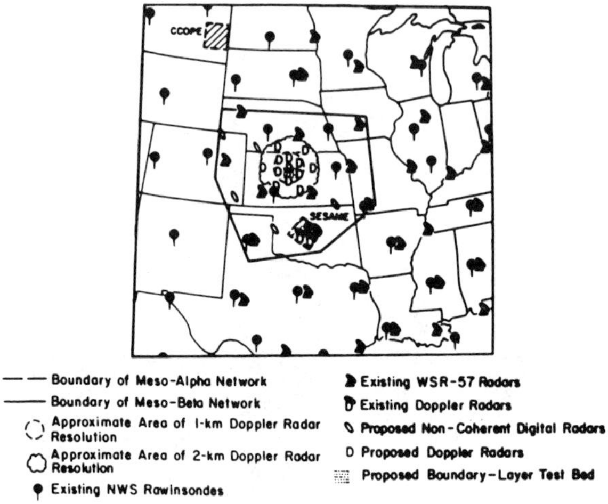

Source: NCAR

Figure 19. Current Storm-Central Observational Network

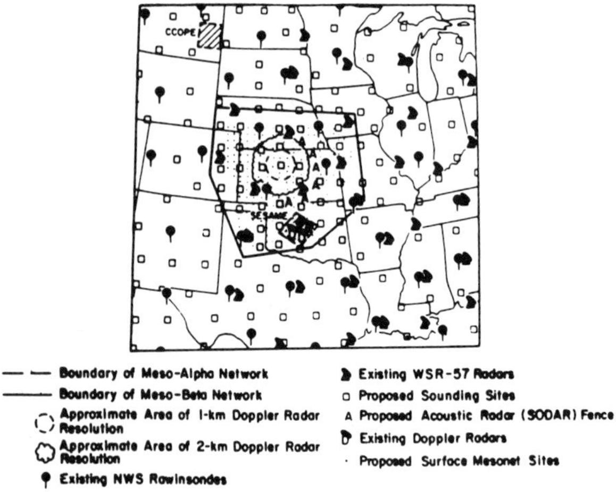

Source: NCAR

Figure 20. Proposed Storm-Central Observational Network

provide high coverage within that area. A few airborne Dopplers will also fly around these circled regions in order to further increase the Doppler coverage. It will then be possible to look at the internal flow characteristics and the life history of these large events, and, as the entire flight of the storm is observed, we will be able to watch how it interacts with its environment. The outer portion of this area, the "meso-alpha network," is the general flow regime, which one way or another, produces the initial disturbance. When the storm becomes active it has a significant feedback effect on its source.

This experiment should help us to learn something of the nature of the interplay between these storms and their source environment, to determine how these storms act cooperatively to maintain themselves for long periods, and to obtain a data base that will allow us to predict which small early disturbances will grow into large configurations. A successful experiment will provide a body of knowledge that should lead to predictability that will fill the mesoscale gap. This formidable problem has not been attempted in the past because of difficult observational and computational obstacles. We now have the ability to perform the necessary tasks and we should get on with the job.

An elaborate undertaking of this kind is very expensive. The proposed project would be performed over a period of about four years and would require a budget split of over 100 million dollars between NOAA, NASA, and NSF. As might be expected, it will be hard to sell the program just now, given the budget difficulties in Washington. We may have to wait a year or so, but this experiment is clearly the next step in the weather forecasting business.

The near-scale and far-scale ideas that I have presented this morning, together with this gleam-in-the-eye experiment that I have just outlined, represent some activities that will improve prediction capabilities in the future. There are many areas in this discipline that are of interest to electrical engineers. Certainly the application of radar, and particularly modern Doppler radar, is one of the key items that has contributed to our present success and is essential for future developments. Weather prediction is an exciting business and we hope to continue to make significant progress. When we return to celebrate the 110th or 200th Cornell E.E. Anniversary, we will find out how we're doing.

Editor's Note: *The New York Times* of Sunday, August 7, 1988, reported a successful demonstration of an advanced Doppler radar microburst detection system at Stapleton Airport in Denver, Colorado. On July 11, 1988, the system alerted airport controllers to warn approaching aircraft of dangerous windshear conditions. Five United Airlines jetliners were kept in a holding pattern until the radar system notified the control tower that the runways were clear. All of the jetliners then landed safely.

THE MST RADAR TECHNIQUE: HISTORICAL BACKGROUND AND POTENTIAL FOR ATMOSPHERIC RESEARCH

Ben B. Balsley

Ben B. Balsley received the B.S. degree in Electrical Engineering from California State Polytechnic University in 1957, and the M.S. and Ph.D degrees in Electrical Engineering from the University of Colorado in 1964 and 1967, respectively. Following a two-year assignment as an electronics engineer and physicist at the International Geophysical Year field station in Trujillo, Peru, he joined the Jicamarca Radio Observatory in Lima, Peru in 1960 as Physicist and Engineer in Ionospheric Studies. In 1969 he was appointed as Physicist in Atmospheric Studies at the Aeronomy Laboratory of the National Oceanic and Atmospheric Administration (NOAA), his present position. He also holds concurrent positions as Adjoint Professor of Electrical Engineering at the University of Colorado and at the Geophysical Institute of the University of Alaska. From the United States Department of Commerce Dr. Balsley has received an Award for Sustained Superior Performance, a Gold Medal Award for Outstanding Performance, and an Outstanding Paper Award. He has also received a Distinguished Alumnus Award from California State Polytechnic University. He is a member of the Scientific Research Society of America, the American Geophysical Union, and the American Meteorological Society. He has 150 publications in ionospheric and atmospheric physics, and has served as Guest Editor of the *Journal of Atmospheric and Terrestrial Physics.*

In my talk today I will focus on some of the earlier aspects of weather prediction mentioned by Bill Hess, with particular attention given to a specific measurement technique that has been developed over the last 12 to 15 years. The principal elements of my discussion are summarized in the following statement:

> Sensitive Doppler radar systems, operating over a broad VHF frequency range of 40 MHz to 2000 MHz, can detect scattering from the optically clear air and thereby can measure winds, waves, turbulence, and atmospheric stability. Such measurements can be made continuously, from just above the ground up to very high altitudes. This technique offers a major new tool for studying the atmosphere and its variations.

Of the radar measurements that Bill described, many referred primarily to observations of particulates in the atmosphere such as raindrops, clouds, and, in some cases, dust. Alternatively, I will be concerned with a subset of those techniques that allows us to look into optically clear air and obtain echoes. One of the new and promising aspects of this procedure is the ability to take data every minute or so, put the information onto magnetic tape, and continue the process for many years, if necessary, giving us a nice high-time-resolution, low-cost method of obtaining atmospheric variables. I will describe the kind of radars that are used for this purpose, briefly outline the pertinent fundamental theory, estimate the capabilities and promise of this technique, and attempt a projection of possible business opportunities.

Over the past 10 to 15 years, clear-air radars have come to be known by a number of essentially interchangeable terms. The most sensitive types are called "MST" radars that are capable of looking at the mesosphere, stratosphere, and troposphere—that is from the ground up to 80 to 100 kilometers. "ST" radars are somewhat less sensitive but are capable of looking at the lower portion of the stratosphere, say up to 20 or 30 kilometers. "Wind profilers" are really ST radars that are used to monitor or profile wind on a continuous basis. The following installations belong to the MST class of radars and have been active in fundamental atmospheric research:

- The Jicamarca facility near Lima, Peru, an extremely sensitive radar, was used in some of the early experimental work in the clear-air wind measuring field. The very large antenna array is some 200 meters on a side and transmits about six MW of peak pulse power. It is the only radar capable of looking continuously all the way through the atmosphere, from near-ground level at two kilometers up to about 100 kilometers. (Note that this total height range is much larger than the first 15 kilometers of atmosphere that is the concern of the typical meteorologist. Altitudes up to slightly above the troposphere are sufficient for most weather studies).

- The Arecibo Observatory in Puerto Rico has been used for a number of special high-resolution studies of the atmosphere. Tor Hagfors will describe that facility later in this program. (Both the Arecibo and Jicamarca radars were designed for other purposes. Their use for atmospheric studies is typical of past procedures that have employed existing installations. The capabilities of the "borrowed" equipment are tested, after which a better system is designed for a particular application).

- The MU radar at the University of Kyoto in Japan is probably the most advanced unit that has come on line. This recently completed facility is a middle and upper-atmospheric radar that was built at a cost of from 12 to 15 million dollars. It is somewhat smaller than the Jicamarca and Arecibo installations and is located in a depression that has been dug in a mountain there. This impressive system will effectively demonstrate the promise of clear-air radars in atmospheric research.

Very large and powerful radars are not necessarily required for atmospheric research. A number of groups throughout the world make use of relatively small installations, formerly called ST radars but now more generally known as profilers. These radars are not like the conventional microwave units with the small rotating dishes that we are used to seeing at an airport. Operating at much lower frequencies, the profiler systems consist of stationary antenna arrays 50 to 100 meters on a side, with simple antenna structures and a small amount of auxiliary equipment for analysis. Although they are fairly inexpensive, consume relatively little power, and are capable of unattended operation, they can be used to perform worthwhile atmospheric research. This particular clear-air technology has advanced rapidly in the last 12 to 15 years, following publication of the initial papers based on use of the Jicamarca radar.

Preliminary work in the United States with these radars suggested more advanced studies, which are now being performed all over the world. The global distribution of stations that are capable of doing clear-air studies is shown in Figure 1. This chart was made about a year ago, so additional stations have since been established. Most of the stations are located in the United States but there are a number in Europe, Australia, and Japan. Additional stations have been proposed for installation throughout the central Pacific that are intended for study of the El Niño phenomena that Bill Hess mentioned in his discussion. These units would perform continuous measurements for several years.

Questions about how echoes can be obtained from clear air, and how they can be applied to atmospheric studies, may be resolved by considering the fundamentals of the pertinent radar system. With any atmospheric radar, we are

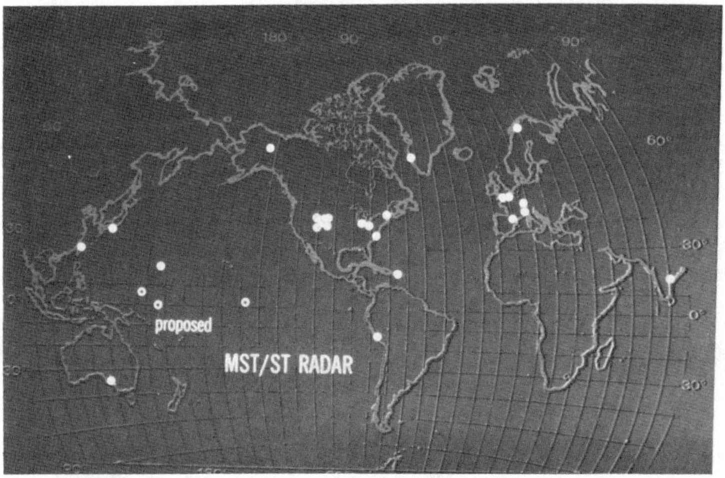

Source: N.O.A.A.

Figure 1. Global Location of Clear-Air Stations

dealing with scattering phenomena resulting from variations in the refractive index of the air that are on the scale of one half-wavelength of the radar in use. At 50 MHz, for example, the corresponding six-meter wavelength limits our concern to three-meter fluctuations in the air. As indicated in Figure 2, a sample volume of the atmosphere could be examined for any particular property such as humidity, temperature, density, and in some cases, electron density. Depending upon the height of the atmospheric sample, statistical variations of one or more of these quantities will occur along the antenna beam. These variations can be expressed as periodic functions. Sampling done on an infinitesimally small scale along the path of the radar transmitted pulse would produce a kind of random variable about some mean. Application of Fourier analysis to this variable would enable determination of the half-wave component of the fluctuation, and the corresponding magnitude of that component would determine the scattering function.

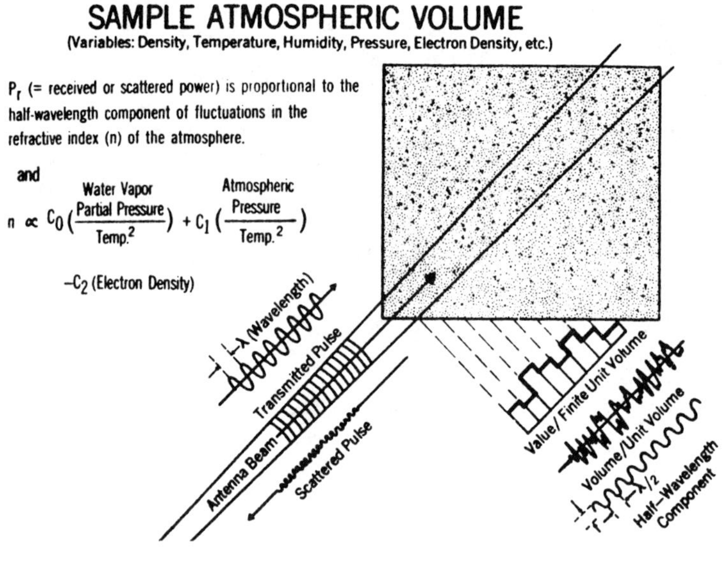

Source: N.O.A.A.

Figure 2. Examination of a Sample Volume of the Atmosphere

Elements related to the nature and cause of the scattering phenomena are depicted in Figure 3. There are a number of different components that contribute to the formation of the echoes. Since humidity is an important factor, water vapor in the first five to seven kilometers provides the best scatter signals. Above that height, temperature and equivalent pressure fluctuations cause changes in the refractive index so that the scattering becomes somewhat weaker because of the decreasing density of the atmosphere up to a height of about 50 kilometers. Above that height, at least during the daytime, the presence of free electrons enhances the scattering phenomena.

Once the scatter signals are in hand it is natural to inquire as to their availability with respect to time of day and various atmospheric conditions. Except for the loss of electron scattering at night above 50 kilometers, the signals typically are not affected in any major way due to these parameters. Useful echoes can be obtained in either clear air or under cloudy conditions, in the middle of the day, or in the middle of the night. Just as in the postal service, the signals are delivered day or night, in sunshine, rain, or snow.

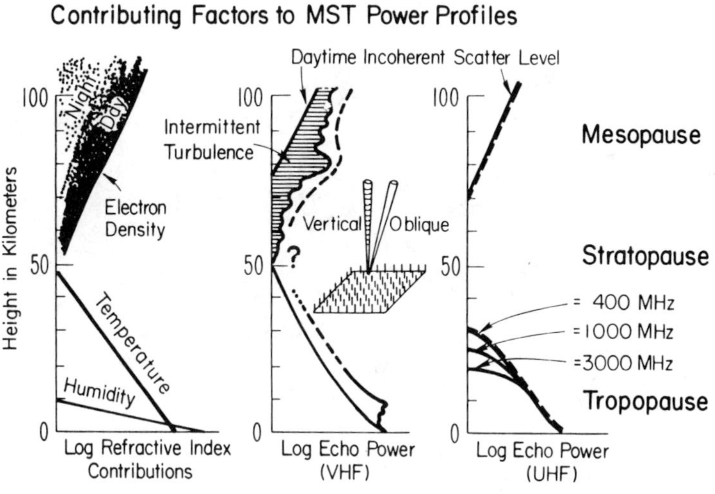

(© 1980 American Geophysical Union) From: K.S. Gage and B.B. Balsley, *Radio Science*, Vol. 15, pp. 243-257.

Figure 3. Elements that Contribute to Scattering Phenomena

The equation shown in Figure 4 illustrates three very important parameters in the determination of detectability. This formula expresses the ability to detect an echo at a given height or a given volume. Although not as complex as it may appear at first glance, the behavior depends on a number of variables that I will not discuss at this time except to point out that the detectability is proportional to the average transmitted power times the antenna area. Thus, the improvement of detectability is based upon the availability of powerful transmitters and large antennas. The only other parameter of immediate interest here is the system noise, the term T_S in the denominator of the equation. As in almost any other type of sensing equipment, the important factor here is the amplitude of the detected signal relative to background noise. With a large antenna and plenty of average power, the desired fluctuations can be detected without difficulty provided that the system noise, from whatever source, is sufficiently small.

As Bill Hess suggested in his discussion, if our sensitive clear-air radars include a Doppler capability, it becomes possible to investigate motions in the atmosphere. The various radars that have been discussed this morning, and others that will be mentioned later, may be subdivided rather crudely into

$$\frac{S_r}{\overline{\Delta S_N}} \propto \frac{\overline{P}_t A_e \eta \sqrt{n_i}}{(T_s + \alpha T_c)\delta f}$$

S_r = amplitude of largest spectral peak of the received signal density after averaging

$\overline{\Delta S_N}$ = mean noise fluctuation level after averaging

$\overline{P}_t$ = average transmitted power

A_e = effective antenna area

η = volume reflectivity for scattering process

n_i = number of spectral averages (incoherent averaging)

T_s = system noise temperature

T_c = cosmic noise temperature

α = transmission line efficiency

δf = signal bandwidth

Source: N.O.A.A.

Figure 4. Simplified Detectability Relationship

coherent *versus* non-coherent types—radars that are capable of detecting Doppler signals and those that are not. For example, many of the old-fashioned radars that are still in use today do not have a Doppler capability. Further subdivision of the Doppler subset reveals that some models have the ability to look at clear air or at hydrometeor structures in the atmosphere. Those radars can be used to measure the desired factors outlined previously: atmospheric winds at all heights, waves at all heights, as well as atmospheric turbulence and stability. Brief examples of how well we can measure each of these parameters will be given in the following sections.

Good characterization of wind parameters requires measurement of both horizontal and vertical components. The standard procedure that has been used for years to examine atmospheric winds on a day-by-day basis involves sending balloons aloft. This method provides a good picture of the coarse behavior of the horizontal winds, but the vertical components cannot be obtained from typical balloon observations. On the other hand, use of Doppler radar techniques allows accurate measurement of both horizontal and vertical components on a continuous basis. As Bill mentioned, however, looking at a volume in one direction with a Doppler radar will not provide a complete wind picture. In order to obtain a full three-dimensional wind vector it is necessary either to look in three separate directions or to scan the beam. Both of these capabilities are illustrated in Figure 5. The sketch on the left shows

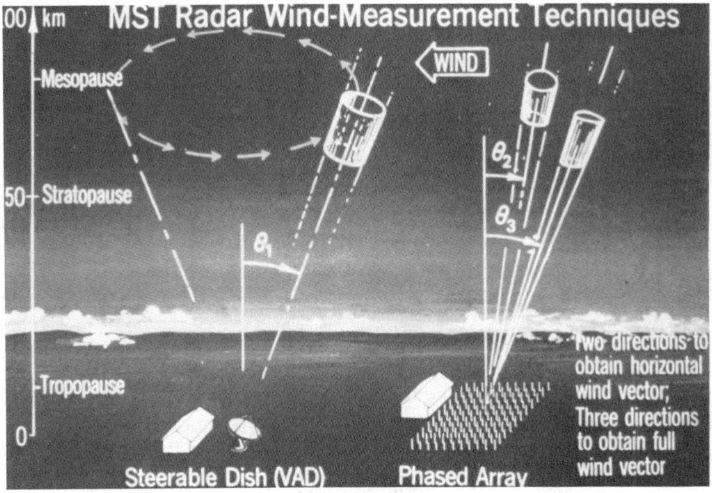

Source: NCAR: *Atmospheric Technology* Vol. 13, 1981

Figure 5. Necessary Capabilities for Three-Dimensional Wind Vectors

that a velocity azimuth display (VAD) can be obtained by continuously rotating the antenna. The resulting Doppler signal allows derivation of the full velocity vector. Alternatively, as shown in the right-hand sketch, either fixed or moving beams may be provided by a phased array. If three fixed beams are operating continuously, a wind measurement can be achieved that is roughly equivalent to the one obtained from a rotating beam.

Our ability to measure the wind is best expressed by a comparison between what people normally accept as the wind field and the results from an actual measurement. Reasonable coincidence between these two values over the available height range would offer convincing evidence of the validity of the wind measurements. Reference to Figure 6 provides a case in point. The simultaneous data in this figure were obtained a few years ago from radar observations made with the Arecibo radar in Puerto Rico and from conventional balloons at the National Weather Service station located some 60 kilometers to the east of the Observatory. A very good agreement is apparent, both

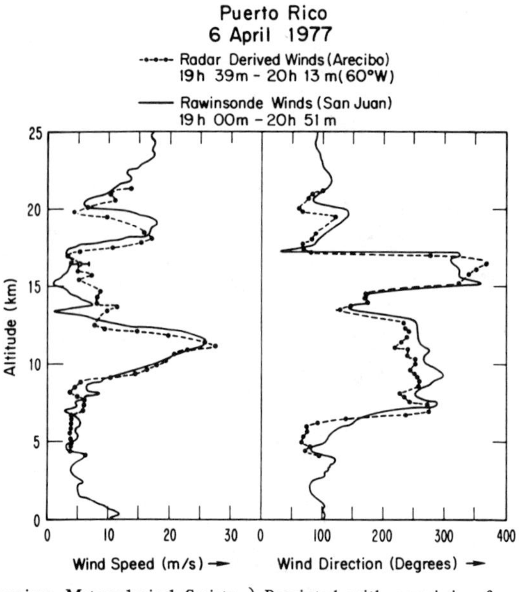

(© 1979 American Meteorological Society) Reprinted with permission from *Journal of Applied Meteorology*, Vol. 18, No. 3, D.T. Farley, B.B. Balsley, W.E. Swartz, and C. La Hoz, March, 1979, pp. 227-230, "Tropical winds measured by the Arecibo radar".

Figure 6. Comparison of Simultaneous Data from Conventional and Radar Measurements

in magnitude of the wind with respect to height, and in wind direction, even though the two measurement techniques were located 60 kilometers apart. The overall conclusion is that we can indeed measure the horizontal wind structure quite well.

If the Doppler technique is used to measure wind profiles on a rapid basis, and if the results are plotted in sequence, as shown in Figure 7, it is possible to obtain a picture of the variability of the wind. These data were obtained a few years ago by another small ST radar in Colorado. The time scale in the figure represents one day between midnight and midnight, followed by the next day. Thus we have a two-day picture of hourly averaged wind profiles in both the east-west and in the north-south directions. The heavier lines indicate the times when the normal balloon launches would have occurred at the Denver Airport. It is evident that most of the detail would have been absolutely lost if the information had been obtained from conventional once-or-twice-daily balloon sightings. The figure gives an impression of strong variability, and indeed, examination of a number of figures of this type reveals that many of the variations can be thought of as wave fluctuations. The atmosphere, in fact, appears to be full of periodic motion ranging from tidal fluctuations over a 24-hour period down to fluctuations on a scale of minutes.

RADAR - DERIVED WIND VELOCITY
Platteville, Colorado

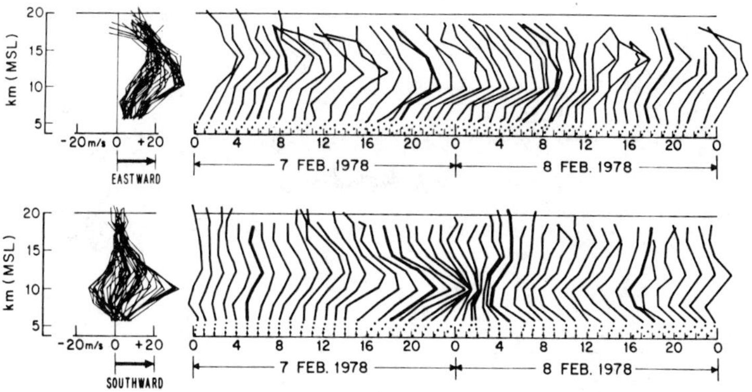

(© 1979 Pergamon Press, plc.) Reprinted with permission from *Journal of Atmospheric and Terrestrial Physics*, Vol. 41, pp. 983-994, W.L. Ecklund, D.A. Carter, and B.B. Balsley, "Continuous measurement of upper atmospheric winds and turbulence using a VHF Doppler radar".

Figure 7. Doppler Radar Measurement of Wind Velocity Profiles

The strong wave-like aspect of the atmospheric fluctuations is demonstrated in Figure 8. This chart is drawn from a sampling of data taken at our central-Pacific radar station at Ponape. The wave-like structure in the atmosphere is shown by the quasi-sinusoidal, short-period fluctuations that occurred right over the Ponape station on May 7 and May 8. This vertically oriented observation furnished data at altitudes ranging from 3.2 kilometers up to 21.2 kilometers, the latter height being the maximum possible capability of the small Ponape station. Comparable information can be obtained on horizontal wind fields.

The very strong kinetic energy contained in the wave motion of the atmosphere has been difficult to observe in the past. Something of the nature of the wave phenomena is presented in Figure 9, a rather intriguing "folksy" picture from a recent text by Colin Hines and colleagues. It gives the impression of waves in the upper air that typically are generated near the bottom of the

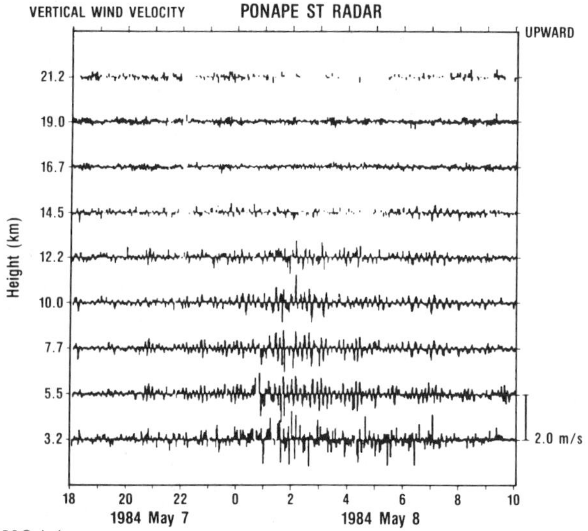

Source: N.O.A.A.

Figure 8. Measurement of the Wave Motion of the Atmosphere at the Ponape Station

atmosphere. As they move upward into regions of decreasing density, they start to grow, and just as in the behavior of ocean white caps, they break when a sufficient amplitude is reached. In the process of breaking they transfer energy from one spot in the atmosphere to another. This process turns out to be strikingly important in understanding the global circulation of energy from the equator to the poles. Both theory and observations of these phenomena have recently come to the forefront in atmospheric studies.

Atmospheric turbulence is illustrated by the fairly striking graph of Figure 10 obtained by a group of German workers at Arecibo in 1981 and reproduced in 1983. Although the darkened areas refer to enhanced detectability or enhanced echo strength, it turns out that the actual enhanced echoes are due, by and large, to the presence of turbulence arising from the dynamic breakdown of motions in the atmosphere. Radar observations of the layered structure between 9 and 21 kilometers suggest regions where enhanced turbulence

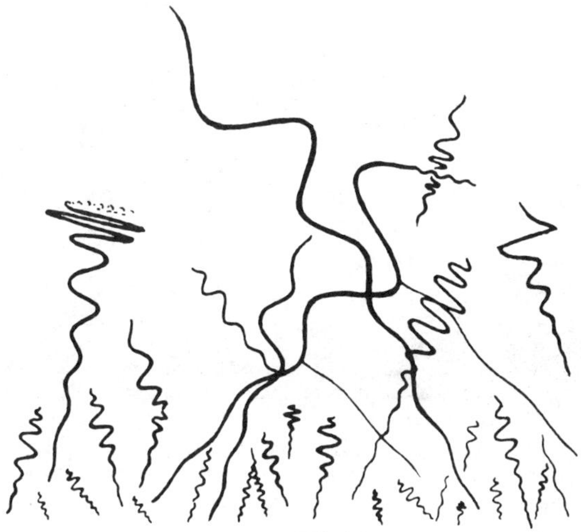

The atmosphere

(© 1974 American Geophysical Union) From: C.O. Hines and Colleagues, *Geophysical Monograph*, No.18, page 194.

Figure 9. An Impression of the Upper Atmosphere in Motion

could occur. Regions of actual turbulence encountered by aircraft flying through the lower portions of these layers do not correspond exactly with the projected turbulent regions although the agreement is very close.

Atmospheric stability is a rather complex subject that, in itself, requires special attention. It is sufficient to comment at this point that the stratosphere (named, in fact, for the stratification of the upper air) is a very stable part of the atmosphere. If these stability conditions can be measured, it becomes possible to obtain temperature profiles of the atmosphere. Very promising work is now in progress on the development of suitable techniques to achieve such measurements.

Thus far, our discussion has been concerned with various kinds of radars and their potential for accomplishing some remarkable breakthroughs in atmospheric research. Turning now to the current status of these techniques, it is proper to ask whether the technique is so well understood that an optimum

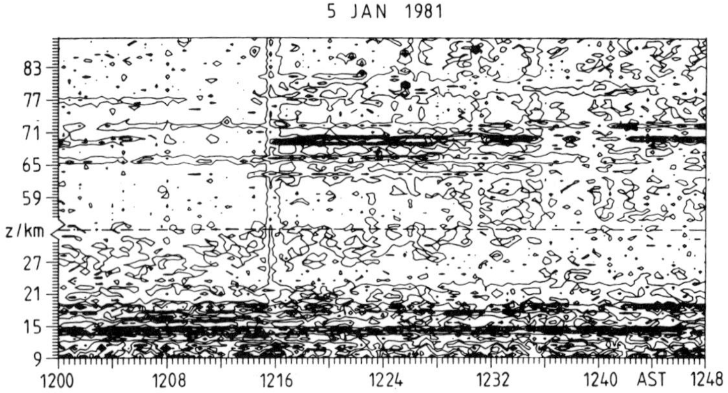

High-resolution height-time contour plot of echo strength in the troposphere, stratosphere and mesosphere. Note enhanced mesospheric echo after 1216 AST following a solar flare. Data obtained at Arecibo Observatory @ 50 MHz.

(© 1981 Pergamon Press, plc.) Reprinted with permission from *Journal of Atmospheric and Terrestrial Physics*, Vol. 43, pp. 789-800, J. Rottger, P. Czechowsky, and G. Schmidt, "First low-power VHF radar observations of stratospheric and mesopheric winds and turbulence at the Arecibo Observatory".

Figure 10. Radar Observations of Atmospheric Turbulence

system has been achieved. The response to this question is profoundly negative. We are in the stages of a preliminary design that could be called a model-A, or perhaps a model-B, configuration of the ultimate system. We have not defined the optimum system nor even the problems to which it can be applied.

Some of the potential difficulties in the design of an optimum clear-air radar system that might arise may be illustrated by means of Figure 11. This chart is a parameterization of the types of radars that could be considered if it is necessary to make observations at various heights. Recall that, in order to obtain echoes from any given altitude, it is necessary to have the proper "average power-aperture product", the name given to the product of average transmitted power times the antenna area (as expressed in the equation of Figure 4). If it is desired to examine conditions at an altitude of about 20 kilometers, for example, the average power-aperture product should be 10^9 watt-meters squared, if a 12 MHz to 15 MHz radar frequency is used. For the same

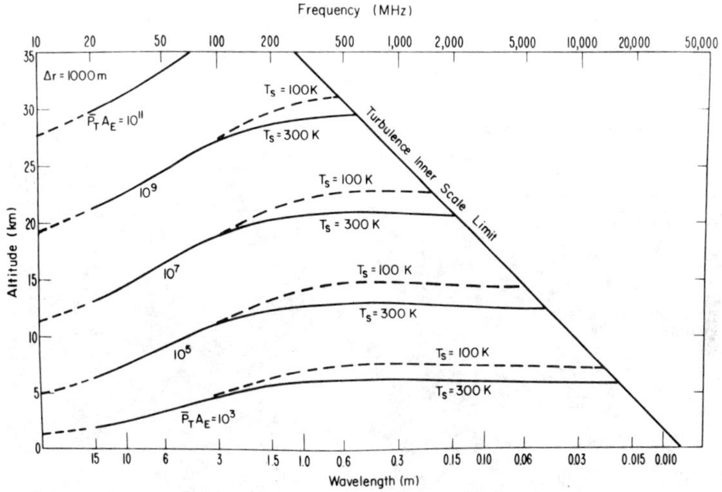

(© 1982 American Meteorological Society) Reprinted with permission from *Bulletin of the AMS* Vol. 63 No. 9, Sept. 1982, pp. 1009-1018, B.B. Balsley and K.S. Gage, "On the use of radars for operational wind profiling".

Figure 11. Desirable Characteristics of Radars for Observations at Various Heights

power-aperture product at, say 50 MHz, it would be possible to look up to about 24 kilometers. Thus, given a particular radar, an appropriate frequency, and an antenna area, Figure 11 can help us determine how high we can see. But there are many other problems that must be considered. For example, it is necessary to define the desired spatial and temporal resolutions, and to specify the rapidity of observations. These parameters are not covered in our present discussion. We have a basic understanding of these factors but we have not reached the point of being able to design the proper radar, nor define the pertinent uses.

The internal components of these radar systems are fairly basic. The antenna configurations, however, are subject to a variety of major design options. For example, a dish-like antenna is used at the Chatanika radar in Alaska. Alternatively, the previously mentioned 50-MHz MU radar in Japan makes use of a phased array of yagi antennas. An even simpler structure is composed of arrays of coaxial cables, appropriately cut every half-wavelength, known as "coco-antennas"—coaxial-colinear antennas. This design turns out to be an inexpensive way to obtain large antenna areas. The latter two arrangements typically provide near-vertically oriented fixed beams, as indicated in Figure 12. In order to obtain all three components of the wind, three different

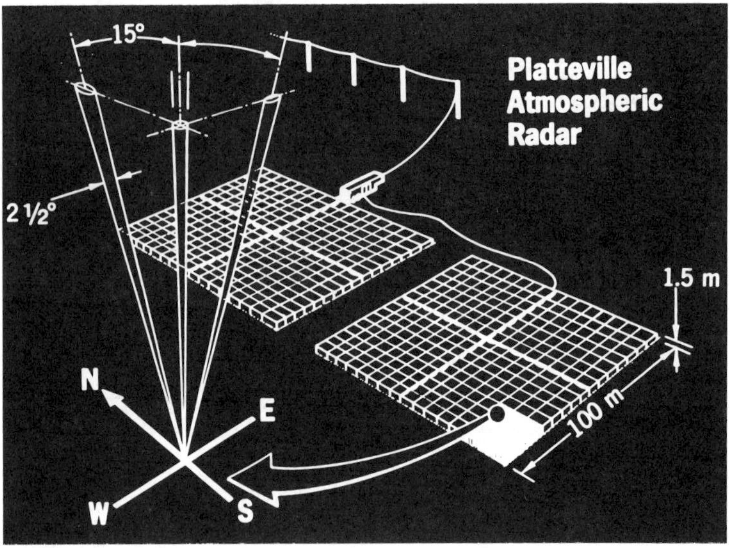

Source: N.O.A.A.

Figure 12. Typical Antenna Structure for Vertically Oriented Fixed Beams

beams are generated by these antennas, one looking toward the north, one toward the east or west, and one being directed vertically. With the beams fixed in these positions, the antenna can operate for many years and provide the same information that would be obtained from a steerable structure. The necessary experimental equipment is also relatively simple and usually consists of a transmitter, a transmitter power supply, the digital components of the system, the receivers, and a tape recorder. Thus a typical installation only requires a very small room for the auxiliaries and a very large antenna.

In addition to the measurement of winds, waves, turbulence, and stability factors, these systems can be used for a number of other activities. I will comment briefly on each of the following applications:

- A national network of profilers (for weather prediction)
- Clear-air turbulence studies
- Rocket-launch wind monitoring
- Tactical uses
- High-seas wind profiling
- Atmospheric sciences
- Pollution studies

The first item, as suggested in Figure 13, proposes a grid of several hundred ST radars, installed throughout the United States, that would greatly improve our capability in the wind-profiling aspects of weather-service tasks. Each unit could consist of a small 100 by 100 meter array, or an even smaller antenna, with a single transmitter. Data from continuous monitoring of winds up to heights of 15 to 18 kilometers could be fed into a central station. These small radars could even be combined with other larger radar installations in order to increase the overall sensitivity of the network. Matching of frequency bands or of radar models would not be necessary since clear-air Doppler radar capability is the only requirement for compatible operation.

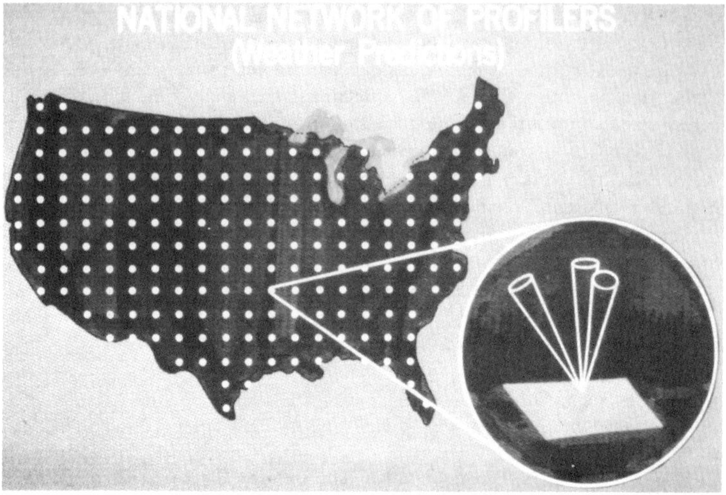

Source: N.O.A.A.

Figure 13. Proposed National Network of ST Radars

A recent NASA bulletin devoted to a discussion of this proposed network contains an important financial comment that may be summarized as follows:

> If a national radar network were to be used only for prediction of the jet-stream location, the complete radar system could pay for itself in three months by the savings that would result from improved flight control of commercial jet-aircraft traffic.

Although this statement is probably an exaggeration, the economics would still be impressive if the complete system could be paid for in this way in a matter of, say three years.

A network of this sort would also be able to monitor areas of extreme clear-air turbulence. Routing of aircraft around such dangerous regions could prevent accidents and many potential aircraft crashes. A good example of this capability can be found in the proposed operation of a system at Cape Canaveral that would monitor the atmosphere continuously for shuttle launches that require good wind measurements. Prior to the availability of this technique it has been necessary to put up a balloon and wait 40 or 50 minutes until it reached sufficient altitude to allow measurement of the really high-altitude winds. By that time the bottom part of the wind profile may have changed drastically for a variety of reasons. With the radar technique, however, rapid changes in high-altitude wind speeds can now be measured on a scale of a minute or two.

One example of a use for tactical purposes would be to embed an antenna into the deck of an aircraft carrier. (Similar arrangements could be made for the other armed services). Current technology is available to allow measurement of winds directly above such an antenna, thereby removing the necessity for releasing balloons that could be detected by enemy forces. Constant monitoring of the wind would thus be available for any aircraft associated with the carrier.

I am particularly interested in the potentially useful ability of these systems to predict wind patterns over the oceans. At present there is a great paucity of knowledge about the wind over oceanic areas. Some limited data can be obtained by means of satellites but much more information is needed. Modified profilers, mounted on fixed platforms in the oceans, could telemeter their observations instantaneously to a central data bank via satellites. Rapid data processing would ensure accurate projections of wind behavior.

Atmospheric research has been conducted on waves, global circulation systems, and turbulence for the last 15 years, with many investigations having been developed from the use of clear-air radars such as the Poker Flat station in Alaska. This system is somewhat smaller than the previously mentioned Jicamarca radar, and was designed specifically for clear-air studies in the mesosphere, stratosphere, and troposphere. The antenna is about 200 meters on a side, the transmitting system has a capability of about six MW of peak pulse power, and a time resolution that is in the range of 1 to 1.5 minutes. Almost continuous operation with this time resolution for the past five or six years has given us an extremely large data base that can now be used for a number of advanced studies.

It is obvious, of course, that clear-air radars can be useful in pollution studies. Small ST units, installed near manufacturing plants that may possibly be causing pollution of the atmosphere, could be used to good advantage in the control of pollution for a given wind direction.

By way of summary, a comparison between conventional weather-balloon technology and ST radar profiler techniques is given in Figure 14. It is important to recognize that the balloon-sounding networks cannot be replaced entirely by the new techniques since the radars can only profile the wind

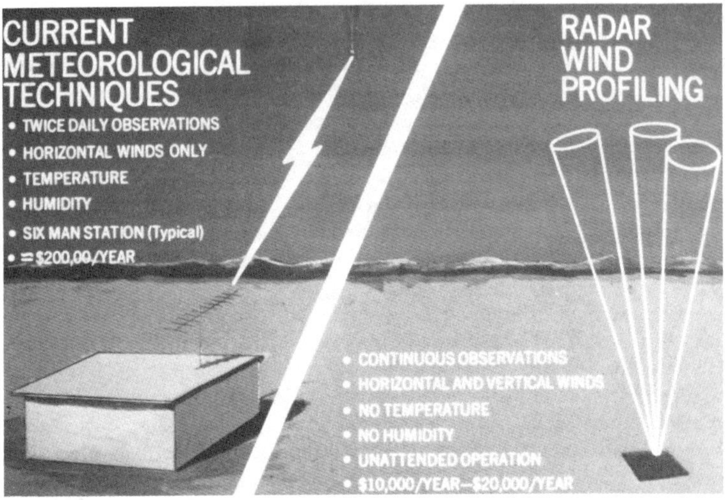

Source: N.O.A.A.

Figure 14. Comparison of Conventional and ST Radar Profiler Techniques

field. Balloons are still required for the measurement of humidity, pressure, and temperature. As mentioned previously, the wind-profiling capability of balloons is limited to measuring only the horizontal components. Typically, balloon soundings are made once or twice daily, with a number of men required at each station to decode the information, operate the system, and transmit the data into some central network. The operational cost of a single system is about $200,000 a year.

In contrast, ST radars are capable of continuous measurement of both horizontal and vertical components of the wind, thereby providing better understanding of these and related phenomena. They cannot, however, obtain temperature information and, for the present at least, humidity measurements are also beyond their capability. The radars operate unattended and typical station and maintenance costs are somewhere between $10,000 and $20,000 a year, representing a considerable economic advantage for the new techniques.

For those who may be interested in the engineering and business aspects of this modern approach to atmospheric studies, Figure 15 contains a fairly complex, albeit simplified, picture of how the profiler technique developed. In the 1930s, 1940s, and 1950s, a number of atmospheric propagation studies were already employing some of the same techniques that are now in use. Also in the 1960s, many ionospheric studies were being performed with the aid of the Doppler radar capability. Digital systems equivalent to some used at present were also being applied in those studies. That background, coupled with the need to improve measurement of winds, to examine the dynamics of wave structures, and to measure pollution, suggested the development of a clear-air radar capability using the Doppler technique.

In the 1970s, we began developing specific systems for wind profilers. Now in the 1980s, regular use of these initial designs has indicated the need for improvements. Generally, acquisition of more powerful, more sensitive,

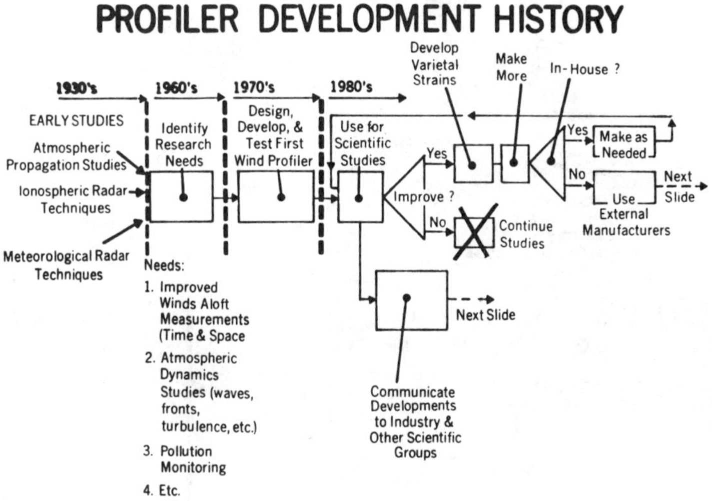

Source: N.O.A.A.

Figure 15. Development of the Profiler Technique

high-resolution radars has occurred within the atmospheric-sciences community by in-house design. It has been found, however, that the in-house development of a number of systems of a single design is too difficult to achieve. In such situations we usually turn to outside contractors for the necessary construction.

This improvement/manufacturing loop is an interesting phenomenon. The beginning stages of any scientific study that is based on the use of a particular system invariably results in a desire for improvement of the system. As the study progresses and new and better units are devised, an inevitable technology transfer occurs: the results of the investigations are conveyed to other scientific groups, both in and out of industry, by means of technical papers, discussions, and seminars of the kind we are having today. In the field of atmospheric research this pattern is evolving at a fairly constant rate.

The influence of technology transfer on the future development of profilers is outlined in Figure 16. In general, the process would start with the

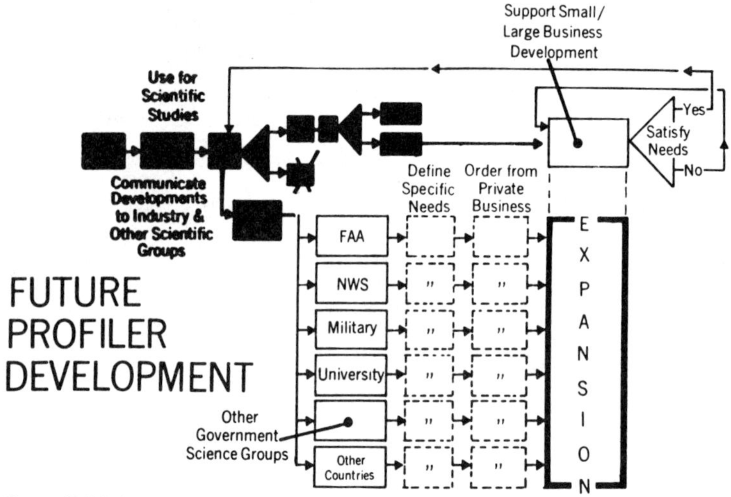

Source: N.O.A.A.

Figure 16. Future Development of Profilers

establishment of a small business to satisfy the current needs of a scientific sponsor. (This step has already been taken in our group). The required technology expansion will probably generate enough business to keep both the sponsor and the manufacturer happy. If, at the same time, the sponsor has transmitted his results to various other scientific groups, one can expect a proliferation of business enterprises, as each new sponsor writes specifications for different antennas, different frequencies, different resolutions, and different capabilities.

This form of expansion is already in progress. Several groups are seeking to satisfy their specific requirements by investigating opportunities to buy various systems from companies that still must be formed. There is a great need at present for the establishment of a production mode for the MST radar field. We have insufficient technology and too few companies to build the radar systems that are needed now and those that are projected for the future.

In this brief summary of the background and potential of atmospheric research, I have tried to express my belief that this is a very exciting, very healthy, and very rapidly expanding technology. I would hesitate to predict where we will be in this field ten years from now.

ATMOSPHERIC REMOTE SENSING: A NEW TECHNOLOGY FOR IMPROVING WEATHER NOWCASTS AND FORECASTS

C. Gordon Little

C. Gordon Little obtained the B.Sc. degree in Physics, with Honors, and the Ph.D. in Radio Astronomy from the University of Manchester in England in 1948 and 1952, respectively. Following a year of duty as an assistant lecturer at the University of Manchester, he joined the faculty of the University of Alaska in 1954 as a professor in the Department of Geophysics, and as Deputy Director of the Geophysics Institute. In mid 1958 he transferred to the United States Department of Commerce National Bureau of Standards Laboratories in Boulder, Colorado. In that organization he held a series of senior posts until 1967 when he became the Director of the Wave Propagation Laboratory of the National Oceanic and Atmospheric Administration (NOAA), his present position. In 1984, Dr. Little was the recipient of the Cleveland Abbe Award of the American Meteorological Society. Over the years he has also received a number of high-level Agency, Department of Commerce, and Federal awards for his contributions to scientific research and departmental management. He is a member of the National Academy of Engineering and a Fellow of the Royal Astronomical Society, the Institute of Electrical and Electronic Engineers, and the American Meteorological Society.

I will introduce this subject by making three points followed by some discussion and examples for each one. First, meteorological services have often been driven by technology in that major advances in the art and science of weather forecasting have required the arrival and exploitation of new technologies. Second, despite advances in applied science, our complex modern society's needs for understanding the atmosphere have outrun the currently used technologies. Third, since weather services can only be as good as the data sets upon which they are based, radical improvements in atmospheric observing capability are required.

As an example of the first point, over a hundred years ago the arrival of the telegraph enabled the meteorologists of that era to prepare the first maps of simultaneous weather conditions across the United States. I imagine that prior to that time it was necessary to exchange letters in order to know the state of the weather on a given date in various parts of the country. Another example would be the advent of the radiosonde, a balloon system that was initiated roughly fifty years ago and is still used as the primary method for obtaining upper-air profiles of wind, temperature, and humidity. This system uses cheap throw-away instrumentation packages that are carried aloft on fairly small balloons. These devices use radio to track the winds and telemeter temperature and humidity information down to the ground.

The computer, of course, is an important new technology that has enabled the meteorologist to use mathematical and physical insights in the arena of numerical weather predictions. Much of the present-day weather forecasting capability across the United States and around the world results from modern computer techniques. The weather satellite, initiated about 25 years ago, also has had a dramatic impact upon meteorology since it made it possible for the first time to obtain information on a truly global basis. A single polar satellite is capable of scanning the entire earth in a period of 12 hours, and multiple satellites can cover the globe at even more frequent intervals. For the temperate and tropical latitudes, the geostationary satellites can explore a region as frequently as every five minutes, thereby allowing existing weather patterns to be monitored almost continuously in time.

The information gap of my second point results from the fact that our needs for better understanding of the atmosphere are much greater now than in the past. An excellent example of this issue may be drawn from Bill Hess's earlier remark this morning about the necessity for minute-by-minute data on windshear and downbursts at an airport. There are many other industries and activities across the country that would receive important benefits from improved weather information. In agriculture, for example, each day thousands of farmers must make costly decisions relative to irrigation, the application of herbicides and insecticides, or the correct time to harvest their crops. These decisions are highly dependent upon detailed information on the current weather and, of course, the predictions for the next few hours or days. Daily operation of our huge electric energy supply industry is geared to the consumption of energy which, in turn, is strongly dependent upon weather conditions. Obviously, the requirements for air conditioning on a hot day must be met by an increase in the generation of electricity by conventional sources and, more recently, by solar and wind-power systems. The efficiency of the electricity supply industry could therefore be improved by better forecasting of the weather.

In our defense programs, many weapons systems are dependent on suitable weather for their correct performance. For example, when we made our abortive attempt to free the hostages from Teheran, there is no doubt that at least part of the failure can be attributed to the lack of weather information. Perhaps the plan would not have succeeded for other reasons, but those helicopters ran into an unanticipated sand storm at 10,000 feet that compounded the difficulties. Since smart weapons systems often require clear air, the ability to provide detailed forecasts of weather conditions on some future battlefield may indeed be critical.

As another example, air pollution and acid rain are both of great concern to the Environmental Protection Agency. In the days of the cave man the presence of these two irritants probably mattered little to the inhabitants of the earth. Now, as our society has become more complex, we find ourselves in much greater need of new methods for obtaining information about the atmosphere.

Data-set quality, the premise of my third point, is best illustrated by use of a term that I call a "nowcast", a report on the current state of the atmosphere at a given time. Certain users of weather information require this service in addition to the widespread standard forecasts. The accuracy and reliability of both nowcasts and forecasts depends heavily upon the density and abundance of the data sets in space and time. In order to achieve significant improvements in weather prediction, I must emphasize that space-time density of the observations must be increased by orders of magnitude rather than by relatively trivial incremental advances of say, five or ten percent.

The need for upper-air data offers a good example. If we are to forecast individual storms, or even some of the mesoscale phenomena that Bill Hess spoke about, it would be desirable to have say, a 50-kilometer grid of wind-temperature and humidity data, rather than the present 500-kilometer grid, thereby introducing a ten-fold increase in data in each linear direction along the ground. Since we are interested in short-duration phenomena for these purposes, the frequency of observation would also have to increase by a factor of ten. Thus the simple linear scale factor of ten increases the amount of data to be obtained by a factor of a thousand. If it is necessary to provide the services that would be desirable at modern airports, a further increase of a thousand would be required in the space-time density of wind-field information, thereby increasing the number of observations by a factor of one million. If possible, we would like these greatly increased observations not only to be more frequent in time and space but also to be more accurate and more representative of the atmosphere. For example, a point sensor, such as an anemometer, may record the wind speed and direction very precisely, but for typical numerical weather prediction purposes, the major interest is in obtaining representative data for a specific volume of atmosphere rather than in precise measurements at a given point.

I come now to my main thesis: that ground-based remote sensing can provide the important and essential breakthroughs in observational capability that are required for major advances in weather services. I will illustrate this proposition by describing four different remote-sensing techniques as follows:

- Production of path averages of near-surface parameters
- Measurement of vertical profiles of wind, temperature, and humidity
- Provision of three-dimensional coverage of the atmosphere
- Operation of the infrared Doppler radar system

The first of these technologies makes use of the scintillation, or twinkling, of an optical or infrared signal that is propagated through the atmosphere. If a laser is used for this purpose and the beam is propagated through the atmosphere for about 10 kilometers, a viewing screen illuminated by the transmitted beam would show a typical moving, shimmering, scintillation pattern caused by temperature eddies carried by the wind that is blowing across the beam. This concept is illustrated in Figure 1. Two detectors placed side-by-side, just a few centimeters apart, sense the diffraction pattern that moves in the same direction as the wind. The observed data are fed to a simple pattern-velocity computer, thereby allowing measurement of the average speed of the wind across the path.

The ability of the system is illustrated in Figure 2 where the upper trace is an optical measurement of the wind speed across the path as a function of time, for data taken over a 24-hour period. The lower trace shows the data obtained by taking the average readings of five anemometers that measured the transverse wind speed across the same path. It is clear that there is an extraordinarily high correlation between the two sets of data.

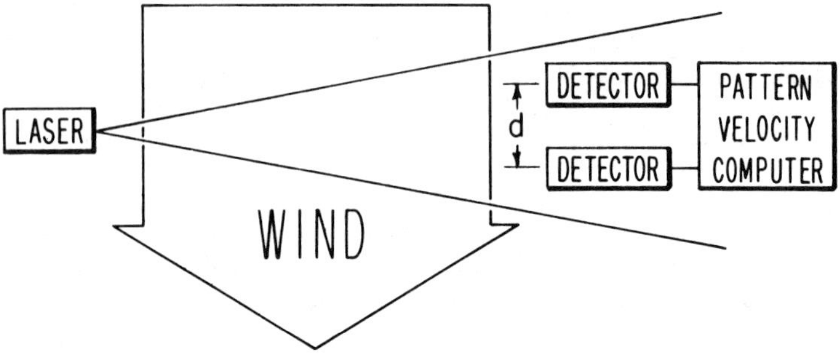

Source: (Courtesy of Mr. G. R. Ochs, NOAA)

Figure 1. Optical Transverse Wind Monitor

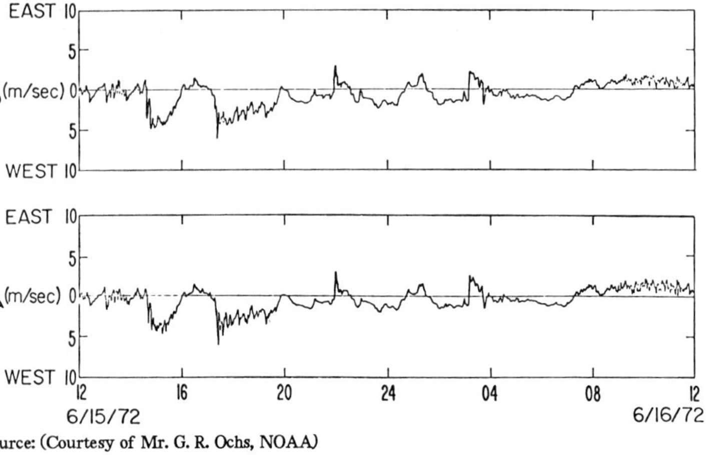

Source: (Courtesy of Mr. G. R. Ochs, NOAA)

Figure 2. 24-Hour comparison of path-averaged transverse winds measured by optical scintillation technique (upper trace) and by a colocated linear array of anemometers (lower trace).

A geographical contour plot of the Four Corners area in northwest New Mexico is shown in Figure 3. The major coal-burning power plant that is operated there, in a region that would otherwise be a pristine wilderness area, has created considerable concern about pollution of the atmosphere. Transverse wind-sensing systems have been used there in irregular terrain to measure the wind-flow patterns. On a larger scale it would be evident that the ground slopes gradually downward from right to left. This figure, however, shows the contours of a ridge, a rocky dam roughly 100 meters or so high, across the general slope. The three lines, A, B, and C represent optical paths parallel to the fault. Line A is at the foot of this escarpment, line B is an optical path on the summit of the ridge, and line C is located along the approximate back side of the ridge.

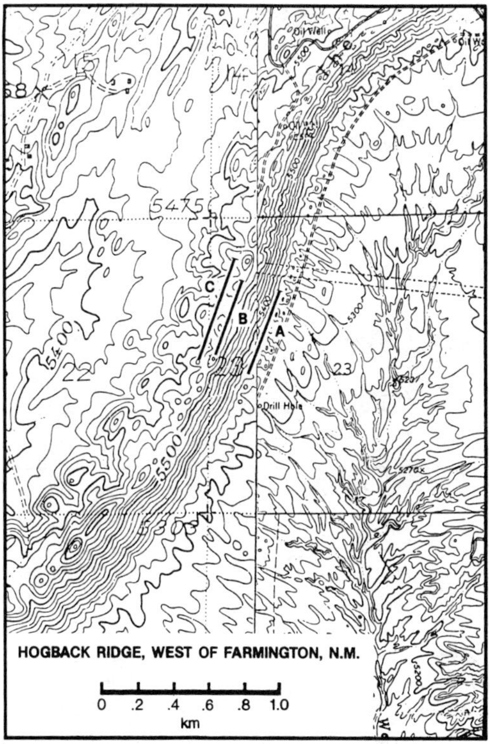

Source: (Courtesy of Mr. R. B. Fritz, NOAA)

Figure 3. Optical Paths Paralleling Ridge Contours in Northwest New Mexico

The plots in Figure 4 are medians of some 20 days of variations in wind speed across the ridge for a typical 24-hour period. The prevailing wind is blowing from right to left during summertime conditions. The upper trace shows variations for the crest of the ridge along path B. For approximately the first 12 hours of the afternoon or early-evening hours, the wind speed is in an up-slope condition. During the evening and early-morning hours, a down-slope condition sets in. The lower trace for line A at the base of the ridge exhibits quite a different pattern. A slight flow that is counter to the prevailing down-slope winds is continuously present through the night-time hours.

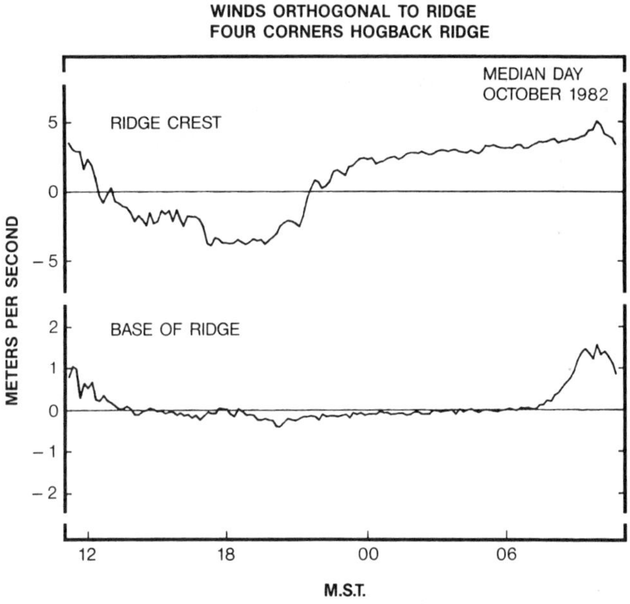

Source: (Courtesy of Mr. R. B. Fritz, NOAA)

Figure 4. Median Diurnal Variation of Wind Speed Component Transverse to Ridge Line

These conditions are illustrated in Figure 5. During the night, as the large-scale down-slope flow of air from right to left is brought up against the ridge, a counter flow develops that is associated with a local down slope along the face of the ridge. This rotor is shown in the figure just above the letter A. Similar conditions exist in many areas of the country. Cities such as Boulder, Colorado and Los Angeles, California are located near mountain ranges so that

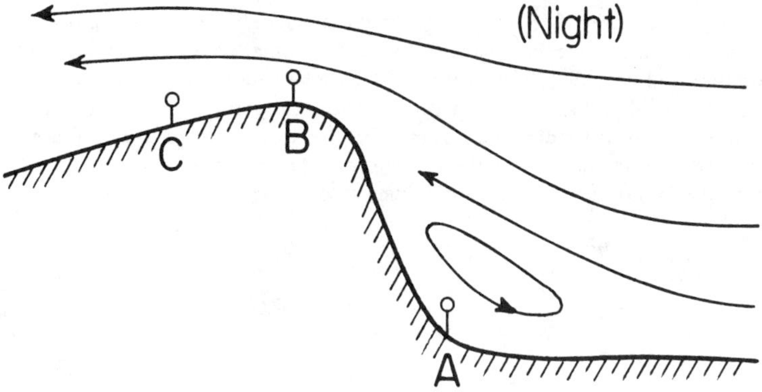

Source: (Courtesy of Mr. R. B. Fritz, NOAA)

Figure 5. Sketch Illustrating Formation of Local Rotor-Induced Counterflow Near Base of Ridge

the prevailing winds may create an eddy of this kind and cause circulation of polluted air. Start-up wind speeds associated with this counter rotation are often very low with typical velocities of perhaps half a meter per second. Since these low speeds cannot be measured by anemometers, the path-averaged techniques are indeed quite valuable, particularly in areas of difficult access such as across a lake or an estuary, or in difficult terrain.

These techniques have been extended well beyond the mere measurement of average transverse wind speed. For example, single-ended systems have been employed to observe the scintillation of the natural scene, rather than of a laser beam, using the contrast in the natural scene to measure the transverse wind speed. Using two more-or-less orthogonal paths, it is possible to determine path-averaged or area-averaged wind speed and direction by taking data along a series of paths in a closed array of triangular or multi-sided form. NOAA scientists have shown that it is possible to measure path-averaged heat flux as well as path-averaged refractive-index turbulence. The technique also measures the size of the smallest eddies and hence the path-averaged momentum flux or drag of the ground on the atmosphere. Analysis of multiple paths allows measurement of the convergence and divergence of the air. The transverse wind speed at different points along the paths can also be obtained by using sophisticated transmitter and receiver arrays. Under conditions of rain it is possible to measure the path-averaged rainfall rate as well as the path-averaged drop-size distribution. Increasing use of this technology is being made in typical air-pollution-type studies since it s quite cheap and the devices are relatively simple.

My second area of technology, generally designated by the term "profiler", encompasses a complex of remote sensors that measure profiles of winds, temperature, and humidity. The diagram of Figure 6 includes the components of a clear-air wind-profiling radar with an associated data processor. The six-channel microwave radiometer indicated in the upper right-hand block measures temperature and humidity by detecting radio waves that are emitted from the atmospheric oxygen and water vapor. Brightness temperature data

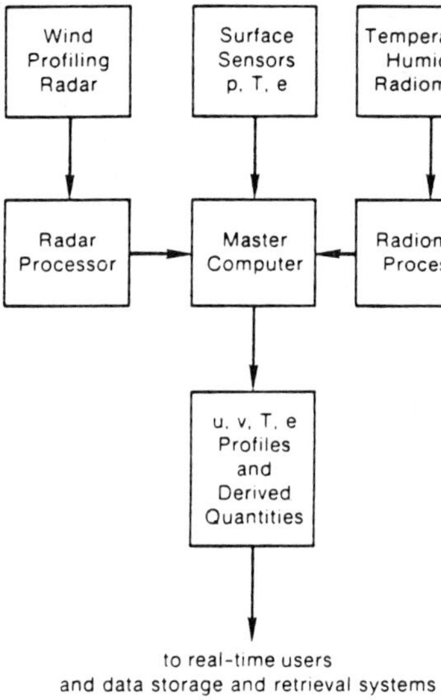

Source: (Courtesy Dr. R. G. Strauch, NOAA)

Figure 6. Components of a Wind/Temperature/Humidity Profiling System

observed on the several channels shown in the diagram are used to calculate vertical profiles of temperature and humidity. The sketch of the beam arrangement of the clear-air radar shown in Figure 7 indicates a typical fixed-beam structure consisting of three beams that are oriented into the zenith, and 15 degrees from the zenith towards the east and the north. A radar system of this type at the Stapleton Airport operates at 900 MHz and uses three horn antennas near the focus of an offset paraboloid to create the three-beam array.

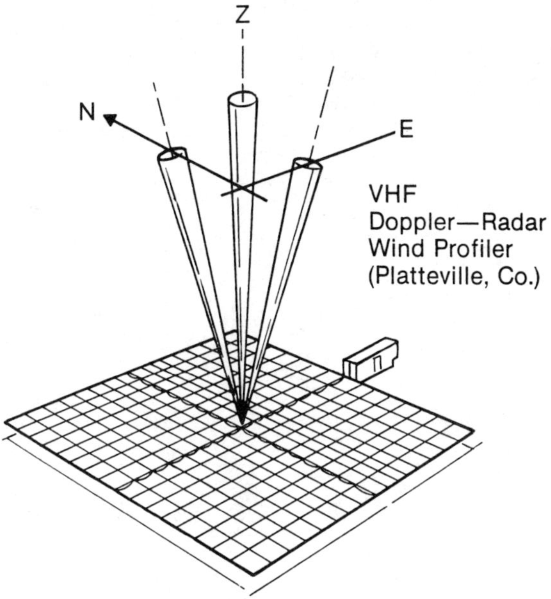

Source: (Courtesy of Dr. R. G. Strauch, NOAA)

Figure 7. Illustration of the Three Fixed Beams of a NOAA Wind- Profiling System

It is possible to obtain some information about the evolution of a weather system by using data obtained from radiosondes. Typically, data points are available only at 12-hour intervals and the wind profiles can change quite dramatically over that period. Specific wind-profiling techniques are capable of supplying better information, as in the wind-barb plots shown in Figure 8. This time/height cross section uses the meteorological forecaster's standard plotting technique, with wind profiles plotted every hour. The time scale starts on the right and covers a 24-hour period. Heights run from sea level to 17 kilometers, as shown on the left. Our radars are located at a ground-level height of 1.6 kilometers. Towards the right of the chart the barbs denote wind speeds of the order of several tens of knots from the west. In the center of the figure a discontinuity marks a transition to lower speeds from the north caused by the passage of a front over the site. Although this chart illustrates the ability of a single wind-profiling system to provide meteorologically significant data for the weather forecaster, it will be necessary to use many radars to define the wind field adequately across the United States.

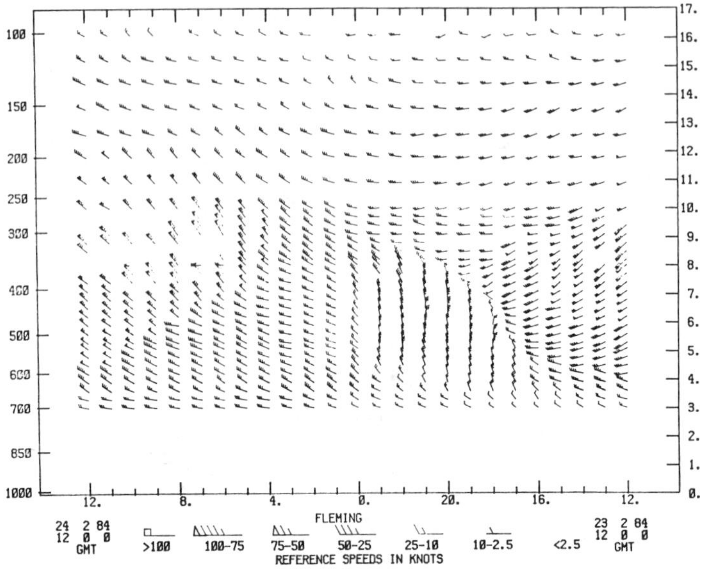

Source: (Courtesy of Dr. R. G. Strauch, NOAA)

Figure 8. Hourly Mean Wind Barbs Plotted versus Height, Showing Passage of a Front

One of the main difficulties with this wind-profiling technique is the extreme variability of the echo intensity. This problem is illustrated in Figure 9, which shows the cumulative probability of the hourly averages of echo intensity from the height range of 8.34 km to 9.21 km above a 50-Mhz radar in northeast Colorado for a 12-month period. The non-linear horizontal probability scale used in Figure 9 gives a straight-line cumulative probability distribution if the parameter is normally distributed. Figure 9 therefore indicates (to a very good approximation) that the logarithm of the echo intensity is normally distributed. Moreover, the slope of the line is such that the 1-percentile and 99-percentile values are 40 db apart. Examination of the distribution indicates that to achieve 90 percent coverage in time, the radar must have a 9 db sensitivity reserve relative to median atmospheric conditions. A 21 db reserve is required if good data are required 99 percent of the time.

Overall, it is fair to say that wind profiling has become a well-established technology. Many hundreds of station-months of data, using various kinds of wind-profiling radars, have been gathered around the world. The results obtained from these installations demonstrate that it is now possible to have essentially continuous monitoring of wind profiles.

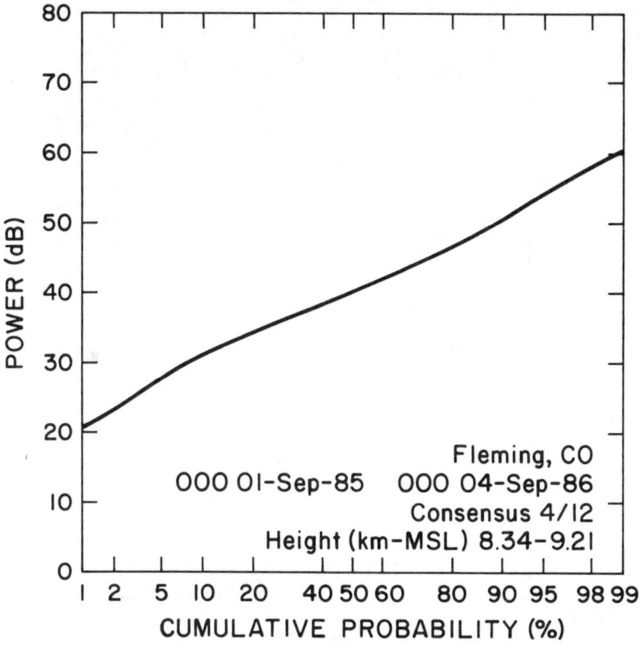

Source: (Courtesy of Dr. A. S. Frisch, NOAA)

Figure 9. Annual Cumulative Distribution of Hourly Median Signal Strengths versus Height for a Clear-Air Wind-Profiling Radar

The weighting functions plotted in Figure 10 illustrate the third technology of this discussion, namely temperature profiling. These functions are seen to be exponentially decreasing with height and are weighted heavily into the bottom half of the atmosphere. If a microwave radiometer is tuned to different frequencies, the received signals come from a correspondingly different range of heights. At 58.8 GHz, for example, the instrument is close to the oxygen absorption band so that the detected radiation is strongly absorbed. Consequently, the sensitivity of the radiometer is limited to the lowest kilometer or so of the atmosphere. On the other hand, at 52.85 GHz, signals are being received mainly from the lowest 500 millibars of the atmosphere. A system based on these principles has been built and has operated for some years in Denver. Vertically incident radio waves from the sky strike a metal plate reflector that transfers the signals to a six-channel microwave radiometer installed in a temperature-controlled environment.

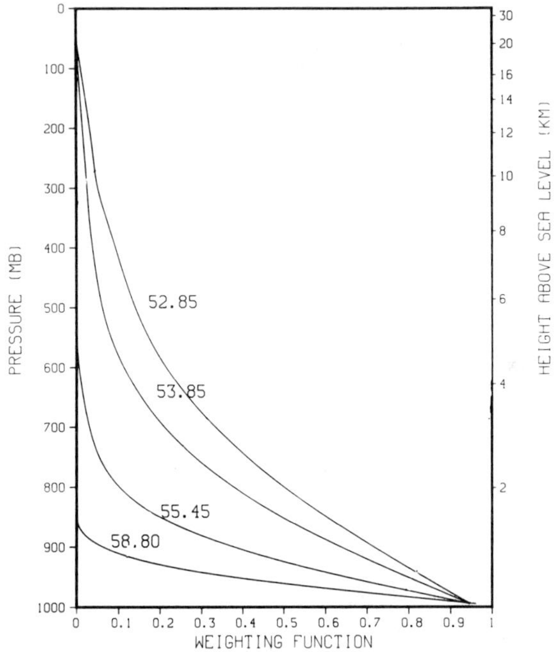

Source: (Courtesy of Dr. E. R. Westwater, NOAA)

Figure 10. Weighting Functions for Various Frequencies used in NOAA Ground-Based Temperature Profiling Radiometer

Temperature profiles in the lower part of the atmosphere can be obtained from these weighting functions, as illustrated in Figure 11. In this chart a series of displaced temperature profiles are plotted as functions of height and time from data obtained at two-hour intervals during the day. The shape of the profile changes as the day advances. The left portion of the figure shows nocturnal radiation-inversion conditions that are caused by warm air aloft and cold air in contact with the ground. As the sun warms the earth, the inversion disappears. By the middle of the day (shown in the center of the chart) a basically adiabatic condition exists with the temperature decreasing uniformly and steadily with height. At the right side of the figure, the plot indicates that as the sun disappeared, the heating stopped, and the ground-based radiation inversion reoccurred. This figure displays actual data taken during a low-altitude inversion over Denver that trapped the contaminants and caused the worst air-pollution event of that winter.

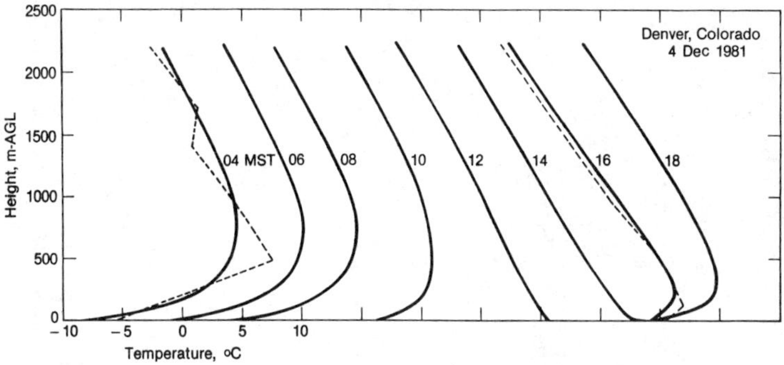

Source: (Courtesy of Mr. M. T. Decker, NOAA)

Figure 11. Displaced Temperature Profiles Showing Disappearance and Reformation of Ground-based Radiation Inversion. (Data obtained automatically by NOAA radiometer at two-hour intervals)

One of the advantages of a temperature-measuring system of this type is the high time resolution that can be obtained. Instead of the radiosonde information being available only once every twelve hours, it is possible to obtain data about every two minutes. The curves shown in Figure 12 are derived from such data plotted at two-minute intervals for two hours, for the lowest few kilometers of the atmosphere. The passage of a weather front is well illustrated in the center of the figure. A very rapid transition from one air mass to another can create a cold front that will move over a site in a matter of two to four minutes. In this particular case, the front came into an existing temperature inversion caused by a pool of cold air in contact with the ground. Since the front was not strong enough to be colder than the already cool air at ground level, it flowed on top of the inversion. Consequently there was negligible effect on the ground temperature for several hours. When this particular front passed over the National Weather Service site, the ground thermometers were unable to detect it but the radiometric system showed its passage in a very convincing manner.

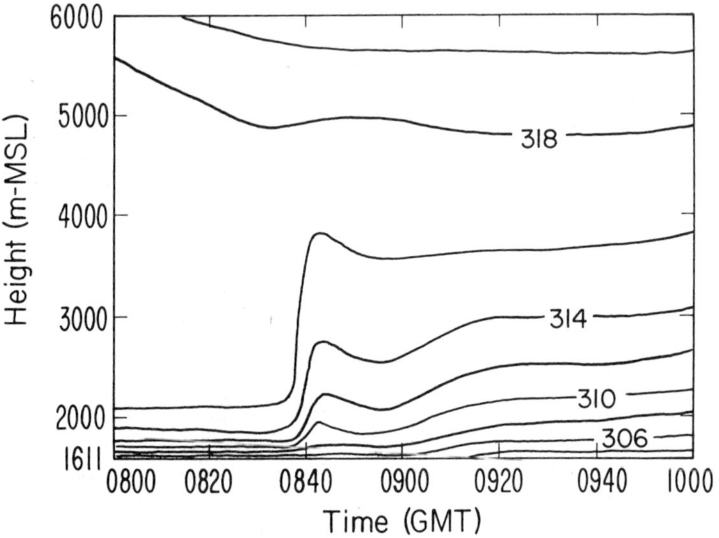

Source: (Courtesy of Mr. M. T. Decker, NOAA)

Figure 12. Passage of Cold Front, Monitored with High Time Resolution by NOAA Radiometer

Weighting functions enable us to measure temperature profiles quite well provided the profiles are relatively smooth. Since the detection equipment operates with only four oxygen/nitrogen channels plus two water-vapor channels, there is insufficient information available for the accurate reproduction of a complicated temperature profile. As illustrated in Figure 13, it is possible to achieve rather good reproductions using both the water-vapor profile and the temperature profile, provided they are basically simple monatonic profiles. The presence of an elevated inversion, however, introduces some difficulty. The ground-based inversion at the bottom of the figure is reproduced quite well but the inversion shown at the top of the figure is weakened. The temperature profiles tend to become very smooth as the inversion occurs at greater heights. The dashed curve in the figure is the radiometer's attempt to reproduce actual conditions, but the resultant curve is a very smooth version of the real temperature inversion. Our instruments just cannot provide the information content that the meteorologist would like to have on the appearance or location of an elevated inversion.

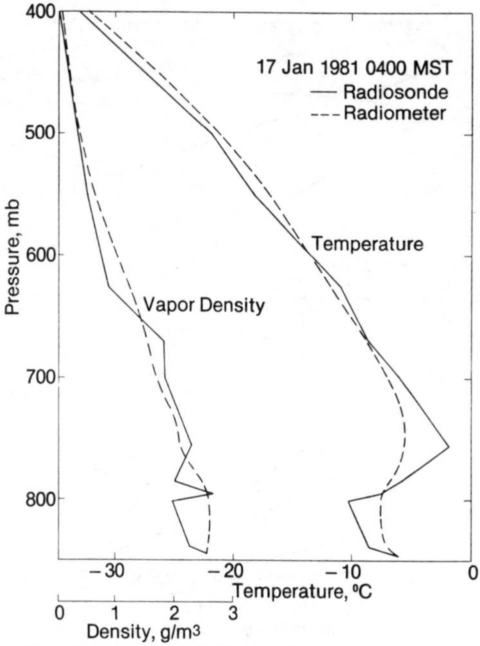

Source: (Courtesy of Dr. E. R. Westwater, NOAA)

Figure 13. Temperature and Humidity Profiles Derived from Radiometric Data, Compared with Radiosonde Profiles

Fortunately, there is a simple and logical way to improve the information content and the quality of the profiles, as illustrated in Figure 14. The weighting functions in this plot are notably different from those of the previous figures. The additional weighting functions are plotted for the same basic frequencies, as obtained from a satellite looking down on the exponentially increasing density of the atmosphere. The dashed curves represent weighting functions that maximize at heights of roughly 600, 300, and 100 millibars, in regions where the ground-based system has lost much of its sensitivity and accuracy. The satellite systems, however, are adversely affected by the radiation coming from the ground. Consequently, they are unable to measure near-ground-level temperatures accurately.

This difficulty may be observed in Figure 15, which shows the error in temperature profiles as measured by the satellites, when compared with corresponding data obtained from radiosondes launched from Denver. The lower right-hand curve in the figure shows a major RMS discrepancy of 6° C.

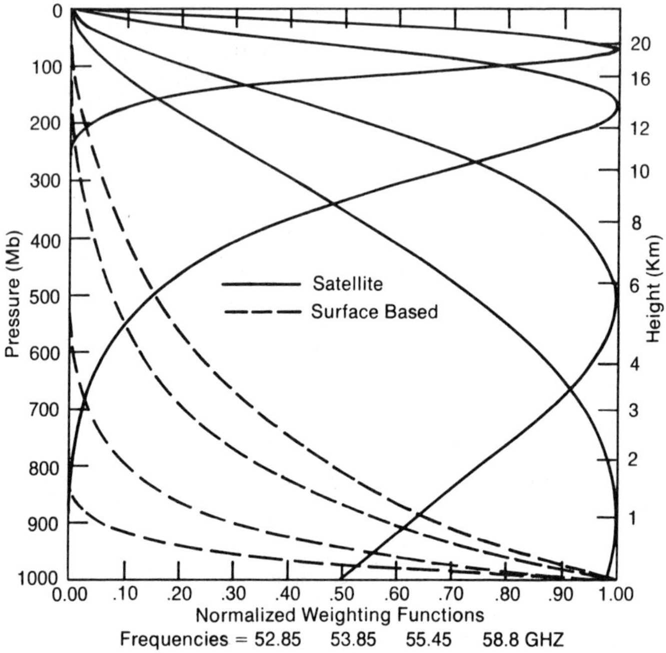

Source: (Courtesy of Dr. E. R. Westwater, NOAA)

Figure 14. Comparison of Satellite-Based and Ground-Based Weighting Functions

On the other hand, the ground-based profiler has much less error, as would be expected from the lower part of the atmosphere. Its accuracy degrades with height and becomes worse than the satellite data in the highest part of the atmosphere. (Such behavior is expected from the exponentially decaying weighting functions of the ground-based system). When the data from these two systems are combined, we find that the composite system is able to provide better data at all heights than either system alone, as indicated in the central portion of Figure 15.

Temperature profiles from ground and space have been studied using geopotential height data, as illustrated in Figure 16. Instead of showing temperature quantities, this plot expresses different pressure surfaces as a function of height. The meteorologists' numerical weather prediction people are very eager to have this parameter. Because of its inability to measure temperature close to the ground, the satellite system has RMS errors of the order of 20 meters at the 700-millibar surface—about three kilometers above sea level. In contrast,

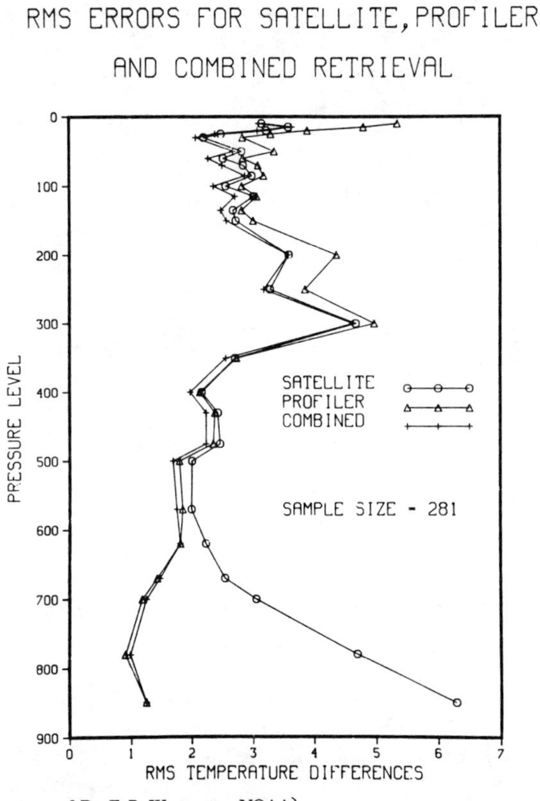

Source: (Courtesy of Dr. E. R. Westwater, NOAA)

Figure 15. Accuracy of satellite and ground-based radiometric profiles versus height, compared with profiles derived by a combination of satellite plus ground-based radiometers

the ground-based system has just a few meters of error at the 700-millibar surface. This error, however, increases with height and eventually exceeds that of the satellite. The dashed curve in the figure is based on the National Weather Service claims for their radiosonde system. The composite system appears to be somewhat better than the radiosonde in the bottom half of the atmosphere and only slightly worse than the radiosonde in the upper half of the atmosphere. Over the years, the Aeronomy Laboratory group has shown that the height of the tropopause can be measured. If that height were to be

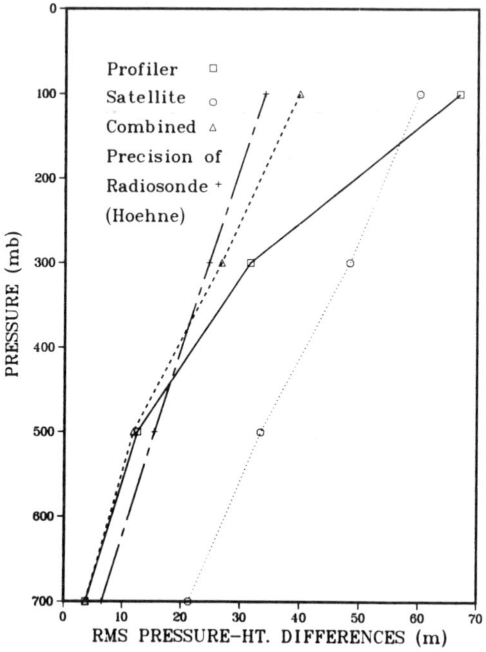

Source: (Courtesy of Dr. E. R. Westwater, NOAA)

Figure 16. Accuracy of measurement of geopotential heights of different pressure surfaces versus height, for satellite radiometer alone, ground-based radiometer alone, and satellite and ground-based data combined together. The accuracy of the current National Weather Service radiosonde system is also shown.

inserted in the data plot, we would anticipate that the composite line would move toward the dashed line and possibly somewhat to the left of it. We would then be able to have continuous, automatic monitoring of the height of the pressure surfaces, with a capability that would equal, and perhaps exceed, the accuracy of the radiosonde.

Two of the six channels in the microwave radiometer are devoted to water content, thereby allowing humidity profiling of the atmosphere. One of these channels is tuned approximately to the water-vapor line at about 22 GHz and provides a measurement of the total amount of water vapor present in the air. The other channel is a so-called "window channel" that determines the quantity of liquid water in the beam. Using surface values of humidity, as

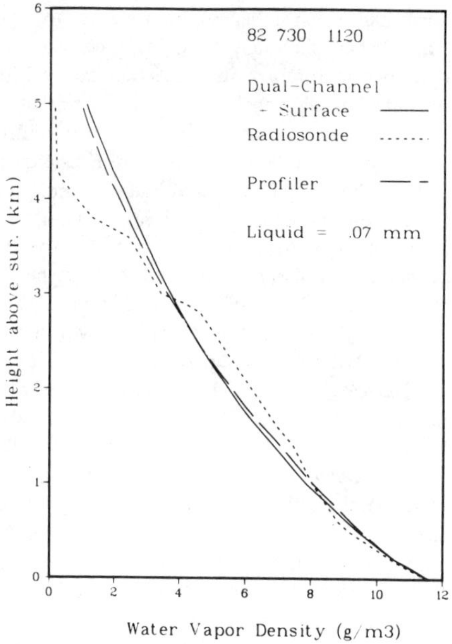

Source: (Courtesy of Dr. E. R. Westwater, NOAA)

Figure 17. Comparison of humidity profiles estimated from ground-based microwave radiometer measurements, with profiles measured simultaneously by radiosonde.

measured by standard National Weather Service instruments, and *a priori* statistics of climatological water-vapor distribution, it is possible to *estimate* the distribution of water vapor in the atmosphere. (It is important to emphasize that this determination is an estimate rather than an actual measurement). These estimates are illustrated in Figure 17, where the dashed curve represents radiosonde data and the solid curve is obtained from the profiler. As before, if the profile is relatively smooth the estimate is reasonably good. If the humidity increases with height to produce a humidity "inversion", the details of the inversion will be smoothed. If the humidity inversion is at a considerable altitude, it would be essentially lost.

Humidity profiling can be improved by combining ground-based radiometric data with other data. As in temperature profiling, satellite measurements of humidity in the upper part of the atmosphere provide better data than can be obtained at the bottom of the murky atmosphere. Again the two data sets will complement each other to improve the composite humidity profile. There is hope that such augmented information will allow the National Weather Service to improve its prediction capabilities. We can also anticipate improvements through the addition of cloud-base and cloud-top heights that will provide reference data on 100-percent humidity levels.

The third technology of this discussion, three-dimensional coverage of the atmosphere, is based on the use of advanced radar techniques. Dr. Hess has pointed out that in addition to obtaining echoes from 10-cm radar that provide an estimate of the amount of rainfall as a function of space, it is worthwhile to study additional parameters of the radar wave such as the Doppler shift. In particular, he emphasized the ability of Doppler radars to monitor convected storms and to provide early identification of the location of the mesocyclones that are often apparent entities within tornadoes. As a result of many years of work by NCAR and NOAA scientists and engineers, a national program called NEXRAD has been established to install over one hundred advanced Doppler radars. These units will replace the 25-year-old National Weather Service radars with equipment based on new technology, and will enable the detection of most tornadoes, particularly the larger ones. They will have the ability to give many minutes of advanced warning of the occurrence of these dramatic and dangerous meteorological phenomena.

Polarization of the Doppler-shift radar wave is another parameter that can be used to obtain the kind of information shown in Figure 18. (This figure is reproduced from an original color-coded slide). In this case, vertical scans of a thunderstorm are being shown. The upper diagram displays the echo intensity of a vertical cross section through the storm with the dark areas (colored red in the original slide) indicating extremely strong echoes. The bottom figure shows, on a logarithmic scale, the differential reflectivity, Z_{dr}, *i. e.* the ratio of the radar echo reflectivities measured simultaneously on horizontal and vertical polarizations. If the targets were spherical, the echo intensity would be equal on both polarizations. During a heavy rainfall, however, unequal echo strengths are recorded because the large drops become distorted as they fall through the air. The lower diagram is a plot of the ratios of these echo strengths, with differential echo intensities being of the order of 1.8 db or larger, for this period.

In the upper diagram, beneath the intense portions of the storm, some white echoes may be seen reaching to the ground. This region has a differential reflectivity of the order of zero db because of the presence of hail. The averaging effect of the tumbling of the hail tends to give quite symmetrical data. This phenomenon was demonstrated during an extremely severe thunderstorm in Denver on June 13, 1984 that caused some 300 million dollars worth of

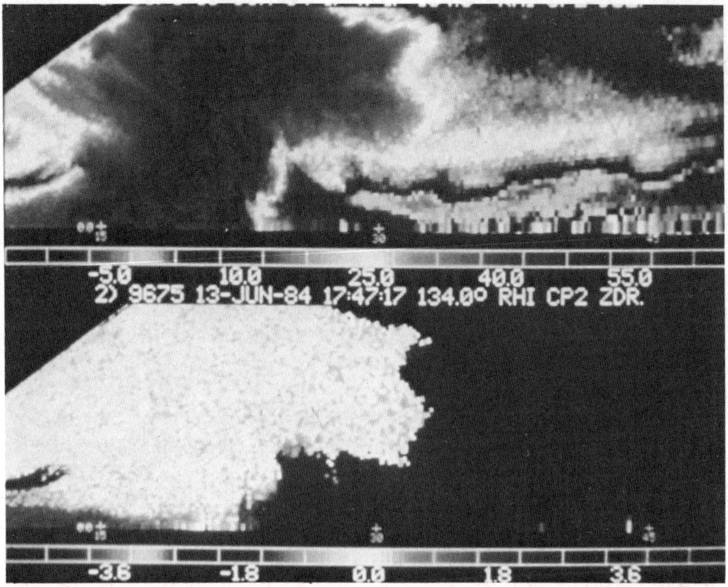

Source: (Courtesy of Dr. R. Carbone, NCAR)

Figure 18. Vertical cross sections through a strong hailstorm over Denver, Colorado. The upper diagram is the echo intensity. The lower diagram is Z_{dr}, the differential radar reflectivity on horizontal and vertical polarizations.

damage. Some areas of the city experienced over four inches of rain and almost five-inch-deep accumulations of hail. There are well-confirmed reports that some of the hail stones were about three inches in diameter. The rain and hail combination blocked the drains and caused extensive flood and hail damage, twenty people were injured, and thousands of automobiles were damaged seriously.

This new technology could be used for hail research and possibly for the detailed prediction of hail storms. This technique has the potential to provide us with the ability to identify regions of hail as well as to determine, without ambiguity, that large hail stones are absent in a particular region of heavy rain. Both heavy rain and large hail stones produce strong echoes; it is very important to develop a procedure whereby the weather forecaster can determine whether those strong echo regions are indeed due to rain and not to hail.

Infrared Doppler radar (known as Doppler "lidar") is the final new technology to be described in this presentation. This new approach complements that of radio-frequency Doppler weather radar in several desirable ways. Normal Doppler radars obtain their meteorological echoes primarily from precipitation. While some powerful radars are able to obtain echoes in clear air (from naturally occurring temperature and moisture fluctuations in the air) they typically require very large antennas that are not readily made steerable. The much shorter infrared wavelengths (shorter by a factor of 10^4 to 10^6) permit much smaller antennas that are easily steered and have much narrower beamwidths. This latter feature can be a critical advantage in a mountainous or urban environment, where the relatively broad beams of standard meteorological radars cause considerable ground clutter and even inoperable conditions. The narrow beam of the infrared lidar (with a cross section of only one meter at 10-km range) makes it ideal for such environments. It has been applied very successfully to studies of air flow in mountain valleys, and to studies of urban air pollution.

In summary, I have tried to make the point that our complex modern society requires highly detailed information on the current and future state of the weather. I believe that the need for this knowledge will increase steadily in response to pressures from many industries, especially aviation, agriculture, energy, and defense. Problems of air pollution, acid rain, and weather modification will require a better understanding of our atmospheric environment. In response to these needs, I have suggested that ground-based atmospheric remote sensors provide major opportunities for improved weather research and services.

A GIANT STEP FOR MANKIND: COMMUNICATIONS

William E. Gordon

William E. Gordon received the B.A. and M.A. degrees from Montclair State College in 1939 and 1942 respectively, and an M.S. degree in Meteorology from New York University in 1946. After several years of teaching and research, he came to Cornell for further graduate study and received the Ph.D. degree in Electrical Engineering in 1953. In the following year he was appointed Associate Professor of Electrical Engineering, and in 1959 was promoted to Professor. Dr. Gordon conceived, planned, designed, and directed the construction of the world's largest radar/radio telescope, the Cornell Arecibo Ionospheric Observatory, and served as Director during its early years. In 1965 he became the Walter R. Read Professor of Engineering. In 1966 he left Cornell to join the faculty at Rice University as Dean of Science and Engineering and Professor of Space Physics and Astronomy and of Electrical Engineering. In 1975 he was named Dean of the Weiss School of Natural Sciences; in July of 1980 he became the Provost and Vice President of Rice University. Dr. Gordon is a member of the National Academy of Sciences and the National Academy of Engineering, a Fellow of the IEEE, and a Fellow of the American Geophysical Union. He was a Guggenheim Fellow in 1972-73. He is the recipient of the Balth van der Pol Award of the International Scientific Radio Union, the American Meteorological Society Medal, the National Academy of Sciences 1984 Arctowski Medal, and the IEEE 1984 Centennial Medal. In addition to these honors, he has held membership in many international, national, state, and local committees and boards.

Editor's Note: In 1986, Dr. Gordon became Distinguished Professor Emeritus at Rice University.

For this lecture, the first of the three for this afternoon that deal with space sciences, I thought about the possibilities for my contribution and decided to focus on communications for three different reasons. One is that much of space science depends on radio communication, usually to deliver information from a sensor in space to a scientist who normally is on the ground, but not necessarily. The problems of delivering that information over distances now defined on the scale of the solar system have stretched the telecommunications art to the point where Cornell engineers and others are being challenged in new ways. Secondly, modern telecommunications are bringing changes of a magnitude that are difficult to imagine--I will say more about these developments later in this talk. Thirdly, and of special importance for this occasion, it was Ezra Cornell who built the first telegraph line that was used by Samuel F. B. Morse in 1844 to carry the message (from Washington to Baltimore) that John Nation mentioned this morning. That was a very important contribution to telecommunications, and it was very important for Cornell University. Ezra Cornell constructed additional transmission lines, became a major stockholder in Western Union, and the success of that venture provided the capital that started Cornell University. So we are all deeply indebted to Western Union and Ezra Cornell.

I like to think that the opportunities for satisfaction and enjoyment in our professions are abundant in spite of our arguments about turf and technical solutions and patent rights and other matters. The fun lies mostly in tackling and solving the tougher problems, talking with our colleagues as we are doing here, beating deadlines, getting products into production or winning contracts and seeing one's product in use. So in the spirit of that kind of enjoyment I will present you with two propositions this afternoon. One I think I can prove, and the second I can only argue, hoping to make you think about it. The purpose is to provoke you so if you're ready to be provoked--it's after lunch, it's siesta time, you'd better get excited about something--I intend to provoke you. But I also intend to give you a chance to reply at the end of my allotted time, because if I can get you mad enough perhaps you will want to let off some steam.

Let us start by focussing principally on the magnitude of change that has been brought about by telecommunications. It is helpful to think about our present age, with the impressive breakthroughs that are occurring all around us, and to compare our society with that of an earlier age when a similar major advance, of which we are all very much aware, made its own compelling impact. My two propositions follow directly from this perspective:

Proposition 1:

> In communication efficiency—the rate at which information is transmitted—the introduction of video communication is a bigger step than the introduction of Gutenberg's press.

Proposition 2:

> The impact of video on society is already greater than that of the printed word despite a 500-year head start for the press.

Proposition 2 is the one that I expect you to argue about. I see some heads waving negatively already. Just get the ammunition ready, you're going to have your chance.

Let me define the terms--particularly video communication--and let us also remind ourselves of the pervasiveness of printing in our lives in order that we may appreciate the magnitude of that second proposition. Perhaps the meaning of video communication can best be examined by means of a quick poll. Please raise your hands in response to the following questions: How many of you have more than one television set at home? (I see that more than half of this group say "yes".) How many of you have a video cassette recorder? (It seems that probably everyone with a TV set--perhaps a few less--have again answered "yes".) How many of you have a video cassette tape library? (We have only a few yeas to this question.) How many of you have a home computer? (Here there are many positive answers.) I conclude that we are in the right place. Now if we take all of these items that you have just indicated by hand signals, and add satellite links, radio links from earth to satellite and back, plus optical fibers for transmitting pictures, pages of text, and telephone conversations--all of that together is what I mean when I say "video communication". If you can think of a better word for it, I'd like to hear it. So video communication is that collection of systems and techniques that covers those items we have just listed, very high-capacity channels of communications.

Without taking a poll, I could ask how many of you have more than a hundred books in your home library and I'm sure virtually every hand would be up. How many of you read more than one newspaper a day or several magazines a month? Most hands would be up. Let me say that there are more letters, reports, catalogues, books, dictionaries, and directories, than you can possibly manage. As a matter of fact, in the USA today, the printing output is the equivalent of one book per day for every man, woman and child. So have you curled up with *your* book today? Did you read *your* book yesterday? Are you ready to read *your* book tomorrow? That's an enormous quantity of print. We're saturated—we're supersaturated! On the other hand, books are nice

Figure 1. A 15th Century Illuminated Manuscript in the Rare Book Collection at Rice University

collector's items and so we have them on display at home. Some of you now have videotapes at home for the same reason—they are collectors items.

Suppose we were to go back about 500 years--that is about 50 years after the invention of Gutenberg's press, just as we are now about 50 years into the video-communication age—we might be meeting in Paris, Florence, or perhaps London instead of in St. Louis. If I were to conduct a poll at such a meeting, as I did a few minutes ago, the questions would be: How many of you own one book? (We probably would have had most of you raising a hand.) How many of you own more than one book? (Some of you would have answered "yes".) Books were attractive then not only for their information but as a piece of furniture, as a prestige item, as a status symbol in the home. Some early books still exist in rare-book collections that encompass a wide range of subjects. For example, the Rice University collection contains an early history of Mexico and a book of erotic French literature that go back almost 500 years. Originally, books were made to look like the handwritten manuscripts that they were replacing, as in the illuminated manuscript depicted in Figure 1. Books were truly a novelty in those days.

TV has now passed through the novelty stage. Whether one could afford an early TV or not, it was fashionable in some areas to have a TV antenna on the roof just to show the neighbors that there might be a TV in the house. A TV displays information, it is often an attractive piece of furniture, and large-screen TV is a definite prestige item.

Early books contributed to intellectual, political, and religious upheaval—as you may recall, Galileo had his problems. The ideas contained in such diverse works as the writings of St. Francis of Assisi, *Days of Sodom,* Erasmus's *In Praise of Folly,* and the handbooks of witchcraft, influenced men in various ways. An invention, of course, may be put to any use.

All of us are familiar with the oft-described "wasteland" of TV and its impact upon our intellectual life. The impact on politics, in fact, started well before TV; it was part of radio broadcasting. FDR and his fireside chats changed the way politicians functioned. Today, office seekers at both state and national levels are packaged attractively for television, and the voters respond to the imagery. Political television commercials say nothing about the issues, they simply present the candidate in the best light and leave an impression that in some sense leads you to the ballot box. The impact of TV on religion extends from theology on the one hand, to shameless huckstering on the other. Of course the medium has many positive aspects. It can show marvelous pictures of the earth taken from the moon, as in Figure 2, provide instant coverage of every disaster, and cover spellbinding Olympic Games that keep you glued to your TV set late into the night.

For those of you who will want to argue that printing has had greater impact than video communication, let me give you one piece of ammunition.

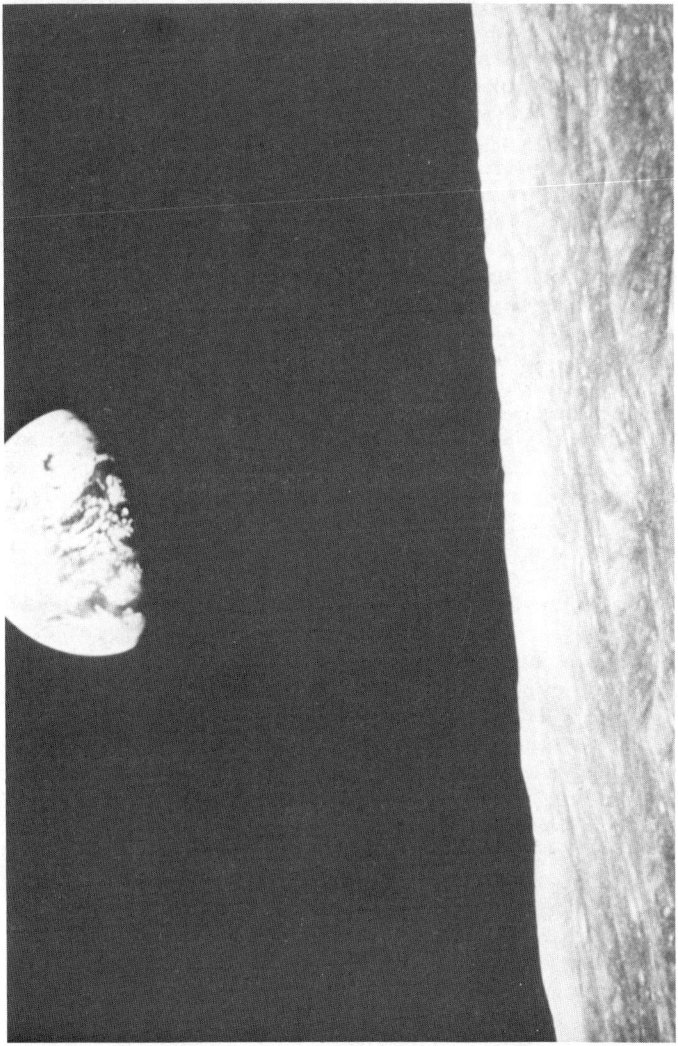

Figure 2. "Earthrise" from the Surface of the Moon

Rabelais equated the invention of printing with that of gunpowder—both were put to use rapidly, both had a leveling tendency, both spread their effects impartially. Rulers acquired cannons and printing presses at about the same time and used both as tools of policy.

Over time, books shifted from being a novelty to becoming an item of public fascination, resulting in the establishment of attractive, profitable, and even indispensable businesses. Similar statements also apply to TV. Where Gutenberg introduced the realm of medieval knowledge to a continent, the news press and television bring details of spectacular events, such as the landing of a man on the moon or the Olympic Games, to a world-wide audience. Today a fascinating cultural reversal is under way. The individual is being offered access to a totality of information that was not possible in the past.

In an industrial civilization such as ours, it has become customary to mark the development of mankind in terms of the wheel, the steam engine, and atomic energy. Alternatively, we might think of these divisions as being: learning to speak, learning to write, learning to print, and learning to use video. Perhaps we should divide history on the basis of communication technique rather than on mechanical technique.

The printing press was the early means for mass production and distribution of ideas and information. Video and TV combined are fast becoming the modern means of communication with an added appeal that is almost hypnotic. What else could hold your attention on New Year's Day from early morning to late at night? What else is so effective a baby-sitter as *Sesame Street*, the cartoon parade, or the availability of even worse fare? There is an attraction to TV that we must learn to use profitably, not just in the business sense of that word, but in the educational sense. We have hardly scratched the surface of that challenge.

Printing is clearly a crucial invention in modern history. Five centuries have been shaped in profound ways by the Gutenberg press. Invented in Europe, it was a potent instrument in affecting the spread of European power and influence. The explosion of print and that other European explosion, the Reformation, belong to the same time frame and are completely entangled. The development of printing cannot be separated from the whole social and cultural process known as the Renaissance. Gutenberg's press was partly a product of the new thinking and partly its instigator. It is easy to see certain patterns that are also evident in our own time, when new ideas about the nature of information are arising precisely when new methods for creating and disseminating words and information are being adopted. Incidently, the Chinese had a printing press, invented before Gutenberg's time, but it was not exploited, in part I think, because of the difficulty of the Chinese picture language; from one thousand to ten thousand symbols—that is printing items—are needed. It simply was not practical.

The switch from stone inscription to handwriting on papyrus signaled new patterns of thought and social organization. The switch from a scribal society to a printing one changed the whole focus of knowledge in the West and created new locations for information in society. The transition from paper to a telecommunications system can hardly prove to be less important, necessitating the development of new skills, new equipment, a new kind of text, and new methods of text storage. As with many other transformations in history, the change becomes evident in government before its impact reaches society in general. In the States, at least, the defense and intelligence communities have already undergone a thorough changeover and well over half of all of the documents relating to intelligence work are now transmitted, filed, and indexed electronically. They pass through specialized networks without generating paper copy and the text, if it needs to be read, is presented on a video terminal. Technical librarianship is one of the next fields likely to undergo substantial transformation and from there the change will spread outward through the educational system.

It is instructive to consider a few of the impacts of printing and video.

1) Printing, in effect, established authors. Prior to printing, manuscripts were simply copied, usually by monks, and the authorship was lost. Once printing came into being, the author's proper place was recognized. Presumably, he also received his proper compensation although that took some time to become established. In TV, greater recognition seems to be given to individuals with a certain appeal who project distinct personalities.

2) Printing had a big impact on written language. There was now a need for grammar and a need for uniform spelling, both tending to create a stabilizing effect on the written word. Similarly, TV has had a big impact on spoken language. Regional dialects tend to be avoided by popular commentators, resulting in the emergence of a nationalized tongue. As an example, consider the map of Indonesia shown in Figure 3. In various sections of this country, spread over a large area of the Western Pacific, something like 250 separate dialects are spoken. The entire country is now served by satellite TV using *the* national language. Simply introducing that satellite has made a major change in the language habits of the people. Nearly 90 percent of the Indonesians now understand and speak the national language in spite of their having been brought up in a large variety of languages.

3) Printing and video are pastimes, they are relaxing, they can be entertaining, they can be exciting or they can just be a way to pass the time. It is easy to become a passive observer for much but not all of

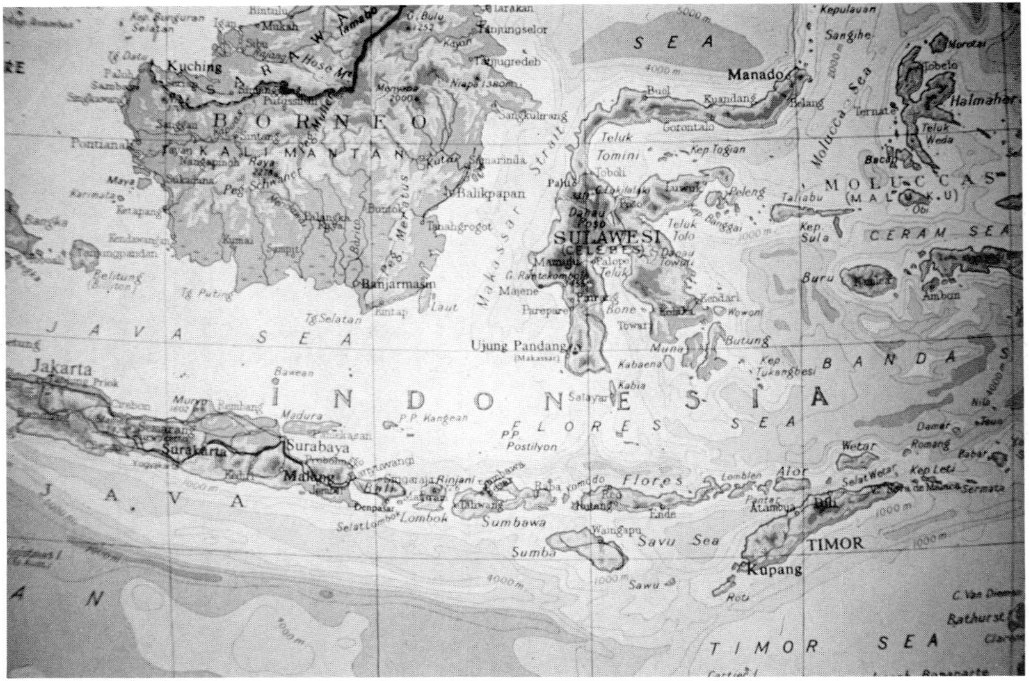

Figure 3. The Land of Indonesia in the Western Pacific

the material. 500 years ago, the entertainment to be found in a tavern might consist of someone reading a book aloud. Upon entering a tavern now, one is confronted with a TV screen showing a prize fight or a football game. These two cases exhibit a very interesting bit of parallel development.

Both kinds of communication suffer from the poverty and the credibility of the information that is being transmitted. Readers and viewers, swallowers of a nonstop flow of words and pictures, do not necessarily believe much of what is being presented but, inevitably, their lives are affected.

Printing output grew rapidly and has reached the previously mentioned present rate of one book per day per person in the U.S. alone. Coping with this massive volume of words requires refinements in storage and retrieval capability. The card catalog of a library can now be recorded on magnetic tapes with all of the information in the whole of the catalog contained on one disc. Consequently, our library-research procedures will be completely changed in the not-too-distant future. Instead of walking up to the wooden cabinet, pulling out the drawer and thumbing through the cards to find the desired information, we will sit down at the computer, type in the reference material, and wait a second or two until the equivalent of the card that would have been found in the catalog appears on the screen. This service will be readily available at terminals in the library, the office, or at home. (It appears that we are heading for a big bargain sale of nice three-by-five-card oak cabinets that will become available when libraries automate their catalogs.) In spite of this rosy picture, it would not be correct to assume that I am implying in any sense that reading will be replaced by watching. Proposition 1 does not lead to that conclusion.

In the earliest days, information was stored largely in man's memory and was supplemented occasionally by drawings on a wall. Information was transmitted verbally from one generation to the next using imprecise words and imperfect but imaginative memories. The key was memory. The impact on memory of the invention of writing was lamented by Socrates whose remarks are equally applicable to printing and TV. "This discovery of yours," says Socrates, "will create forgetfulness in the learners' souls because they will not use their memories; they will trust in the external written characters." It is quite likely that at this moment most of us have in our pockets (our "active files") some notes as reminders for what we must do tonight or what we must do tomorrow–a direct indication that we no longer have complete trust in our memories.

Let me say a few things about modern communications before I prove the efficiency proposition. Communication is the glue that holds society together. Modern communications systems and styles create various problems and tensions but never before in history has mankind's survival and well-being hung so precariously on accurate and swift communication. Television, satellites, lasers, and optical-fiber links are technologies that are less than 50 years old. The great hero of telecommunications, Marconi, performed his magic less than 100 years ago when he sent signals across the Atlantic in 1901; today we are critically dependent on these communication techniques. People need global news, even in unedited or undigested form, because everyone is affected by what is happening all over the world. Corporations have far-reaching information-transmission needs. The roof of a company building is likely to contain antennas of various kinds, including a dish pointing at a satellite. Governments need to communicate with each other and with their constituencies.

The Jules Verne of communication is Arthur Clarke (Figure 4). You may know that in 1945 Arthur Clarke suggested that any two points on the surface of the earth could be linked by a communication channel that used orbiting satellites. In the lower part of Figure 4, the small circle in the middle is supposed to be the earth. There are three satellites in what was then called "Clarke's orbit" (known today as "geosynchronous orbit") that essentially cover the full sphere of the earth. By relaying signals from a point on the ground to one or more of the satellites, it would be possible to communicate with any other point on the earth. Since Sputnik was not placed into orbit until the late 50's, Clarke was nearly 15 years ahead of the time when we had anything in space, let alone having something in geosynchronous orbit. He was more than 25 years ahead of the orbiting communications satellites, particularly those in geosynchronous orbit. The profound effects of the convergence of communication and computing technologies have brought Clarke's 40-year-old predictions to reality, causing previous communications techniques to change almost completely in the 80's:

- Cable TV seems established although it may face a struggle with direct-satellite broadcasting to the individual viewer. That service is ready to go as soon as somebody is convinced that there is money to be made with it.
- Video cassette recorders will be in one out of every two homes by 1990, providing a new source of home entertainment, new profits for the film industry, and new educational opportunities ranging from the "how-to" kinds of hobbies to courses that universities such as Cornell might rent to show in your plant or in the comfort of your home.
- The video disc, once thought to be a popular substitute for the video cassette, is pretty much dropping out of the entertainment field but may turn up as a means of replacing stacks of books in the library. It is hard to imagine shrinking the libraries at Cornell into something that would look

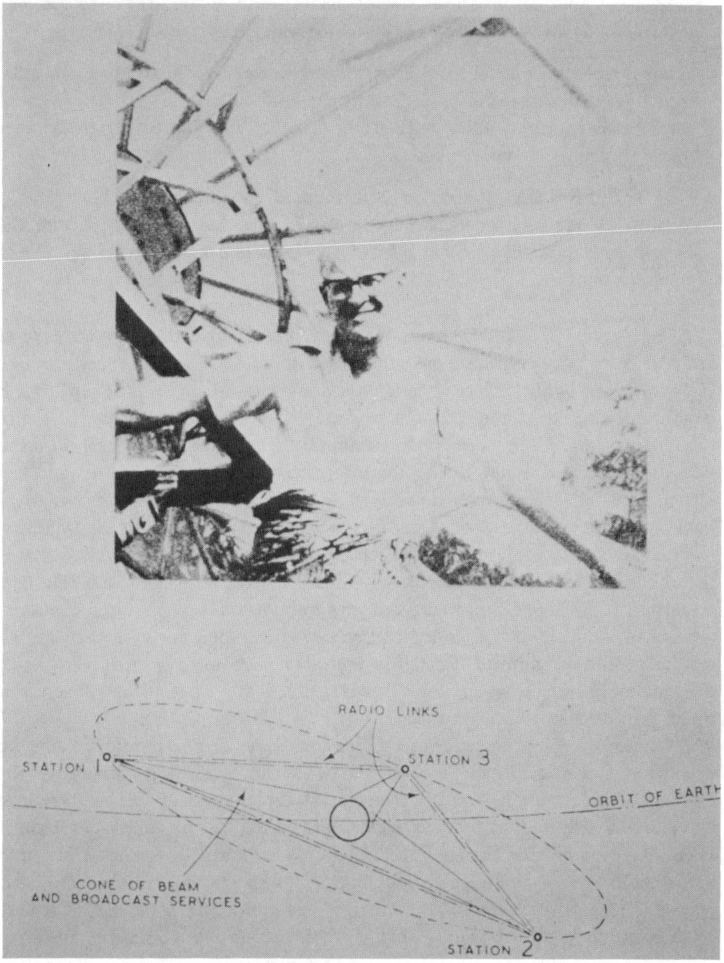

Figure 4. Arthur C. Clarke and "Clarke's Orbit"

like a big juke box filled with laser discs instead of long-play records, but that is likely to be the future role for the video disc.

- Transmission on optical fiber is competing fiercely with satellite transmission. We are within range of sending the equivalent of about 40 TV channels for a hundred miles without a repeater. For short-haul traffic that capability will be hard to beat.

- High-definition TV is ready to be introduced. Soon after you have bought a new TV set you can expect to be offered one that is better because the screen will probably fit flat against the wall and the picture will be of higher quality.

The matter of communication efficiency is well illustrated by the rate at which I have been transmitting information to you as I speak—something like 100 words per minute, more or less. You know the old saying that a picture is worth a thousand words. That is wrong. A picture today is worth many thousands of words, perhaps 10^4 to 10^5 words. It is important to keep in mind that the question is not whether the information can be absorbed or used but rather how fast it is delivered. If I write a letter and it arrives overnight, I have delivered about a word a minute, depending on the length of the letter. If I mail a book overnight, the rate would be something like a hundred words per minute when it is subdivided. If I mail a video disc, then the rate goes from something like a hundred words per minute to a few hundred thousand words per minute! If I communicate via satellite, then the rate may be ten million words per minute. Certainly there is no argument on this point and I did not see people wagging their heads when I said that video efficiency is much greater than that of printing.

I spoke once on this subject at Rice and some bright physicist came up afterward and said to me, "Do you realize that in a painting masterpiece the amount of information is about the same as the amount of information in a symphony?" I leave that for you to calculate on the back of an envelope sometime. But I do have an assignment for you: What is the information rate that is available to you—available to your mind? As an example, think about your eye which has a resolution of the order of one hundredth of a degree. It may have a hundred-degree field, although that is probably optimistic, and that gets the number of resolution cells up to something like 10^8. Allowing for how fast your eye can respond, the information rate might be well above a million bits per second. I am sure that estimate is high since there must be all kinds of filters and processes in the visual system that simply get in the way. I am told that the ear of a young person can probably hear up to 20 kilohertz. Since the aural canals contain on the order of 100 working receptors, even by ear the information rate is of the order of a million bits per second, although again it seems to me that the processing and the filtering must cause substantial reduction in communication efficiency. So there are two parts to your assignment:

(a) To whom do you address the question as to what branch of science deals with how fast the information can be absorbed?

(b) What are the real limits to how fast information can be absorbed—not processed but simply absorbed?

I return now to Proposition 2 and the matter of total impact on society, of video compared to that of the printed word. In passing I have mentioned many similar impacts that occurred at similar stages in the history of printing and of video. Let me give you just two more. First, the *Wall Street Journal* and the *New York Times*, at least in Houston and, I presume, in St. Louis, are now delivered at breakfast time to your doorstep because of the availability of satellite communication combined with a local printing press. Second, the youngsters who go to school these days and then go on to enter universities, have been exposed to a variety of experiences via TV that none of us had as we grew up. They have observed more than we could have done at their age of what goes on in this world in all kinds of situations—good and bad. This wealth of experience, so different from our own backgrounds, is certain to have profound effects upon their education and their future careers.

For my final comparison I will read from Act 3, Scene 7, of *The Tragedy of King Lear* by William Shakespeare starting a few lines into the scene. What I want you to do is to imagine what is going on. Then I will show you a videotape of the Royal Shakespeare Company doing the same scene so you may compare the imagery acquired from just listening, to that seen on the screen. So we are in Gloucester's castle, Cornwall, Regan and the servants are present. Cornwall ......

[In this scene, Gloucester is brought in, bound to a chair and accused of treason by Cornwall and Regan. A servant attempts to intervene but is slain. Cornwall then plucks out Gloucester's eyes, crying "Out vile jelly! Where is thy lustre now?" In the video scenes, all of the physical acts of violence are shown but the eye mutilation occurs with Gloucester's back to the viewer.]

It is hard not to comment, in view of all of the violence now available on TV, that Shakespeare was one of the most accomplished producers of violence in the theatre.

This exercise is an obvious example of the kind of visual imagery that is available on TV. Here we are in the year 550 of the printing press and almost in the 50th year of video communication. My judgement is that the present-day impact of video is at least equal to that of printing, and my proposition to you is that it is greater.

In conclusion, our ability to use video communication profitably has a long way to go and should serve as a challenge to us in whatever profession or industry we find ourselves.

In a sense, the only choice that history does not make for us is that between optimism and pessimism. Whether we have a big or a little government, whether we become a gregarious or lonely society, individualistic or regimented, is largely the result of the way we choose to use the technology that is available to us. The new electronic storage systems for information, interlinked globally, have the potential to offer a renewed sense of the unity of the world and of the accessibility of knowledge. We are in a brave new world!

Those of you who have ammunition for the discussion are now invited to roll out the cannons and start firing.

THE ARECIBO OBSERVATORY: PAST ACHIEVEMENTS AND FUTURE PROSPECTS IN THE EXPLORATION OF THE IONOSPHERE AND THE SOLAR SYSTEM

Tor Hagfors

Tor Hagfors received the Master of Engineering degree from the University of Trondheim, Norway in 1955 and the Ph.D. degree from the University of Oslo in 1959. From 1955-1959 and from 1961-1963 he was a scientist at the Norwegian Defense Research Establishment (NDRE). During the 1959-1960 period he was a Research Associate at Stanford University in California. He was a staff member at MIT Lincoln Laboratory from 1963 to 1966, served as Director of the Jicamarca Observatory in Lima, Peru from 1966-1969, and then returned to Lincoln Laboratory for another three-year period until 1971 when he became Director of Operations at the Arecibo Observatory in Puerto Rico. In 1973 he became Professor of Electrical Engineering at the University of Trondheim and also served as Director of the European Incoherent Scatter (EISCAT) Radar Observatory from 1975 until 1982. From 1977-1982 he was also a member of the Fachbeirat Max-Planck Institut fur Astronomie in Lindau-Hanz, Germany. In 1982 he was appointed to his present position as Director of the National Astronomy and Ionosphere Center (NAIC) and Professor of Electrical Engineering and Astronomy at Cornell University. Dr Hagfors has been directly involved in the ionospheric heating program, in planning cometary tail occultation experiments, and in devising new methods for the measurement of scattering from plasma oscillations in the atmosphere. He is a member of the American Astronomical Union, the American Geophysical Union, and the International Union for Radio Science. He is a Senior Member of IEEE and holds membership in a number of European professional societies.

The Arecibo Observatory has been in operation for more than 20 years. It was originally conceived by Professor W.E. Gordon as a device for exploring the upper atmosphere. The original ideas were presented at a colloquium at Cornell in the late 1950's by Professor Gordon and were, as far as I understand, widely accepted at the time. Fortunately there was some misconception about the process of scattering. At the time, it was expected that one would see scattering from free electrons in the ionosphere. In reality one observes scattering from density waves. This was not realized by Dr. Gordon, and, fortunately for us, he concluded that a very much larger antenna would be required than what actually is required for the purpose originally intended for Arecibo. On this basis, the facility was built about ten times as large as it needed to be, and that was extremely fortunate because the installation became the largest single reflecting telescope in the world and is now being used extensively and very fruitfully for purposes not originally envisaged.

Before considering the achievements of the Observatory during its existence, it may be of interest to examine a few pictures of the Observatory as it appears today. Figure 1 is an aerial view and, as you can see, it is neatly nested in the hills of Puerto Rico. It was placed there because the nature of the terrain minimized the amount of excavation required for its construction.

Source: N.A.I.C.

Figure 1. Aerial View of the Arecibo Observatory

Source: N.A.I.C.

Figure 2. Close-Up of the Observatory

I have been told that the site was selected by choosing a map of appropriate scale and sliding a quarter along the map until a coincidence was obtained.

Figure 2 shows a close-up of the Observatory. The reflector is fixed and consists of 38,000 panels which can all be adjusted individually. The feed is mounted on a carriage house which can move along a bridge or azimuth arm which in turn can rotate on a circular track. The angular position of the azimuth arm determines the azimuth, and the position of the carriage house on the azimuth arm determines the zenith angle of the beam. Because of the spherical geometry of the reflector, steps must be taken to correct for the spherical aberrations. At present, this is done by the use of line feeds, a leaky wave-guide providing illumination of the right amplitude and phase on the aperture. Figure 3 shows a carriage house with line feeds. Another carriage house with a similar line feed is usually positioned at the opposite side of the azimuth arm to maintain balance–about 450 feet above the reflecting surface. Originally the surface was composed of a wire mesh that limited the Observatory to frequencies slightly less than 1 GHz because of the surface inaccuracies. About 12 years ago the first major upgrading of the Observatory took place with the installation of the present surface panels. With these panels, the surface accuracy is 3.8 millimeters RMS which means that the reflecting surface

Source: N.A.I.C.

Figure 3. Carriage House with Line Feeds

can support frequencies as high as 4 GHz. This modification turned the Observatory into a major radio-astronomy facility. In particular, the capability of observing the neutral-hydrogen line at 21 cm made it an extremely exciting installation for radio-astronomy purposes.

The Observatory was built by Cornell University with funding from ARPA and the Air Force. It has since been transferred to the National Science Foundation (NSF) but is still operated by Cornell University as a national observatory.

Over ten years have passed since the Observatory was transformed into a major telescope. We are now in the process of continuing the upgrading and we have, during the last four years, made extensive plans for the next stages. These plans involve further expansion of the frequency coverage and a large increase of the sensitivity. I shall return to these plans presently.

Most of the radio astronomy work at the Observatory does not pertain to the solar system and may not strictly be classified as "Atmospheric and Space Science," the topic of this symposium. Nevertheless, radio astronomy is the most important part of our program. It takes fully 75 percent of the observing time, and I cannot resist the temptation of showing one or two examples of this research.

Figure 4 is a spectrum of a so-called rotation curve of a galaxy. One item of interest is the Doppler shift, because the mean Doppler shift gives the distance to the galaxy through the Hubble law. Another is the spectrum shape because it supplies information about the distribution of emitting matter in a rotating disk since the Doppler shifts from the approaching and the receding

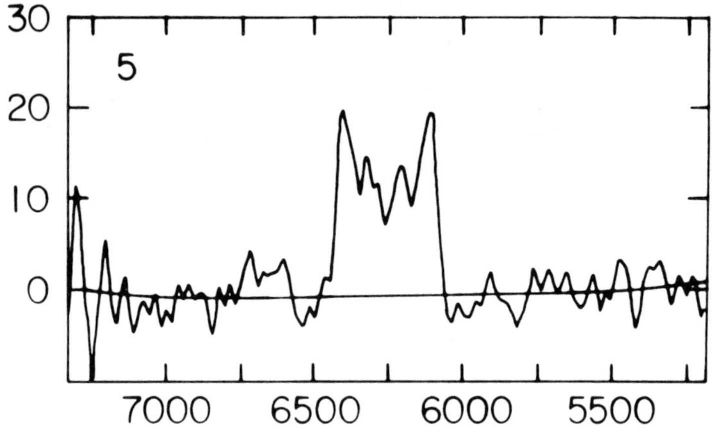

Source: N.A.I.C.

Figure 4. Rotation curve of galaxy. The horizontal axis shows the velocity in km/s. The velocity difference of 400 km/s between the two edges of the galaxy is due to rotation. The vertical axis is relative power received (R. Giovanelli, private communication).

sides are different. Clearly much can be learned about the structure of distant galaxies from the study of such rotation curves. A considerable part of the work in Arecibo is related to galaxies of this kind, and several thousands of them have been classified. Some of them are so remote that the original 1420 MHz frequency is shifted to 400 MHz, indicating their origin at a very early time of the universe. The statistical study of such galaxies reveals that they often occur in superclusters, and are not evenly distributed throughout the universe as one might have suspected.

Another type of object that has attracted much interest lately, and that is well suited for study at Arecibo, are millisecond pulsars. All of the three known millisecond radio pulsars have been detected at Arecibo. Figure 5 shows the detection of the first millisecond pulsar at a pulse frequency of 642 Hz. The rotating neutron star, giving rise to the pulses much like a lighthouse beacon, has a mass larger than that of the sun and is rotating at a rate much like that of a typical kitchen blender. This pulsar was discovered in 1982 and has since then been monitored regularly from Arecibo. It is a very interesting and possibly technologically useful object because the rotation rate is exceedingly constant. We monitor the pulsar against the best atomic clocks available, and it seems that the stability of this pulsar will soon exceed the stability of these clocks. It may well be, therefore, that millisecond pulsars will come to play a vital role in time-keeping. It is not enough to use only a single pulsar to establish a time reference, and partly for this reason we have conducted intensive searches for more, and have so far identified two additional ones. For purely scientific purposes, these millisecond pulsars may be used for studies of scintillation effects in the interstellar medium in our own galaxy or for the study of gravitational effects in the solar system. The solar system is

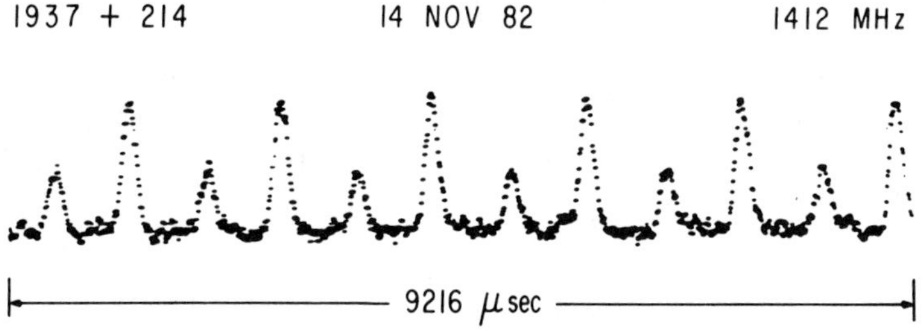

Source: N.A.I.C.

Figure 5. Millisecond Pulsar Detection, Pulse Power versus Time (discussed by D.C. Backer, S. R. Kulkarni and J. H. Taylor, *Nature*, 301, 314, 1983).

exposed to long-period gravitational waves which may cause minute perturbations that may be detectable through the observations of pulsars.

Figure 6 shows another example from the pulsar work at the Observatory. In this case, the pulsar period varies in a periodic manner which is consistent with the pulsar being in orbit around a companion. The period of the orbital motion, according to general relativity, should increase very slowly with time because of emission of gravitational waves. The agreement shown in the illustration is extremely good.

Radar Astronomy is a field where the Observatory is pre-eminent. This field was started after the end of World War II when radar systems had become sensitive enough to detect the moon, the first astronomical radar target. Figure 7 shows the power scattered back from the moon as a function of the delay. Observe the very strong echo at the nearest point and the extremely fast decline of echo power beyond that point. The moon, therefore, looks like a slightly roughened, smooth sphere. If we were able to view it with radar-sensitive eyes, we would see the moon as a very bright spot at the closest point at the center of the lunar disk, with an extremely dark area extending toward the rim. A study of curves such as shown in Figure 7, and similar ones from planetary targets, permits a deduction of the RMS slope of the surface.

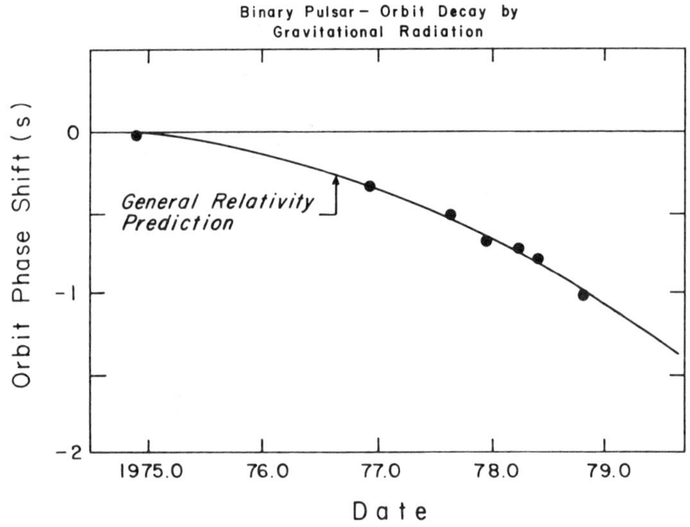

(© 1982 The American Astronomical Society) Reprinted by permission.

Figure 6. Pulsar Orbit Period Slow-Down due to Gravitational Wave Radiation (J. H. Taylor and J. M. Weisberg, *Astrophysical J.*, 253, 908, 1982)

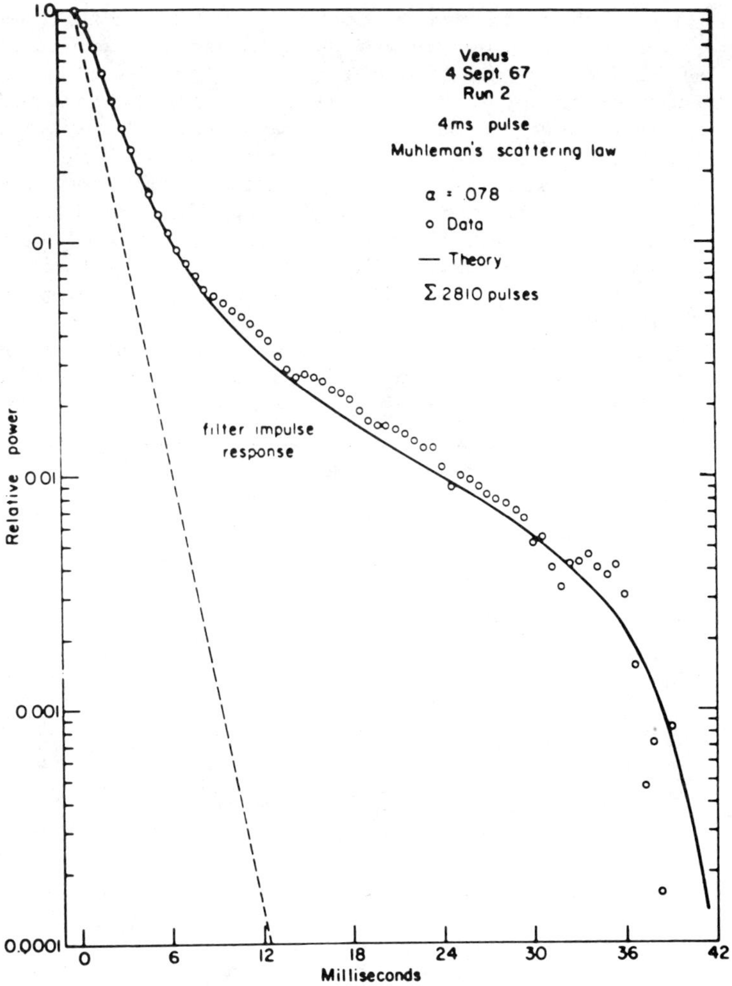

(©1963 American Geophysical Union)

Figure 7. Echo-Power Versus Delay for Lunar Radar Returns (J. V. Evans and G. H. Pettengill, *J. Geophys. Res.*, 68, 423, 1963)

Figure 8 shows an interesting Arecibo discovery. When the first echoes were obtained from Mercury and the frequency spectra of these returns were analyzed, it was found that the planet could not possibly be in bound rotation about the sun, as everybody believed at the time. The great surprise was the discovery of a 59-day rotation period rather than the expected 88-day period (see Figure 8). It was determined that the planetary rotation was bound to 2/3 of the orbital period. This means that on successive passages through perihelion, the planet turns opposite sides toward the sun. The erroneous rotation period of 88 days had been determined by optical astronomers who had discarded observations every other revolution because of "bad seeing!" The determination of the rotation period of Mercury was one of the early successes of radar astronomy.

So far I have given examples of studies involving either range resolution or frequency resolution. It is often possible to combine these two types of observations to obtain two-dimensional resolution on a planetary surface.

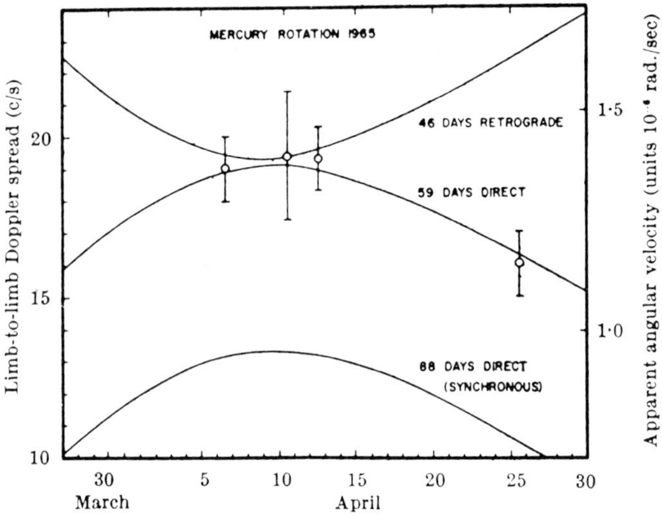

(© 1965 Macmillan Journals Ltd.) Reprinted by permission from *Nature*, Vol. 206, No. 4990, p. 1240.

Figure 8. Plot of the apparent rotational angular velocity of the planet Mercury versus date for several values of rotation during the inferior conjunction of April 1965. The values inferred from the measurements are shown with their estimated errors. (G. H. Pettengill and R. B. Dyce, *Nature*, 206, 1240, 1965)

Figure 9 explains how this is done. Transmission of a short pulse, and the gating out of a window of the same length isolate the power returned from a narrow ring centered on the sub-earth point. Spectral analysis isolates slices of the target that are all parallel to a plane containing the apparent rotation axis and the radius vector between the center of the target and the earth. The two sets of rings intersect at two points indicated by δA allowing a two-fold ambiguous map to be constructed. The ambiguity can either be solved if the antenna beam is narrow enough to distinguish the ambiguous points, or it can be resolved if an interferometer is available with the right fringe resolution. Such an interferometer, with a 10 km north-south baseline is available in Puerto Rico.

Figure 10 shows some early maps of the reflectivity of the moon obtained at 70 cm wavelength by the range-Doppler technique using the beam resolution to resolve the ambiguity. The two maps show the so-called depolarized and polarized radiation, that is the two orthogonally polarized circular components obtained by illuminating the moon by circularly polarized radiation. In the latest lunar maps made in Arecibo, the resolution was about 4 km.

Even more spectacular results have been obtained from the mapping of Venus. Figure 11 shows an example of the application of this technique to

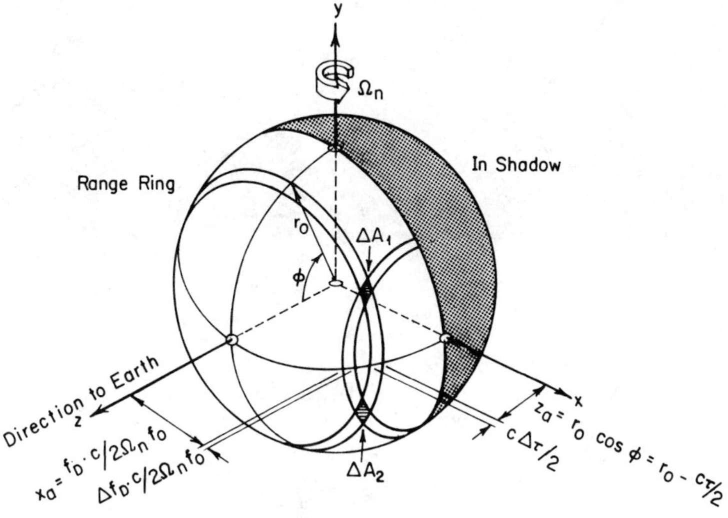

Source: N.A.I.C.

Figure 9. Delay-Doppler mapping showing the slicing up of a planetary body into range rings and Doppler rings.

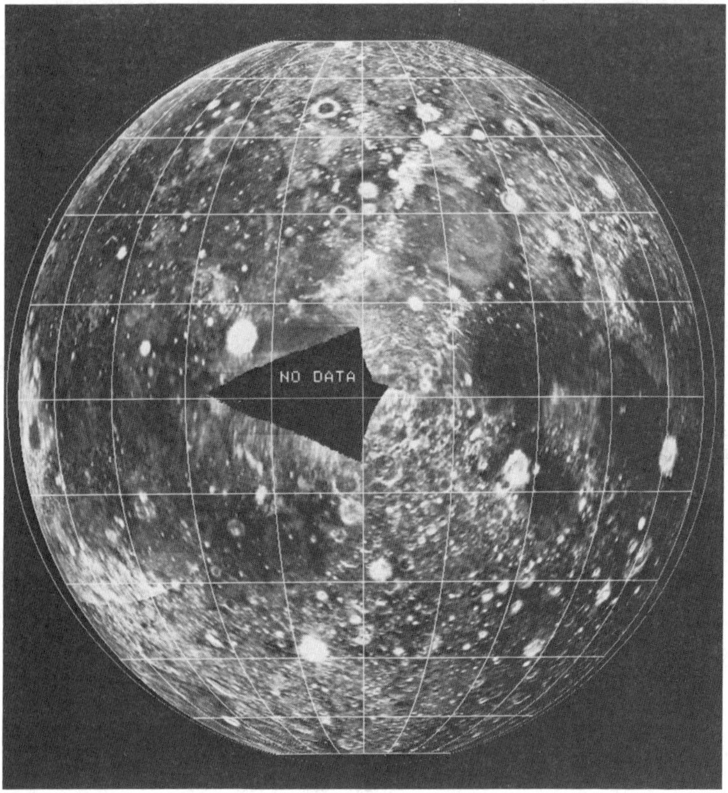

Source: N.A.I.C.

Figure 10. Radar Reflectivity Map of the Moon (J. Thompson, private communication)

map Venus at a wavelength of 12.8 cm, using the interferometer to resolve the ambiguity. The particular area shown is the Maxwell-Montes region. The resolution is 2 to 3 km, comparable to the resolution obtained with the side-looking radars on board the Soviet Venera spacecraft. These results have been of considerable interest to geologists as they attempt to determine the kind of activity that has shaped the surface of the planet.

Figure 12 is an example of the topography measured on Mercury by simply ranging to the sub-earth point on the planet. The depth of some of the craters is particularly impressive. On the basis of such information it is now

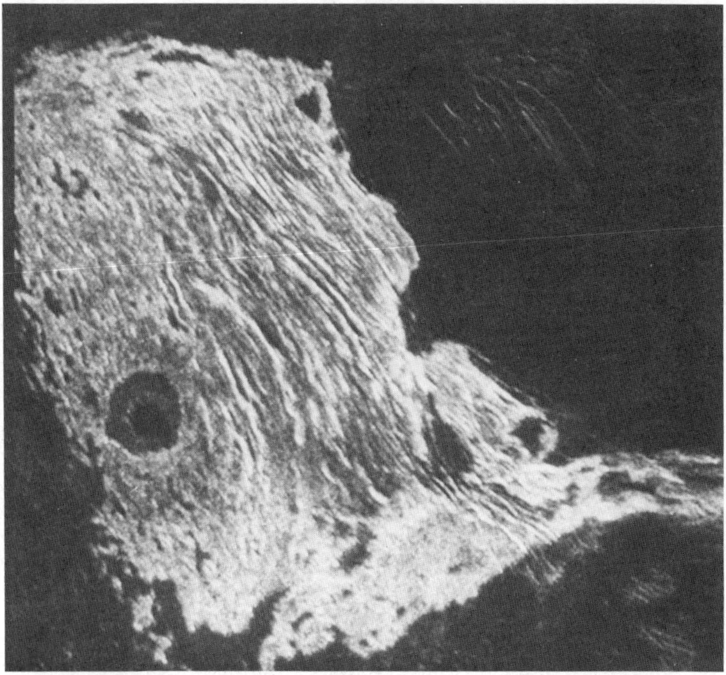

Source: N.A.I.C.

Figure 11. Radar Reflectivity Map of the Maxwell-Montes Region on Venus (D. B. Campbell, private communication)

the consensus that there has been some tectonic activity on Mercury. Similar observations have also been made on Mars.

An extremely interesting radar observation was made about two years ago of the comet IRAS-Araki-Alcock. The spectrum of the return is shown in Figure 13. The very narrow spike in the spectrum clearly comes from the nucleus. This in itself is of considerable interest because there has been uncertainty even as to whether comets have a solid nucleus at all. Even more interesting is the broad spectral feature that indicates that there is a cloud of cm-sized particles surrounding this nucleus. A few other comets have also been observed. Attempts will be made to obtain echoes from Halley at the end of this year.

There have been observations of three of the satellites of Jupiter with extremely strange results: Ganymede, Callisto, and Europa all have extremely

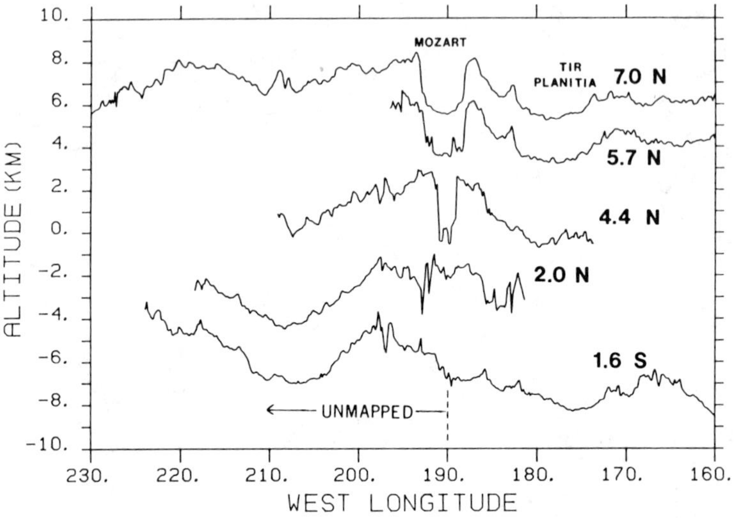

(© 1986 American Geophysical Union)

Figure 12. Radar Topography Map of Mercury (J. K. Harmon, D. B. Campbell, D. L. Bindschadler, J. W. Head, and I. I. Shapiro, *J. Geophys. Res.* Vol. 91, No. B-1, pp. 385-401, Jan. 10, 1986)

large radar cross sections, larger than smooth metallic spheres of similar diameters; they all return circularly polarized incident waves with the opposite polarization of that of any other body in the solar system; they do not show the specular glare at the sub-earth point which other bodies do. There are several competing theories to explain these bizarre properties, one of which involves large refractive index variation in objects embedded in fluffy ice or snow. Clearly extensive studies of these objects will occur when Jupiter again becomes observable in Arecibo in 1987.

Other objects of particular interest for future studies are Io, Titan, and numerous asteroids. The latter may play an important part in the exploration and industrialization of the solar system because they are more accessible to mining than is the moon.

Let me now turn to observations of the ionosphere, the original target of exploration of the Observatory. Figure 14 shows a theoretical frequency spectrum corresponding to scattering from thermal-density fluctuations in the

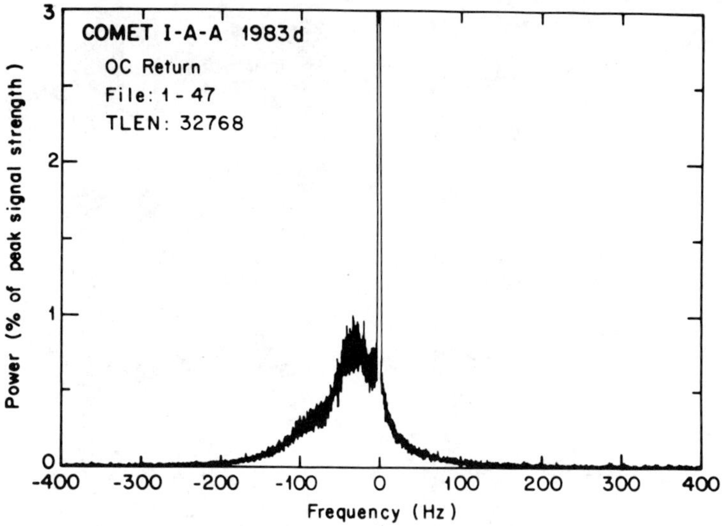

Source: N.A.I.C.

Figure 13. Radar Returns from the Comet IRAS-Araki-Alcock (D. B. Campbell, private communication)

ionosphere. Such spectra form the basis of the diagnostic use in ionospheric research of Arecibo, and of other observatories such as Jicamarca in Peru, Millstone Hill in Massachusetts, Sondre Stromfjord in Greenland, and EISCAT in Northern Scandinavia. The central part of the spectra reflects the ionic temperature and composition, and the plasma line which is offset by the plasma frequency corresponds to thermally excited Langmuir oscillations, that is longitudinal oscillations of the electrons without participation by the ions. These spectra can be predicted theoretically with great precision and represent one of the most successful results based on linear plasma theory. I will now discuss a few examples of measurements obtained in Arecibo.

Figure 15 shows electron-density profiles over a wide height range, including the peak of the F-layer. This type of information represents what was originally sought when the Observatory was first constructed. The illustration shows a sequence of such profiles. The sequence starts some time after sunset and goes on through the night. The development of the layers in the morning at sunrise can be clearly seen. Peculiarities such as very thin layers

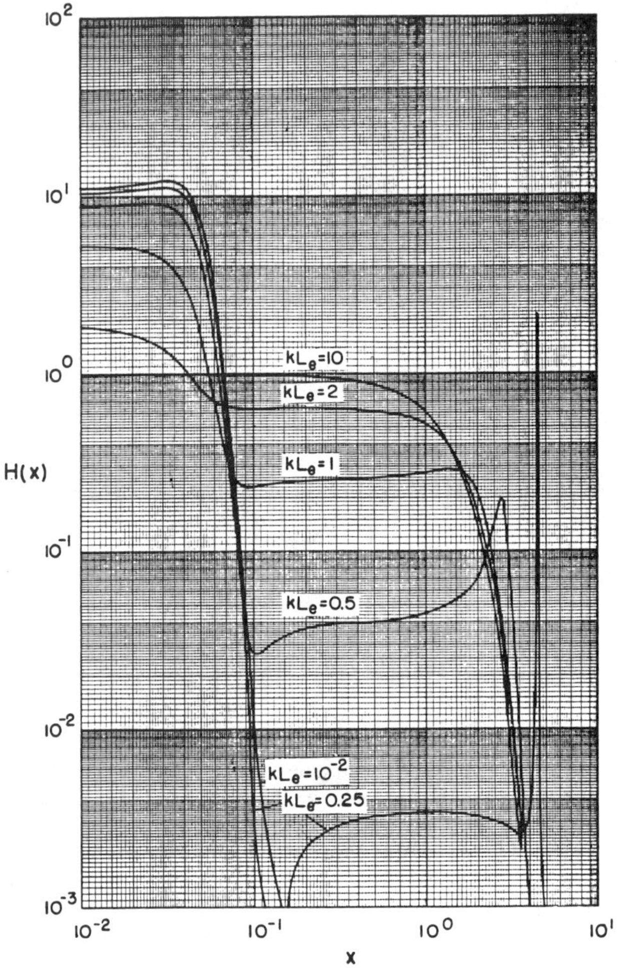

(© 1961 American Geophysical Union)

Figure 14. Theoretical ion-line frequency spectrum. The horizontal axis is the normalized frequency axis. The curve parameter is the ratio of Debye length to spatial wavelength times 2. (T. Hagfors, *J. Geophys. Res.* 66, 1699, 1961)

at about 100 km altitude can be seen, probably due to metallic ions having a long lifetime and creating so-called sporadic activity.

Temperatures, temperature ratios, composition, electric fields, neutral winds, and densities have been extensively studied in Arecibo, and to a large extent as a result of these studies, the properties of the lower middle ionosphere are quite well known by now.

Most of the information that has been derived from such observations has so far been obtained from the central part of the spectrum, the ion line. The plasma line occurs at approximately the local plasma frequency of the

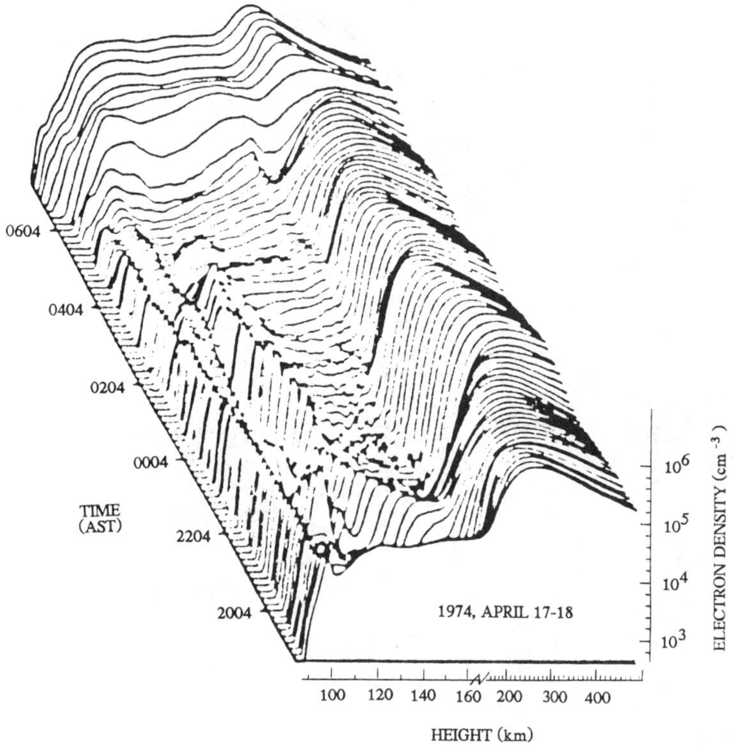

Source: N.A.I.C.

Figure 15. Time-Sequence of Electron Density Profiles (W. Harper, private communication)

ionosphere and gives a direct measure of the density. Since it represents the local plasma density, however, the frequency of this spectral feature changes rapidly through the ionosphere. This makes it rather difficult to detect because as soon as the volume is expanded to increase the amount of returned echo, the bandwidth will increase because of this variation. It is, however, desirable to study this feature because it depends on a large number of physical parameters of great interest, such as the electron temperature, the ionospheric vertical current density, the tail of the electron velocity distribution apart from the electron density. In order to increase the detectability, we are currently

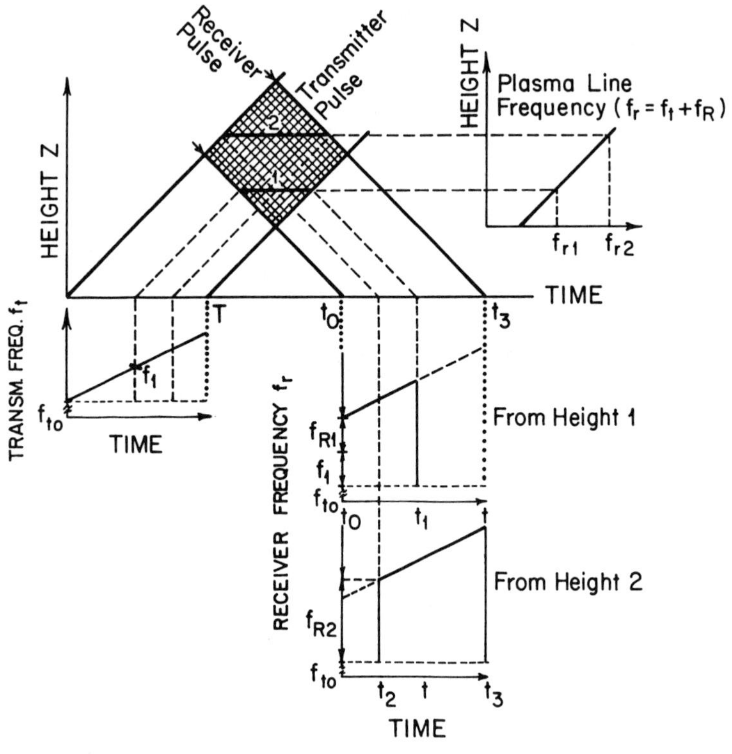

(© 1984 American Geophysical Union)

Figure 16. Chirped Plasma Line Observation Technique (T. Hagfors, W. S. Birkmayer, and M. Sulzer, *J. Geophys. Res.* 89, A8, 6841, 1984)

developing a technique explained in Figure 16, that involves changing the transmitted frequency at a rate corresponding to the gradient of the electron density in the ionosphere. By this technique, we can make plasma-line detection in a small fraction of a second. We can even detect it in a single pulse. Clearly we can begin to look for transient phenomena such as variation in the electric field in the ionosphere caused, for instance, by thunderstorms. We have also been able to apply this observation technique during high frequency ionospheric modification experiments and have been able to identify such exotic nonlinear phenomena as solitons.

Ionospheric modification experiments, or heating experiments, have been conducted extensively in Arecibo, much encouraged by the enthusiasm of Professor Gordon. We have exposed the ionosphere to a very strong high-frequency wave and have excited parametric instabilities, solitons, accelerated electrons leading to airglow, field-aligned large and small irregularities, and other phenomena. Some of the phenomena we observe are of considerable interest in many aspects of physics. The heating experiments also have possible engineering applications. For example, if there is a current in the ionosphere, the heating of the plasma may lead to changes in the collision frequency, or in the electron density. In either case, there may be a modulation of the conductivity at some ELF frequency which may be transmitted from the ionosphere in the form of electromagnetic waves which could be used for communication purposes.

What about the future of the Observatory? Clearly much has been accomplished and only a fragment of the activity has been reported here. Before describing the upgrading plans, it is necessary to list some of the shortcomings of the Observatory as it is today. One of these is the increase in system temperature with zenith angle shown in Figure 17. With a spherical reflector, the pointing of the beam is accomplished by moving the area illuminated on the reflector. Usually this area is about 700 feet in diameter. When pointing to the largest zenith angles, a fraction of this area is spilling over the reflector edge and is interacting with the hills surrounding the antenna. When that happens, thermal noise from the surrounding ground is picked up and the performance of the telescope deteriorates seriously. For instance, if we were to observe the Galilean satellites as soon as Jupiter rises above the horizon, they would appear at large zenith angles where the system noise is high. If observing an object like Titan, the travel time is about two hours and as the maximum zenith angle is 20°, observations will only be possible at high zenith angles to the west where the noise performance is poor. We plan to eliminate this extra system noise contribution by erecting a 60-foot screen around the dish as shown in Figure 18. The screen will cause rays that normally spill over the edge of the reflector and interact with the ground to deflect into the cold sky.

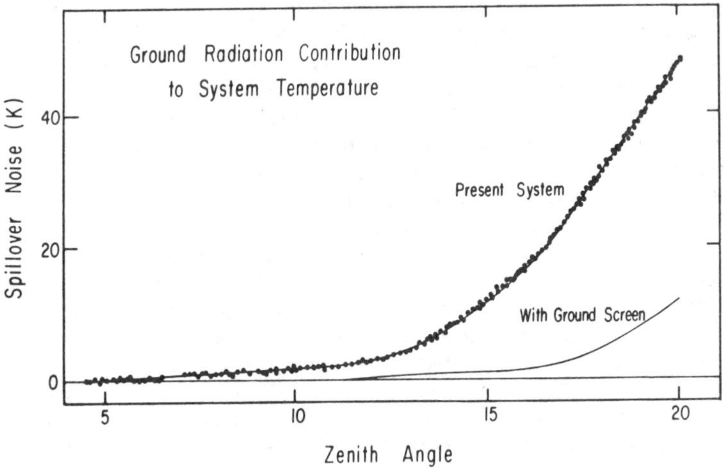

Source: N.A.I.C.

Figure 17. Increase of System Temperature with Zenith Angle

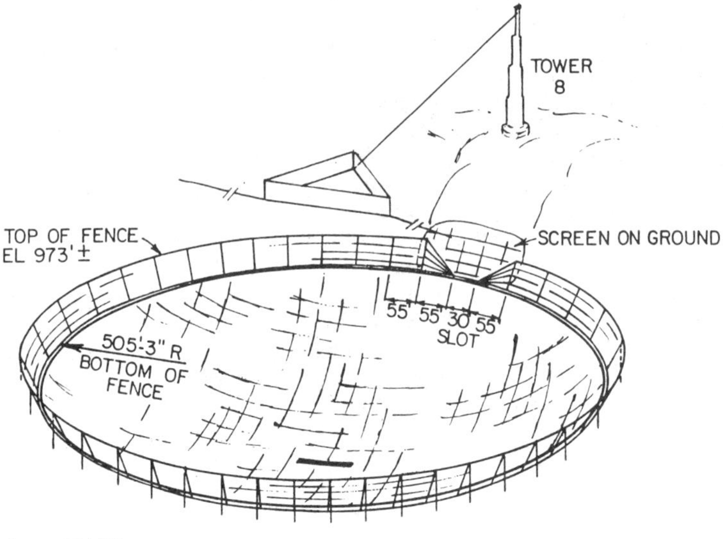

Source: N.A.I.C.

Figure 18. Ground Screen

The line feeds which are used to remove the spherical aberration are long, lossy traveling-wave structures of very limited bandwidth. In observations of a system of spectral lines or in observations of the plasma line in ionospheric experiments, this limited bandwidth seriously constrains the types of studies that can be conducted. The reflector illumination provided by the line feeds is far from optimum, and in addition, the loss in the feeds contributes to the overall system loss. To overcome these difficulties, we are planning to install a feed consisting of a conventional horn and two reflecting surfaces (see Figure 19). The final structure which must be suspended from the azimuth arm is expected to look as shown in Figure 20. The radome surrounding the secondary and tertiary reflectors will be covered by metal sheets, and the rays will pass through the radome only through an opening in the lower part. This will reduce interference and clutter because both are caused by direct line-of-sight paths to the present feeds.

CROSS SECTION THROUGH SUBREFLECTORS WITH RAYS

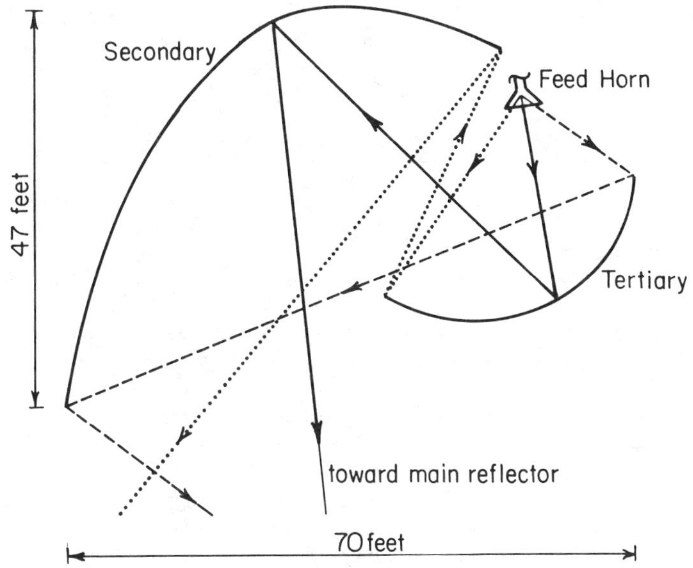

Source: N.A.I.C.

Figure 19. Geometry of Dual Reflector Feed System

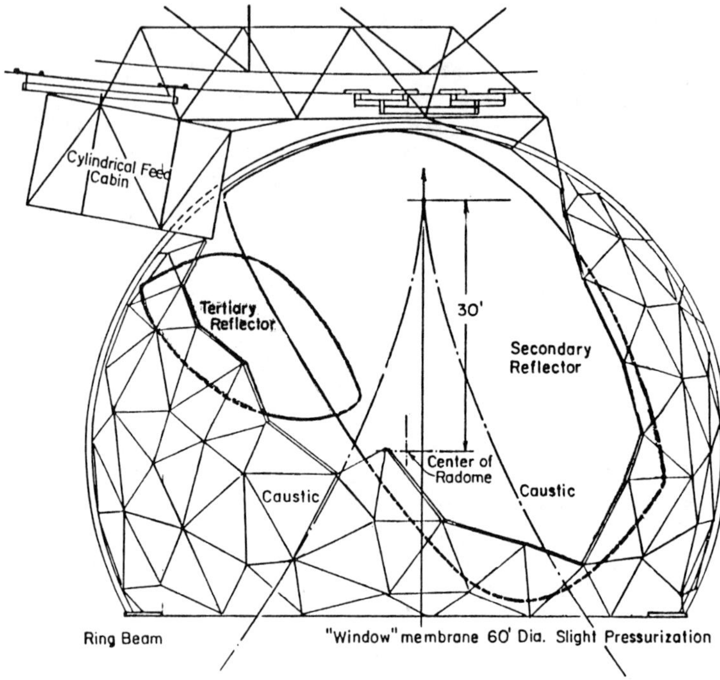

Source: N.A.I.C.

Figure 20. Gregorian Feed System

In addition to these two improvements, we are also resetting the main reflector to slightly less than 2.5 mm accuracy to increase the frequency coverage.

When all these improvements are completed, the frequency coverage will be extended to a wavelength of 3.75 cm, the sensitivity for radio-astronomy experiments will be increased by a factor between 3 and 10, the sensitivity for radar-astronomy experiments will be increased by a factor of between 6 and 40, depending on the zenith angle, and the interference will be substantially reduced. The total cost of all these improvements will amount to between 10 and 15 million dollars, about 1.5 times the annual operating budget, representing a very small fraction of the total investment in the Observatory to the present time.

The upgrades could be completed in a time span of four to five years. So far, the upgrades are in the advanced planning stage and no funds have been made available for the project by NSF as yet. We believe, however, that the chances for NSF approval of the project are excellent in view of the very large improvement in capabilities at moderate cost.

EXPLORING THE OUTER SOLAR SYSTEM: VOYAGER, GALILEO, AND BEYOND

Joseph F. Veverka

Joseph F. Veverka received the B.Sc. and the M.Sc. degrees from Queen's University, Kingston, Ontario in 1964 and 1965, respectively, and the M.A. and Ph.D. degrees in Astronomy from Harvard University in 1970. He came to the Center for Radiophysics and Space Research at Cornell University in 1970 as a Research Associate. In 1972 he was promoted to Senior Research Associate and held that position until 1974 when he was appointed Assistant Professor in the Department of Astronomy at Cornell. In 1977 he was promoted to Associate Professor, and in 1982 he became Professor of Astronomy, his present position. At Cornell he is Director of the Spacecraft Planetary Imaging Facility and a member of the Laboratory for Planetary Studies. Dr. Veverka holds membership in the Space Science Advisory Council, the Solar System Exploration Committee, and the Planetary and Lunar Exploration Committee of the National Academy of Sciences. He has been a member of several NASA teams concerned with imaging science on the Mariner 9 and Viking missions to Mars, and the Voyager and Galileo satellites. In 1979 he was awarded the NASA Medal for Exceptional Scientific Achievement for his investigations of the moons of Mars, and the NASA Group Achievement Award for the Voyager Flight Team. He is a member of the Division of Planetary Sciences of the American Astronomical Society, the American Geophysical Union, the Meteoritical Society, the Royal Astronomy Society of Canada, the International Astronomical Society, and is a Fellow of the American Association for the Advancement of Science.

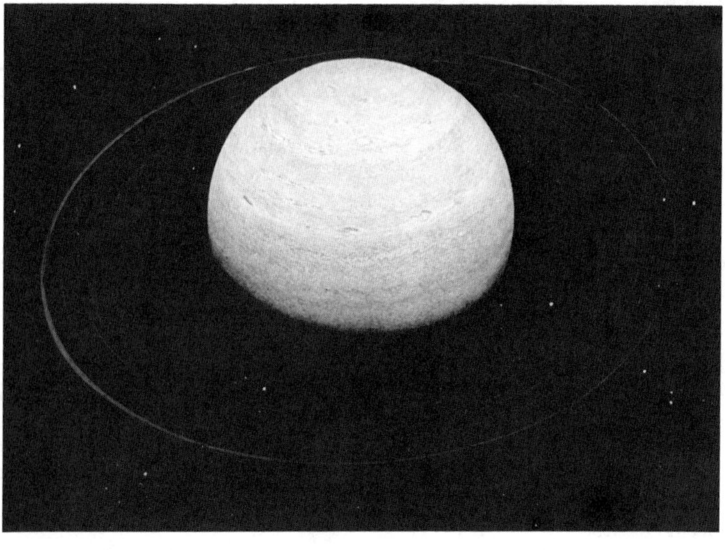

Pre-Voyager artist's concept of Uranus, showing the planet's extreme tilt and the nine very faint narrow rings.

INTRODUCTION

Over the past two decades, the United States program of planetary exploration has been spectacularly successful. We have investigated every one of the earth-like planets: Mercury, Venus, and Mars. Our astronauts have travelled to the moon and brought back pieces of its surface for detailed analysis on Earth. Our spacecraft have even ventured across the belt of asteroids to the realm of the giant planets: Jupiter and Saturn. One of them, Voyager 2, is now on a trajectory that will take it past Uranus in January 1986. I would like to describe some of these ongoing voyages of exploration, enterprises in which we at Cornell and our colleagues across the country have been involved.

THE VOYAGER SPACECRAFT

There is no better place to begin than with the currently ongoing Voyager mission. Two Voyager spacecraft were launched in the Fall of 1977, with Jupiter as their initial target (Figure 1). Both got there successfully in 1979 and went on to Saturn. Voyager 1 arrived at Saturn in 1980 and swung by on a trajectory that took it out of the plane of the planets (the ecliptic plane).

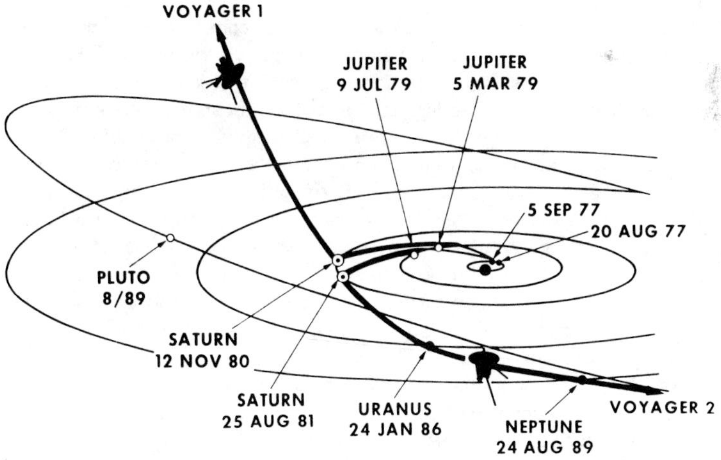

Figure 1. Voyager Trajectories to the Outer Planets

This maneuver was intentional and is making it possible for Voyager 1 to study, for the first time, the magnetic field and charged particle environment outside the plane of the planets. Voyager 2, after its encounter with Saturn in 1981, stayed in the plane of the planets and is heading for Uranus, with a scheduled arrival date of January 24, 1986. After that, Voyager 2 will be on track to meet the next planet further out, Neptune, in late August of 1989. My talk will begin with a brief summary of what Voyager 2 will be looking for at Uranus, and beyond that, at Neptune, and will finish with a discussion of other NASA missions planned for the next decade.

The first obvious but important fact to stress about the outer planets is that compared to the Earth, they are very far away from the Sun. Uranus is about 19 times as far away, and Neptune about 30. Therefore, at Uranus the Sun appears about $(19)^2 = 360$ times fainter than to us on Earth, and at Neptune solar radiation is 900 times fainter than it is at the Earth's orbit. Consequently, it is very cold out there; whereas a typical temperature on the Earth might be around 10°C or 280°K, a typical temperature at Neptune is closer to -210°C or 60°K. Not only will water freeze, but other volatiles such as ammonia and methane as well. In such a frigid environment, a major concern is to maintain the spacecraft at a healthy temperature. So far away from the Sun, solar panels are useless. The spacecraft carries radioactive plutonium, which generates the electricity needed to keep Voyager warm and operating (Figure 2).

Communications are another problem. From Uranus, even at the speed of light, it takes about 2.8 hours for a signal to get to Earth. Thus, to send a command to the spacecraft and get an answer back requires over 5 hours. To operate successfully under these conditions requires a very sophisticated spacecraft. Not only is Voyager a complicated spacecraft, it is also very large. In mass it is almost a thousand kilograms. The high gain antenna has a radius of about 2 meters, big enough for a person to sit in quite comfortably. Such a large antenna is needed to communicate over the vast distances involved. Other essential components include smart computers which allow the spacecraft to operate autonomously when necessary and make sure that all the science instruments carry out the complicated measurement sequences designed for them, and finally that these data are sent back to Earth in a clear and ungarbled manner. Voyager carries a dozen different science instruments—cameras to take pictures of planets, satellites, and rings; spectrometers to measure temperature and the compositions of atmospheres and surfaces; and a variety of particles and fields instruments that measure magnetic fields and charge particle environments. In short, Voyager is a smart, sophisticated and very complicated spacecraft that can take care of itself at distances of more than 3 million kilometers from its home base: Earth.

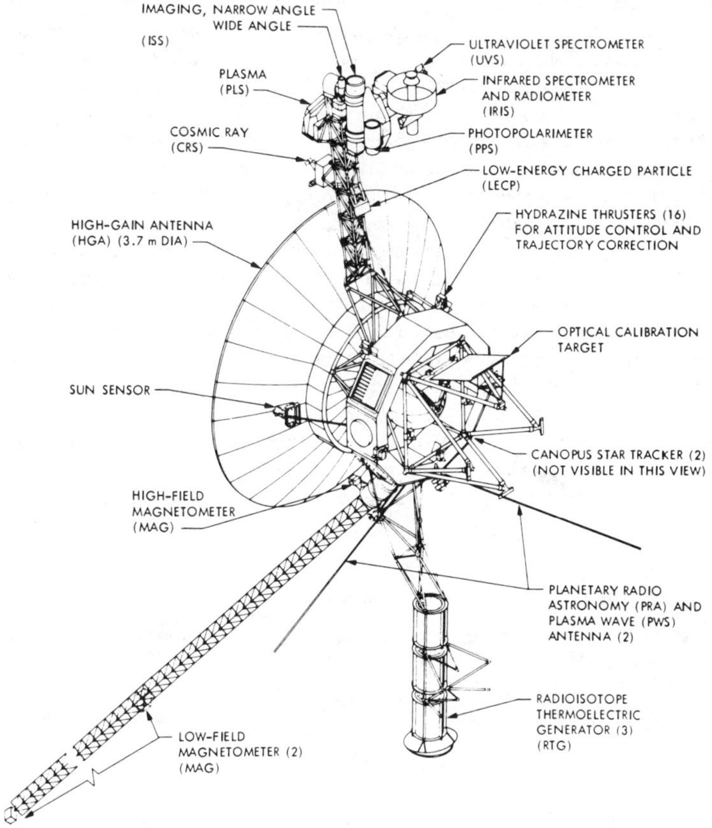

Figure 2. Schematic Diagram of the Voyager Spacecraft

URANUS AND NEPTUNE: WHAT DO WE KNOW?

Since Uranus and Neptune are so distant, they are extemely difficult to study from the Earth. Hence, we know precious little about them, much less than we knew about Jupiter or Saturn before Voyager. This lack of detailed knowledge has led to some common misconceptions. A particularly common one is that the four planets: Jupiter, Saturn, Uranus, and Neptune, are all very similar. Even from Earth, we can gather enough data to know better (Table 1). One reason we can be sure that these two planets are nothing like Jupiter or Saturn is that we know their average densities. If a planet has a satellite, from its motion around the planet and using Kepler's Third Law, one can determine the planet's mass. In this way, it has been found that Uranus and Neptune are each about 15 times as massive as is the Earth. We also know the sizes of the planets, even though being distant objects their disks appear very small, even in the largest telescope. But occasionally, as seen from Earth, Uranus or Neptune are seen to pass in front of a distant star. If one observes such an occultation, one sees the star disappear behind the planet, only to reappear a little later on the opposite side of the planet's disk. From an accurate measurement of the time during which the star was occulted, and from knowledge of the two orbits–that of Uranus and that of the Earth–one can calculate the absolute radius of the planet. Both Uranus and Neptune have radii which are approximately four times bigger than that of the Earth. From the radius one can readily compute a volume, and the mass divided by the

Planet	Distance from Sun (AU)	Orbital Period (yr)	Equatorial Radius (km)	($R_\oplus$)	Mass ($M_\oplus$)	Average Density (g/cm^3)	Spin Period (hr)
Earth	1	1	6,378	1	1	5.5	23.93
Jupiter	5.20	11.9	71,500	11.2	318	1.3	9.9
Saturn	9.55	29.4	60,300	9.5	95	0.7	10.6
Uranus	19.2	84.0	26,200	4.1	14.5	1.2	16-17?
Neptune	30.1	164.8	25,300	4.0	17.2	1.6	17-18?

Table I: Outer Planets Compared to Earth

volume gives the average density. The result is that the densities are such that one cannot make these two planets of the same kind of stuff as Jupiter and Saturn: mostly hydrogen and helium, like the Sun. One needs to add a larger proportion of heavier material, most likely carbon, nitrogen, oxygen—the elements that are the components of ices such as water ice, methane ice, and ammonia ice. So one important fact is that in terms of composition, Uranus and Neptune have much less hydrogen and helium than Jupiter and Saturn.

Uranus and Neptune should not be thought of as miniature Jupiters and Saturns. Neither are they anything like the Earth. Compared to our planet, they have massive atmospheres. From spectroscopy we know that these are made mostly of hydrogen and contain hydrocarbons such as methane. The spectra of Uranus or Neptune are very similar. Most of the absorption features seen between 0.5 and about 1 micron are due to methane (CH_4), but there is also evidence of large amounts of hydrogen. Helium, although spectroscopically not detectable from Earth, must also be present.

Another strong indication that Uranus and Neptune are different from Jupiter and Saturn is their appearance. Through a telescope, these two planets have a definite greenish/bluish hue, quite different from the orangey tints of Jupiter and Saturn.

Now why is it that atmospheres compositionally seemingly so similar to those of Jupiter and Saturn look so different from the outside? A likely explanation is that the difference in appearance results from the relatively colder temperatures of the upper atmospheres of Uranus and Neptune. The big difference between Uranus and Neptune on the one hand, and Jupiter and Saturn on the other, may be that Jupiter and Saturn never get cold enough in their upper atmospheres for methane and other hydrocarbon gases to condense into clouds.

On Jupiter the whitish clouds that we see in the "zones" are believed to be ammonia crystals; the reddish clouds making up the "belts" are guessed to be something like ammonium hydrosulfate (NH_4SH). A typical model of the atmosphere of Jupiter has three levels of clouds: the ammonia cloud, the reddish cloud down below which is probably ammonium hydrosulfate, and below that a water cloud. Saturn is very similar but because this planet is farther away from the Sun, its upper atmosphere is colder and more ammonia condenses, forming a thicker ammonia cloud; thus, on Saturn one cannot see the colored clouds below as well as in the case of Jupiter. Following the trend out to Uranus and Neptune, the reddish clouds might be there all right, but they would be so deep that we can't see them. The atmospheres of Uranus and Neptune are cold enough not only for all of the ammonia to be frozen out in the high atmosphere but for methane to freeze out as well, adding still another high-altitude whitish cloud. These hazes could well prevent our seeing any colored clouds which might be farther below. [An additional factor is that due to the colder temperatures the clouds in the atmospheres of Uranus and Neptune occur at "deeper" levels of the atmosphere than on Jupiter and Saturn: there is more atmospheric gas between us and the cloud layers. Thus, increased scattering by the intervening gas molecules makes it harder to see the clouds below.] There are important measurements that Voyager can carry out to test such ideas. Especially crucial are accurate temperature determinations as a function of height in the upper atmosphere.

A more sophisticated misconception about Uranus and Neptune is that the two planets are twins. We know from the mean densities that this also cannot be true. The mean density of Uranus is about 1.3 grams per cubic centimeter; that of Neptune is almost 1.6 (Table 1). That slight difference is very significant. On the inside these planets must be quite different. A confirmation of this fundamental difference comes from the infrared radiation emitted by the planets. As in the cases of Jupiter and Saturn (but probably for a different reason), the heat emitted is not just reflected sunlight. Jupiter, Saturn, and Neptune have appreciable sources of internal heat that is escaping into space. Uranus does not appear to have a similar heat source.

Uranus has one peculiarity which is worth stressing at this point. Most planets spin in such a way that their rotation pole is more or less perpendicular to the plane of the orbit. The Earth's pole is tilted about 23° from the perpendicular to this plane; that of Neptune, by about 29°. But Uranus is flipped way over on its side! We don't really know how Uranus got into this state (any more than we can explain the spin of any other planet), but one implication is that as Uranus orbits the Sun once every 84 years there are two times when one of the poles is pointing almost exactly at the Sun, while the other pole is totally in darkness (Figure 3). Right now Uranus is in a configuration where the south pole is facing directly toward us and toward the Sun. From the point of view of Voyager, this means that, unlike the case for Jupiter and Saturn, where the spacecraft flew by more or less in the plane of the satellites and rings, the ring and satellite system of Uranus will be encountered at almost right angles by Voyager 2. Hence, we cannot fly by the satellites in succession, and only one hemisphere of the planet or any satellite will be in daylight during the brief period of the flyby.

The orientations of the spin axes of Uranus and Neptune are known reasonably well. We observe satellites orbiting the planets, and assume that the spin axes of the planets are approximately at right angles to the orbital planes of the satellites. But we don't know the precise spin periods. From the Earth, it is extremely difficult to detect any features on these planets. The best telescopic pictures of Uranus and Neptune are not very revealing. In the best views of Uranus, only faint smudges are discernible (Figure 4). Neptune is more interesting: faint variable cloud markings can sometimes be detected. Some of these have been tracked by Bradford Smith and Rich Terrile, who

Southern Summer

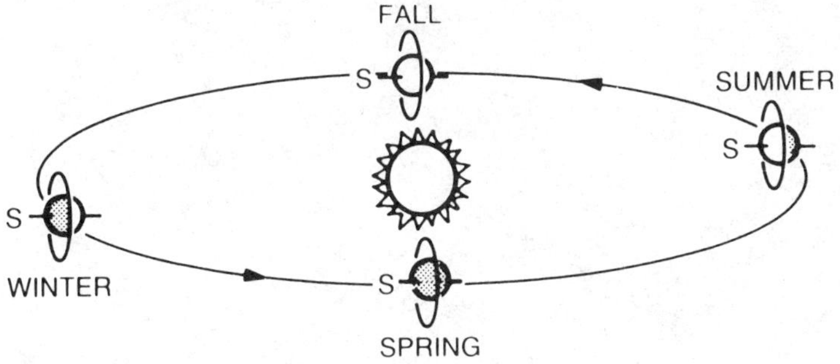

Figure 3. Seasons on highly tilted Uranus. Voyager flew by the planet during summer in the southern hemisphere, the situation depicted at extreme right.

inferred a rotation period for Neptune of between 17 and 18 hours from such data. The markings are believed to be methane clouds high up in the atmosphere. The appearance of Uranus is considerably less patchy than that of Neptune, for reasons which we don't understand but which should become evident as Voyager 2 has flown past both planets. The spin period of Uranus has been estimated by less direct spectroscopic methods: the best current guess is that its period is between 16 and 17 hours. For both Uranus and Neptune, the estimates of spin periods are very uncertain, but accurate determinations are expected from the Voyager data.

We are also looking forward to finding out what causes the fundamental difference between Uranus and Neptune mentioned briefly above: the fact that temperatures in the upper atmosphere of Neptune are similar to those at comparable levels in the atmosphere of Uranus, even though Neptune is one and a half times farther away from the Sun. The fact that the temperatures are very similar as are the albedos of the two planets (the fraction of incident sunlight that is reflected) suggests that something peculiar is happening on Neptune. Neptune is much warmer than it ought to be, considering incident solar energy alone. Thus, like Jupiter and Saturn, Neptune must have an internal source of heat. To understand what is happening, we need a more refined measurement of the ratio of heat that's coming from the interior to that which is derived from absorbed light. Such measurements of the "heat balance" of Uranus and Neptune are key goals of the Voyager flybys.

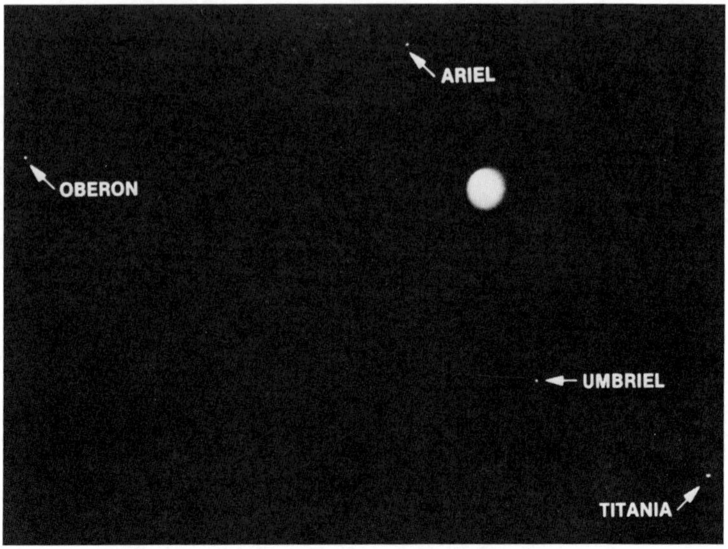

Figure 4. A Voyager 2 view of Uranus and four of its satellites. This view, obtained on July 15, 1985, when the spacecraft was still 247 million km from the planet, shows detail at a level comparable to that in the best photographs obtained from Earth. Note that the rings and the inner satellite Miranda cannot be seen in this picture.

The indication that Uranus is heated only by absorbed sunlight, whereas Neptune has an additional internal heat source, is a significant clue to the interiors of the two objects being significantly different. Models of the interior of both Uranus and Neptune are rudimentary (Figure 5), a situation which is not surprising given our limited knowledge of these planets. Suggestions are that these planets are about one-third (by radius) made of rocky substances (the kind of stuff that makes up the Earth), a third radius made of icy materials such as water ice (H_2O), and methane (CH_4), ammonia (NH_3), and the outer one-third consisting of an envelope of predominantly hydrogen-helium but with other gases such as methane and ammonia. One can speculate what the temperature-pressure distribution would be in such a body, with the interesting result that significant pressure ionization is expected in the "icy" layer. The result must be increased electrical conductivity and if these layers are fluid as seems likely, there is a strong possibility that Uranus and Neptune may have significant magnetic fields. At present, we do not know whether

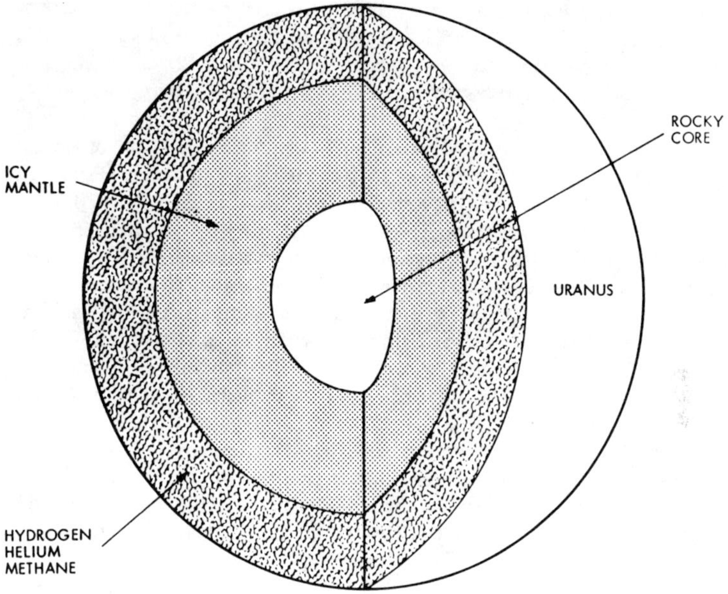

Figure 5. A Sketch of the Possible Interior Structure of Uranus

either planet has a magnetic field and a magnetosphere of trapped charged particles. Finding out is another major objective for Voyager. With a complement of more than half a dozen relevant instruments, the spacecraft is admirably instrumented to answer such questions.

SATELLITES

At the present time, we know that Uranus has five satellites (Table 2); we suspect that there are more. It is expected that all five of the known satellites orbit within the magnetosphere, a situation similar to that of the large satellites of Jupiter. Thus, it is likely that the surfaces are exposed to significant fluxes of high-energy protons and electrons, which can produce significant chemical and physical modifications. In the best photographs that can be obtained from Earth, the five satellites–Miranda, Ariel, Umbriel, Titania, and Oberon–appear as mere dots (Figure 4). Until a few years ago all we knew about them was their orbits. Recently a number of new techniques have made it possible to determine their actual physical sizes; and from their mutual gravitational effects on each other's orbits, it has been possible to deduce their masses. From the masses and radii, we can calculate mean densities and obtain insights into the constitution of these moons (Table 2). All of these determinations–radii, masses, and mean densities–will be refined considerably by Voyager.

Compared to our Moon (radius = 1,738 km), all of the satellites of Uranus are small; the largest, Oberon, is only 800 kilometers in radius (Figure 6). Their general characteristics are comparable to some of the icy satellites of Saturn. The densities are those expected for satellites made of water ice with some rocks mixed in, and there is spectral evidence that water ice is present on the surfaces. (These satellites are too small to have atmospheres.) The spectrum of water frost (H_20) has two prominent, distinctive absorption bands at 1.5 and 2.0 microns, which show up prominently in spectra of the Uranus satellites. Thus, to first order we are justified in thinking of Uranus' five moons as icy bodies--500 to 1600 km across–with water ice as probably the dominant volatile.

The Voyager encounter will provide the first close look at the surfaces. At last, we will be able to see what kind of geologic evolution has taken place during the past 4.6 billion years. Since Voyager will be coming in at right angles to the plane in which the satellites orbit, the decision has been made to fly between the orbit of the nearest satellite, Miranda, and Uranus and its rings (Figure 7). Such a trajectory will provide the best possible look at the planet and its rings and at its innermost known satellites. The other moons will be observed at greater distances and therefore in less detail. What do we expect to see? One possibility is that these small icy satellites all look like Saturn's Mimas–a heavily cratered ice ball some 200 km in radius (Figure 8). There is little evidence that Mimas has done anything but been subjected to external bombardment by meteoroids. That's an awfully dull way to have spent the last 4.6 millions years!

But Mimas has a close neighbor, Enceladus, which although not much bigger, has had a very different history. You don't have to be too much of a geologist to figure out that something interesting has happened to Enceladus (Figure 9). Extensive areas on the satellites have no craters, suggesting relatively recent geologic activity. In fact, there is considerable evidence that Enceladus remains warm enough near the surface for some water to still erupt at the present time. For example, it is likely that Enceladus is the source of ice particles that populate Saturn's diffuse E-ring which is densest at the orbit of Enceladus (Figure 10). To sustain geologic activity one needs energy. Of the various possibilities the only one that seems to work is tidal heating: Enceladus orbits close to a very big planet. A satellite in a slightly non-circular orbit around a planet finds itself at times closer to the center of the planet than at other times. When closer, tides distort the shape more than

	Distance from Planet (R♅)	Orbital Period (days)	Radius* (km)	Mass† (10^{23} gm)	Average Density (g/cm^3)
Miranda	5.0	1.41	250 ± 110	1.7 ± 1.7	?
Ariel	7.3	2.52	665 ± 65	15.6 ± 3.5	1.3 ± 0.5
Umbriel	10.2	4.14	555 ± 50	10.0 ± 4.2	1.4 ± 0.6
Titania	16.7	8.71	800 ± 60	59.0 ± 7.0	2.7 ± 0.6
Oberon	21.7	13.46	815 ± 70	60.0 ± 7.0	2.6 ± 0.6

Using R♅ = 26,200 km.

*Determined by R. Hamilton Brown.

†Determined by Ch. Veillet.

Table II. Uranus Satellites (Pre-Voyager Data)

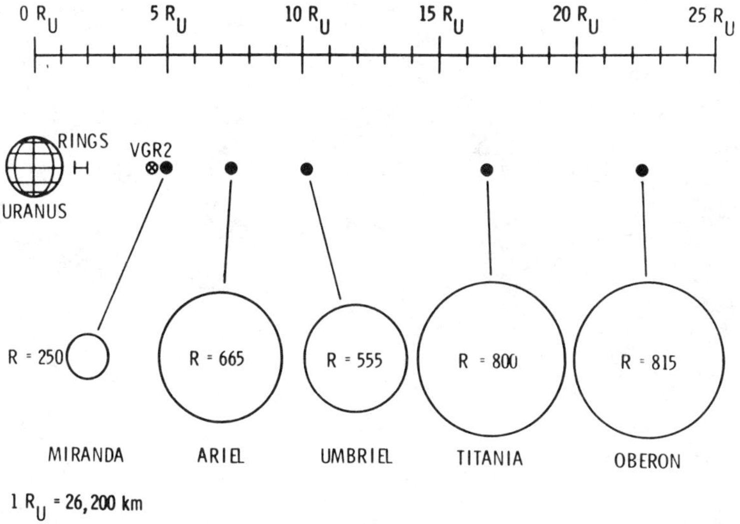

Figure 6. Summary of our pre-Voyager knowledge of the physical characteristics of the satellites of Uranus (see also data in Table II).

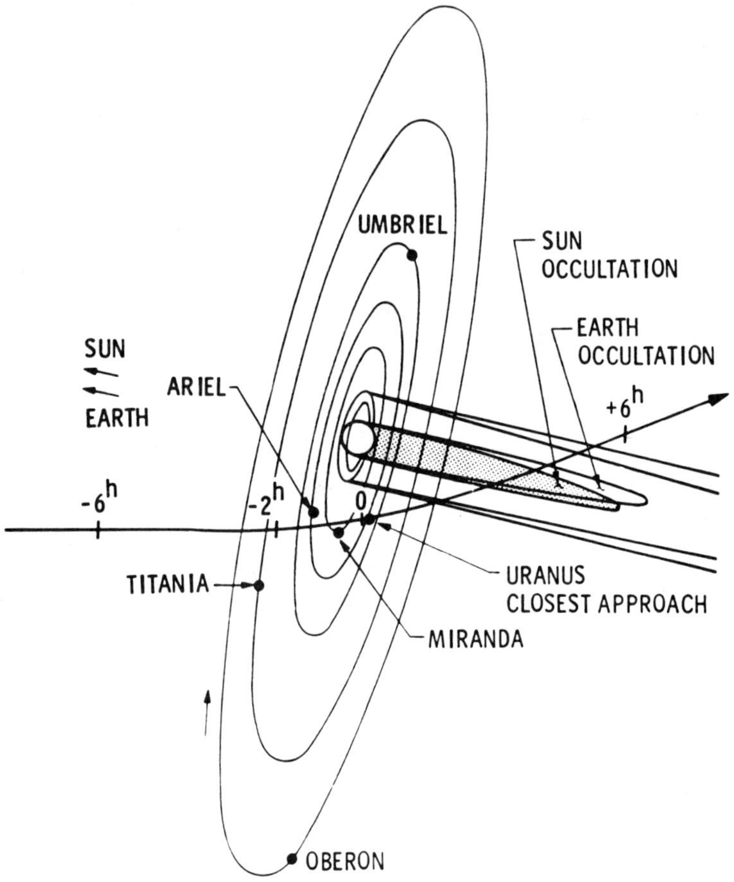

Figure 7. Voyager's flyby trajectory at Uranus. The spacecraft is to come within 29,000 km of Miranda. The closest approach to Uranus will be 107,000 km on January 24 (10 am PST).

Figure 8. A Voyager view of Saturn's inner satellite Mimas. The profusely cratered surface of this icy object suggests that impact cratering has been the major geologic process during the satellite's history.

when it is further out. So during each orbit the satellite is being flexed back and forth and since there is friction, some of the flexing energy gets dissipated into heat. It's somewhat similar to making a spoon warm by bending it back and forth in your hands. Tidal heating may be the answer in the case of Enceladus. The problem is that we don't understand tidal heating very well. The same equations that can be used to predict tidal heating for Enceladus predict even stronger tidal heating for Mimas. Mimas is closer to Saturn and its orbit is even more non-circular than is that of Enceladus. Yet Voyager images don't betray any evidence of internal activity on Mimas. That's a serious problem. Given the uncertain state of the theory, it is difficult to use it with any certainty to predict whether or not any of the satellites of Uranus have undergone significant tidal heating. It has been suggested that Ariel may

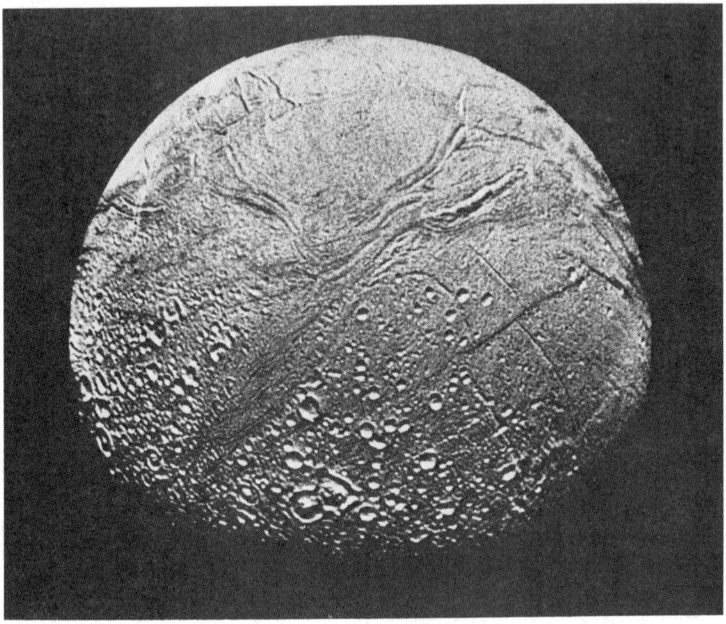

Figure 9. Saturn's satellite Enceladus, imaged by Voyager 2. Note the remarkable variation of crater density on the surface and the groove-like sinuous features, which suggest internal geologic processes.

have, but we really don't know. That is why we are so anxious to see what Voyager will reveal next January. Will Ariel look more like Enceladus, or more like Mimas? Or will it be entirely different?

RINGS

We know that Uranus is surrounded by nine very narrow rings (Figure 11). These were discovered during careful observations of an occultation of a star by the planet in 1977. As Uranus approached the star's position, sharp dips in the intensity of the starlight were recorded—dips that were matched by others observed once the disk of Uranus had cleared the star. The only reasonable explanation was that each pair of symmetric dips represented the temporary blocking of the starlight by a ring. The rings are very narrow. The

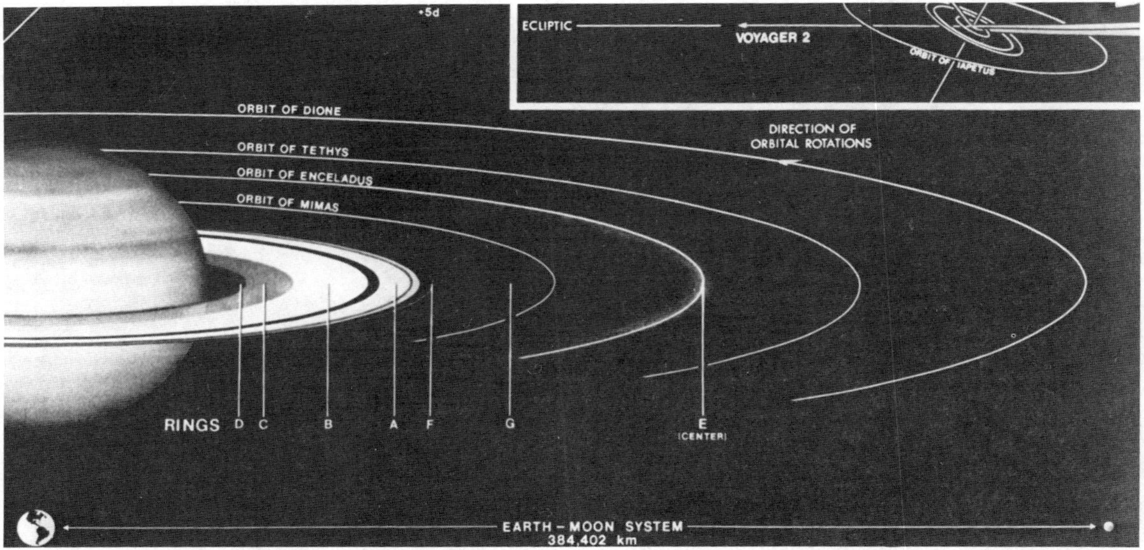

Figure 10. Schematic diagram of the Saturn ring and satellite system. Note the location of the narrow Ring F just outside Ring A, as well as the fact that the diffuse Ring B appears to be centered on the orbit of Enceladus. One speculation is that geologic activity on Enceladus continues to produce the ice grains that make up Ring F.

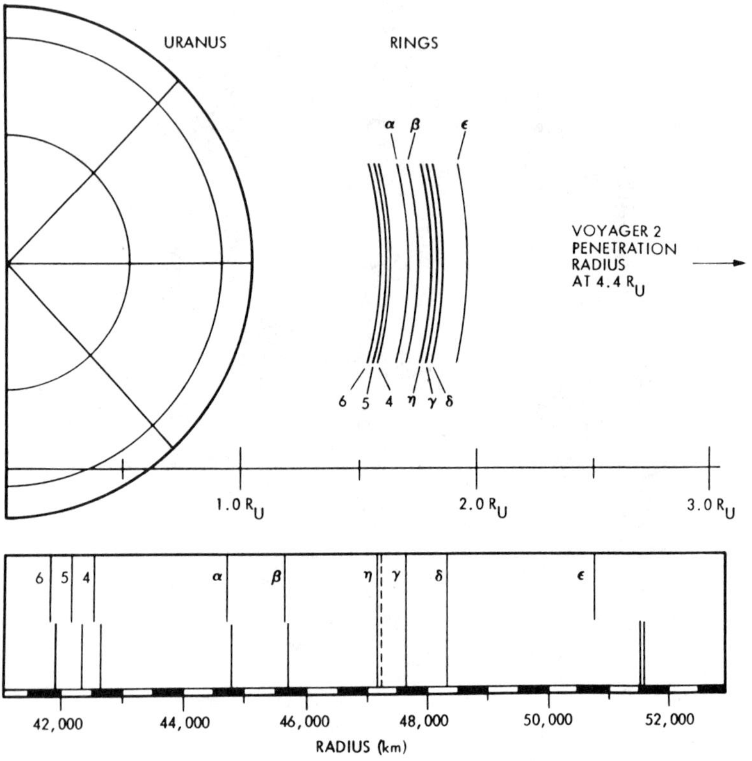

Figure 11. Schematic of our pre-Voyager knowledge of the geometry of the nine narrow rings of Uranus. As shown in the bottom panel, some of the rings (especially Epsilon) are noticeably non-circular. (Shown are the maximum and minimum distances from the planet's center.)

shortest dips imply rings only 10-20 km across. The widest suggests a ring some 100 km in extent. Since this discovery, the puzzle has been to explain why the rings of Uranus are confined in such narrow bands. We know rings are made up of individual small particles, each a minuscule satellite of the planet. Various processes—for example, occasional collisions—are at work in such rings. Most of these processes have a tendency to spread the rings out. Thus, if one starts with a very narrow ring, with time the ring will spread out and become diffuse. Therefore, to explain what we see we have two choices. Either the rings formed very recently (not a very palatable assumption) or there is a process that is keeping the ring particles from spreading. A similar situation was encountered in the case of an unusual ring of Saturn—the narrow F-ring (Figure 12). Just outside the main rings that one sees through a telescope, spacecraft discovered a narrow ring no more than a hundred kilometers wide. The proposal was made by Peter Goldreich and Scott Tremaine

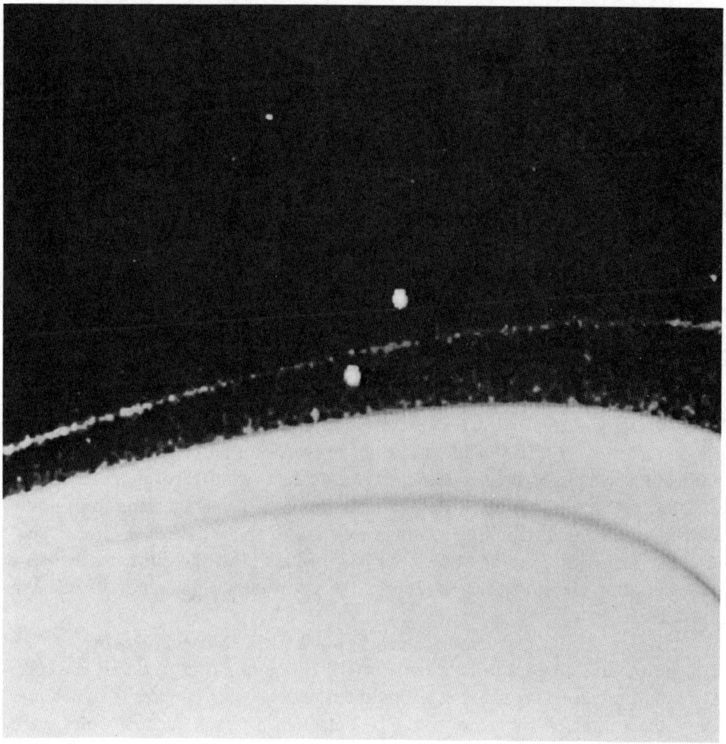

Figure 12. The two shepherding satellites that are believed to confine Saturn's narrow Ring F are seen in this Voyager image. It is speculated that similar shepherding satellites are confining the particles in the rings of Uranus to narrow lanes.

that one way of keeping a ring confined was by means of two shepherding satellites. If there is a satellite just inside the ring and another satellite just outside, one can calculate that the inner satellite tends to push the ring material out, while the outer satellite pushes it in. Thus, by means of this simultaneous pushing out and pushing in, the ring is kept confined. In the case of Saturn, Voyager 2 looked for the two predicted shepherding satellites and found them (Figure 12)! If this explanation applies in the case of Uranus—there are nine narrow rings—we should expect to see a lot of shepherding satellites. Since they are expected to be too small to be visible from the Earth, we have to wait till Voyager's visit to test the theory.

BEYOND URANUS: NEPTUNE

After its flyby of Uranus, Voyager 2 will be on a trajectory that will take it close to Neptune in August 1989. Our exploration goals at Neptune are similar to those at Uranus. Does the planet have an internal heat source? A magnetic field? How fast does it spin? There have been recent reports based on occultation observations that the planet may have a ring. We know that Neptune has at least two satellites, and one of them, Triton, must rank as one of the most interesting objects in the solar system.

Triton has been the subject of a lot of mischief. Realizing that one is dealing with a very distant object which is extremely difficult to study from Earth, it is easy to understand that one can make up a lot of good stories. Even the best photographs show Triton as nothing more than a speck of light close to Neptune. From its brightness it is clear that Triton is a sizeable body. But how big is it, really? Until recently, it has not been possible to determine the diameter of Triton. This fact has not prevented a lot of guessing and there is an old textbook tradition—unsupported by fact—that "Triton is the largest satellite in the solar system." Recent measurements using the same techniques that have been applied in determining the radii of the satellites of Uranus suggest that Triton is about the size of our own Moon. But the estimates are very uncertain and we must look to Voyager to eventually tell us how big Triton really is.

Another vexing problem concerns Triton's mass. Since Triton orbits close to Neptune and appears to be a large satellite, one might expect that the orbital motion of the planet and satellite about their center of mass (i.e., not precisely the center of Neptune) could be detected. In fact, such wobbles were measured decades ago and used to determine the satellite's mass. Unfortunately, at least two mass determinations exist, and they are not very consistent. Taking a mean value and assuming that Triton is the size of our Moon leads to an unreasonably high mean density. At the present that is a major Triton problem. Fortunately, Voyager will fly very close to Triton, and

Figure 13. An artist's concept of what Voyager 2 will find when it flies by Triton in August 1989. Vast Neptune looms over this frigid ice-scape that may include an ocean of liquid nitrogen.

readily determine the mass, radius, and mean density very accurately. For the first time, we will have reliable evidence as to the internal makeup of this satellite.

Since Triton has always been perceived as a large satellite, there have been many suggestions that it should have an atmosphere. These have been confirmed by recent spectroscopy. The near-infrared spectrum of Triton was found to be rich in absorption features: some due to methane gas, CH_4, and others to methane ice. Most of the spectrum can be understood if there is abundant frozen methane on the surface of Triton and some methane gas in the atmosphere: not an unreasonable situation at the cold temperatures (50°-60°K) that prevail on Triton. But there is a prominent feature in the spectrum at 2.16 microns which does not fit methane. Dale Cruikshank and his colleagues at the University of Hawaii, who have analyzed the spectrum of Triton in detail, interpret this 2.16 μm absorption as due to nitrogen, but nitrogen in a strange form: liquid nitrogen! According to this interpretation of the spectrum, Triton is an object with a lot of methane ice and a lot of liquid nitrogen on the surface, and a little bit of methane gas in the atmosphere. A very, very strange object indeed, with oceans of liquid nitrogen and perhaps icebergs of frozen methane floating in them (Figure 13). Voyager is ideally instrumented to send back high resolution images of Triton and to make more detailed measurements of its spectrum. Within five years we will know whether this strange concept of Triton is fact or fantasy.

AFTER VOYAGER

By the end of the decade the two Voyagers will have done remarkable things: they will have explored Jupiter, Saturn, Uranus, and Neptune, almost half of the known planets in the solar system. Neither will get very close to Pluto; unfortunately there are no current plans to send spacecraft to this remote planet.

At the time that Voyager 2 will be encountering Uranus next January, most of the world will be preoccupied with the return of Halley's comet. Halley returns every 76 years; its last visit was in 1910. Certainly Halley always attracts much attention, but this time around scientific preoccupation with the comet will be even more intense. A veritable flotilla of spacecraft is being readied to investigate a comet. The Japanese, the Europeans, and Soviets each have spacecraft that will encounter the comet. A serious problem is that Halley comes in through the inner solar system at very high velocity. Thus, all of these missions will be very high speed fly-throughs. They will provide our first look at what the heart of a comet–its nucleus (presumed to be a chunk of dirty water ice a few miles across)–really looks like. They will also obtain the first detailed analyses of the chemistry of the gases and dust that is liberated as the ices on the nucleus surfaces are vaporized by sunlight. But comets are known to be very complex and a much more thorough study of them is essential. Although the United States does not have a mission to Halley, it has plans for a follow-up mission which would study a comet in considerably more detail than is possible during a very high speed flyby. The spacecraft is designed to match velocities with a comet, and get within tens of kilometers of the nucleus (Figure 14). In fact, the plan is to orbit the comet

Figure 14. The next mission to a comet. The NASA Comet Rendezvous/ Asteroid Flyby spacecraft hopes to rendezvous with a comet nucleus and land an instrumented package on its surface during the 1990's.

nucleus and study it in great detail for several years as its distance from the Sun changes. In addition to the usual remote sensing instruments, this spacecraft would carry a "penetrator," a device which would implant key analytical instruments into the nucleus surface. The mission—called CRAF for Comet Rendezvous Asteroid Flyby—would fly by an asteroid on its way to its final target: the nucleus of some short-period comet such as Tempel 2 (Figure 15). (Short-period comets have orbits more or less within the orbit of Jupiter. Tempel 2 has a period of only 5.2 years, compared to Halley's 76 years.) Currently CRAF is planned for the mid-1990's.

Much closer in the future is the scheduled launch of Galileo to Jupiter in May 1986. Galileo is a follow-on to Voyager and consists of two parts which are launched together but separate before arrival at Jupiter. One is the Probe, which will be dropped into the clouds of Jupiter; the other is the Orbiter, which will fly around the planet and study Jupiter and its satellites. Galileo is the first planetary spacecraft scheduled for launch by the Space Shuttle.

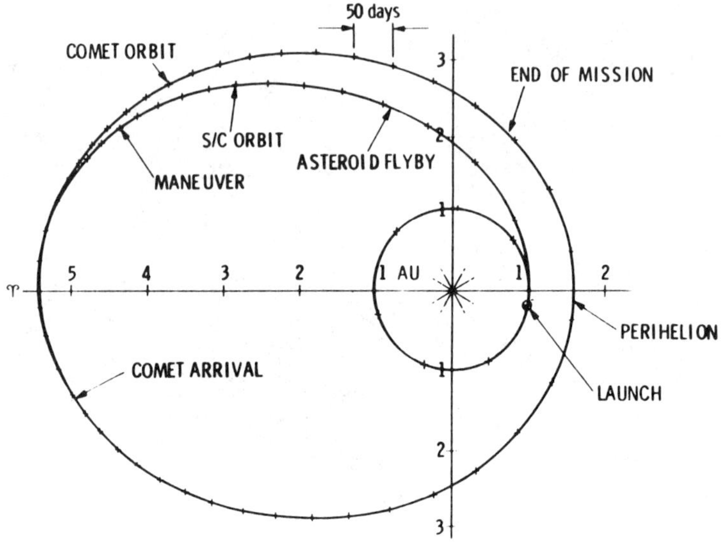

Figure 15. A representative trajectory for the proposed Comet Rendezvous/ Asteroid Flyby Mission. Note that on its way to the comet the spacecraft would fly by an asteriod, and that it would stay with the comet through at least half of an orbital period. Comet activity is expected to increase dramatically as the comet approaches the Sun.

The main purpose of the Probe is to investigate the composition of the atmosphere and clouds and the structure of the atmosphere above and below the visible clouds. The Orbiter will provide detailed observations of Jupiter and its satellites for several years. Four of the satellites of Jupiter are large enough to perturb the spacecraft sufficiently. Thus, once Galileo is in orbit about Jupiter, the moons can be used to alter its path in such a way as to effect close flybys of all four large satellites.

Galileo will send back images with a resolution a hundred times better than Voyager. Even more importantly, the spacecraft carries an instrument which on a comparable scale, measures spectral reflectance, from which information on composition can be derived. Thus, not only will we see the morphology of surface features at very high resolution, but we will be able to tell what materials they are made of. For example, we should be able to test once and for all the extent to which the volcanic surface of Io is covered with sulfur and sulfur compounds.

CONCLUSION

The exploration of our solar system has proceeded greatly beyond the situation that planetary astronomy was in 30 years ago, when we relied entirely on observations made using telescopes here on Earth. The information we had, and the questions that could be asked and answered were all very limited. Due to advances in rocket technology we can now take our instruments beyond the confines of the Earth to the very places we want to study. Equally important have been the advances in electronics. Because of these advances one can build and operate spacecraft as sophisticated as Voyager which, for a period of ten years, can take care of themselves, execute complicated sequences at a distance of billions of kilometers, and send back to us the precious scientific data they have gathered. It is because of these advances in rocket technology and in electronics that we have participated, and will continue to participate, in what has been one of the most spectacular intellectual adventures in our history: the direct exploration of the solar system.

EPILOG

The tragic loss of the space shuttle Challenger and its crew in January 1986 has delayed the execution of future plans described above. For example, Galileo, which was to have been launched in May 1986, has been postponed until 1991.

Any difficult enterprise must expect to encounter temporary setbacks. Currently NASA is working very hard to get back on track and continue its very successful exploration of our solar system and of the universe beyond.

Author's Note: All illustrations in this article are from NASA or from NASA's Jet Propulsion Laboratory, and were produced originally for use by the Voyager Project.

FUTURE DIRECTIONS IN ELECTRICAL ENGINEERING

ENERGY FOR THE FUTURE

PROCEEDINGS of the

Cornell Electrical Engineering Centennial Symposium

at Washington, D. C.–April 25, 1985

Simpson Linke, editor

Promethean Press
Ithaca, New York

PROCEEDINGS OF THE
CORNELL ELECTRICAL ENGINEERING CENTENNIAL SYMPOSIUM:
FUTURE DIRECTIONS IN ELECTRICAL ENGINEERING-
ENERGY FOR THE FUTURE

Copyright 1988
by the School of Electrical Engineering
Cornell University, Ithaca, New York
All rights reserved

Published by Promethean Press
Ithaca, New York

Printed in the United States of America

October, 1988

CONTENTS

PREFACE

The celebration of the Centennial Year of the School of Electrical Engineering at Cornell University featured a series of symposia based on the central theme of "Future Directions in Electrical Engineering". Each seminar concentrated on a major facet of the discipline, with individual topics presented by outstanding specialists in each field. The symposia were held at sites around the country that are widely known as centers of activity for each particular subject area, thereby providing significant updates on pertinent electrical engineering technologies at appropriate locations, and allowing a large number of Cornell Electrical Engineering alumni to participate in the celebration in a meaningful way.

The ultimate success of the Symposium Series must be credited to the dedicated efforts of the faculty, symposia, and local alumni chairmen who were responsible for planning the programs, engaging the participants, and attending to local arrangements within specific time constraints. The following schedule of events attests to the overall magnitude of the task and also displays the broad range of topics that were presented:

March 15, 1985—Microelectronics—Stanford, Calif.

April 1, 1985—Computers—E. Windsor, N. J.

April 10, 1985—Communications—Boston, Mass.

April 17, 1985—Atmospheric and Space Sciences—St. Louis, Mo.

April 25, 1985—Energy for the Future—Washington, D. C.

June 12, 1985—Societal Impacts and Future Directions—Ithaca, N. Y.

Each session was videotaped to facilitate preparation of the proceedings of each meeting and for possible use of the individual presentations in classroom seminars and elsewhere. The audio portions of the tapes were transcribed into preliminary text and coordinated with figures obtained from the video images and from hard copy provided by some of the authors. Each presentation was edited with a view toward achieving continuity and coherence without

compromising the natural style of the author. In several instances it was necessary to introduce substantial changes in format and phrasing in order to achieve this goal; in others, the hard copy supplied by the authors made the task easier. In all cases, the final versions were sent to the authors for their criticisms, corrections, and final approval. I am grateful to all of the contributors for their excellent cooperation and assistance in bringing their presentations to this present form.

I wish to thank Professor Edward D. Wolf for his enthusiastic interest in this project and for authorizing financial support from the National Research and Resource Facility for Submicron Structures to assist in the publication of the Symposia Series *Proceedings*. Additional funding from the College of Engineering and the School of Electrical Engineering is also gratefully acknowledged.

It would not have been possible to produce this work without the skillful secretarial assistance of Catherine Kuhl, Deborah Overslaugh, and Linda Struzinsky who performed the difficult task of transcribing the tapes into typed versions of the unedited manuscripts. The expert guidance provided by Mrs. Diana Nathanielsz on format and production of this volume is very much appreciated.

I am particularly grateful to Catherine Kuhl for her diligence and care in forming the camera-ready copies of several of the manuscripts on that versatile new instrument that is so illustrative of "future directions in electrical engineering" — the laser printer.

Simpson Linke

Ithaca, New York

INTRODUCTION

John A. Nation

Director, School of Electrical Engineering

The celebration of our Centennial Year affords a unique opportunity to review past accomplishments, to evaluate our present circumstances, and to make some reasonable estimates of prospects for the coming years. A glance at the history of the University reveals that the Cornell *Register* for 1883 contained a "Special Notice" from President Andrew Dickson White announcing the establishment of a "Course in Electrical Engineering." President White had convinced the University's Trustee Executive Committee to create the new program "provided such course occasions no additional expense."

The early curriculum was rigorous and in addition to the traditional engineering topics of mathematics, physics, mechanics, mechanical drawing, and shop work, required rhetoric, French and German. The senior year also included courses in several Cornell specialties such as photometry and electric lamps, telegraph instruments, lines and cables, the dynamo, and electric motors. The first baccalaureate degrees in electrical engineering in the United States were awarded by Cornell (as well as by M.I.T.) in 1885, and the first Ph.D. in the new discipline was conferred by President White at Cornell in the same year. Thus we are truly celebrating 100 years of electrical engineering education.

One of the major figures in the E.E. School of those days was Professor Harris J. Ryan who ran the Electrical Engineering Department essentially as a one-man faculty for 17 years and is credited with having trained almost 1000 electrical engineers during that period. Ryan left Cornell in 1905 to become head of the Electrical Engineering Department at Stanford. It is interesting to note that 12 of the first 13 members of the Stanford faculty were educated at Cornell.

This brief summary of the beginning of electrical engineering instruction at Cornell would not be complete without mention of the internationally known Professor Vladimir Karapetoff who joined the faculty following Professor Ryan's departure and made many contributions to early electrical engineering theory and analysis in several textbooks and many technical articles. Something of the flavor of electrical engineering studies of the time is suggested by the following excerpt from Karapetoff's *Electrical Laboratory Notes*, published in 1906:

> "In coming to the laboratory, bring with you a slide rule, an inch rule or tape, a speed counter, a screw

> driver and a pair of plyers [sic]. This will save you time and trouble of looking for them or borrowing them. Do not forget to have a pocket knife for skinning off wire; a bicycle wrench is also sometimes very handy to have."

In these days of hand calculators and personal computers, it is rare indeed to find a student who has even seen a slide rule, much less to have used one.

Classical electrical engineering education at Cornell was devoted principally to electric machinery, power systems, telecommunications, and early electronics. As the technology developed from the 20's through the 40's, the curriculum was, of course, continually upgraded as new and advanced concepts were introduced, but the basic format remained essentially unchanged until the mid-50's when research began to play a central role in the E.E. School, coincident with the move from Franklin Hall to the newly constructed Phillips Hall. The following figure among the highlights of our research development over the years:

> An early solar telescope project initiated an era of Radio Astronomy research that culminated in the establishment of the world's largest radio-radar telescope at Arecibo, Puerto Rico, now known as the National Astronomy and Ionosphere Center.
>
> A research program in the new field of Physical Electronics gradually expanded into the present National Research and Resource Facility for Submicron Structures housed in the Knight Laboratory adjacent to Phillips Hall.
>
> The Old High Voltage Laboratory on Mitchell Street, converted to an Energy Laboratory in 1960, served as the staging area for the Cornell Extra-High-Voltage Cable Test Program and housed E.E. research on high-voltage vacuum switches and arc physics until the formation of the interdisciplinary Laboratory of Plasma Studies in 1966. This facility, together with adjacent installations, now contains the largest particle-beam program in any university in the world and has made many important contributions to the understanding of intense electron and ion beams, particularly as they relate to inertial confinement thermonuclear fusion experiments.
>
> The national energy crisis in the early 70's created renewed interest in power and energy research within the School leading to the combination of power-system elements, control theory, and new computer facilities, that now constitutes the unique Kettering Power Systems

Laboratory for the analysis of modern complex power-generation and transmission networks.

Two recent events attest to the high regard that the University and the E.E. School enjoy in the national research community:

1. The Program in Microscience and Technology was established in 1982 and receives about $2 million annually through the Center of Excellence Program of the Semiconductor Research Corporation.

2. The NSF-sponsored Cornell Center for Theory and Simulation in Science and Engineering, with its associated supercomputer production facility, will augment and expand research capability in the School and will encourage greater interdisciplinary activities with other scientific units on campus.

One of our strengths in recent years, as the Electrical Engineering School has evolved from a purely teaching organization into a teaching and research institution, has been our ability to attract good new faculty. A positive measure of this quality is in our receipt of Presidential Young Investigator Awards from the National Science Foundation in the first year of the establishment of these awards. The E.E. School received 5 of the total of 26 awards made available to electrical engineering departments in the nation. Our success in attracting these high-quality young people bodes well for the continued future growth of the School.

Remarkable changes in electrical engineering education have occurred in the School within the past 25 years as the influence and impact of new faculty members and their research have brought modern concepts to curriculum and laboratories that had been essentially identical to those of World War II days. The advent of computers has been particularly significant. We have been very fortunate in the last decade in the support that we have received from industrial corporations. Their generous donations of equipment including computers, work stations, and state-of-the-art measurement devices have been of great value in upgrading our laboratories to present-day standards.

On the other hand, the current huge interest in computer engineering and electrical engineering has introduced major enrollment problems in the School. Using 1965 as a year of reference, the size of the faculty has remained fixed at about 40 members whereas the number of undergraduate students has tripled. Our senior class in 1985-86 numbered 186 students. We have 265 graduates students in the School with about 145-150 MS/PhD students and the remainder in the Master of Engineering Program, the latter being the second largest professional degree program in the University. This year we graduated

30 MS/PhD students and 88 Master of Engineering degree candidates. An encouraging factor is that 12 of these advanced-degree students were women. Today about 15 percent of the students in the E.E. School are women and 25 percent are minority students.

The increase in student enrollment is coupled with a research effort that is growing at a compound rate in excess of 10 percent per year in constant dollars. Clearly we are approaching the limits of our present capability. It is hard to realize that within 30 years of our initial occupancy of Phillips Hall we are now bursting at the seams. Further, the building which was then state-of-the-art, is now an antiquated facility in terms of present-day technology, with little in the way of the clean-air environments, air conditioning, and other amenities that are necessary for the proper functioning of modern equipment.

Plans are in progress to build a new engineering complex on campus in which electrical engineering will have substantially more space than in the past. When these new facilities are completed, our faculty will be able to work at its full potential. We expect to see a relatively modest growth in the number of faculty members. Since a faculty lifetime post is forty years, and student-interest time cycles are rather shorter than that, the response to large student increases in enrollments and interest in electrical engineering cannot be followed instantly by new faculty positions. Attempts will be made to maintain our undergraduate size at about its present level with some growth allowed in the number of our graduate students. It is certain that there will be many more interactions with industry in the near future, and these will take on a much broader form as the years go by. For example, an exchange program may be arranged with industrial laboratories so that their scientists will come to work in the School for a period while a faculty member will have the opportunity to spend corresponding time in industrial environments.

This brief perspective of past, present, and future encourages us to begin our second century with a true sense of excitement. We may be justly proud of the important contributions that our School of Electrical Engineering has made to many of the developments that are reported in these *Proceedings*. As we celebrate these past achievements, we can look forward with confidence to producing equally significant additions to our challenging discipline.

PART 5

ENERGY FOR THE FUTURE

PROCEEDINGS of the

Cornell Electrical Engineering Centennial Symposium

at Washington, D. C.–April 25, 1985

"I am pleased to open this fifth symposium of the Cornell Electrical Engineering Centennial Series on Future Directions in Electrical Engineering. The program for today will focus on prospects for meeting the nation's energy needs for the remainder of this century and beyond. These topics are indeed appropriate for this celebration since electrical engineering education at its inception, and for many years afterward, gave principal attention to the field of energy.

"At this time I wish to acknowledge the wholehearted cooperation of our General Chairman, Donald Kerr, and to express my appreciation to all of our speakers for their many significant contributions to this symposium. I am grateful to my colleague, Professor Bob Thomas, for his valuable assistance as Faculty Co-Chairman of this event. I also thank our Local Arrangements Chairman, Joe Stregack, for his careful attention to and skillful execution of the many details that make for a successful meeting.

"Now for the official welcome to this gathering, it is my pleasure to introduce Professor Joseph Ballantyne, former Director of our School of Electrical Engineering, who conceived and planned this Centennial Celebration in the final year of his tenure as Director before assuming his present position as Vice-President for Research at Cornell."

–Professor Ravindra N. Sudan, Symposium Faculty Co-Chairman
From Opening Remarks, April 25, 1985
Washington, D. C.

"I am pleased to welcome you this morning to the fifth symposium in our cross-country series of meetings that celebrate the centennial of electrical engineering at Cornell and in the United States. This coming June will mark the 100th anniversary of the awarding of this country's first electrical engineering degrees by MIT and Cornell. We are particularly happy, therefore, that Dr. Gerald Wilson, Dean of Engineering at MIT, will participate in today's discussions.

"I believe that Electrical Engineering, as a profession, has been spectacularly successful in carrying out its mission to find ways to apply electricity for the benefit of society. In almost all aspects of our daily lives, from access to worldwide communication, to vehicular transport, to production of food and services, and even to entertainment, electrical engineering has played, and continues to play, a prominent role. (In reference to communication, it is interesting to note that today, April 25th, is the 111th anniversary of the birth of Guglielmo Marconi in Bologna, Italy). In recent years, our abilities to deal with complexities in science, technology, and the marketplace have been greatly enhanced by the advent of the computer, inherently an electrical engineering development.

"Applications of electricity are centrally dependent upon the continuous availability of electrical energy. It is entirely appropriate, therefore, that this meeting today will concentrate on the future production and supply of our vital energy needs. I look forward to hearing from the distinguished gentlemen who have graciously consented to help us examine that future.

"It is with pleasure now that I introduce our Symposium Chairman, Dr. Donald M. Kerr, who obtained his undergraduate and graduate degrees from Cornell, all in electrical engineering. As a student he was a National Merit Scholar, a Ford Foundation Fellow, and a James Clark Maxwell Fellow. After obtaining his Ph.D. degree in 1966, he joined the staff of the Los Alamos National Laboratory and held various technical and administrative positions at the Laboratory and in Washington, D.C. until 1979 when he became Director of the Laboratory. He has served on many research and development committees and advisory boards in the fields of energy, plasma physics, atmospheric physics, and national security policy, and has contributed research papers in all of these areas".

–Professor Joseph M. Ballantyne, V. P. for Research, Cornell University
From Introductory Remarks, April 24, 1985
Washington, D. C.

Editor's Note: In 1985, Dr. Kerr left Los Alamos to assume his present position as Senior Vice President of E. G. & G. Inc. in Wellesley, Massachusetts.

"All of us present today who have worked together on energy-related problems have observed the considerable development of the field of energy over the last decade. It is certainly important at this time to take stock of what we are doing. In pursuing this task, our speakers will develop one common theme: the need for stability of purpose in the conduct of long-term energy endeavors, thereby providing an attractive scientific and technological environment in the future that will move students to concentrate their study and research in this field.

"In this city it is always appropriate to begin with institutions and then move on to process, so we will start with a discussion of international collaboration in energy, an area in which our government has developed considerable interest in recent years. We have ample evidence of the great intellectual and technical benefits to be gained from these activities although there are some cynics who might not agree. There is a particularly compelling need to share energy technology with other countries that are not industrialized, as in Latin America, because of the need to displace dollar-supported oil imports with indigenous resources.

"Following these international concerns, the panel will concentrate on our national interests in the future of oil and gas production, the prospects for fusion power, and the trends in electric-power technology from the viewpoint of the Electric Power Research Institute. Attention will then be given to one of the most important topics of all: the encouragement of interest in, and the education of, the people who will make these wonderfully challenging energy expectations come to pass. Finally we will have a review of the part that the U. S. Department of Energy is playing in the development and coordination of all of these activities.

"At this point I wish to thank Ravi Sudan for his outstanding efforts in the planning of this meeting and arranging the program. I also thank our distinguished speakers who have consented to give so generously of their time today to this broad discussion of the whole field of energy."

–Donald M. Kerr, Symposium Chairman
From Introductory Remarks, April 25, 1985
Washington, D. C.

INTERNATIONAL COLLABORATION IN ENERGY RESEARCH AND DEVELOPMENT: AN IMPERATIVE OR AN ILLUSION?

Eric H. Willis

Eric H. Willis received the B.Sc. degree from the University of London in 1947, and the M.A. and Ph.D degrees in Geophysics from Cambridge University in 1958. From 1957 he was Assistant Director of Research for radiocarbon dating at Cambridge University until his appointment as Vice President and Director of Research for Teledyne Isotopes in 1964. He joined the Advanced Research Projects Agency in 1970 as Director of Research in nuclear monitoring and remained in that position until 1975 when he became associated with U.S. ERDA, first as Director of the Geothermal Energy Division, followed by appointment as Assistant Administrator of Institutional Relations. In 1977 he became the Deputy Assistant Secretary of Energy Technology of the U.S. Department of Energy and held that post until 1980, when he was appointed Director of the Office for Energy Research and Development of the International Energy Agency (IEA) in Paris, France, his present position. Dr. Willis has performed research in radiocarbon dating, atmospheric chemistry, seismic verification problems associated with underground-nuclear-test detection, and exploitation of geothermal energy from hydrothermal, geopressured, and hot-dry-rock resources. He received the Meritorious Civilian Service Medal from the Secretary of Defense in 1974. Dr. Willis is a member of the Geochemical Society and of the American Geophysical Union.

Editor's Note: In 1985 Dr. Willis left IEA to become Principal Partner in the Meridian Corporation, a management and engineering consulting firm in Alexandria, Virginia.

INTRODUCTION

Much has been said, and even more written, on the subject of international collaboration in energy research. It would seem such a natural thing to pool resources, share costs and reap mutual benefits in the common good, namely of achieving energy security in the industrialized economies and developing countries alike. In this way each country might reap the greatest gain at the least cost. While this simple thesis drives what international collaboration exists, it can hardly be said to have reached the dimensions of a contagious and compelling movement. Clearly, there are factors at work which, for one reason or another, render the thesis either unattractive as a viable proposition in itself or as a weakening of national competitive positions. A conflict thus emerges. On the one hand the need to pool resources in a period of almost universal financial stringency appears to render international cooperation an imperative – on the other hand the impediments to its realization could render it a cynical illusion. The present time becomes a time of choice.

A few minutes reflection on the course of collaboration since the first oil crisis of 1973 is instructive. There are several international fora for such collaboration, some dealing exclusively with single technology areas such as the Nuclear Energy Agency (NEA), others dealing with relatively unsophisticated renewable technologies for developing countries (UNDP). More recently the Economic Summit process has embraced the pursuit of high technology, specifically fusion energy. Since only the International Energy Agency (IEA) has tried to embrace the whole spectrum of energy technologies as a basis for international collaboration, its experience over the past ten years is particularly revealing in how the process has worked in varying political and economic conditions.

THE FIRST PHASE - 1974 TO 1980

The IEA was formed by a political act in 1974, namely the International Energy Program, and it has remained primarily a political entity since that time based upon the premise of oil sharing in an emergency. Research and development activities carried out under its aegis inescapably carry with them a certain political connotation, although the political scene has certainly changed in a fundamental sense since the inception of the Agency.

Initially, there was a desire to achieve visibility for its activities to demonstrate the ability of member countries, acting in concert and showing collective cohesion, to tackle the problem of obtaining alternative energy supplies through technical innovation. It was hoped that all countries large and small would be able to match short-term commitments in emergency oil-sharing with the long-term interest in advanced technology. In short, it was an act of political faith, untested by the slings and arrows of outrageous fortune of vested interests, national or industrial.

The venture got off to a good start. The political will to achieve success ensured that capable individuals with considerable drive and insight were put to the task by member governments. Further, the ability to attract funding of significant proportions was seldom a problem. National budgets for energy R&D were increasing fairly rapidly, and funds earmarked for international activities posed no threat to the funding of domestic programs. This latter point proved to be critical later. The United States in particular was poised at the beginning of a major domestic energy R&D program after the demise of the Atomic Energy Commission gave birth to the Energy Research and Development Administration. The considerable U.S. investment in a wide spectrum of energy R&D provided, in turn, a major impetus for IEA collaborative ventures, many initiated by the U.S. which was often the major financial contributor. This support had the beneficial effect of allowing smaller countries with limited scientific infrastructures to benefit from participation in projects they could not otherwise have contemplated. In this manner, a portfolio of some 50 projects was brought into being, ranging from a few million dollars each to the 70-80 million dollars represented by the Almeria Solar Power Tower project in Spain and the Grimethorpe Pressurized Fluidized Bed project in England. Country participation was broad and the overall quality of the programs was generally good despite the general perception that politically driven technical ventures have inherent built-in constraints, particularly if political agreements have not been matched by the enthusiastic support and vested self interest of the technical community.

THE SECOND PHASE - 1981 TO 1982

The latter half of the 1970's saw the second major rise in the price of oil, giving further impetus to energy R&D and in particular to a drive to secure alternative sources of energy by the early commercialization of synthetic fuels. "Commercialization" was the keynote of this period, without perhaps there being a real understanding of the nature of the marketplace and its receptivity to new technologies. Nevertheless, conventional wisdom saw a steadily increasing oil price for the foreseeable future and a sanguine, even naive, view of the incipient competitive position of new technologies.

For a number of reasons, this euphoric investment in the commercialization of new technologies was short lived. A certain cynicism was being evoked by huge cost overruns that were becoming all too frequent. The magnitude of these overruns was such that it cast grave doubt on the adequacy of the technical process and its planning. Of particular note was the cancellation of the Solvent Refined Coal Project (SRC-II) in early 1982. This cancellation stemmed primarily from the philosophy of the new U.S. Administration which held, reasonably, that government should be non-interventionist and that market forces would themselves establish the priorities for the energy supply/demand mix. Cancellation of the project would probably have occurred anyway as the three participants, the U.S., Japan, and Germany, contemplated the fourfold rise in their obligations in a more sober light. Nevertheless, the cancellation heralded a new dimension in the progress of energy R&D – one of caution in embarking upon international commitments. Soon to be followed was a general and major de-escalation of the U.S. Government energy programs, leading inevitably to less U.S. money being available for joint ventures in the international field.

There are some serious lessons to be learned from this period, especially from abrupt and unheralded cancellations of prior commitments. For whatever good reason and however inevitable, unilateral non-negotiated actions by a Member Government affecting the programs of others can have far-reaching and totally unintended results. International collaboration, experience now tells us, is not an instinctive reaction of governments. Less so is it the reaction of bureaucrats, legislators, and program managers with domestic constituencies to satisfy. International collaboration, and the climate in which it flourishes, is a delicate fabric woven laboriously and with much care. It takes years to create but can be shattered in an instant. Thus the real casualty of the SRC-II episode lay not in the demise of a large project, per se, but in the tarnishing of the climate of confidence which had been a necessary prerequisite to negotiate the complex agreement in the first place. The critical value of an international framework for negotiating collaborative ventures is often badly underestimated – the subtlety of the necessary climate of confidence rarely appreciated.

At first, the downward trend of the U.S. Energy Program seems to have caught most other IEA countries unawares, particularly since it was such a direct reversal of the interventionist policy practiced by its predecessor administration. With the principal pay master sitting on the side lines, the other players were understandably bewildered. At that time, in early 1981, the effects of recession, although talked about, had not really bit into western economies. Soon, however, other worries replaced the realization that the U.S. was no longer the locomotive driving the collaborative research program of the IEA by virtue of its ubiquitous funding. The emergence in other Capitals of budget constraints caused by recession and increased welfare payments due to rising unemployment, was forcing a re-thinking of both domestic programs and international commitments alike.

Thus in late 1981 and into 1982, a perceptible slackening of the momentum for collaborative projects was in evidence. Projects tended to become either task-shared in character rather than jointly funded, and smaller in scale. The larger projects, $50 million or more, found little support for planned second phases, and instead more emphasis was placed on less ambitious basic science and information-type projects. Such projects are not at all without merit or even less than first rate in quality – they simply cost less in times of financial stringency. International collaboration was passing into the doldrums, and moving, perhaps, to its lowest point.

THE THIRD PHASE - 1982 TO 1984

Towards the end of 1982, the combination of conservation measures, recession, a mild winter, oil company stock draw-downs, and the development of indigenous oil supplies within the Organization for Economic Cooperation and Development (OECD) countries, had reduced demand for imported oil to the point where the world price of oil could no longer sustain its $34 per barrel floor price. Since U.S. inflation had already caused a 15% drop in its real price since 1979, the $34 per barrel of 1983 now represented a decrease in real terms – its drop to $26 per barrel had the further effect of reinforcing the inflation-related drop in real terms. Thus the most effective stimulant to energy R&D, the high price of oil, was considerably lessened. It should be noted, with a certain irony, that the effective price of oil still remained at a high level which would have been considered catastrophic at the inception of the IEA's energy R&D programs – such is the capacity of society to adjust. Indeed, for the European members of the IEA, the price of oil was 70% higher than in 1980 in terms of European currencies due to the dramatic change in dollar exchange rates, since oil prices continued to be denominated in dollars (see Figure 1). For Europe at that time it seemed ingenuous to suggest that the incentive to use energy more efficiently had been diminished by the slackened oil markets – for many countries the real cost of oil imports had increased substantially with consequences both to their balance of payments and their economies. Nevertheless, the psychological effect of the nominal drop in the dollar-denominated oil price, a resilience to high prices, and a common perception that the price would stay at or around the current apparently affordable prices for the foreseeable future, made the search for alternative energy sources less urgent than they had seemed only four years previously when the perception of a high oil price, climbing rapidly higher with each passing year, had caused the search to become almost frenetic.

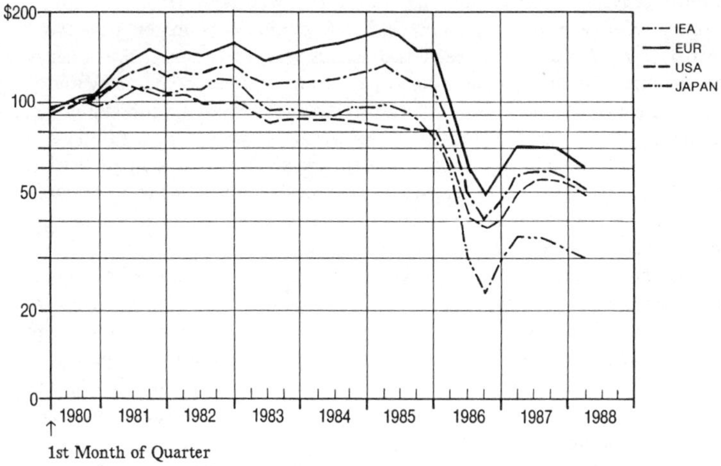

Source: *Energy Prices and Taxes*, International Energy Agency, First Quarter, 1988.

Figure 1. Index of Crude Oil Import Prices in National Currencies (1980 = 100)

THE FOURTH PHASE - 1985 TO THE PRESENT *

The following years saw the single most important boost to the World Economy in ten years – the drop in the world price of oil. In particular, for European economies that had labored under the twin burdens of $34 per barrel oil *and* an overpriced U.S. Dollar, the relief was reinforced by the precipitous drop in the value of the dollar starting in February, 1985 and continuing through late 1987. The end of the Gulf War in August, 1988, and the need for both Iran and Iraq to obtain revenues to restore their economies, probably implies that a regime of lower oil prices will prevail in the short term into the mid 90s.

Energy Technology R&D, under severe financial stress between 1982 and 1984, would have probably seen a dramatic extension of its downturn since its

*Editor's Note: In view of the dramatic changes that have occurred since April, 1985, Dr. Willis has added this updated section to his discussion and has made appropriate changes in Figure 1.

erstwhile driving impetus, the high oil price, was now largely removed. Strangely, energy research has not suffered as much as might have been expected if its fortunes had been tied exclusively to the oil price. Energy research is now being driven as much by environmental concerns, particularly those associated with acid rain, the carbon-dioxide-induced greenhouse effect, and the effects of the Chernobyl nuclear accident, which have provided a fresh impetus to certain segments of the energy R&D portfolio. In particular, the need for the clean use of coal has caused considerable sums to be invested in the search for remedial technologies. The greenhouse effect is a haunting spectre whose existence has yet to be demonstrated while its inevitability seems accepted, and it threatens to be both significant and widespread in its climate implications. We are faced with a cruel paradox – on the one hand we may want to burn coal cleanly, and on the other hand we may not want to burn it at all. A purely technological, and some might even say logical, response would indicate a resurgence of the nuclear option. Unfortunately, the aftermath of Chernobyl together with the unresolved nuclear-waste issue have further dampened its appeal. Thus the dilemma will continue for some while – the probable but undemonstrated irreversible climatic changes *versus* an unpopular non-fossil-fuel energy source. Unfortunately, publics seldom react to undemonstrated and intangible threats until they materialize. Only the clear and unequivocal manifestation of the effects of increased carbon-dioxide concentration will suffice to change present prejudices against nuclear energy – the tragedy could well be that by the time the climate effects are clear enough for their cause to be recognized, technical fixes will be unable to stem the tide.

AND FROM NOW ON?

While the overall energy picture has undoubtedly changed significantly in recent years, and generally for the better, one important fact remains undiminished. Concerns for energy security and the drive to achieve it have not been dispelled – the problem of energy security remains with us. It remains also true that energy technology will contribute materially to attaining this security and with it the economic prosperity objectives of IEA Member governments. Technology applications can provide diversified sources of energy supply and also permit that supply to be used most effectively. It is inherently a long-term effort, and long-term needs cannot be overly subjected to short-term considerations if the potential contribution of energy technology is to be realized.

Long-term energy technologies, however, require sustained effort and continued financial support – some of this support could involve significant outlays. Unfortunately, the need for sustained effort comes at a time when present trends in national budgets indicate that financial support is unlikely to increase, and may even decrease. Thus a real conflict is arising between continued need for energy technology R&D on the one hand, and financial stringency and lack of immediate incentive on the other. It is the thesis of this paper that this conflict can be resolved in part by achieving the same or greater technological and energy security results at reduced cost through international collaboration.

Although the confluence of events points to the need for increased collaboration, such collaboration within the foreseen budgetary constraints implies political decisions that cannot be contemplated by the technical community alone. Technical need, in and of itself, is not a sufficient driving force to achieve the level of collaboration needed to resolve the conflict referred to above – it must be reinforced by the decisions of policy makers at the highest level. Because the conflict is unlikely to recede but actually promises to become more acute, it follows that now is the propitious time to consider a new and well-directed thrust for international collaboration with political-level support. It is an opportunity which may not present itself for long.

Lest one get prematurely and unduly euphoric at the prospect of solving financial and technical problems by extending hands across the sea, it has to be acknowledged that international collaboration has some obvious and practical drawbacks that require additional management skills, careful planning of the project from the outset, and appropriate contractual arrangements to reap the very demonstrable advantages to be gained. It is important to recognize the increased chances of success the further the technology is removed from the competitive inhibitions of the marketplace. This is particularly the case for ventures involving basic science and technologies with no immediate commercial application, where patents, proprietary rights, and other commercially competitive factors are not inhibitors.

International collaboration can only thrive if it is firmly grounded in, and is subordinate to, the process of national decision-making. This must be a basic premise if the suggested process of early consultation is to succeed. A corollary to this statement is that budget savings can only be realized if the international collaboration is in consonance with national objectives. It also follows that not all countries will wish to participate in every collaborative venture, even when and if they have been party to the consultation process. The size of an economy, and thus its technical capability and need, will determine the nature of its collaborative interests – this is particularly true for smaller economies which, by virtue of resource limitations, are not in a position to contemplate large-scale and expensive research facilities even in the absence of the current financial stringencies.

Collaboration works best when national needs are brought into confluence and relate to a specific technical area – the process of early consultation can help identify such areas in which the confluence of ideas promises success. In this way, the expression of national need (sometimes expressed by the news media as the "top down" approach) is reinforced by specific ideas and initiatives from the technical community (the "bottom-up" approach).

What then should be the criteria for international collaboration? It should be seriously contemplated if:

- reduction in technical and economic risk is evident to each participant;
- cost to individual national budgets is lowered, permitting areas of research to be pursued that would be otherwise unaffordable;
- duplication of effort is limited without suppressing competition;
- approaches to technical problems are focussed on broad intellectual efforts that make full use of available scientific talent;
- progress is made possible in technologies where "next step" costs might preclude proceeding at all on a national level;
- problems with unique transboundary implications can be tackled comprehensively.

MODALITIES OF COLLABORATIVE ACTIVITIES

In what form can ventures be put together on the international scene? The combination of possible modalities seems, at first sight, to be both endless and daunting. Nevertheless, they fall into some well defined pathways. They are:

(a) *Bilateral:* Where two parties agree, usually to the exclusion of others, on a course of action mutually beneficial to them, and generally focussed upon specific research topics or information exchange. Bilateral agreements are easier to implement since concerns of the two parties are the only ones to be considered in negotiations.

(b) *Multi-bilateral:* Where one country, usually with large and attractive resources, executes bilateral agreements with a number of other countries in the same technical area or project. Retention of project control effectively resides with the initiating country and the other partners may or may not interact with each other in the process. This is sometimes known as the "hub-and-spoke" model – it is useful when the relative contributions of the partners to the project are more disparate.

(c) *Interconnected Multi-bilateral:* Where the different bilateral parties establish relationships among themselves vis-à-vis each other or vis-à-vis the initiating country. This could be viewed as adding a "rim" to the "hub-and-spoke" model.

(d) *Multilateral:* When a group of countries interested in a particular technical area or project agree, on a broadly equal basis, to pursue a joint activity either *ad hoc* or within the framework of an existing international body such as the IEA. In the latter case, the framework of existing relationships and commitments could facilitate negotiations. One or more partners may be designated "lead countries" by the others, but overall management is exercised broadly by a body appointed by the partners for that purpose. This mode is the one generally adopted in the IEA. While subject to certain disadvantages inherent in multilateral undertakings, experience shows that potential problems can be accommodated by careful preparation and attention to detail in the contractual and management arrangements.

Other modalities could be devised. The modality best suited to perform a given R&D activity (program or project) will depend greatly on its nature and size, and the number and capacities of potentially interested parties.

Apart from the modalities of association, there are two categories of arranging the funding. They are:

(a) *Task-shared projects:* The parties enter into an agreement in a specific technical area whereby each party undertakes to conduct one or more of a closely coordinated series of activities as a contribution to a concerted plan of research so that:

- The task of each participant is clearly delineated and is part of an integrated whole. Duplication of effort is thus avoided.
- Consensus is obtained through a shared view of objectives, goals and means, but each country is free to pursue its particular task(s) under its own control;
- Budgetary expenditures continue to reside in the country of origin and no funds are transferred.
- Complications of exchange rate fluctuations are avoided.
- Experience in operations and research results are shared by all parties – the whole becomes greater than the sum of the parts.

(b) *Joint-funded projects:* The parties agree on one single project and its technical objectives, and agree to share the costs. Certain principles have to be understood and agreed to by the parties, namely:

- National governments must be prepared to accept the possibility of a net outflow of funds and technology – the principle of "juste retour" is not an inflexible rule but neither can it be entirely ignored.
- Commitments once made must be honored, together with continued access to markets.
- Arrangements that are most effective, inexpensive, and equitable start by participation of the parties at the conceptual and design phases, not by invitations later to participate in "fait accompli."
- Principles of proprietary rights and licensing arrangements must be carefully prepared. This becomes particularly sensitive when more immediate commercial advantages are perceived.
- Designation of a lead country can be made by the participants for a given project, but as a general principle an equitable distribution of project leadership among partners must be evident.

TECHNOLOGIES FOR COLLABORATION - PLACING THE BETS

Having discussed the vicissitudes of international collaboration and the criteria for its use, one then has to grasp the nettle firmly – collaboration cannot be discussed in the abstract – there comes a point when one has to be specific. What are the technical areas which could lend themselves to international collaboration, in one or other of the forms described previously? The IEA has recently completed an Energy Technology Policy Study in an attempt, *inter alia,* to make a systematic analysis of the rationale that is needed for developing specific technologies, and the way in which they lend themselves to international collaboration.

Coal resources will outlast all other fossil-fuel resources – they are more extensive and are better distributed within IEA economies than all other fossil-fuel supplies. Coal is a resource both as a combustible fuel and as a feedstock for the petrochemical and synthetic-fuel industries. By comparison with oil technologies, its use has been relatively unsophisticated, its R&D relatively neglected. If it is to realize its full potential in IEA economies, its combustion or conversion must be achieved so that pollutants are minimized. Not only must the environment be safeguarded but it is important to ensure that the pollutant-removal process before, during, and after combustion imposes a tolerable financial burden on society – it must be affordable and easy to maintain. Combustion processes should minimize the formation of oxides of nitrogen and sulphur. Governments should give the highest priority to the remedial technologies for air-pollution abatement. Basic coal sciences are increasing in sophistication, and as they do so they will open up a wider range of possibilities for its use. Productivity, safety, and efficiency of mining operations also remain key factors in the use of coal. Thus it seems profitable to develop coal research towards:

- enhancing the fundamental knowledge base resulting from years of relative neglect of basic coal science;
- understanding the phenomena of combustion to improve efficiency and address the formation of pollutants and the development of more advanced total-control technology;
- improving the economics of techniques to utilize coal more efficiently and cleanly;
- developing utilization and/or disposal techniques for waste generated by pollution-abatement measures; and
- advancing mining technology to ensure improved safety and efficiency of production.

The abundant resources of coal can further be used in the production of synthetic alternatives to oil and gas through the various technologies for coal liquefaction and gasification. In the development of these technologies, attention must also be paid to ensuring environmental acceptability. An abundant supply of fossil-fuel resources is also contained in both *tar sands and oil shales* – they are more abundant than normal oil and gas reserves but less than coal. They exist in many IEA countries. While interest revived in these resources in the 1970's, it diminished in the 1980's under the growing realization that the product would be expensive at a time when oil prices were not anticipated to continue rising. Thus governments face the dilemma posed by *synthetic fuels* generally, namely how to increase the viability of the option presented by these resources through well-directed research leading to a firm and sophisticated technology base without necessarily proceeding to the commercial or even demonstration scale. This is a major policy challenge. It follows, then, that the priorities in this area are:

- achieving further understanding of basic processes (e.g. reaction kinetics, catalysis);
- producing durable components;
- engaging in a continuous pursuit of process technologies both of sufficient diversity to permit use of synthetic fuels products and to a scale adequate to permit decisions based upon commercial considerations;

- deriving advanced cost-estimation techniques to increase the reliability of projections for future investments;
- overcoming phase separation and engine corrosion from the combustion of alcohol;
- developing new gasifiers, reactors and underground-gasification technology to enhance both the technology and the economics of coal-conversion processes; and
- devising hybrid systems, namely, coal and natural-gas feedstocks and coal conversion/electricity production.

All of these areas, besides being priorities, could lend themselves to international collaborative ventures.

Oil and to a lesser extent *gas* continue to dominate energy use in IEA countries and the vast infrastructures in place for producing, transporting and marketing them underline the likelihood of their ongoing importance and the necessity of discovering and producing these resources to the maximum extent. Improved exploration and drilling techniques and, particularly for offshore resources, advances in design and maintenance of offshore structures are necessary to permit fullest resource recovery. Enhanced oil recovery (EOR) technologies make possible the production of substantial additional amounts of oil, not extractable through conventional methods, from existing oil fields by injecting steam, CO_2, gas, and flooding agents into the oil-bearing deposits. While responsibility for the development of these technologies has historically been that of the producers, governments are increasingly aware of the impact of EOR development on the formulation of their depletion and oil-leasing policies and on security of oil supply. As a practical matter, the research opportunities involve:

- further development of the basic exploration sciences, drilling technology, reservoir technology and enhanced oil recovery;
- improved design, maintenance and inspection of offshore structures, and associated distribution systems;
- further research on the identification and extraction of unconventional gas resources; and
- improved techniques for health, safety and environmental protection.

International collaboration is much more difficult in this domain, dominated as it is by large-scale commercial interests. Nevertheless, governments could pursue collaboration in their own interests, including the basic technology areas.

Since *nuclear power* for IEA countries as a group is a key ingredient in a balanced electricity supply system and potentially as a source for district heating, thereby further reducing dependency on oil and gas, its degree of penetration into the market is a source of continued concern. Concerns about the health and safety aspects of reactor operation, waste management and ultimate disposal have led, in some countries, to fears that it will be unable to play the full role its potential had previously indicated. Its failure to do so could have serious consequences. These concerns can only be assuaged by a continuing and convincing record of both safe operation and sustained progress on the major issues of public concern translated into appropriate legislation and regulation. Public confidence in the ability and will of their governments to protect their interest in the health and safety aspects of nuclear power will as much determine its future market penetration as will its economics. Governments will have to decide and begin to take action on development of national nuclear waste-management strategies, giving consideration to an international component. Further developments are also expected with respect to advanced thermal reactors, including compact light-water-reactor core designs, and an improved high-temperature gas-cooled reactor (HTGR), all of which may lead to improved cost and operational performance for the next generation of nuclear power plants. Thus the priorities would include:

- a comprehensive program to resolve the remaining issues in nuclear-waste management;
- continued improvement in reactor safety;
- examination of ways to improve the next generation of thermal reactors, such as through improved fuel efficiency, improved plant reliability, enhanced inherent reactor safety, and facility standardization;
- an evaluation of the merits of developing advanced thermal reactors, including the merits and feasibility of developing "compact" simpler, more forgiving and shop-fabricated plants that might be attractive to certain operators; and
- development of new designs and associated technology, and analysis of economic and market potential for the HTGR.

International arrangements are often likely to be bilateral, although the NEA (Paris) and the IAEA (Vienna) conduct multilateral research programs.

Breeder technology holds different prospects for different countries. For some countries, not richly endowed with indigenous energy resources and reliant on imported fuels, it offers the prospect for the first time of an indigenous fuel source and the opportunity to be less beholden to imports – in a word, it is a matter of their energy supply security. This is not necessarily the view of governments whose countries are better endowed at present with natural energy resources. The economies of breeder technology that currently dominate the thinking in several countries do not compare favorably with those of the current generation of nuclear reactors. Research must be directed towards lowering the cost and ensuring commensurate health and safety in the operation of breeder reactors. The prospect of extending the effective life of uranium over many hundred years, the avoidance of vulnerability to imports and the possibility of reduced generating costs will, however, continue to make breeder-technology development an attractive option in many countries. It is prudent, therefore, that breeder research:

- moves toward the technology required for demonstrating commercial-scale operation at reduced capital costs;
- addresses the problems and economics of reprocessing of highly radioactive spent fuel.
- explores the economic potential and feasibility of new reactor concepts, as well as innovative technical approaches; and
- continues to be oriented to close collaboration, pooling of resources, exchange of experience, and the development of common safety criteria at the international level.

Fusion power, unlike conventional nuclear power or even breeder technology, remains as yet unproven as a controlled energy source. The achievement of a sustained fusion-reaction through intense magnetic confinement remains the future challenge for fusion-energy research. Once a sustained thermonuclear reaction has been achieved, further major challenges will be faced in successfully engineering the technology to the point of practical application. While the current machines represent important steps along the road, they will need to be followed by other equally important steps in the next decade before it can be said with conviction that an energy option is truly available. Decisions on the next steps, based on the lessons from the present research, will require much thought and debate. Governments will be faced with forceful arguments and difficult choices involving large investments. They should not however lose sight of the immense potential this technology offers even in light of the imponderable obstacles and the long time frame that confronts them. The technical evidence presently points in the direction of success, and to do other than to support the technology to the point of proof of technical feasibility would not do justice to future generations. The point of proof will be roughly at the end of this century: governments should not reduce their support at this time below a level where this proof cannot be reasonably achieved. They should also consider, on an urgent basis, the possibilities afforded by enhanced international co-operation to reduce their financial burden and utilize resources more efficiently. Questions of prime importance in the fusion field are:

- determination of the direction that the "next step" should take as a logical progression from the present large tokamak machines;
- development of supporting technologies, such as superconducting magnet research, materials technology and tritium handling; and
- provision of active opportunities for international collaboration to share costs, risks, and benefits of investment in new large and costly experiments and supporting technology development.

Renewable Energy Sources may be extremely important regionally, especially for those economies that possess few other indigenous resources with the exception of well-established hydropower, fuelwood and geothermal. They do not promise, however, a major contribution to the energy supply/demand balance within this century when considered over IEA countries as a whole. Taken together, the full range of renewable energy sources represents a very large potential – if only a small fraction of this potential were to be realized, the contribution would materially affect the energy supply mix of IEA countries. The financial structure of the present renewables industry is typical of a new and emerging industry – its capacity to engage in R&D is quite limited and many of its current products are not economic without favorable incentives for introduction. The past ten years of experience have largely narrowed the range of technologies being pursued to the most promising, and governments should support those technologies that can harness their natural resources of sun, biomass, wind, wave, and geothermal energy reliably, economically and in an essentially benign manner. The long-term nature of this endeavor should be re-emphasized, and sustained government support for its development will be required to realize its potential – the corollary is that the investment should continue to be sharply focussed. It can be said then, that renewable resources:

- represent a very large potential. While some technologies have reached commercial maturity (solar heating, wind, biomass, and geothermal), those whose contribution is expected to be largest (solar photovoltaics) will need long-term development;
- would benefit from government-supported R&D to address critical cost reduction and reliability problems;
- need better data on locally available resources and an improved understanding of end-use patterns and their adaptability to renewable energy possibilities;

- may need some development of effective storage systems as part of a total system concept; and
- could make important contributions to regional development.

The "*technology base*" embodies those scientific principles and interdisciplinary issues that nourish the development of ideas for later specific technology applications. While such research has proceeded vigorously in the nuclear fission and fusion sectors, concomitant efforts in the basic sciences related to coal, energy-saving, and renewables have not been forthcoming. The process of combustion, for instance, is a highly complex mechanism that was poorly understood before the advent of modern diagnostic techniques born of sophisticated laser technology – the consequences of this understanding will be far reaching. There is a similar need to improve the understanding of the behavior of materials – in many cases technologies are limited in their progress simply by the lack of appropriate materials. The pay-off in one sector from an investment in another is a recurrent feature of modern science. The investment can be relatively modest, but the leverage is great. Governments should ensure that the natural desire to see a technology enter the marketplace does not divert attention from the continued generation of those ideas from which technological development stems. As a general point, it must be stressed that:

- Research and science-policy funding generally should reflect the importance of basic energy science, including materials science, reaction kinetics, geophysics relating to coal and oil and gas structures, and biogenetics.

ELECTRICAL SYSTEMS TECHNOLOGY

- Electrical technologies should focus on improvement in the efficiency of conversion, transmission, load-levelling, and storage.
- Improvements in electricity storage must include improved batteries, which will influence the further deployment of some renewable energy technologies and electric vehicles.

CONCLUSION

We have seen that there is no lack of technical opportunities that satisfy the twin requirements of technical urgency and international interest. As national budgets have shrunk, and are still shrinking in some countries, the critical choice to be made is between proceeding on an international basis for some technologies or facing the very real possibility of not proceeding at all.

Fusion is clearly a case in point. Making choices in this field is difficult since there are currently three excellent programs running in parallel in Japan, Europe, and the U.S. National prestige and the national need to enhance expertise are all involved. But one must seriously wonder on what basis either group can go it alone to the next step, or how research can be funded at its current levels for the next forty years before there is a glimmer of power on line. Society cannot lose faith in this promise for the future.

The past ten years of IEA activities have generally been in response to *ad hoc* stimuli and represented targets of opportunity, and the results, on the whole, must be considered successful. It is questionable, however, whether the *ad hoc* and relatively modest approach of the past, and the existing level of collaboration that resulted, is sufficient to meet the changing circumstances of today and will build a foundation for the future. What appears to be needed now is a more efficient process to plan and coordinate R&D activities taking place bilaterally among IEA countries. This could lead to genuine collaboration and execution in a systematic fashion. Such collaboration can be particularly suitable when it:

- increases efficiency of R&D and the return on each party's investment;
- involves R&D relating to technologies with trans-national implications, particularly in the environmental, health and safety areas, such as the short-term problems of airborne pollution and long-term effects of increased atmospheric carbon dioxide;
- depends upon a common understanding of the technical imperatives and leads to the identification and a definition of future experimental needs;
- permits proceeding with expensive and/or high-risk new undertakings that would be difficult or impossible for individual countries to do on their own;
- allows more rapid technical progress than otherwise possible.

The prospect of doing more together than could be done alone clearly contributes to the concept of energy security for all the IEA countries. This prospect becomes the imperative in a shrinking world – narrow self interest and sheer inertia could render it an illusion. The stakes are high, and the time-window for action could be short. The first steps are clear – they comprise the dual activities of joint planning of future research needs and at the same time the generation of a climate of confidence. Without either, we will live with an illusion.

THE DEVELOPING COUNTRIES AND THE CONTINUING ENERGY CRISIS

Jack Vanderryn

Jack Vanderryn received the B.A. and M.S. degrees and the Ph.D. degree in Chemistry from Lehigh University in 1951, 1952, and 1955 respectively. Following his graduate study he was an Assistant Professor of Chemistry at Virginia Polytechnic Institute until 1958 when he joined the Research and Development Division of the U. S. AEC Oak Ridge Operations Office. In 1962 he was transferred to AEC in Washington, D.C. where he held a series of technical and administrative positions within that organization and its successor, ERDA, until 1977. From 1967 to 1971 he was scientific adviser of the U. S. Mission to the International Atomic Energy Agency in Vienna, Austria. In 1977 he joined the U. S. Department of Energy as Director of the Office of International Research and Development Programs. In 1982 he assumed his present position as Agency Director for Energy and Natural Resources of the Agency for International Development (AID). Dr. Vanderryn's research interests include: science policy development, energy technology development, international cooperation in energy research and development, and administration of research and development.

INTRODUCTION

Energy is a key ingredient for the economic development of the world's poorest nations. With high population growth, increased energy use in economic activity, and the shift from traditional non-commercial fuels to petroleum and electricity, demand for energy in developing countries is growing faster than anywhere in the industrialized world.

Between 1980 and 2000, the populations of countries benefitting from the U.S. foreign assistance program will continue to grow rapidly, on average over 2%, and in some of the poorest African countries, over 4% annually. The urban population is growing twice that fast. In 1982, roughly one in four persons in the Third World lived in a city; by 2000 that percentage will rise to one in three. This growth will require 8 to 12% more in fossil fuels per year as compared to virtually zero growth in the U.S.

Most of this increasing least-developed-country (LDC) demand for energy will be met by imported oil in the near- and mid-term since traditional fuels, such as wood and agricultural residues, are becoming increasingly scarce and development of increased supplies of indigenous conventional and alternative energy supplies require more time and capital to develop. Fuel and power-distribution systems are inadequate, and significant supply from alternative fuels requires long lead times, improved infrastructure, stronger national commitments, and increased private sector involvement.

The pace of exploration for oil in the oil-importing developing countries relative to the rest of the world's exploratory activity has declined in recent years. Oil exploration in these countries became more concentrated in those already producing petroleum, especially Argentina, Brazil and India. Exploration has been hampered by political uncertainties, economic policy disincentives discouraging to investment by the multinational companies, and lack of local infrastructure.

To pay for the imported energy, developing countries are diverting scarce capital and foreign currency from other urgent investments. This policy, coupled with servicing the debt crisis, leaves little for investment in much needed increased agricultural production (especially in Africa), industrial development, and other critical development needs. And, while the U.S. is now paying less for oil, the developing countries continue to pay more. Because of declining exchange rates relative to the dollar, energy prices in local currencies of developing countries are still increasing, not declining. It is expected that oil-importing LDC's will spend $75 billion to import fuels in 1985, up *50%* from $50 billion in 1980. For example, since 1981 the real price of fuel oil has increased 40% in the Sudan, 140% in Peru, and 170% in Ecuador. Twelve months ago, Jamaica paid J$86 for a barrel of oil. Now, with 32% devaluation of their currency, they pay J$126.70 per barrel. A January 1985 attempt by the Jamaican government to pass on some of their higher oil cost to consumers, led to widespread rioting. Many other developing countries face similar problems.

Too large a share of developing countries' foreign exchange is drained off to pay for imported oil. In 1980, 22 countries assisted by the U.S. Agency for International Development (A.I.D.) spent over 40% of their foreign exchange earnings to pay for oil, 25 spent between 20 and 40 percent, and 10 countries spent less than 20 percent. Regretfully, the situation is no better today.

This energy crisis provides a series of challenges to the U.S. and the rest of the world. With respect to our own national interests we have:

(1) The humanitarian challenge to improve the well-being and earning capacity of the world's poor.
(2) The challenge to our political and national security interest from the destabilizing effect of this energy crisis on the internal economic, social and political fabric of the developing countries.
(3) The challenge to global monetary stability and economic recovery resulting from the LDC's debt crisis.
(4) The challenge to our commercial banking system with significant exposure to loans in developing countries and the challenge to U.S. corporations that are dependent on LDC markets.

Without adequate attention to these energy problems, economic growth in the Third World will be stifled. And it is well established that growth in per capita GNP and commercial energy consumption (fossil fuels, hydropower, nuclear, etc.) are inextricably linked, especially in those countries in the lower half of the per capita income scale who are recipients of U.S. foreign economic assistance. Studies have shown that a 1% rise in LDC income will require a 1.3% rise in fuel consumption; without fuel or increased energy inputs that growth will not occur.

How are we dealing with these problems? What actions can be taken in the economic sphere, using technological solutions, policy reforms, and institutional actions.

DONOR ENERGY ASSISTANCE TO DEVELOPING COUNTRIES

Numerous bilateral and multilateral donors are providing financial and technical assistance to the developing countries to support energy infrastructure development, increase the production of indigenous energy supplies, provide training and support institutional development, assist in the analysis and formulation of national energy policies and legislation, and support research and technology demonstrations and information dissemination, thereby assisting in both urban and rural development.

Official energy development assistance by all donors, except for the Communist countries, totaled $3.5 billion in 1982, the last year for which comprehensive data is available. A breakdown of these commitments is given in Table 1.

TABLE 1

ODA COMMITMENTS FOR ENERGY DEVELOPMENT, BY INDIVIDUAL DAC MEMBERS AND MULTILATERAL DEVELOPMENT FINANCING INSTITUTIONS AND OPEC DONORS, 1976-1982

	$ Million		
	1980	1981	1982
I. *DAC MEMBERS*	*2,239*	*2,111*	*2,045*
Australia	7	6	9
Austria	22	131	20
Belgium	3	8	2
Canada	93	160	88
Denmark	37	2	1
Finland	5	3	4
France	157	188	173
Germany	277	589	244
Italy	2	23	80
Japan	902	613	720
Netherlands	74	32	42
New Zealand	5	5	8
Norway	26	15	15
Sweden	44	26	30
Switzerland	2	8	2
United Kingdom	217	59	345
United States (FY)	150	154	212
EEC	(216)	89	37
II. *MULTILATERAL INSTITUTIONS*	*1,259*	*403*	*1,017*
IDA	1,101	123	801
IDB (Fund Spec. Operat.)	42	127	107
AfDB (Spec. Fund)	9	9	35
AsDB (Spec. Fund)	107	144	74
III. *OPEC*	*542*	*507*	*487*
Bilateral	434	400	320
Multilateral	108	107	167
IV. *OVERALL TOTAL*	*4,040*	*3,021*	*3,549*

Source: OECD

TABLE 2

DAC COUNTRIES' BILATERAL ODA COMMITMENTS FOR ENERGY DEVELOPMENT AS A SHARE OF BILATERAL ODA COMMITMENTS

	Percent of ODA		ODA to Energy
	1979-80	1981-82	as % of GNP, 1982
DAC COUNTRIES	*8.2*	*8.8*	*0.30*
Australia	2.3	1.3	0.06
Austria	12.4	28.9	0.29
Belgium			
Canada	24.2	13.6	0.31
Denmark	10.4	0.6	0.02
Finland	3.0	3.0	0.08
France	2.7	4.2	0.33
Germany	7.8	13.5	0.37
Italy	1.0	9.2	0.23
Japan	23.1	18.9	0.68
Netherlands	4.9	3.7	0.31
New Zealand	8.4	13.1	0.34
Norway	10.4	5.8	0.32
Sweden	4.2	5.1	0.31
Switzerland	1.6	2.2	0.02
United Kingdom	15.0	19.3	0.73
United States (FY)	3.4	3.3	0.07

The share of development assistance devoted to energy by the different donors is given in Table 2. Of the bilateral development assistance in energy, 53% goes to the low-income countries; these countries, however, received only 16% of the energy export credits provided by the 20 industrialized member countries of the Organization for Economic Cooperation and Development OECD.

TABLE 3

EXTERNAL SUPPORT FOR ENERGY DEVELOPMENT
BY SUB-SECTOR, 1979-1982 AVERAGE
Provided by OECD Countries

Energy Sub-Sector	
1. Energy planning and technical co-operation	14
2. Oil	8
3. Gas	2
4. Coal	9
5. Nuclear	-
6. Hydro-power production	28
7. Electricity transmission	19
8. Fuelwood, charcoal, geothermal, alcohol, biogas, solar	2
9. Other unallocated	17
Total	100

The assistance from OECD countries is distributed by energy sector as indicated in Table 3.

Of the 36 least-developed countries, India and Bangladesh have received by far the most energy assistance. Other countries with 1980 GNP per capita under $600 that have received significant assistance include Indonesia, Egypt, Pakistan, Sri Lanka, Liberia and Burma. For those countries with 1980 GNP per capita between $600 and $1200, Colombia, Ivory Coast, Nigeria, Philippines and Thailand were the largest energy assistance recipients.

TABLE 4

A.I.D. ENERGY PROGRAM LEVELS
Development Assistance and
Economic Support Funds
($ millions)

	FY 1984	Estimated *FY 1985*	Requested *FY 1986*
Fuelwood	21.1	14.3	25.7
Renewables	54.5	65.1	50.3
Fossil	56.9	19.9	9.5
Power	98.2	33.3	99.3
Total	230.7	132.6	184.8

Energy assistance support by the United States, through our foreign assistance program administered by A.I.D. is summarized in Table 4. In Fiscal Year 1986, the program is very heavily concentrated in a few countries – Pakistan ($140 million in FY 1986), Egypt, India and the Philippines – giving too little attention to much needed increased energy supplies for Africa and Latin America. Under the direction of the Los Alamos National Laboratory, however, we have just initiated a $10 million program of technical assistance to the Central American countries. This effort concentrates on geothermal development, energy planning assistance, and preliminary investigations of peat and other energy resources.

Most donor assistance, as indicated in Tables 3 and 4, is focussed on capital assistance for the electric power sector. A.I.D., and to some extent other donors, increasingly has emphasized the need for attacking other aspects of the problem – the need for revised national energy policies; strengthening infrastructure and institutional capability; increased training of LDC personnel; and technology transfer and research. I will discuss these areas using program and technology headings.

ENERGY PLANNING AND POLICY DEVELOPMENT

Public and private sector leaders in less developed countries have a relatively good understanding of the key energy constraints to development in their countries, but they have too little knowledge of solutions available to eliminate those constraints. Since 1973, U.S. industry has shown remarkable ingenuity in identifying our domestic energy problems and in finding solutions. This ingenuity has grown out of serious problem analysis and a careful consideration of available investment options. Energy planning and policy development assistance aims at transferring such expertise and ingenuity to LDC's.

Results of this knowledge transfer can be remarkable. For example, based on the results of an A.I.D.-sponsored analysis, the Government of Liberia closed down an inefficient major oil refinery, saving $15-20 million per year – 2% of Liberia's GNP. Government revenues from taxes on petroleum products have increased, yet consumers are saving about 10% at the gas pump. In Jamaica, an A.I.D. team discovered that 150 million U.S. dollars could be saved per year in imported oil costs (out of a total of $180 million spent) by using the bagasse from its sugar production rather than oil to generate power; Prime Minister Seaga has asked A.I.D. for a feasibility demonstration. I will return to this development later.

The developing countries need to give high priority to reforming their policies that discourage rational solutions to energy difficulties. Critical policy reform areas include energy price subsidies that inhibit development of indigenous resources and encourage waste, trade policies that discriminate against import of needed exploration and development equipment, and investment policies required to encourage formation of local and foreign capital to establish energy infrastructure. U.S. A.I.D. is providing technical assistance to help accomplish such changes. We are calling upon national laboratories of the Department of Energy and other public and private organizations in the U.S. that are experienced in these matters to participate in these efforts. To date, more than 10 countries have been assisted; models are being developed for all A.I.D.-assisted countries for use in building their own planning abilities and to better rationalize policy for energy development.

ENERGY CONSERVATION

Energy conservation is the cheapest and quickest solution to the oil import problems of developing countries. The potential for increased energy efficiency in the Third World is large and can be captured quickly. When viewed as an energy "resource," it costs less than other supply options. Conservation also yields substantial secondary benefits such as softening the impact of removing energy price subsidies. By 1990 developing countries could save 15 percent of their 1981 level of commercial energy consumption without inhibiting economic development. The industrial and power sectors are big energy users. In most A.I.D.-assisted countries, these two sectors account for 40 to 60 percent of all commercial energy used. Roughly 30 percent of all industry and power conservation measures can be implemented within 1 year, 50 percent after 2 years, and almost all within 4 years. Short-term conservation programs can achieve energy savings of up to 20 percent in transportation energy use, although increased transportation-sector energy efficiency in the developing world will not be implemented until difficult institutional barriers and vested interests are overcome.

To save a barrel of oil per year, at the current cost of $25-30 per barrel, requires the expenditure of only $12 to $15 on conservation measures; this one-time expense eliminates the fuel expenditure year after year for the life of the project. In countries with subsidized energy price structures, conservation helps to ease the transition to higher energy prices. Moreover, conservation can improve the management and hence profitability of private firms.

The conservation program at U.S. A.I.D. provides technical assistance to developing countries to identify the significant conservation opportunities, and places priority on projects that will leverage larger investments by the World Bank and other public and private financing institutions. Major A.I.D. programs are underway in Jamaica, the Dominican Republic, Central America and the Sudan. Efforts have also been initiated in Sri Lanka, Ecuador, Pakistan, Egypt and the Philippines. Over 25 other developing countries have either requested assistance or have opportunities to save considerable amounts of energy.

INDIGENOUS FOSSIL FUELS TO SUBSTITUTE FOR IMPORTED OIL

Over 50 A.I.D.-assisted countries have serious trade imbalances caused by imported oil; 13 of these countries have significant gas reserves that are either underexplored or underutilized; an additional 22 countries have substantial coal reserves that are not being used to any appreciable extent. Other countries have oil reserves that have not been evaluated or explored and a few have untapped peat reserves.

For example, Pakistan has enough coal to create a new power generation industry, thereby doubling present generation capacity, and converting the country's energy-intensive cement industry to locally mined coals instead of imported fuel. This effort would not only save billions of dollars in foreign exchange due to reduced oil imports, but would also create a new domestic industry and many new jobs. We are encouraging Pakistan to undertake this development through the private sector, rather than expanding government-owned enterprises or creating new parastatal organizations. In Thailand, A.I.D. and the World Bank are exploring the possibility of using unused domestic natural gas to substitute for imported oil for that country's exploding motor vehicle fleet. A project in the Philippines has found that using coal water mixes, derived from locally mined coal, could replace significant amounts of imported oil now used for power generation. These approaches are applicable to many other countries.

In order to utilize such approaches, many developing countries need assistance to determine the nature and extent of their indigenous fossil fuels and to design ways to use them for economic development. This function can be important for foreign assistance agencies such as U.S. A.I.D., and serves as a stimulus and essential precursor to much needed large-capital investment by the private sector. Countries need help in organizing existing geologic data, establishing national geologic surveys and similar institutions and, more importantly, in developing a favorable policy climate and practices that will stimulate private investment.

With respect to oil production in the developing world, in the period 1970 to 1979 the share of world oil produced by the multinational companies decreased from 94% to 45%. In an effort to reap greater benefits for their countries, LDC governments have increasingly established national or parastatal oil companies, thereby discouraging private investment. It is not at all clear that these moves have been to the overall economic benefit of Third World countries. The national oil companies tend to lack technical capability, sound economic practices, and efficient management. Our assistance efforts should encourage them to increase private-sector technical and financial participation. One approach is to help governments develop or re-examine terms for exploration and production contracts that would provide improved financial returns to both the country and private investors in a positive sum game. Starting next month, A.I.D. is offering a newly-designed six-week training course on economic and financial management in oil and gas operations, with the hope that developing country personnel who take the course will be able to improve their skills in contract negotiations and in attracting private investment.

RENEWABLE ENERGY INCLUDING BIOMASS

Virtually all developing countries have large untapped sources of renewable energy that have the long-term potential to supply significant portions of the energy needs of those countries. Hydroelectric, biomass, geothermal, solar, and wind energy may provide sustainable sources, especially for agricultural production and small or medium-scale enterprise development in rural areas.

Their use will require careful technical, economic, social and institutional analysis for each application, and sufficient demonstration to prove their cost effectiveness.

- **Small hydro:** Small hydropower systems are ready for commercial application in the developing world. In India small hydroelectric systems could provide up to 10,000 megawatts of power, one-third of the current national electricity generation. In the Philippines, 4,000 megawatts of small hydropower are possible, equivalent to the present total installed electric-power capacity. We are currently examining 40 A.I.D.-assisted countries where small hydro can significantly foster rural development.

- **Solar:** Much has been said in recent years about the potential for solar energy in the developing world. The potential uses of photovoltaics for stand-alone dispersed low-power sources in rural areas have been widely publicized. A.I.D. and others have supported demonstrations of photovoltaic-powered systems for irrigation pumping, medical refrigeration, village lighting and similar applications over the past 3-4 years. Our experience with these photovoltaic systems is currently being evaluated. Wind machines are receiving increasing attention in the developing world. Solar thermal technologies have not yet found significant application although there is probably considerable potential for low-temperature process-heat sources for industrial processing.

- **Biomass:** Today 14% of the world's primary energy is derived from biomass – equivalent to 20 million barrels of oil per day. The predominant use is in rural areas where one-half of the world's population lives. Ethiopia and Nepal derive nearly *all* of their total energy from biomass – from fuelwood and agricultural residues. In other countries their dependence is large – Kenya derives three-fourths of its energy from biomass, India about one-half, and Egypt about one-quarter.

These so-called "traditional" fuels are disappearing rapidly. At present they are largely tapped for household uses, and often provide a "free" energy source for the rural poor. Unfortunately, efforts to expand supplies have only had marginal success. The continuing increase in population and the resulting increase in demand for agricultural land, combined with increasing demands for wood, not only as a fuel but also for construction and other products, are exacerbating the problem. Even if we were to completely discount the use of wood as the primary energy source for 2 billion of the world's population and were able to introduce alternative fuels, it would still be necessary to preserve the woodlands. Forests are essential to survival because of their impact on soil stabilization and agricultural productivity, on water supply through their effect on regulating and buffering annual water flow in a watershed, and on climate.

The solution to the problem is two-fold. First, increased research is needed on deriving more energy from biomass, both with respect to improved biomass-production methods including development of new crop varieties using biotechnology, as well as new, improved and more efficient energy-conversion techniques. The approaches are many – ranging from dedicated fuelwood plantations on lands too marginal for agricultural production, to increasing the fiber production in sugar cane and converting it to electricity by combustion in a utility boiler, to anaerobic conversion of biomass to useful fuel, as well as to production of alcohol from sugar and starch crops in situations that are economically feasible but do not discourage food production or increase food prices in areas of food shortages. Such approaches involve difficult and complex social and economic policy tradeoffs for many countries. The question of how to achieve both food and biomass-fuel production locally and on a sustainable basis leads to the second essential solution: firmer establishment of priorities for agriculture, forestry and energy and a better understanding of their close interrelationships. This policy requires improving developing-country institutions in these areas and strengthening their interdisciplinary approach. These efforts must be coupled with adequate attention to national economic policies that provide the appropriate incentives and market signals.

One example of the potential of biomass for energy production is the often-mentioned situation of alcohol production in Brazil. Another is the present feasibility study/demonstration of energy-cane production in Jamaica. By shifting plantation practices to maximize fiber production from the cane while simultaneously increasing sugar production, and burning the fiber (bagasse) in utility boilers, the potential exists for cutting Jamaica oil imports by 80 percent. This process would sharply reduce foreign exchange expenditures for oil, create additional jobs, and revitalize the sugar industry. There may be 40 other developing countries where a similar cane-energy approach could be taken.

Several steps must be taken, however, to help assure an adequate role for renewable energy in the supply mix of developing countries:

(1) Increased attention must be given to the economic, social, infrastructure, and institutional aspects of renewable energy systems to help assure their widespread application. Local entrepreneurs must be involved. U.S. industry should develop joint ventures with the LDC private sector to establish local systems. The most important end-use applications, such as irrigation pumping, should receive the focus of attention and be matched

with the most appropriate energy source technology rather than seeking a suitable end use for a new technology. Adequate training and development of public and private institutional capacities to support renewable energy-system development and application is essential.

(2) The LDC economic and policy climates need to be improved so that renewables can compete fairly with other energy sources. Tariffs and other barriers to importing foreign technology need re-examination in many developing countries. Tax incentives and encouragement for research and development expenditures need to be developed. These and other measures, as well as LDC experiences with renewables to date, will be examined at a renewable-energy roundtable meeting with key developing country energy-policy makers, to be held in Sacramento, California on May 28-30, 1985.

(3) Sound, evaluated, objective information on renewable-energy systems, their economics and performance characteristics, must be available to LDC energy planners, decision makers, and practitioners in both the public and private sectors. In the past three to five years, renewable-energy systems have been oversold. They are now being considered more realistically and the search is on to seek their appropriate place in the developing countries. Establishment of networks linking developing-country renewable-energy practitioners and institutions to collaborate on R&D, develop a common methodology for evaluation and monitoring of renewable systems, conduct joint state-of-the-art evaluations, as well as to address policy and implementation issues, are being supported by A.I.D. and other donor agencies. Networks are being established on bioenergy, wind, use of cookstoves, biomass gasifiers, and fuelwood and forestry research. A.I.D. is increasing its efforts to provide evaluated information on renewable-energy systems in published form, coupled with on-the-spot technical assistance and training.

CONCLUSION

The energy crisis in the developing countries threatens their economic stability and development. The industrialized countries, and especially the U.S., have a major humanitarian, economic, political, and national-security stake in alleviating the crisis and providing assistance. The problem requires active involvement of both the U.S. public and private sectors, each having an important and appropriate role to play. The effort must take place on several levels:

- Establish policy dialogues with the developing countries and revise energy pricing, taxation, technology-import regulations and measures to accelerate private investment

- Furnish technical assistance and technology transfer to provide LDC decision makers with the information and experience they need to make their energy-system investment choices

- Develop the LDC human resource and institutional capacity to manage energy systems that are required for economic development

- Assist in these efforts by increasing and more informed participation by the U.S. technical community so as to better utilize the vast wealth of information and experience in our private and public institutions, and

- Devote an increased share of the U.S. foreign assistance program to energy assistance for the Third World, commensurate with the key role energy plays in Third World economic development and U.S. interests therein.

EPILOG

Since the preparation of this paper, the world energy situation has continued to be fluid, oil prices have remained low, but the overall thrust of the presentation is valid. The energy supply shortages in developing countries continue, and demand is rising much faster than in the industrialized world. A major constraint is the inability of developing-country governments to devote a still larger share of their development budgets to energy infrastructure. This difficulty has focussed attention on the need for much more efficient energy systems in the developing world, both for energy production as well as for use. Increased efficiency is also required in the management and fiscal operation of these energy systems. Ownership and operation of facilities by the private sector would help to increase efficiency and reduce governmental capital outlays.

The recent increasing concerns about global and developing-country environmental issues (global climate warming, deforestation, environmental pollution) also argue for much more emphasis on improved energy efficiency and adoption of more environmentally acceptable combustion technologies, such as fluidized-bed combustors. The U.S. foreign-assistance program, and, we hope, those of bilateral and multilateral assistance agencies, will increasingly reflect these concerns in its activities.

FUTURE OF OIL AND GAS PRODUCTION IN THE UNITED STATES

Robert L. Hirsch

Robert L. Hirsch received the B.S. degree in Mechanical Engineering from the University of Illinois in 1958, the M.S. degree in Nuclear Engineering from the University of Michigan in 1959, and the Ph.D. in Nuclear Engineering from the University of Illinois in 1964. After four years at ITT Industrial Laboratories, he joined the U.S. Atomic Energy Commission (AEC) in 1968 as a Senior Physicist in controlled thermonuclear reactions research and served as Division Director of that group from 1972 until 1976 when he became Assistant Administrator of solar, thermal, and advanced energy systems in U.S. ERDA. He joined the EXXON Corporation in 1977 as Deputy Manager of the Science and Technology Department and became the Manager of its Synthetic Fuels Research Laboratory in 1980. In 1983 he was appointed as Vice President and Manager of Exploration and Production Research of ARCO Oil and Gas Company, his present position. Dr. Hirsch has been a member of several national and international committees on energy policy including, the US-USSR Joint Committee on Peaceful Uses of Atomic Energy, and the International Fusion Research Council. He is Chairman of the Board of Directors of ARCO Power Technologies, Inc., and was a member of the Advisory Board of the Princeton Plasma Physics Laboratory. Over the years he has received several Special Achievement Awards from AEC, ERDA, and NASA, and in 1983 he received an Outstanding Technical Achievement Award from the American Nuclear Society. He is a Fellow of the AAAS and a member of the American Petroleum Institute, the American Physical Society, the Society of Petroleum Engineers of the AIME, and the New York Academy of Sciences.

INTRODUCTION

My purpose today is to share some of my views on the future of oil and gas production in the United States. I will be upbeat because of what I will call new technological oil and gas production, which I believe will be brought forward within a decade. In spite of my basic optimism, I do my forecasting with great humility because I well recognize how chancy a proposition I have undertaken. Consider for a moment some notable predictions from the past:

> "The world tells its own tale and in its general decadence bears adequate witness that it is approaching its end. There is less rain in winter to encourage the growth of seeds; springtime is not now so enjoyable or autumn so fruitful; everything in these days is rushing to its doom."

That forecast was made by Cyprian of Carthage in A.D. 250. And what about this forecast of the future:

> "The growing south-pole ice cap has become a stealthy, silent, and relentless force of nature . . . The creeping peril, deadly menace, and the divinely ordained executioner of our civilization . . . The earth of today may quite readily be compared to a top-heavy, dying out, wobbling, spinning top, getting ready to fall over on its side."

That prediction was made by H. A. Brown in 1911. Brown's theory of an imminent "poleshift" is still believed. A few years ago, a group in Stelle, Illinois, was planning to build a lighter-than-air craft to rise above the cataclysm, which they anticipate on May 5, 2000.

Finally, I am mindful of Samuel Goldwyn's admonition:

> "Never make forecasts, especially about the future!"

My intention is to provide you a better prediction than these. I am well aware, however, of the past great misjudgments of better men than I in this field, so I shall proceed with suitable caution.

THE MARKETPLACE AND OIL PRICES

We cannot discuss oil and gas production intelligently without having a good understanding of the energy marketplace, so that is where I will start. Up until 1973, the world was blessed with abundant energy supply at low costs. The subject of energy futures was of interest to a relative few in industry and government. The Arab oil embargo changed all of that, and the term "energy crisis" became part of our everyday jargon. Most of us had to endure long waiting lines to get gasoline for our cars and all of us suffered through energy price escalations that totaled more than a factor of ten for petroleum in particular.

Today the situation is dramatically different. Energy supplies are plentiful, and energy prices are actually coming down. The rising oil prices of the 1970s spurred increased world-wide exploration, and after a while, new supplies resulting from those efforts began to flow. This increase, coupled with a significant decrease in demand due to conservation and economic recession, created a downward pressure on prices, which virtually everyone is enjoying today.

As you well know, the oil and gas industry is very big business. Annual world-wide oil and gas revenues approach a trillion dollars. Annual oil and gas industry investments involve many hundreds of billions of dollars, and government incomes from the world-wide industry are huge. Both of these factors create a large premium for industry and governments to project future oil prices with some degree of accuracy, but our collective record is pretty bad. Nevertheless, if I am going to give you a reasonable projection for the future of oil production, then I will have to base it on some kind of oil-price projection.

Oil-price forecasting is extremely difficult for many reasons. The world-wide energy system is composed of capital machinery and facilities with a replacement value that would be many tens of trillions of dollars. Lead times for large-scale equipment and development of new oil and gas discoveries are measured in large fractions of a decade. Such a system is not only difficult to understand but slow to change. These facts coupled with a myriad of economic, political, and physical variables mean that energy production and usage is very difficult to track and to judge. As a result of all of these difficulties, our overall record of oil-price forecasting has been pretty miserable.

Figure 1 shows a number of past price forecasts by industry and government. I won't attempt to discuss this "spaghetti" graph because it would take too long. Rather, I ask you to simply note the extreme variations wherein even the forecast direction of prices was wrong, let alone the magnitudes. Because price forecasters are smarter today, their forecasts now are usually smothered with caveats, but there are still no guarantees that our current outlook is any better than those of the past.

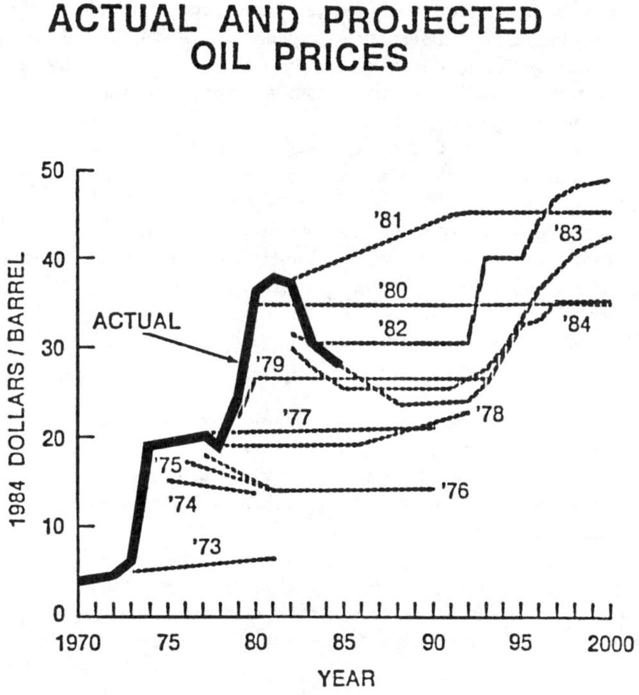

Source: ARCO

Figure 1. Oil-Price Forecasts by Industry and Government

The situation in oil-price forecasting is well summarized in a recent study: [1] "The costs of incorrectly anticipating long-term oil price behavior have in the last few years proved staggering, both for companies and for countries. . . . In 1980-81 alone, on the order of half a trillion dollars was

[1] "The Future of Oil Prices: The Perils of Prophecy." Cambridge Energy Research Associates and Arthur Anderson and Co., 1984.

invested around the world on the assumption that oil prices would continue to rise throughout the rest of the century. Some of this investment remains highly viable under today's very different expectations. A significant part of it will, however, add up to be one of the most expensive business errors ever. The perils of prophecy are already evident in the current troubles of the oil and gas and oil services industries throughout the world; in the growing number of non-performing loans in North America, which have caused or threaten to cause the collapse of major financial institutions; and in the mountainous debt of oil-exporting countries.

" . . . Since modern, long-term oil-price forecasting began in the early 1970s, consensus has been the norm . . . and, in each case, the consensus proved incorrect."

"There is today nearly unanimous agreement about the probable future picture of oil prices. Nearly everyone . . . interviewed [in their study] expects falling real [oil] prices for the next several years, followed by a period of rising real prices."

"This nearly unanimous consensus is troublesome, rather than comforting, for two reasons:

- First, similar agreement has existed four times in the recent past--and in each case proved incorrect.
- Second, forces at work in today's world oil market threaten and could undermine the price trajectory on which most people are operating today."

In spite of this warning, for the purposes of this presentation, I will adopt today's conventional wisdom about future oil prices, namely falling real prices for the next several years followed by a period of rising real prices in the 1990s.

The practical implications of this outlook are that domestic oil and gas companies will continue to tighten up their operations and hold steady if not decrease their near-term exploration activities. This will mean slowly decreasing U.S. production, probably through the end of this decade. During that period, however, I expect that the larger, more progressive oil companies will be preparing for the technological oil production of the 1990s.

BASICS OF OIL PRODUCTION

Before I describe these possibilities, I need to tell you a little about some of the basics of oil and gas production.

To begin with, oil and gas are trapped in microscopic spaces in porous rock underground, like water in a wet sponge, except that the pores are usually too small to see with the naked eye. Usually water is trapped with the oil and gas. Often the oil "floats" on top of the water.

Without going into great detail, I must tell you that the rocks underground were formed by sedimentary processes involving deposition of sands, silts, clays, and organic matter over hundreds of millions of years. These sediments were progressively buried in varying depths. As they were buried, they were subjected to increasing heat, pressures, and a variety of stresses that distorted, moved, and broke them in all manner of ways. The net result is a subsurface that varies from some with nice regular layers with gentle topography to other areas that are completely folded and fractured into an incredible jumble of broken, distorted, and overturned strata. Parts of many of these structures contain oil and gas.

When a hole is drilled into oil-or-gas-bearing rock, the oil or gas trapped in that rock usually flows through the pore spaces to the well and then to the surface because of the great pressure that exists at depth under ground. Oil brought to the surface in this manner is called "primary production."

After a period of primary production, the underground pressure decreases, and so we usually start water injection into oil reservoirs to maintain pressure and to push more oil out. That procedure is called "secondary recovery."

These two production steps capture roughly 20-40 percent of the oil originally in place. In other words, we end up leaving 60-80 percent of the oil in the ground after primary and secondary production!

ENHANCED OIL RECOVERY (EOR)

Varying amounts of that remaining oil can still be captured using enhanced recovery techniques. Some of these involve pumping various kinds of gases into the reservoirs or injecting special chemical surfactants similar to soaps to wash the oil out. We don't do a lot of this now because with today's technology the costs are too high. But there exist hundreds of billions of barrels of known oil in the ground in reservoirs that we have pretty well understood and described, and there is no question that significant amounts of that oil can be produced through enhanced recovery techniques, when we have better technology and/or better economics.

EOR processes are extremely complex. While a great deal of research has helped us to understand these processes, we have much yet to learn. Picture, if you will, a porous solid rock composed of sand or carbonate grains of varying compositions. These grains are cemented together to varying degrees such that there are thousands of pore spaces per cubic centimeter and a large number of channels connecting the pores. Oil, water, and gas are trapped in the pores. Many of the oil droplets are too big to move through the microscopic channels or are surrounded and immobilized by water and are therefore trapped. These droplets interact with the sand or carbonate grains in varying ways because of the range of mineral compositions and chemical and physical forces involved.

As many of you know, we are in the midst of a revolution in surface-science instrumentation and understanding these days. Those new tools are allowing us to begin to unravel the complexity of catalysis, and they are also being brought to bear on the problem of oil trapping. In parallel, we do complex computer modeling and laboratory core floods whereby we force varying amounts of oil, water, gas, and chemicals through cylinders of real reservoir rock to obtain data on phase behavior, gross motions of fluids, etc. All of this is helping us to develop better and better understanding and to slowly improve our methods of enhanced oil recovery.

We have a number of workable processes in the laboratory now, and some have been tested in real field situations where the rocks have widely varying properties, including interlaced impermeable shale layers, fractures, and large-scale breakage, called faults. Results are technically very promising. But the economics for many processes are simply not attractive at today's oil prices and with today's oil-price outlook. Therein is one reason why I began with a discussion of the oil-price outlook because it has a major impact on how oil companies plan activities such as enhanced oil recovery projects.

So what do I think will happen? I believe that laboratory research with new surface-science tools and innovations in chemistry will give a much better understanding of these processes in the next 5-10 years, and that will lead us to lower cost techniques for EOR. As that work progresses and as the oil-price outlook improves, there will come a time when massive EOR projects will make sound economic sense. So EOR is then one kind of "technological oil" that I expect will begin to flow in large quantities in the 1990s in the United States. That flow of EOR oil could reach 2-3 million barrels of oil per day according to a recent National Petroleum Council Study.

HEAVY OIL

There is another kind of oil just sitting there waiting for us. It is very heavy and either will not flow to the surface on its own or does not flow at a high enough rate to be economical to produce at today's prices. In Northern Alaska alone, we know of 30-40 billion barrels of such heavy oil that is waiting for better technology and acceptable economics. There are roughly 50 billion additional barrels of heavy oil in the lower 48 states, some of which is now being produced using steam injection to heat and mobilize the oil. ARCO and others in our industry are hard at work trying to characterize heavy oil in greater detail, to understand how it interacts with its environment, and to develop economical techniques to produce it.

Heat is an obvious tool to help recover heavy oil because the viscosity of these oils decreases rapidly, and these oils thereby flow much more easily as their temperature is increased. But heat means burning something, which is expensive and adversely impacts economics. Much of the science and technology that I believe will lead to better recovery techniques for light, mobile oils will help us to better understand heavy-oil-recovery options. But the problem will be further complicated by the form, rate, and quality of the heat source or heat-transfer media that we use.

The heavy-oil problem has other dimensions. A large amount of heavy oil has been found in the United States over the years, and much of that has been simply ignored and abandoned because of the poor quality of the oil and the poor economics of recovery using existing techniques. Nevertheless, we are making great progress in the laboratory, and I fully expect development of a number of new processes for heavy-oil recovery. These developments coupled with better future project economics will make increased heavy-oil production yet another stream of new oil production in the 1990s.

MICROBIAL ENHANCED OIL RECOVERY

Before I leave the subject of enhanced oil recovery, I must make note of a gleam in the researcher's eye due to a process called microbial enhanced oil recovery.

The idea is that we may be able to use microbes to help us recover hard-to-produce oil through very different mechanisms. Some crude field experiments were done using this approach years ago, but results were mixed at best. Today, there is a revolution in bioengineering, and the understanding and tools developed therein may provide us with a method for enhanced oil recovery.

This approach will likely be very complex because there are a wide range of ways that microbes can interact with oils underground. Microbes can, in principle, "chew" the oil and produce lighter, more mobile oil molecules, or they can feed on oil and produce chemicals or gases that can enhance recovery. I am much less certain where these techniques will lead us, but they do offer very interesting long-term possibilities for yet other enhanced recovery processes.

FINDING MORE OIL AND GAS IN THE UNITED STATES

Let me turn next to the problem of finding more oil and gas in the United States. The point I want to make is that there is much, much more to be discovered, and improved technology will be a major factor in at least part of that effort. I will discuss two topics that bear on this problem. The first relates to improvements in seismic technology and the second deals with oil and gas in so-called stratigraphic traps.

Seismic Prospecting

The primary tool for imaging the subsurface before drilling is with sound waves *via* the so-called seismic technique. Sound sources can be vibrator trucks, dynamite charges, or air guns for underwater applications. Special detectors called "geophones" sense the reflected sound waves, very large computers process the data, and specially trained people manipulate and interpret what is measured.

The typical onshore seismic process involves a series of coordinated steps. A promising area for prospecting is selected through a variety of studies, often involving complex geological concepts and tools. Next, a number of seismic trucks, usually three to six positioned in an appropriate linear pattern, "shake" the ground with heavy vibrators, generating coherent sound waves that propagate deep into the earth. As those sound waves strike discontinuities in subsurface rock strata, some reflection occurs. Part of those reflections travel to the surface and are picked up by long strings of sensors (the geophones) spread on the ground, usually in a linear array. Geophone data is digitized and stored for later computer manipulation and analysis. It is the temporal and spatial recording of reflected sound waves that give us a picture of the subsurface before we drill.

This procedure may sound relatively simple but it is extremely complex in practice. The reflected signals are very weak. The transit time of sound waves depends on the velocity of sound in the rocks, which varies considerably in ways that we cannot anticipate because we do not often know the rock types in the area. Reflector surfaces often involve extremely complex geometries, which reflect sound in many directions. There are many sources of both coherent and incoherent noise that can mask the good data. The sound frequencies employed in this process present an additional problem. Varying degrees of resolution of the underground layers are obtained since the thicknesses of the layers may be much larger or much smaller than the wavelengths of the pertinent sound waves.

In spite of this remarkable complexity, the seismic method works well in many, but not in all, of the cases of interest in oil and gas prospecting. The seismic technique has helped us to find most of our oil and gas so far, but it has also missed enormous quantities because of its limited capabilities.

Table 1 shows how much seismic resolution has improved in the last 10 years. We feel that much more is possible and so in the next 5-10 years we expect major further improvements, as shown. These improvements will

SEISMIC TECHNIQUE

	10 YEARS AGO	*NOW*	*FUTURE*
RESOLUTION AT 4 KM	60 M	40 M	10 M
STRUCTURAL ACCURACY	± 100 M	± 50 M	± 5 M
SIGNAL / NOISE	2.5 / 1	10 / 1	1000 / 1

Table 1. Improvement in Seismic Resolution

allow us to discover significant amounts of oil that the industry has passed over in the past.

Because of time constraints, I must purposely simplify what in reality is a multi-faceted problem that involves a myriad of geological inputs and the use of other tools besides the seismic process. Progress across the spectrum of tools must accompany improvements in seismic technology if the industry is to take full advantage of the expected progress in seismic methods.

Stratigraphic Traps

Let me now turn to a relatively untapped source of oil and gas, namely so-called stratigraphic-traps. For an underground oil or gas trap to be of practical interest, it must have a number of important features, one of which is that it must be a good container. If that were not the case, the oil and gas that might tend to accumulate would simply leak away. The two major types of oil and gas traps are known as "structural" and "stratigraphic", with many variations thereof also possible.

Since structural traps are generally characterized by relatively distinct boundaries that often can reflect seismic sound waves rather cleanly, the use of seismic techniques can provide good results. On the other hand, the sound reflections of stratigraphic traps are usually so subtle that today's seismic technology is not good enough to "see" those traps clearly. Nevertheless, in many cases we have found large amounts of stratigraphically trapped oil by accident when structural traps were the principal target. The most notable large strat trap in this country is the multi-billion-barrel East Texas field.

Many geologists estimate that there is as much oil yet to be discovered in stratigraphic traps in the United States as has already been found in structural traps. Furthermore, there are still a fair number of structural traps out there yet to be found. It will be very difficult to develop the technologies that we need in order to find stratigraphic oil and gas, but I believe that we will master this task in the next 5-10 years. When that happens, I expect a significant, subsequent increase in U. S. oil and gas production from subtle stratigraphic traps. Production of hydrocarbons from stratigraphic traps will be economically similar to production from structural traps. Therefore, a significant amount of strat-trap production can occur relatively independent of oil/gas prices.

"Political" Oil and Gas Resources

Before I try to sum all of this up and provide a "fearless forecast," I must make note of what I might call "political oil and gas" resources. This oil and gas lies in areas that have not been made available for prospecting in the United States. The reasons why these regions have been withheld is most often related to environmental protection, which is clearly a high national priority. Nevertheless, the vast onshore and offshore lands that have been withheld have been estimated to contain more than a hundred billion barrels of oil or gas equivalent, certainly a major national resource.

Many people argue that lower dependence on foreign sources of oil and gas is a high national priority, on a par with protection of virgin lands. Oil industry specialists feel that these lands could be prospected and the oil could be be produced with minimal environmental insult. But our federal and state governments have chosen not to disturb most of these areas, so the contained resources remain idle and undiscovered.

Whether, or how fast, we allow oil and gas prospecting in these areas is a matter of political judgment, and therefore, I will make no predictions. It is worthwhile noting, however, that in the past we have chosen to limit U. S. production from such areas for perfectly valid reasons, which may or may not change in the future.

THE REST OF THE WORLD

What I have said thus far has been related to our situation in this country, which is the most heavily prospected area in the world. Clearly, if I am correct in my optimism regarding increased production in the United States, many of my arguments apply also to the rest of the world.

On that basis, we should expect significant world-wide oil and gas production for a hundred years or more. If that prediction is true and, as I expect, the cost of the more expensive oil from enchanced recovery will not be too high, then the world will be able to depend on oil and gas for a long, long time. That is good news in many respects and perhaps bad news for those who look to oil shortages or much higher prices to stimulate shifts to other technologies.

"FEARLESS FORECAST"

Before concluding, I will try to combine my previous remarks with some additional factors for my own "fearless forecast" of the energy future. First, I believe that the world will rely on oil, gas, and coal as its primary sources of energy for much more than the next hundred years. I also believe that U. S. oil and gas production can be maintained at reasonable levels for many, many decades through improved technologies and enlightened government policies.

In my view, after nuclear plants now under construction are completed, nuclear-power production in the United States is unlikely to grow for a decade or more. The Three-Mile-Island reactor accident is a dramatic illustration of the unsnarlable regulation and financial exposure that afflicts the industry at present. Elsewhere in the world the importance of nuclear power will grow because other countries have taken a more rational approach to management and regulation.

Next, solar energy will grow but will only make a modest world-wide contribution. The problem will be cost and the fact that sunshine is inherently intermittent.

I expect that fusion will become commercial early in the 21st century, but it will not make a major energy contribution until the middle of that century. This prediction is based simply on the fact that it takes fifty or so years to build lots of anything in the energy field, whether it is fusion, solar, or whatever.

CONCLUDING REMARKS

The domestic oil and gas industry is currently going through a period of difficult readjustment, brought about in part by overspending related to the overoptimistic price forecasts of the late 1970s and early 1980s. Other contributing factors include the activities of the so-called corporate raiders, and various changing economic realities. Many companies are in the process of changing their cultures, their ways of doing business, and the size of their staffs, as part of this changeover.

I expect these readjustments to be essentially complete in three to five years. During that time, the better companies will be developing the tools necessary to produce the technological oil and gas that I have described. That oil and gas should begin to be evident in the United States in the early to mid-1990s and, I believe, will provide us an extremely important national resource for the 21st century.

EPILOG

Following the presentation of this paper in April, 1985 a number of significant changes have occurred. World oil prices began an unprecedented collapse in late 1985. By the second half of 1986, prices had fallen to roughly half of their early 1980s levels thereby causing dramatic personnel layoffs in the U. S. oil and gas industry along with the bankruptcy of a great many companies. As a result, the U. S. oil and gas exploration and production (E&P) infrastructure was reduced by roughly one third, oil and gas R&D was reduced by 20-25 percent, and a large part of what remained was rechanneled toward shorter-term, lower-risk activities. All of these circumstances will slow the development of new technological oil to a great extent.

The latest generally held oil-price forecast is for low real prices ($17-$18 per barrel) to continue until perhaps the mid 1990s after which a real-price escalation is likely to occur. Consequently, U. S. oil production is expected to drop significantly in the years ahead due to these oil industry difficulties and price projections. The most likely effect will be a steadily increased U. S. dependence on imported oil. Since there appears to be ample supplies of oil worldwide for the foreseeable future, the next U. S. "energy crisis" is unlikely to involve oil shortages. Rather, the next "energy crisis" is likely to be related to how our country manages an imported oil bill that may approach 100 billion 1988 dollars/year by the year 2000.

PROGRESS TOWARD PRACTICAL FUSION POWER

Harold P. Furth

Harold P. Furth received the A.B. degree from Harvard University in 1951, engaged in post-graduate work at Cornell University in 1951-52, and received the Ph.D. degree in Physics from Harvard in 1960. He was a Physicist with the University of California Lawrence Radiation Laboratory from 1956 to 1965, and was a Group Leader there from 1965 until 1967 when he joined the Faculty at Princeton University as Professor of Astrophysical Sciences. In 1981 he was also appointed Director of the Plasma Physics Laboratory at Princeton, his present position. Dr. Furth is an internationally recognized leader in research on plasma physics and controlled thermonuclear reactions. He received the E. O. Lawrence Award of the AEC in 1974, the J. C. Maxwell Prize in 1983, and the Joseph Priestly Award from Dickenson College in 1985. He has been a member of the Board of Editors of *Physics of Fluids* and *Reviews of Modern Physics* and continues to serve on the editorial boards of *Nuclear Fusion, Plasma Physics and Controlled Fusion*, and the *Journal of Fusion Energy*. He is a member of the National Academy of Sciences, and is a Fellow of the American Physical Society.

A discussion of practical fusion power might as well start with the primary energy source in our neighborhood, so Figure 1 shows a piece of the surface of the sun. It is moderately hot–about 5,000 degrees–and radiates energy, which is a good thing, because much of what has happened on earth, including us, is very much dependent on that radiation. In spite of that, rather little thought has been given, over the history of mankind, to what makes the sun shine or whether it will continue to shine tomorrow. Of course there were occasional periods of concern when the sun unaccountably went off for a minute. I am reminded of the gasoline rationing at the pumps, which for a while set people to thinking that perhaps the supply was not infinite after all. Now the lesson seems to have been forgotten again.

In the case of the sun, it wasn't until about the end of the 19th century that some people became smart enough about physics to be able to see there was a problem. When they sat down and calculated the cooling-off time of the sun on the basis of its expenditure of radiative energy, they found it was very, very much shorter than the apparent history of the earth. Clearly something was wrong. A little later Albert Einstein came along with his

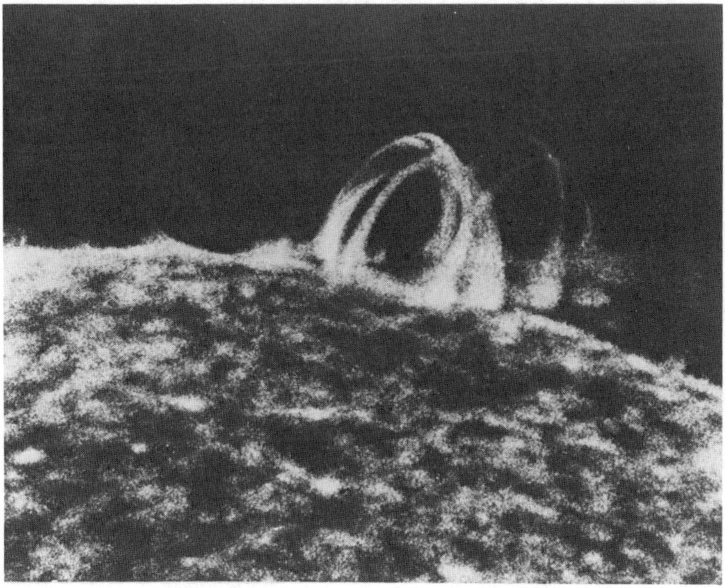

Figure 1. The Surface of the Sun

famous equation E = mc^2. Given this clue, by the end of the twenties, physicists began to think that perhaps they understood what is happening in the sun: hydrogen is being fused to make helium, and the extra mass that disappears in this process is keeping the sun hot. To prove that such things are possible required direct laboratory experiments. The first successful demonstration was performed in Lord Rutherford's Cavendish Laboratory in Cambridge.

In the early thirties, I think it was actually 1934, Lord Rutherford and his colleagues bombarded deuterium with deuterium, representing at that time the cutting edge of high-energy physics. For this purpose they had linear accelerators operating at 100,000 volts so the particle energy was 100 keV. In the deuterium-deuterium reaction, as shown in Figure 2, they found some new elementary particles namely a lighter form of helium, He_3, and a heavier form of hydrogen called tritium (T). Also, quite astonishingly, they found that something like 50 times more energy was released than had been put into the reaction. That made a big impression. The *New York Times* had a front-page article reporting the event, not quite saying that the energy problem was

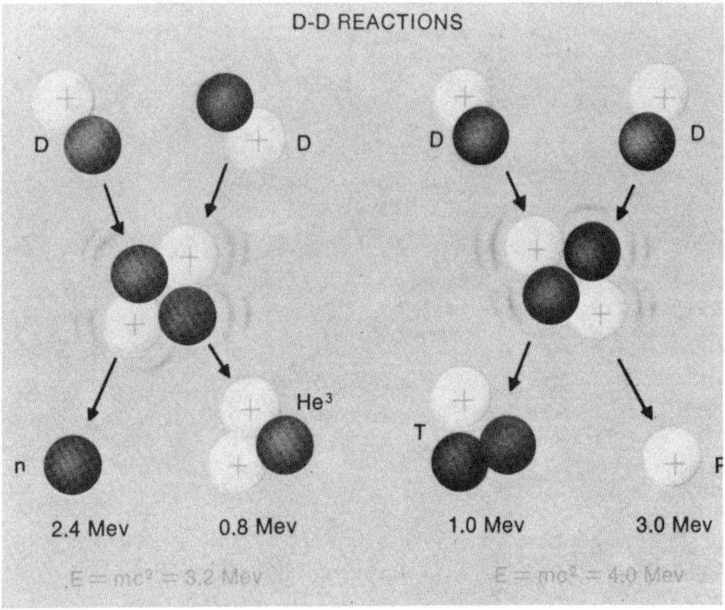

Figure 2. Deuterium-Deuterium Reactions

solved forever, but not failing to suggest to intelligent readers that something very important had just happened.

Lord Rutherford himself was not overly excited about the practical implications of his results, and one can see why. Here is this remarkable 50-folding of the investment in a particular reaction, but Lord Rutherford well knew that for every particle that his accelerator made react there were hundreds of thousands that did not fuse but just scattered and degraded their energy into ordinary heat. So it was perfectly clear that using an accelerator was not a sensible way at all to produce net fusion power; it was just a way of degrading the electric input power. That is why I think he used the word, "moonshine" when asked about the great electrical power resource that might spring from nuclear reactions.

To make fusion power, one cannot simply shoot particles at particles. One has to take into account that the scattering cross-section is much bigger at all energies than that required for fusion. Consequently, it is necessary to create an environment much like that of the sun where particles can bounce off each other over and over without losing their energy. In other words, it is necessary for the entire environment to be hot. The average kinetic energy has to correspond to the amount of energy desired to make reactions. That means temperatures of at least some tens of kiloelectron volts, corresponding to some hundreds of millions of degrees, are needed. The sun doesn't actually work quite that hard. It loafs along at a mere 14 million degrees, but that is because it is a very good "thermos bottle"—a big object sitting in a nice vacuum, so it does not have to be a very vigorous power source. On earth, as may be guessed, if an attempt is made to maintain a temperature of hundreds of millions of degrees in a cold environment, then thermal insulation is rather problematical. It is necessary to run at hundreds of millions of degrees so the fuel will burn faster and thereby compensate for the heat leakage.

One way to make life a little easier is not to burn deuterium alone but to burn deuterium with tritium, as shown in Figure 3. Tritium was discovered in the D-D reactions shown in Figure 2. It does not exist in nature, because it decays after 12 years. The reaction process, however, can be made self-sufficient in tritium. Fusing deuterium with tritium creates helium as a principal ash, with one neutron left over. This neutron can then be used to split lithium, thereby breeding back the tritium and thus continuing the nuclear reaction cycle. Ideally one can breed about a triton and a half per fusion reaction, so there is no great shortage here; it is a very fast-breeding cycle. The helium, in turn, is nice because about 20 percent of the energy comes off in the kinetic energy of its nucleus, which tends to be trapped within the fusion fuel and helps to heat it. If the fuel can be surrounded with sufficiently good heat insulation, it can literally begin to burn, much like the familiar materials that can be ignited with a match. Once they burn, it is not necessary to keep

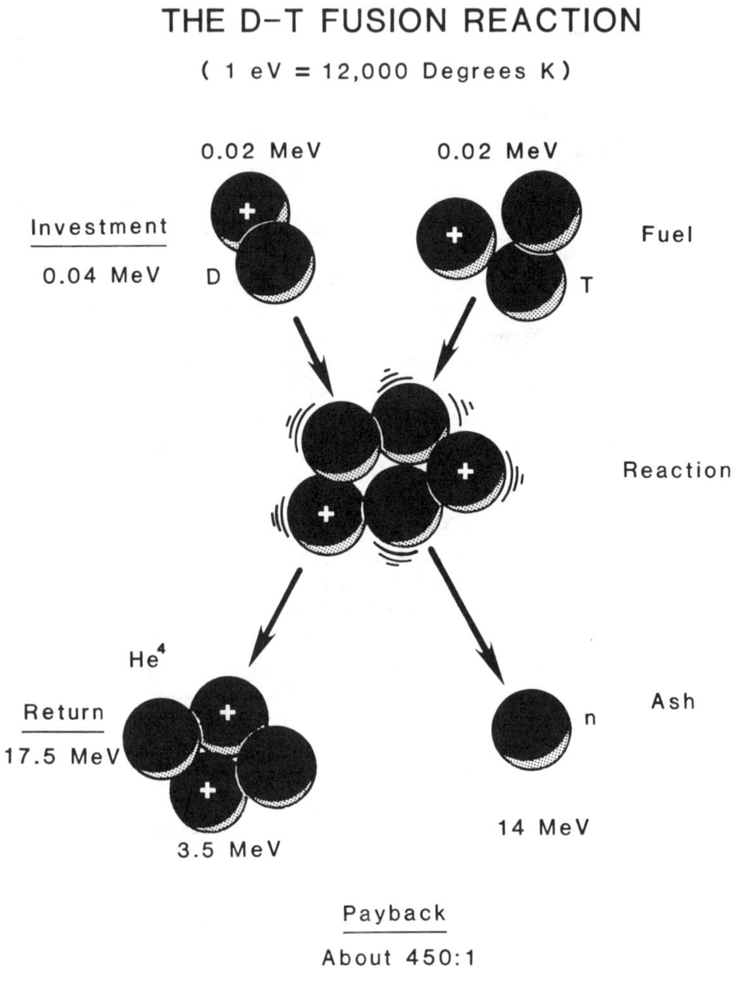

Figure 3. Deuterium-Tritium Reaction

heating them. That is what happens here. The fuel must be heated up initially to a hundred million degrees or so, but thereafter the production of energetic helium nuclei by the fusion reaction will keep the fuel hot if the heat leakage is slow enough. This is really quite a fine process. The basic fuel, deuterium, is enormously abundant and negligibly cheap. The big problem that we are attacking is how to make a contraption that will burn this wonderful fuel without requiring such a huge capital cost that the interest charge becomes greater than the cost of burning oil.

A comparison between a fossil-fuel power plant and a fusion power plant is shown in Figure 4. In terms of material going in and out, fusion is enormously attractive. The data are given for a thousand-megawatt electric plant. If the fuel is coal, many thousands of tons are required and great quantities of effluent are dumped into the atmosphere, helping also to contribute to the CO_2 problem. From that point of view it is splendid to contemplate a pound of deuterium with a few pounds of lithium going in, and a few pounds of ordinary helium coming out, with no direct radioactive ash at all. The neutron that is shown in Figure 3, however, will pass through the structural material of the fusion reactor and activate it. Naturally the first wall will not be made of material with terribly nasty long-lived radioactivity. The radioactive waste problem for fusion reactors will be much easier than that for fission

DAILY FUEL CONSUMPTION AND WASTE PRODUCTION

	COAL PLANT	D–T FUSION PLANT
FUEL		1 lb. D_2
	9000 T. Coal	3 lb. Li^6
		(1 1/2 lb. T_2)
WASTE	30,000 T. CO_2	
	600 T. SO_2	4 lb. He^4
	500 T. Ash	
	80 T. NO_2	

Figure 4. Comparison of Fossil and Fusion Power Plants

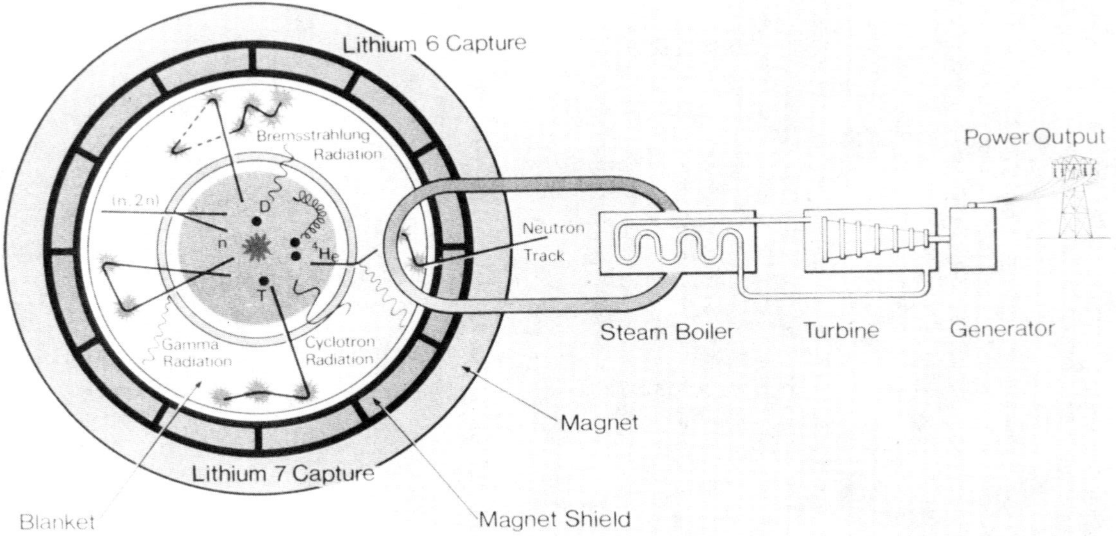

D-T Fusion Reactor

Figure 5. Elementary Sketch of a Fusion Reactor

reactors, where the fragments resulting from the splitting of the uranium nucleus can be rather unpleasant. In the immediate outlook, if reasonable kinds of steel are used, the first-wall radioactivity would persist for about 100 years. It would not be necessary to use geological burial, and the material could be recycled eventually. The use of vanadium would further reduce the storage time. If industry becomes clever enough to devise a ceramic first wall, the radioactivity could die down literally in weeks. All in all, the environmental prospects of fusion seem exceptionally benign even if we are not yet smart enough to accomplish everything in terms of present-day technology.

A very abstract form of fusion reactor is shown in Figure 5. Here is the hot stuff cooking, neutrons flying out, depositing most of their energy in the blanket which is quite cool—less than a thousand degrees. The blanket breeds tritium from lithium and also changes its absorbed heat into steam, the steam turns the turbine, and the turbine makes electricity.

Earlier I mentioned that we need a good enough "thermos bottle" so that the reaction will keep itself burning. It is like being in a damp muddy place trying to build a fire that tends to go out when the match is removed. In order to raise the fuel to the desired temperature of hundreds of millions of degrees, suitable heat insulation is required. Figure 6 shows the conditions for reaching the "break-even" point where the fusion-power output equals the input heating power. In order to achieve these conditions, the confinement time, τ, of the heat content, the density, n, of the fuel, and the fuel temperature, T, establish the criteria for the deuterium-tritium reaction. Since fusion is a binary reaction, less time is required if the density of the fuel can be increased. Hence, it is the so-called $n\tau$ product that must have a specific magnitude. If we want the fuel to "burn"—that is, to be self-heating—then $n\tau$ has to be about five times higher than for break-even.

CONDITION FOR BREAK-EVEN WITH D-T REACTION

$$n\tau > 10^{14}\ \mathrm{cm}^{-3}\ \mathrm{sec}$$

$$\mathrm{AT}\quad T = 10\ \mathrm{keV} = 10^{8}\ {}^{\circ}\mathrm{K}$$

Figure 6. Approximate Criteria for Break-Even Conditions

At this point there are several approaches that can be taken. Operation in the steady state places a limitation on the density of the fuel. It must not be too dense, because something that is a hundred million degrees hot and also very dense is essentially a bomb. Ideally, the density should be about a hundred thousand times less than that of the atmosphere, and the fuel then has about the same pressure as hot steam. On this basis, the typical density is a few times 10^{14} particles per cubic centimeter, and the required heat confinement time is then about a second.

Inertial fusion is one alternative solution that I think is also quite likely to succeed. The idea, as indicated in Figure 7, is to blow off the surface of a fuel pellet of multi-millimeter dimensions. The reaction from this blow-off squeezes the fuel into super-dense, hot matter. The fuel "burns" for a very short time—about a nanosecond—but the associated product of density and time, $n\tau$, is sufficient to attain the ignition point. The driver could either be a laser beam, an electron beam, or an ion beam, and the reactor would look something like the device shown in Figure 8. The little pellets are injected into the central cavity at the rate of a few per second and are heated to create the fusion reaction. The lithium blanket that is needed to breed back the

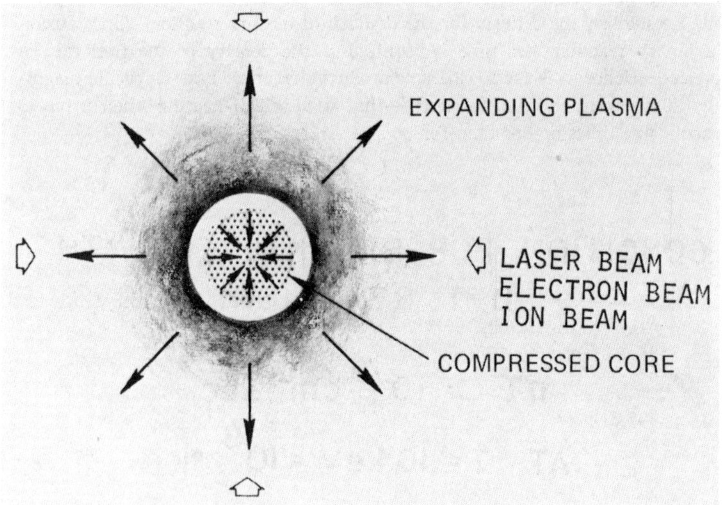

Figure 7. Inertial Fusion Concept

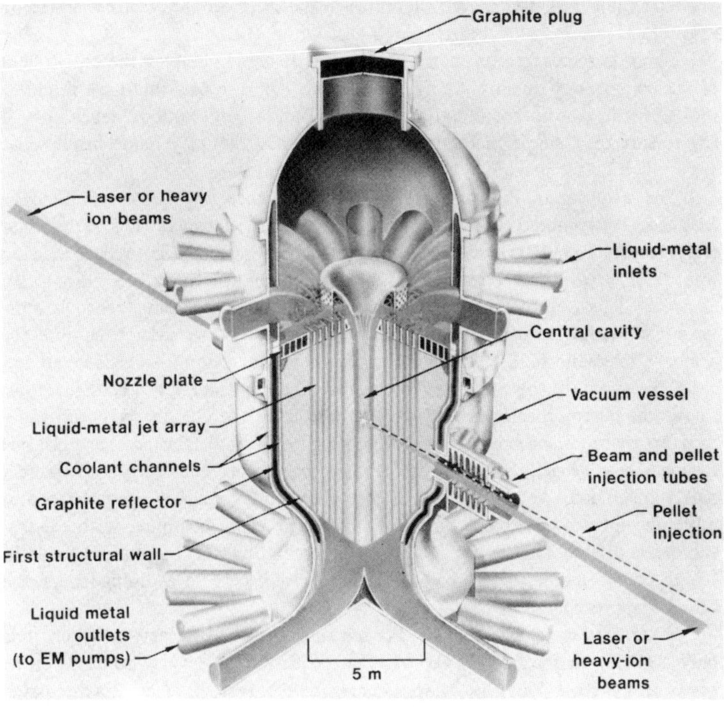

Figure 8. Projected Inertial Fusion Reactor

tritium is in the form of a liquid-metal curtain that could also serve as a shock absorber.

Inertial fusion is clearly very different from magnetic fusion. While both approaches still have many unexplored technological problems, as well as potentials for coming to grief on the economics, the difficulties are of a quite different sort. It would seem extraordinarily unlikely that in the case of inertial fusion one would encounter precisely the same snags as in magnetic fusion. Such technological diversity contributes to the richness and vitality of fusion research.

There is a great temptation at this point to try and cover the entire field of fusion without missing a single approach. For instance, there are fascinating things to be said about mu-meson fusion and other exotic schemes. I will try to stay on a single main track, however, and gradually home in on what we are doing at Princeton.

The material in Figure 9 allows me to advance toward my main topic. Magnetic confinement is based on the observation that even at a rather low temperature, like 10,000 degrees, the atoms in a gaseous fuel will break up into their constituent electrons and nuclei, both of which are electrically charged. This state of matter is called a plasma. In a plasma, there is some hope of obtaining heat insulation, because charged particles cannot move freely *across* a magnetic field. The next concern is what to do about escape out the ends. Fortunately there are clever ways of pinching the ends in configurations known as mirror machines, and one can also bend the chamber around into a torus so there are no ends. I will talk mainly about the latter approach, but let me just say briefly that the fine art of pinching ends has become extraordinarily ingenious. As indicated in Figure 10, complex magnetic geometries and multiple mirrors have been developed in recent years, complete with electrostatic potentials. This is one of the most fascinating and innovative fields in fusion at the moment; but I will confine myself here to a discussion of the toroidal approach.

Several years ago, the *New Yorker* magazine had a cartoon where one lady is saying to another: "Well, what *do* you say to a person who tells you he is working with a doughnut-shaped energy field?" Perhaps after this discussion you will know what to say. Figure 11 shows a typical doughnut-shaped plasma that is confined by means of magnetic surfaces composed of helical magnetic field lines, one of which may be seen in the picture. The plasma tends to stick to those field lines but not perfectly. It can make instabilities and collective motions. In particular, it tends to move out along the major radius, into regions of weaker magnetic field. Our task is to make and heat such doughnut-shaped plasmas and try to provide good enough heat insulation, even though we are faced with collective leakages of heat and hot matter.

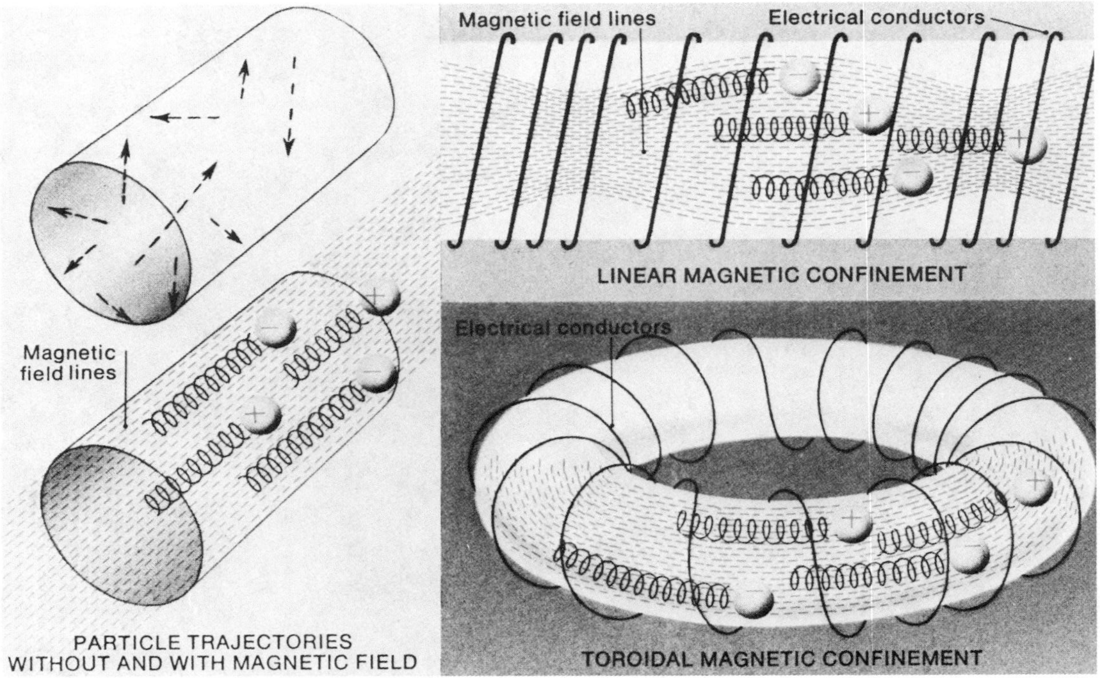

Figure 9. Schemes for Magnetic Confinement

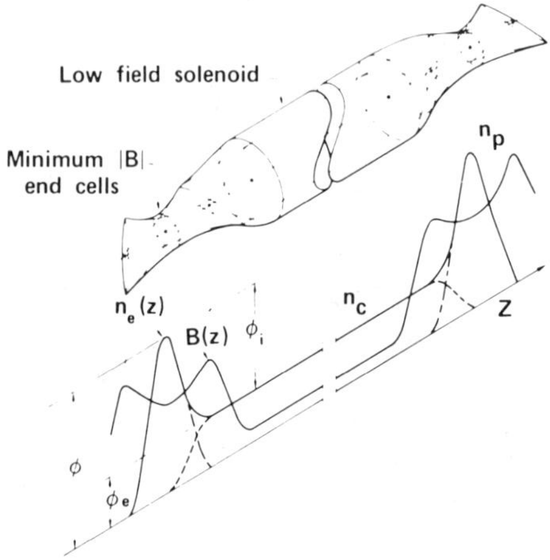

Figure 10. Geometry of a Simple Tandem Mirror Machine

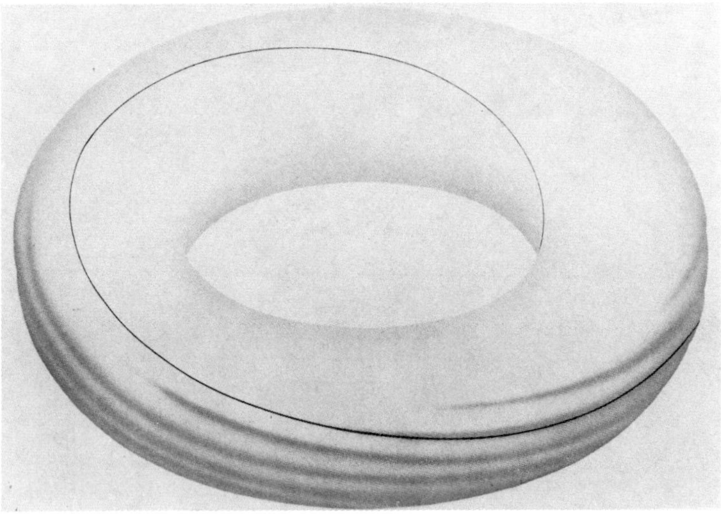

Figure 11. A Doughnut-Shaped Plasma with Flutelike Surface Instabilities

The basic idea that seems to work best at the moment is called the "tokamak." It was invented in 1950 by Igor Tamm and Andrei Sakharov. Figure 12 is a classified Soviet document from the 1950's which has had its titles translated to English but otherwise is the same. It is a doughnut sitting on its side showing the minor cross section. The article that describes this device is remarkably perceptive in terms of the important physics problems, but its projections are actually rather pessimistic. It is looking toward a pure-deuterium reactor with about 900 megawatts of gross thermal power, in other words, less than 300 megawatts of electric power. The diameter is 24 meters, so the major radius is 12 meters, and the field is 50 kilogauss. This is a rather large and expensive object for the production of so little power. If somebody had been sharp enough at the time to complain about that, perhaps Mr.

$$H_0 = 50\,000 \text{ G}$$

$$\left.\begin{aligned} d &= 4 \text{ м} \\ \partial &= 8 \text{ м} \\ D &= 24 \text{ м} \end{aligned}\right\} \qquad \begin{aligned} &\text{Volume} = 0{,}96 \times 10^{9} \text{ см}^3 \\ &\text{Area} = 0.96 \times 10^{7} \text{ см}^2 \end{aligned}$$

$$n_0 = 3.0 \times 10^{14} \text{ см}^{-3}$$

$$T_0 = 100 \text{ keV}$$

Release of thermonuclear power 17.6×10^{6} erg/cm^3 sec.

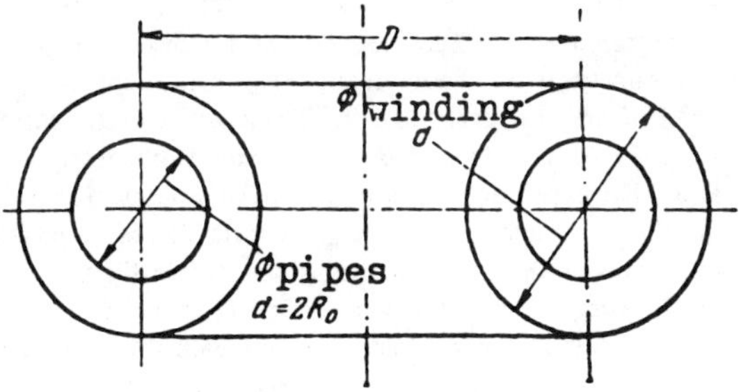

$$W = 8.8 \times 10^{15} \text{ erg/sec} = 880{,}000 \text{ kW} .$$

Figure 12. Early Tokamak Reactor Concept

Sakharov would have been discouraged. Fortunately, no one seems to have complained, so the Soviet Union launched an ultimately very successful line of tokamak experiments.

Skipping over a lot of difficult early history, we come to the tokamak machine shown in Figure 13, which was the first really great success in fusion. In 1968-69, a doughnut-shaped experiment called T-3, at the Kurchatov Institute in Moscow, gave some fairly astonishing results: temperatures of five million degrees and confinement times of 5 milliseconds were achieved. People fell off their chairs when they heard the news. We could hardly believe the reported results. It was quite evident that this success would change the whole course of future research. At this point, the United States, along with the rest of the world, started working intensively on tokamaks.

The following five figures display U.S. progress on these machines. Figure 14 is a sketch of a machine called TFTR, which began to be built around 1975-76. Here again is a doughnut-shaped chamber, with toroidal magnetic field coils that are supposed to provide about fifty kilogauss. These primary windings are designed to induce a current of two-and-a-half million amperes in the plasma which would thereby be heated to ten million degrees or more. Subsequent injection of intense beams of neutral atoms will bring the plasma above the 100-million-degree level. The hope is that when a 27-megawatt beam is shot in, out will come about 27 megawatts of fusion products, in at least an approximate break-even reaction, during a pulse that should last for about a second. The projected 27 megawatts is not bad for a power producer, but this is not really a reactor. It is simply an experiment that is supposed to produce the reactor-plasma state for a second every ten minutes or so.

Figure 15 shows the appearance of the TFTR facility at the beginning of 1982. That scene troubled us, because we had promised Congress that by Christmas of the same year we would begin experimental operations. Miraculously, at 3 a.m. on Christmas eve, 1982, the TFTR was actually running. Figure 16 shows the TFTR as it looks today. The area is cluttered with lots of the kind of junk that experimentalists like to have around. Figure 17 shows the device from the inside and also gives a good indication of its physical size.

The present status of the experiment is illustrated in Figure 18. The full design field of 52 kilogauss and the full design current of 2.5 megamperes have been achieved. It is amusing to note that to obtain this enormous current of two-and-a-half million amperes required exactly one volt, which suggests that this hot plasma is a good conductor, and it is indeed. Dale Meade, head of the TFTR physics group, will present a summary paper on April 26, 1985, at the American Physical Society meeting in Washington, D.C. outlining the most recent progress. I don't want to anticipate his information here. However, one item of considerable interest should be mentioned. When we ran at only one million amperes, and heated with only 5.5 megawatts of neutral beam, which

Figure 13. View of the T-3 Tokamak Experiment

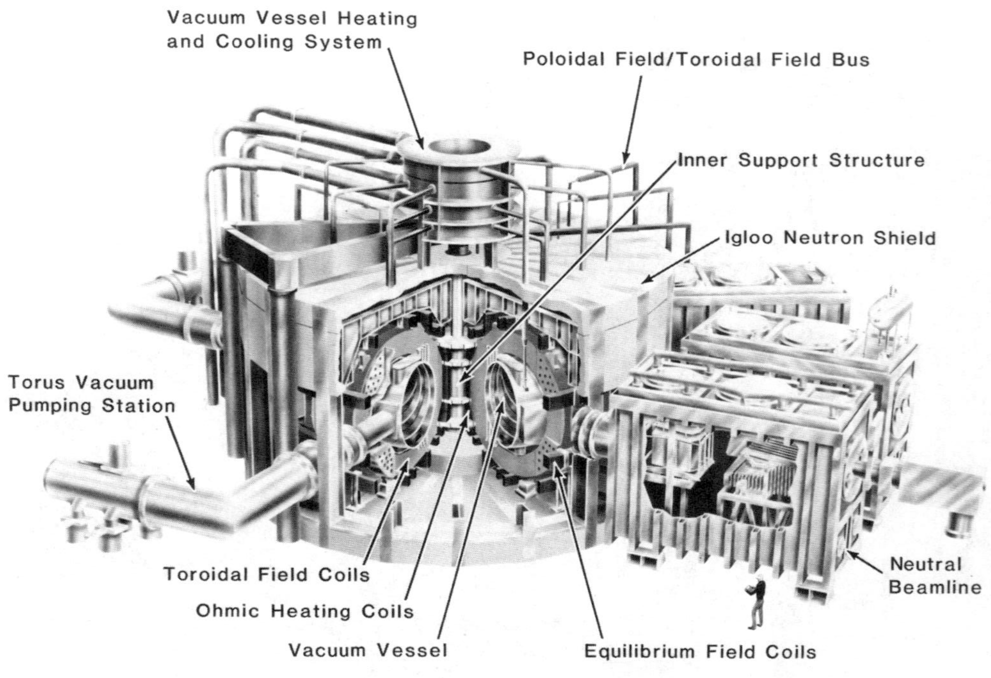

Figure 14. Tokamak Fusion Test Reactor (TFTR)

Figure 15. Early View of TFTR Facility

Figure 16. View of Completed TFTR Machine

Figure 17. Inside View of TFTR Machine

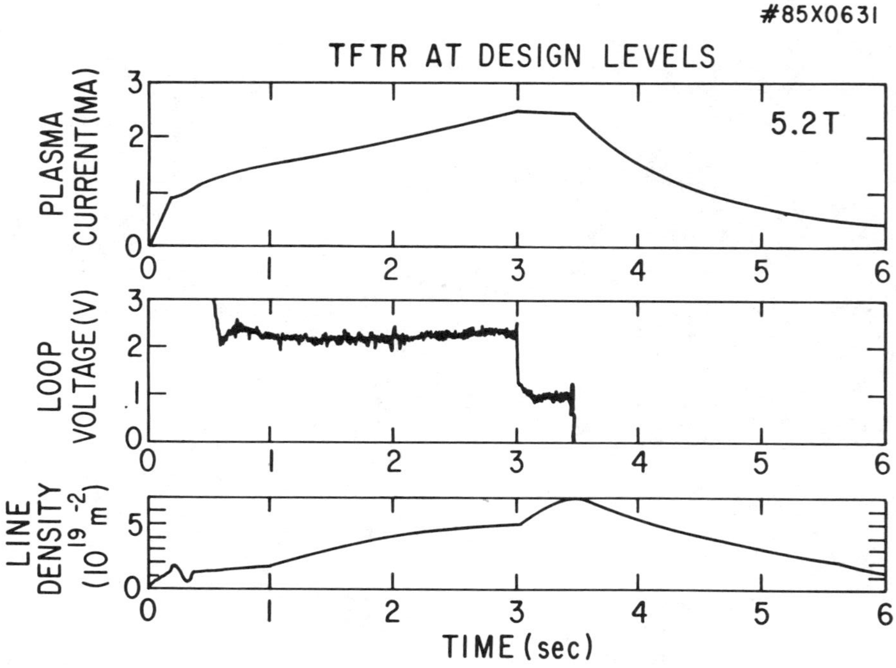

Figure 18. TFTR Operating Parameters

is our present maximum capability, we measured the peak ion temperature shown in Figure 19. Here we have about ten kiloelectron volts and that is over 100 million degrees—which would be the first time in tokamak research.

Naturally, when something good like that occurs, many questions arise. For example, we are measuring here the temperature of a small titanium admixture in the deuterium plasma, but is the deuterium at the same temperature or, as seems likely, is it a bit colder? What is the next step necessary to improve the experiment? Since these results were obtained with less than a fifth of our planned heating power and with only 40 percent of the design plasma current, we seem to have made quite an auspicious start.

In terms of the grand road map of fusion research, we may well ask where tokamak experiments stand at present. In Figure 20, we are at present aiming for the break-even curves. Ultimately we would like to sit on the curve marked "ignition" where the alpha particles of the reaction will keep

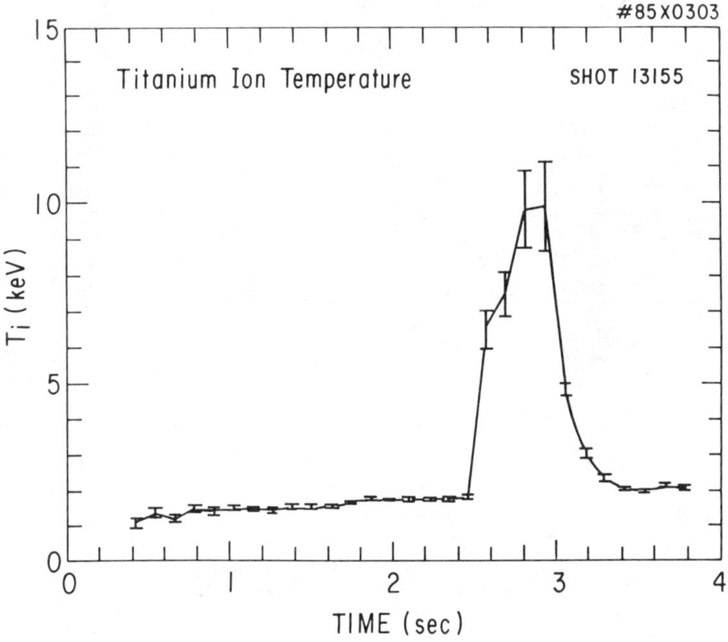

Figure 19. Ion-Temperature Increase During Neutral-Beam Heating in TFTR

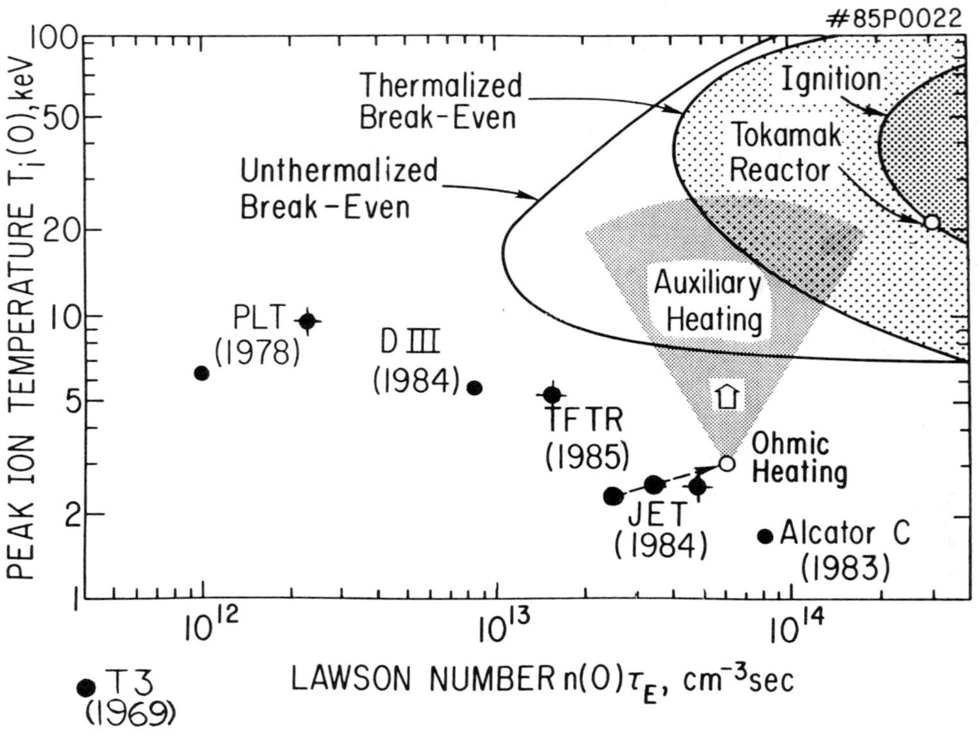

Figure 20. Progress of Tokamak Experiments Toward the Reactor Goal (Crosses Mark TFTR Data Points)

the fuel hot, burning just like the sun. That is what we would like to have in a reactor. The dot in the lower left corner of the figure is the relative position of the decisive breakthrough made by the T-3 device of the Soviet Union in 1969. The other dots show that tokamaks have been making pretty good progress meanwhile. The TFTR experiment has now assumed a front-line position, marked by crosses in the figure, and seems to be moving forward at a great rate. The immediate goal of the TFTR is to enter the region marked "unthermalized break-even" in Figure 20, rather than attempt to reach the neighborhood of the ignition region. At this time, it appears rather likely that we will be able to do just that.

Assuming that break-even conditions can be reached, it is reasonable to ask if it is worthwhile. All we are doing is taking electric power and turning it into 14 MeV neutrons—but, a device of that sort is needed fairly early-on in order to develop the nuclear technology of fusion. Increasingly as the plasma physics succeeds, the nuclear engineering problems loom large, and so we would like some engineering test facility that can produce an appreciable number of megawatts of fusion neutrons per square meter. The interactions of the neutrons with the first wall and the blanket will allow us to learn of all the dreadful materials problems that must be solved before a practical reactor can be developed. A device very much like TFTR in size and current would make quite a good neutron source. Further, if we are content with an energy multiplication around one, we could run this kind of tokamak in steady state: A piece of interesting incidental information that we picked up in the recent TFTR experiments is that when you shoot in the beam, it can maintain a steady-state current. Therefore, one sort of "useful application" of TFTR could be to build a steady-state machine with parameters similar to those given in Figure 21, and use it to help develop nuclear fusion technology.

The next three figures show that the potential of tokamak fusion experiments has not gone unnoticed in the rest of the world. Figure 22 is a view of the Joint European Torus, JET, in Culham, England. It is considerably bigger than TFTR, about twice as tall, with a little lower magnetic field. In fact, the aperture for the plasma, shown in Figure 23, is about as big as would be needed for a real commercial reactor. The major radius will need to be spread out, so that a blanket can be inserted, but the actual plasma aperture is already at full size. The JET team is doing well with this machine, lagging behind us a bit in getting the plasma hot, but in terms of long confinement times we have 400 milliseconds and they have 800 milliseconds at a somewhat lower density. The 800 milliseconds is already very close to the previously mentioned 1000 milliseconds that are required for confinement. The Japanese entry, the JT60, is shown in Figure 24. When this picture was taken it was still under construction in Mito—which is a little north of Tokyo. It went into operation on April 8 and seems to be a very strong contender, with some

#85A0032

STEADY-STATE TCT/CIT RADIATION TEST FACILITY

MAJOR RADIUS	3.0 Meters
MINOR RADIUS	0.75 Meters
ELONGATION	1.7
MAX. FIELD AT TF COILS	8.5 Tesla
FIELD AT PLASMA AXIS	3.7 Tesla
PLASMA CURRENT	2.7 MA
NEUTRAL BEAM ENERGY	100 KEV (D°) 150 KEV (T°)
NEUTRAL BEAM POWER	110 MW
AVERAGE ELECTRON TEMPERATURE	6.0 KEV
AVERAGE ELECTRON DENSITY	$5 \times 10^{13} cm^{-3}$
$\bar{n}_e \tau_{Ee}$	$5 \times 10^{12} cm^{-3}$ sec
BETA	0.045
Z_{eff}	1.5
FUSION GAIN, Q	1.2
FUSION NEUTRON POWER	97 MW
NEUTRON WALL LOADING (At Midplane)	0.75 MW/m^2

Figure 21. Projected Parameters of a Steady-State Reactor Based on the TFTR Experiments

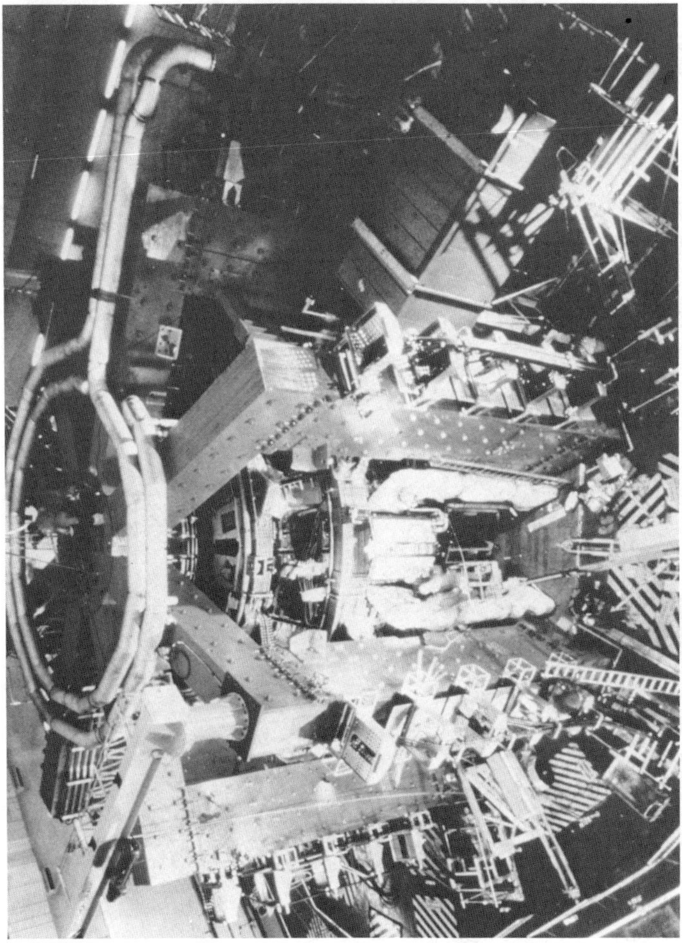

Figure 22. View of the European JET Experiment

Figure 23. Inside View of the JET Machine

Figure 24. View of the Japanese JT60 Experiment

advanced technological touches. The Soviet entry, the T-15, is shown in Figure 25. It is the most advanced of all, in a sense, because it has superconducting coils which we will want to have in the ultimate tokamak reactor. On the other hand, it will not go on the air until 1987. Only TFTR and JET are designed to use the D-T reaction. As for the TFTR, the plan for the latter part of next year is to demonstrate deuterium plasma conditions that will be suitable for break-even in deuterium-tritium plasmas. A few years later, in 1988, we will actually be able to put in tritium, hopefully reach break-even conditions, and put out the 27 megawatts of fusion power as originally targeted.

The machines that we have just discussed all came about in a mood of international competition that prevailed during the mid '70s. Many things have changed now, budgets are tight, and international cooperation, as we heard earlier today, is regarded as a very good thing by the heads of state. Consequently thoughts have turned to the question, whether we really need four of everything or could perhaps build just a single international machine in the future. As early as 1979, a joint project for the design of a next-step reactor called INTOR was started under the auspices of the International Atomic Energy Agency, with participation by the Soviet Union, Japan, Europe, and the United States. The INTOR design is shown in Figure 26. The design activity has flourished, much serious work has been done, a lot of knocking of heads together has occurred, and, remarkably, a convergence of thinking has emerged, so that now all those rather diverse parties are pretty much agreed on what would be a reasonable design.

The participating governments were not all that serious about the prospects for actually building this machine together, so purely national versions of the INTOR plan were always maintained. The Europeans have a machine called NET and say they are prepared to go it alone and start in the early '90s. The Japanese have a similar plan, the FER, and claim to be ready to go in 1990. The United States led the way in this trend by announcing quite some years ago, in 1980, that it was prepared to go it alone. A design project was started and an act of Congress was passed. As the energy crisis went away and the budget crisis appeared, those prospects have faded and the general orientation of the United States fusion program has diverged quite a bit from the rest of the world.

The sort of commercial reactor the fusion programs of the rest of the world are aiming at is pretty much like the design shown in Figure 27. There is no little man in the picture, so the scale is not clear; however, from Sakharov's 12-meter radius, we are down to about six. Such a device would make about a thousand megawatts electric. This is the standard concept of the most probable first-generation commercial fusion reactor, as it prevails in many parts of the world. Quite a bit of attention is being paid to the details

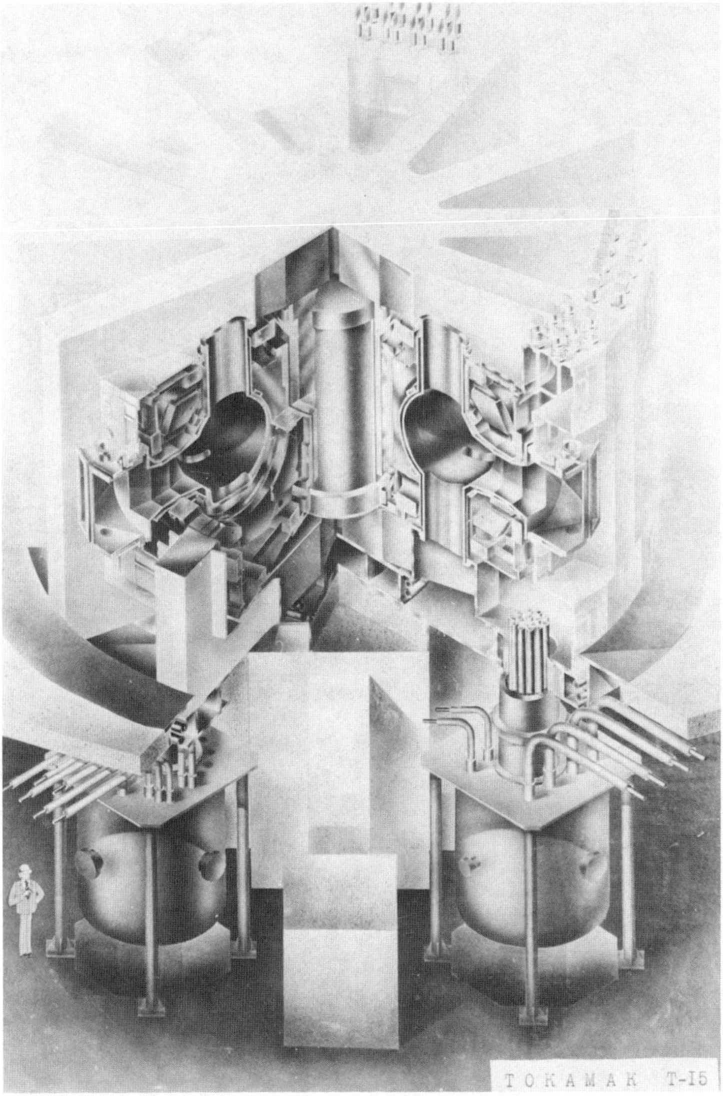

Figure 25. Schematic View of the Soviet T-15 Experiment

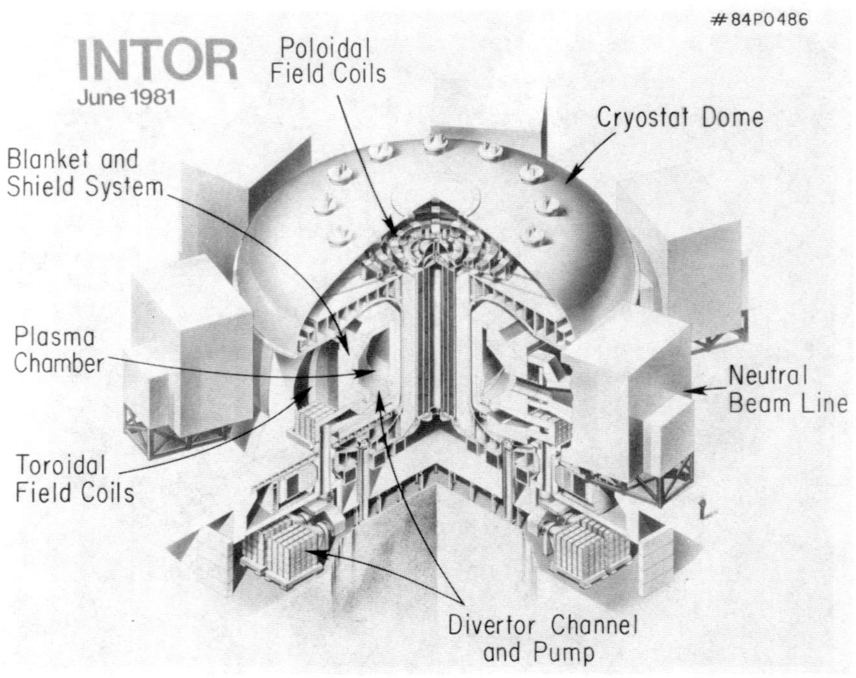

Figure 26. International Design for an Engineering Test Reactor (INTOR)

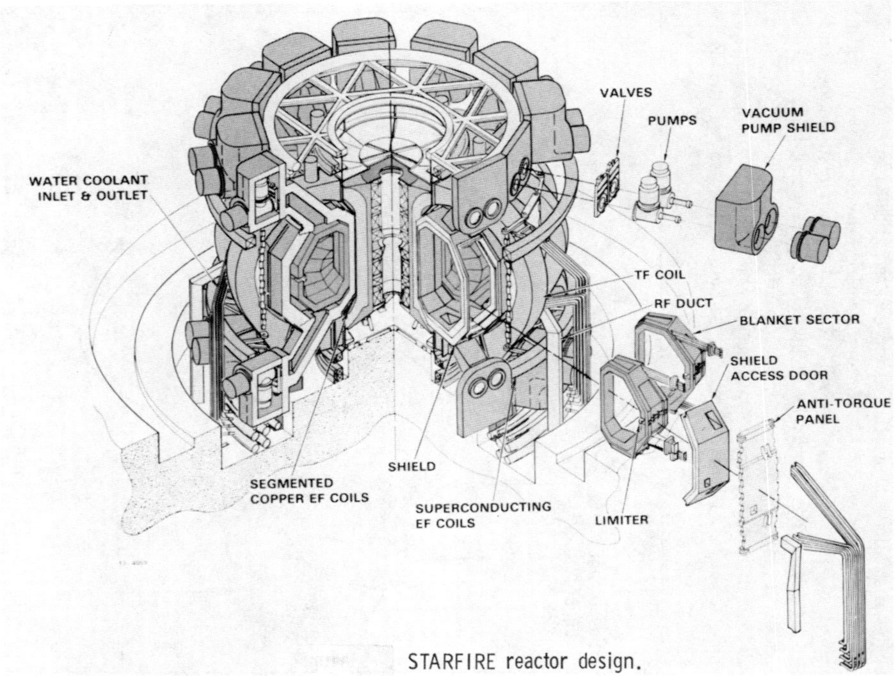

STARFIRE reactor design.

Figure 27. A Current Design for a Commercial Tokamak Reactor

of how such a device would be made, how it could be taken apart, and so forth, but clearly the entire engineering aspect is still in its infancy.

In the United States, the early thinking was to go in the same direction with an INTOR-type engineering test reactor. Then, as circumstances changed, we thought of smaller and smaller enterprises, as indicated in Figure 28. To some extent, these tokamaks are getting smaller because they are getting smarter: All of these machines are still expected to ignite. In another sense, the trend has been driven purely by shrinking budgets, which have caused us to put off the engineering part of our tasks, particularly the nuclear engineering, and look more and more toward just making a physics experiment to study how to ignite plasmas. Similarly, perhaps stimulated by the fission

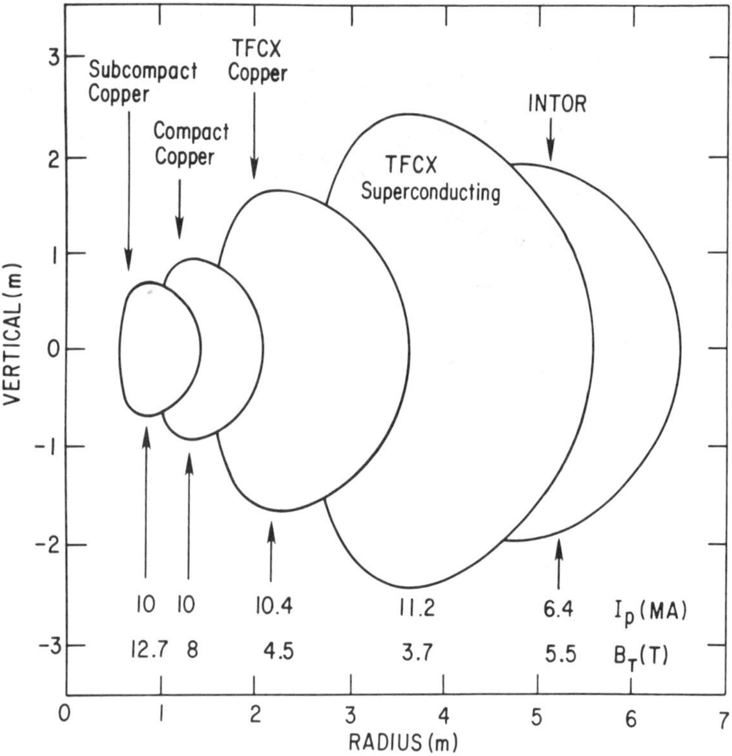

Figure 28. Comparison of Dimensions of Proposed Next-Step Experiments

Major Features of a 350 MW_e High Beta Tokamak with a Bean-Shaped Plasma

Fusion power	1000 MW
Average beta	25%
Major radius	3.6 m
Minor radius	0.9 m
Plasma Current	4.3 MA
Maximum TF coil magnetic field	8.2 T
TF coil magnetic field at plasma axis	3.7 T
Operational pulse length	0.5 hr
Operational recirculating power fraction	15%
Net plant efficiency	30%
Cost of electricity (84 $)	
• Stand-alone	71 mills/kW_eH
• Multiplexed	53 mills/kW_eH

Figure 29. Parameters of TRW Reactor Design

experience the U.S. is now looking toward smaller commercial fusion-reactor units. Figure 29 contains the parameters of one that has been worked on by TRW, with a projected output of 350 megawatts electric. Here we are getting down to the level of power output that Sakharov had in mind, but now it may be seen that the major radius is no longer 12 meters but only 3.6. If such a machine were in the world all by itself, its electric output energy would be estimated to cost 71 mills per kilowatt hour. If four of them were to work together, however, in a common power plant, the cost would come down to 53 mills. This seems like a much smarter scheme.

As for time scale, I would agree with Bob Hirsch's projections. We can reach ignition essentially in the mid '90s and that seems like a good project for the United States to do, since we have now shifted in that direction. If the Europeans and the Japanese are very much in earnest, and if we work with them, I think at least one of those machines could be started in the early '90s and completed in the year 2000. We would then have a real working

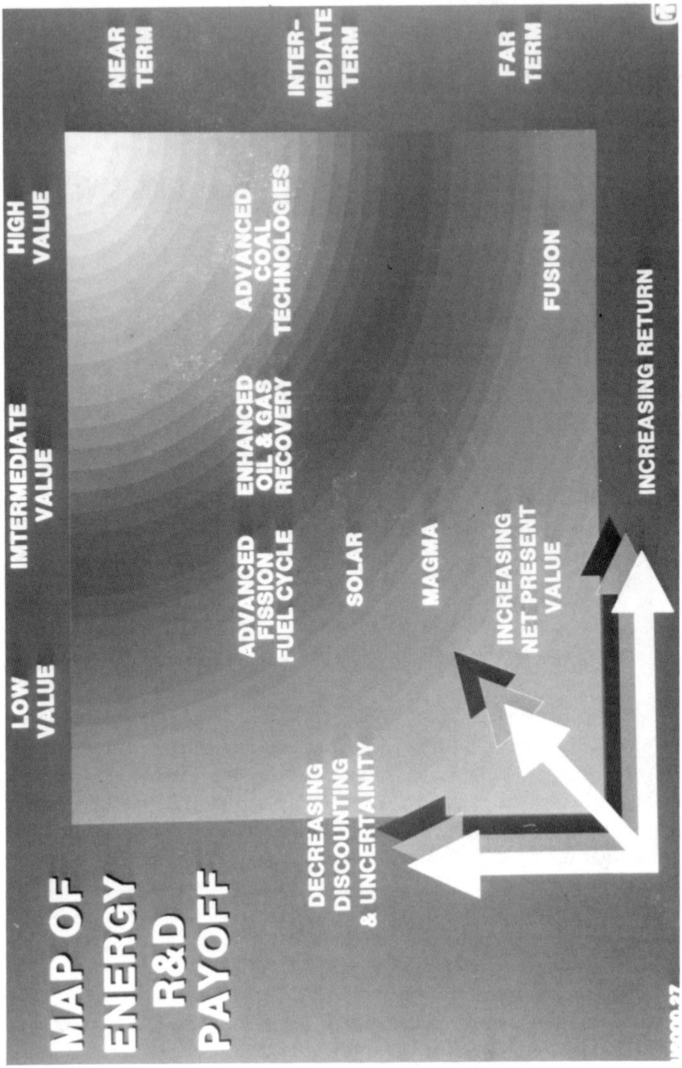

Figure 30. Relative Status of Energy-Research Activities

fusion reactor. The ignition experiment and the more ambitious nuclear engineering-oriented experiment would fit very well together and would help each other. During subsequent decades, commercially suitable machines will then be developed and built.

If the Japanese and the Europeans play the main role in the engineering of the standard tokamak reactor, there will be pressure on the United States to invent something a little smarter during the late '90s; so we, in turn, may build a next-generation reactor aimed at the commercial market. Be that as it may, it is not unreasonable that by the year 2020 someone will have a fusion reactor to sell, and, as Bob Hirsch suggested, it only takes thirty years after that for acceptance by the utilities. Thus by the year 2050 there could be a significant contribution by fusion to the world's energy needs. That is not bad timing because, while I am sure the ingenuity of the oil people will manage to stretch their resources, by 2050 they will be pretty far along.

Finally, Figure 30 is something I stole from George Dacey, the director of the Sandia Laboratories. Although this chart may be slightly controversial with respect to the location of each item, I think it states the basic situation in energy research very well. The vertical axis defines how close we are to actual success in a given process. On this scale, fusion ranks rather low. The horizontal axis defines the quality of a particular process, and there fusion ranks very high. Obviously, we need to concentrate on sensible things such as enhancing oil and gas recovery in the near term, and it would be folly to suggest that emphasis should be removed from these activities in favor of long-range projects. On the other hand, it is also perfectly clear that, sooner or later, oil, gas, and coal will run out, so there is no long-term prospect for them at all, and something must be done to meet that eventuality.

Returning to the chart, if we magnify the fusion corner of Figure 30, similar considerations will apply. Tokamaks are closest to reality. Other fusion concepts such as plasma "smoke-ring" devices are much further from reality but hold very exciting promise. Again, it is necessary to strike a reasonable balance in supporting what appears to be feasible in the near term, while not neglecting projects that may be the carriers of some great promise for the future.

Author's Note: All illustrations in this article are from the Plasma Physics Laboratory at Princeton University.

TRENDS IN ELECTRIC-POWER TECHNOLOGY–THEIR CAUSES AND EFFECTS

Milton Klein

Milton Klein received the B.S. degree in Chemical Engineering from Washington University in St. Louis, Missouri in 1944 and the M.B.A. degree from Harvard University in 1950. From 1950 to 1974 Mr. Klein served in the U.S. Government in several energy-related activities including a post as Assistant General Manager of the U.S. Atomic Energy Commission where he was responsible for both nuclear-reactor and non-nuclear energy programs of that agency. He was also the Manager of the NASA-AEC Space Nuclear Systems Office where he directed the programs of both agencies for the development and production of nuclear-powered generators for space missions such as Apollo, Pioneer, Viking, and Voyager, and the development of nuclear-powered rocket engines. In 1974 Mr. Klein became Associate Technical Director of Energy Programs for the MITRE Corporation, and in 1976 he was appointed Director of Research, Development, and Technology Applications at the International Energy Agency (IEA) in Paris, France. In April, 1980, he assumed his present position as Group Vice-President of the Electric Power Research Institute (EPRI) with responsibilities for industry relations, planning, and information services. Mr. Klein is the author of many articles on energy technologies, serves on several space and energy committees, and is an adviser to the IEA on electricity issues.

It is certainly no revelation to Cornell and its School of Electrical Engineering that electricity is a very special form of energy, one that has been revolutionary in its impact. Even people outside the Cornell Electrical Engineering School know it, at least if they think about it a moment; from the 50 percent or so of our population who have the freedom to choose either to work outside the home or inside it as a result of the electrical appliances in the home, to those in the work place, from the steelworker at an electric arc furnace, to everyone whose life is touched by the computer.

Not so well understood, however, is the fact that, properly applied, electricity is an energy-conserving way to use energy. Engineers know, and many of the lay public have been taught, that modern generating plants have about 35-40 percent thermal conversion efficiency. But that doesn't tell the efficiency story at all. The special character of electricity, its flexibility, the ability to control it with precision, to direct it remotely virtually anywhere, makes high, productive energy efficiency possible. Electricity has been a vital ingredient in improving the productivity of our economy.

We have historic and continuing evidence that electricity has saved energy. Figure 1 shows the energy intensity of the U.S. economy over the last century. Since 1920 the general trend has been downward, the energy used per constant dollar of GNP declining about 50 percent.

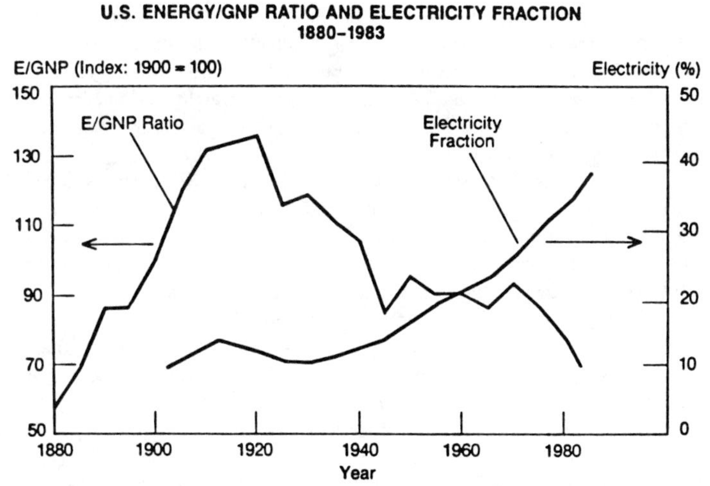

Source: EPRI

Figure 1. Energy Intensity of U.S. Economy in the Last Century

On the other hand, since about 1920 electricity intensity has been on an upward trend. So, while we have been using more electricity to produce our goods and services, we have been actually saving energy, strong circumstantial evidence of the productive efficiency of electricity. Even since the 1973-74 and 1978-79 oil difficulties, electricity use has generally continued to increase, while total energy use has leveled off or declined.

We all know of examples of this effect. Historically, electric motors vastly improved the efficiency of factory operations over the earlier mechanical systems--steam or water-driven devices. It is happening today. In steelmaking, the most competitive mills in an otherwise troubled domestic industry use electric arc or induction heated furnaces and electric finishing. On a total energy basis, the overall production energy is about 40 million BTU per ton of steel shipped from integrated mills with conventional steel furnaces, compared to about 10 million BTU per ton of steel shipped from electric mills with continuous casting. These savings result principally from the ability to use more scrap metal in the melt and from less processing loss and recycle. Construction costs of the electric mini-mill are also substantially less per ton of capacity that the conventional integrated mill. There are similar stories in other process industries.

But what of the future? I suppose that the natural thing to do at a centennial celebration is offer predictions of what the next 100 years will bring. I started to do that, but then decided you wouldn't believe me anyway. In fact, technology forecasting is a hazardous occupation. We don't even know what our resources will be, much less how much of them there may be, as evidenced by uranium a century ago. But advances in technology made them available. We are very good at extrapolating in straight-line fashion from the recent past to the infinite future, but a straight line is not the way the future evolves.

Nevertheless, it is useful to take stock of where we are and see the direction we are heading. We are on the threshold of major changes in the technology for generating and using power. And it's important to both utilities and the nation that we proceed over that threshold. Through much of the first 100 years of power engineering, it consisted of understanding, tracing, and controlling electrons and applying mechanical engineering skills to the processes involved. In the 1980's we are in a different era. The electricity system now requires all the high-techniques of the day. It now involves complex chemical and chemical engineering knowledge. It requires the analytical and control framework that sophisticated computers permit. In fact, a goodly fraction of the output of the Electric Power Research Institute (EPRI) consists of software--models and codes for research and system operation. It involves nuclear engineering. We need to apply new civil engineering and construction techniques to improve the constructibility in this most capital intensive of industries. So, engineering of utility systems is one of the challenging fields of today.

To understand why, let us take a look at the most important influences on utility planning, particularly as they relate to technology requirements, and then touch upon the technological trends which result from these influences. In these remarks, my perspective will be that of the Electric Power Research Institute, though certainly the Federal Government, principally through the Department of Energy, also plays an important role particularly in basic research and long-term developments.

There are five primary influences, some affecting the short-term and others the long-term:

- the changing fuel mix and probable long-term upward trend in fossil energy prices;
- the uncertain future rate of growth in electricity consumption;
- a regulatory system which tends to serve as a disincentive for new investment;
- the view that the consumer will play a growing role in defining the scope and value of service;
- the strong environmental ethic in this country.

Let us look at each of these factors. First, the influence of energy prices. Until the last decade, electricity prices continually declined. Since then, electricity costs, like all energy prices, have been increasing, although less than those of primary fuels over the last 12 years. As a capital-intensive industry, inflation and high interest rates have a special effect on utility costs. And, environmental and safety requirements also add. To counter these trends, utilities are placing increased emphasis on reducing costs or at least minimizing increases.

Second, the easy predictability of an approximately seven-percent annual growth in consumption that characterized the three decades after World War II is no longer with us. One can easily find heated arguments about the growth rate to be expected in the future. Population growth continues. Electricity penetration into society continues. For many years, electricity demand has tracked GNP quite well, including during the last decade. But is that relationship changing? What is the effect of the changes in U.S. industry? What will be the long-term effects of higher electricity costs?

Since past early experience and even the experience of recent years has shown that the use of electricity bears a close relationship to economic growth, there is little doubt that the demand for electric energy will continue to grow if we are to have a healthy economy. Electricity seems to be an essential ingredient of economic growth. Whether it will grow one-for-one with GNP is, of course, uncertain. But even with modest electricity growth, say two percent annually compared to the four-to-five-percent growth of recent years, additional capacity of about 100,000 MW above that now in place and under construction will be needed by 2000. Each additional one percent adds about another 100,000 MW. This compares to about 600,000 MW of current capacity. At the same time, older plants are reaching the age at which at least some will need to be replaced.

But building new plants poses problems as a result of another of the major influencing factors—the impact of the regulatory system on utility operations. I will not try to deal with all the regulatory features. Some of them have the effect of adding to the planning difficulties—the Public Utilities Regulatory Policies Act (PURPA) and the Fuel Use Act are examples. But ratemaking has had the most profound influence. The regulatory system of 49 separate state commissions copes easily when rates are decreasing as they did for the first 90 years of the power supply system, not so well on the up-side. So, in many cases rates have not kept up with costs, particularly when new construction is involved, and the result has been to reduce the financial return to the stockholders and difficulty in financing new plants. The strongest utilities financially are those without major construction programs. There is quite naturally, therefore, an incentive to avoid new major investments.

As the consumer reacts to increased costs and the utilities attempt to provide reliable power with minimum investment, utilities recognize that the consumer evaluates electricity in terms of the service it provides, and that electricity demand will be dependent on optimum, high-value service. Thus, increasing attention is being given to the customer side of the meter—conservation and end-use management—as a service strategy, encouraging consideration of conservation options on an equal basis with construction of new facilities.

The last of the influences is the environmental ethic. All industrial and human activities pose environmental issues. In the case of the electricity system, the technological demands are ubiquitous. The constantly changing and uncertain nature of the requirements, through law or regulation, amplifies the issues. And the impact of nuclear regulation is an extreme case which has added to the financial woes of some of the utilities.

Taken together, these influences have led to an R&D strategy that:

1) emphasizes technology development to extend the life and increase the productivity of existing plant in order to minimize new plant investment;
2) provides the technology for load management and efficient end-use of electricity;
3) provides system options, because of the uncertain rate of demand growth, that can be added in smaller capacity increments with minimum sacrifice of economy of scale;
4) focuses on generation concepts that make use of the most plentiful and least costly of our energy resources, with due regard to the resources that have important potential even though regional in character.

In all cases the dual criteria of minimizing power costs while meeting society's environmental requirements are fundamental.

To illustrate how this strategy translates into technology terms and trends, I will, in turn, touch upon generation technology, transmission and distribution technology, and technology developments in the use of electricity.

First let us consider generation. In the near term, to get as much reliable power as possible from the existing plant and to do so with high efficiency, there is considerable emphasis on instrumentation for control and diagnostics, non-destructive testing techniques, and on developing improved understanding of plant chemistry and materials-aging effects.

For example, the chemistry of flue-gas desulfurization systems is a critical factor in the reliability and costs of those systems. Research to identify the optimum materials and reagent chemistry is proving to be perhaps the most important means of improving scrubber performance. Water-chemistry control is vital to the reliability and durability of steam generators in pressurized water reactors. Material selection to minimize cobalt in piping has been found to have a major effect on minimizing radiation exposure to nuclear plant workers.

Newly developed instruments for diagnosis of plant condition and control include:

- an automated system for continuously monitoring and tracking water chemistry;
- a torsional vibration monitoring system. This is a device that continuously monitors turbine generator shafts to assess fatigue-life accumulation resulting from system disturbances. It can be used to provide an alarm to prevent catastrophic damage to the turbine generator;

- a continuous analyzer of the characteristics of coal being fed to a power plant, using neutron activation, in order to determine the most economical way of managing a utility's generation system (Figure 2);
- an automated pipe-crack detector, to monitor pipes in nuclear power plants in order to avoid unanticipated shut-downs (Figure 3);
- a miniature linear accelerator to inspect, in place, thick metal sections for flaws. Such inspections are required periodically in nuclear power plants; to perform them without removing the components is a substantial cost-saver (Figure 4).

These are among the innovations that will raise the performance of existing plants or extend their useful life with minimum degradation in performance.

But what of future power generation technology? For the forseeable future, coal and uranium are likely to be the fuels that can provide the cheapest power. Both have difficulties, coal with its environmental consequences and uranium with the public concerns about safety and disposal of radioactive waste. Nevertheless, we have or can forsee technology that can allow the expanded use of these fuels in an environmentally sound and safe way.

Source: EPRI

Figure 2. Continuous Coal Characteristics Analyzer

Source: EPRI

Figure 3: An Automated Pipe-Crack Detector

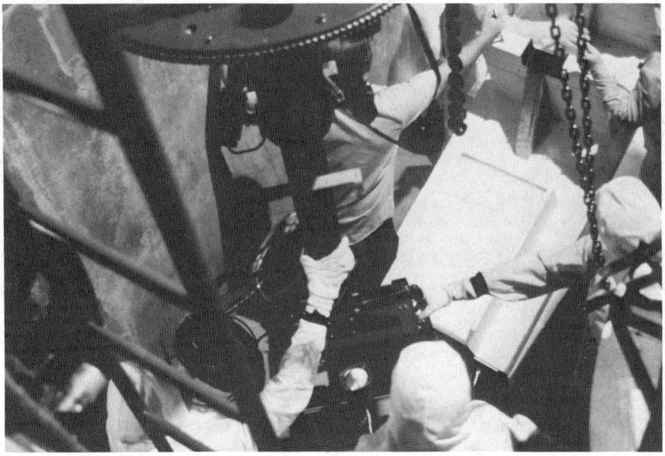

Source: EPRI

Figure 4: Thick-Metal Flaw Detector

Today's technology for burning coal and minimizing environmental problems involves cleaning the coal before use and employing wet scrubbers to remove sulfur dioxide in the flue gas, and electrostatic precipitators in the stacks for particulate removal. But there are ways now being developed that over the next several decades will replace today's systems with better economy and lower emissions.

One such method is through burning coal in a fluidized bed (FBC) which also contains limestone or equivalent calcium-containing material. A 20 MW FBC is operating on the TVA system as pictured in Figure 5. In this bed of coal and limestone, made to act like a fluid by flowing air through it, sulfur is captured by the calcium with high efficiency. Also, the combustion temperature is lower, so that less nitrogen oxide is formed. And a fluidized bed boiler can accommodate a wide range of fuels—not only differing coals but even municipal refuse or old railroad ties and tires. Finally, FBC plants promise to be smaller and less expensive than today's technology. Several larger plants are in initial stages. The largest of them is a 160 MW(e) plant sponsored by EPRI, TVA, Duke Power, the state of Kentucky, and others, which is currently in design.

Source: EPRI

Figure 5: A 20-MW Fluidized-Bed Coal Combustor

Another advanced coal-utilization technology for electricity generation involves the production of a low-to-medium BTU gas in an integrated system with a combined gas-turbine and steam-turbine power generation system, the so-called integrated gasification-combined-cycle. Sulfur is very effectively removed in the gasification step and the combined cycle, of course, enhances conversion efficiency.

There are several gasifier approaches developed by different companies. The largest experimental plant currently in operation is the Coolwater Plant in Southern California. It uses a Texaco gasifier and is shown in Figure 6. This 120 MW(e) plant is principally sponsored by EPRI, Southern California Edison, Texaco, GE, Bechtel, and a Japanese consortium. This first-of-a-kind plant was started-up last May and has exhibited remarkable performance, equaling or bettering the availability of conventional coal fired plants, with emissions well within standards.

Source: EPRI

Figure 6: An Experimental Coal-Gasification Plant

We can also expect advances in important auxiliary technologies. One example is the development of improved methods of particulate removal. Another is the development of improved approaches for cleaning coal before use. EPRI has built a facility for developing and testing improved cleaning methods, at Homer City, Pennsylvania (Figure 7).

Source: EPRI

Figure 7: Testing Facility for Improved Cleaning of Coal

Turning to nuclear power, we encounter an enigma (Figure 8). Nuclear power can be one of the least expensive ways of producing power. Most objective assessments conclude that it is one of the most environmentally benign. And it has a good safety record—no member of the public has been injured either in this country or, so far as we know, abroad.* Yet, no new nuclear plants have been ordered in this country for many years and none are anticipated for some time to come. From a utility perspective, the financial risks are too great, despite the fact that the early plants are among the cheapest

*This paper was delivered before the Chernobyl accident in the USSR. The Chernobyl reactor design was based on a significantly different concept than that of the light-water reactors commonly used in other countries.

power producers. Even plants currently being built will probably be economic in their lifetime, and in other countries, nuclear power remains a viable, economic and preferable means of power production.

Source: EPRI

Figure 8: Nuclear Power Reactors

I don't intend to try to explain how we came to this point in the U.S. Rather, I will simply assert that it is in the national interest that this form of energy be available, and therefore it is important to have R&D to provide a basis for such a future.

The experience gained through now many years of reactor operation has pointed the way to design improvements for light-water-reactor (LWR) plants, the most widely used types. There is work underway to incorporate this experience in advanced LWR designs. Development efforts are also underway on high-temperature gas-cooled reactors. For the more distant future, breeder reactors will be necessary to make full use of uranium and thorium resources. So there is R&D in the U.S. and in other countries to this end.

Earlier I commented on the efforts to achieve as high reliability and low cost as possible for both coal and nuclear plants. For nuclear plants, there is another whole dimension–the continuing analysis to assure their safety. Of course, a reliable plant is a safe plant. But the continuing assessment has brought into focus the possibility that prior estimates of radioactive material release from light-water-cooled reactors of the type commonly used in the United States, Western Europe and the Far East have been unduly conservative, even in the event of a very severe accident. The release may be much lower than previously assumed. The experience at Three Mile Island, at other plant accidents, and laboratory data all tend to confirm model work that has shown that the possible public hazard for this class of reactor is so low that there may be no or at most few injuries even in a severe accident. Research and analysis on this issue continues.

While coal or uranium is likely to be the fuel for most new generating capacity for some time to come, we can anticipate growing use, in special circumstances, of other energy sources. Geothermal in the form of dry steam is producing about 1200 MW(e) in California and is being used in other countries as well. EPRI, in cooperation with San Diego Gas & Electric and the Department of Energy, is developing binary-cycle systems to use effectively the more plentiful, lower temperature hydrothermal resources. A 45 MW(e) plant is being constructed at Heber in California (Figure 9).

Source: EPRI

Figure 9: A 45-MW Geothermal Binary-Cycle Plant

A large number of wind machines, of sizes ranging from a few kilowatts to several megawatts, have been deployed. In California, more than 3,000, most in the 50-kW(e) range, have been installed. Figure 10 shows one of the wind farms. All have been heavily subsidized through tax credits or by direct government funding. Our assessment of the technology and experience leads us to conclude that economic wind power is possible and that the optimum size of the unit is probably in the 500-kW(e) range.

Source: EPRI

Figure 10: Wind-Turbine-Generator Farm in California

We also expect solar electric systems to have a specialized role in power generation in the future. It is difficult to predict when, but progress is certainly being made in photovoltaics (Figure 11). Cells with efficiency of about 16-18 percent in one-sun arrays and about 25 percent in concentrating arrays, and at acceptable costs, are necessary to be economically competitive at future alternative energy prices. For one-sun cells a suitable combination of cell efficiency and cost is still to be attained. The best concentrating cells have achieved over 20 percent efficiency and it is believed they can be manufactured to achieve costs that will ultimately be competitive.

Source: EPRI

Figure 11: Photovoltaic-Generator Installation

Source: EPRI

Figure 12: 4.5-MW Phosphoric-Acid Fuel Cell

In the long term, fusion, discussed by Dr. Furth, may be a principal power source. The utility industry through EPRI is maintaining a careful watch on progress and is participating in a limited way in some of its system and engineering aspects.

A variety of advances in energy conversion systems is foreseen. A 4.5 MW(e) phosphoric-acid fuel cell installation is being tested in Tokyo (Figure 12). Other approaches are also being explored. The fuel cell has the potential for high conversion efficiency in small power increments. MHD is at an earlier stage of development, but also has the potential for high efficiency. And one day we will have generators with superconducting windings.

I mentioned earlier that the uncertainty of future rates of growth in electricity use has led to increasing interest in smaller capacity increments than the 800-1200 MW(e) plant additions which have been typical in recent years. One of the features of several of the technologies of future interest is their suitability for relatively small units with little loss of economy of scale. Both gasification-combined cycle and fluidized-bed-combustion systems appear economically feasible in 200 to 300 MW(e) increments. Even nuclear plants may be more attractive in sizes smaller than those of the current plants. Obviously, fuel cells, solar, and wind systems are very modular in nature. In addition to the advantage of smaller increments in periods of uncertain load growth, they also permit more of the systems to be shop fabricated, a distinct advantage in terms of cost and quality control, and reduce the financing and rate-shock impacts on the utilities and their customers.

The increasing use of cogeneration is incremental in character, but originates from other considerations. The technology involved tends not to be at the cutting edge. The eventual importance of cogeneration on the electric power system remains to be seen.

But increasing the number of points of generation through smaller central-station plants, wind machines, cogeneration installations, or small hydro places a new dimension on the transmission and distribution system. Indeed, advances in technology associated with electricity delivery can be very important, since about 25 percent of the costs of delivered power and one-third of utility capital investment is involved (Figure 13). And the reliability of the T&D networks is critical to make up for the relative unreliability of power plants.

Apart from relatively mundane devices to reduce costs, such as improved means of laying underground cable, reducing costs of transmission towers through improved design methods, and extending the life of wood poles, there

Source: EPRI

Figure 13: High-Voltage Transmission and Distribution Substation

are some major advances being made. Amorphous metal in transformer cores can reduce the core losses as much as 80 percent. Solid-state devices under development could, among others, produce circuit breakers with no moving parts, improving reliability and eliminating switching disturbances.

Increasing ac voltages above the current 765-kV peak would allow us to put more power through rights-of-way, particularly as obtaining new rights-of-way becomes more and more difficult. More use of high-voltage dc is expected as utilities seek more cost effective means to transport large blocks of power between utilities and power pools.

Eliminating or damping line disturbances and extending network-control capabilities to accommodate increasingly complex transmission systems requires new hardware and software. One example is the NGH damping system, shown in Figure 14, designed to counter subsynchronous resonance that can occur on very long transmission lines. These oscillations lead to mechanical stresses that can and indeed have shattered generator shafts (Figure 15).

Source: EPRI

Figure 14: Transmission-Line-Disturbance Damping System

Source: EPRI

Figure 15: Generator Shaft Destroyed by Synchronous Resonance Effects

Control of transmission systems is highly automated. Extending automation to the control of distribution is a next advance. Automated distribution could provide improved substation transformer loading, detect and isolate fault conditions, lead to quicker restoration of power in storms as well as provide load management and meter reading functions.

This brings me to the customer side of the meter. We will continue to see new uses of electricity. I mentioned some in steel-making earlier. A program to apply plasma jets to steel fabrication is underway as a further extension of electricity use. Lasers are finding their way into previously mechanical operations. One such application is the manufacture of specialty saw blades as shown in Figure 16.

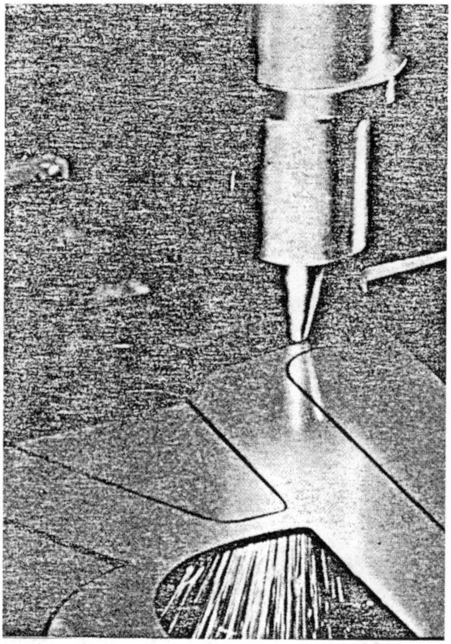

Source: EPRI

Figure 16: Application of Lasers to Saw-Blade Manufacture

Source: EPRI

Figure 17: Heat-Storage Furnace

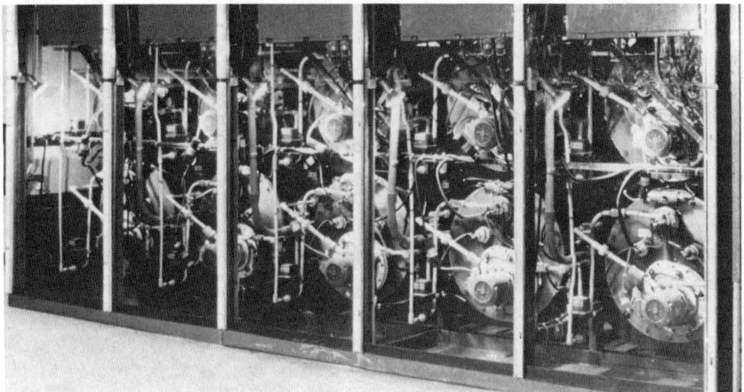

Source: EPRI

Figure 18: Zinc-Chloride Battery Test

Further use of heat pumps in commercial and industrial as well as residential applications is occurring. Heat-storage furnaces, widely used in Europe, are being improved to meet U.S. needs and will help in the leveling of utility daily loads (Figure 17). Battery development for load management is also the key to the electric vehicle. The zinc-chloride battery, pictured here in test, is one example (Figure 18). The technology using improved lead-acid batteries is near at hand for certain types of electric vehicles but is not yet at hand for personal transportation vehicles.

Finally, I would like to touch upon a few environmental issues. As I mentioned earlier, environmental considerations are pervasive. Among the most important advantages of gasification-combined-cycle and fluidized-bed combustion are their ability to use coal cleanly. The environmental and safety requirements of nuclear power are virtually the "name-of-the-game" in nuclear power R&D. Gaining experience with the environmental effects of wind machines is one of the purposes of those activities. At EPRI, our work goes not only to environmental control systems, but also to ecological effects. Thus, we have a major program relating to acid deposition, performing research on atmospheric transport and chemistry and on ecological effects in forests and lakes. We also have for many years been supporting research on high-voltage effects and effects of toxic materials. Through such work, we are trying to provide a scientific basis for the policy decisions that can all too easily be based on emotion and political expediency, and ultimately be counter-productive.

I have tried to present a sampling rather than an exhaustive description of the directions utility R&D is taking. There is one more characterization of significance. Most of our work is applied research and development. We have a limited program of exploratory and basic research. We are very conscious of the importance of such work, and are trying to increase our level of support. However, without meaning to enter the heated budget discussions in this town, I must note that we have depended on the Federal Government to be the principal supporter of such work. Without such support, we have a gap of national consequence.

Indeed, the technological health, in total, of the electricity enterprise in this country is of national consequence. It will continue to be essential to our national well-being. It is a vibrant activity, even though its high-tech character doesn't get the media attention that computers get. It poses the type of challenges that require continuous infusion of good young engineers and scientists. So Cornell's Electrical Engineering School can be as important in the next 100 years as it has been for the last.

FUTURE OF ELECTRIC-UTILITY RESEARCH IN EDUCATIONAL INSTITUTIONS

Gerald L. Wilson

Gerald L. Wilson received the S.B. and S.M. degrees in 1961 and 1963, respectively, and the Sc.D in Mechanical Engineering and Magnetohydrodynamics in 1965, all from the Massachusetts Institute of Technology. From 1965 through 1975 he was an Assistant Professor and then Associate Professor of Electrical Engineering at MIT. In 1975 he was promoted to Professor of Electrical Engineering and appointed Director of the Electric Power Systems Laboratory. In 1978 he became Head of the Department of Electrical Engineering and Computer Science at MIT and remained in that post until 1981 when he became Dean Of Engineering at MIT, his present position. Dr. Wilson held a Ford Foundation Fellowship in 1965-66, and was a consultant with Dynatech Corporation and the American Electric Power Service Corporation in the 1965-67 period. His research interests include magnetohydrodynamic energy conversion, electrohydrodynamics, electromechanics, and electric-power engineering. Dr. Wilson is a member of the National Academy of Engineering and is a Fellow of the IEEE.

Several months ago, I was asked to speak at this wonderful occasion. When the topic for this talk "The Future of Electric Utility Research in Educational Institutions" was suggested, a single question popped immediately into my mind. Is there one?--Is there a future for our best students in the electric utility industry? Or is this an industry which is going to get by with pretty much the status quo and therefore with whatever students they can attract? A good university is not going to design research and educational programs and hire faculty in a field whose vitality and vision of the future rises and falls with each eclipse of the sun. The past relationship between the electric utility industry and universities has been checkered at best. Soon after the work of Thomas Edison, universities and the industry were working closely together to research and develop the generation, transmission, distribution, and utilization of this wonderful new source of energy. You cannot look at any history of Cornell's one-hundred-year-old department, whose anniversary we celebrate today, without learning of students and faculty who worked together to invent the technology for bringing electric energy into the service of mankind.

During the period just after the second world war and into the fifties, the interest in research in electric power tapered off drastically, due to lack of support, lack of vision, and lack of the perceived need for major innovation. It was an industry that had experienced its revolution, had matured and would grow through evolutionary not revolutionary steps. At MIT, Gordon Brown removed electric power research and education from the electrical engineering curriculum and replaced it with a new curriculum for the future, a curriculum that emphasized the teaching of fundamentals to our undergraduates.

In the late 1960's, Philip Sporn came to MIT and to other institutions and expressed his deep concern that a leading Institute of Technology had no commitment or programs related to electric-power technology. He chided us, cajoled us, encouraged us, and later supported us in the manner that was so unique to Philip Sporn. Our program and others grew and flourished during the ensuing decade.

Our students were challenged by the advent of 765 kV transmission, by the growth of central-station plants from 600 to 1400 MVA, by the need for new relaying techniques aided with the development of the microprocessor, by the development of system controls using stochastic and adaptive-control techniques. They learned that engineering in the electric utility environment was an opportunity to bring a wide range of disciplines, to integrate a knowledge of mechanical, chemical, electrical, and computer techniques into an environment of complex, technical, environmental, and political constraints, all for the public good.

In the mid seventies, we were turning talented students away for lack of support and faculty time. We taught our students that the field of electric energy was not limited to one or two disciplines. It was a field in which we could see the contribution of our efforts benefiting society; which is after all the essence of engineering--using the benefits of science and technology for the public good.

Today, however, we are at a crossroads. The utility field is attracting very few of our brightest students. The challenges and visions of the future are not as evident. At present the industry is spending only four-tenths of one percent of sales on research and development, while the computer and semiconductor industries are spending fifteen times that amount, and the aerospace, office equipment, and leisure-equipment industries are spending twelve times as much.

In some ways these are bleak times for all of us. This country is in the midst of a national dilemma. We are losing an international competitive race for high-technology markets. The Japanese and Europeans are employing our technology more carefully, more thoroughly, and more creatively to challenge our economy at its very foundations. "Made in U.S.A." has become a warning label for the consumer. Seventy-four percent of the American public believe the Japanese are equal or superior to us in technology and science. The Japanese have not limited their competitive inroads to high-volume manufacturing operations. The U.S. construction industry is feeling the competitive pressures. Dams, bridges, and highways will soon bear little signs, innocuous of course, that read "Made by Japan." They have started with major components for the electric utility industry, including transformers, circuit breakers, and generators. How soon before the Japanese propose power plants that are markedly competitive with U.S.-built plants? These will be modern, automated units that are equipped with well-conceived sophisticated controls, that emit minimum amounts of SO_x and NO_x, and have very low capital and operating costs; built for us but "Made by Japan."

In many ways the U.S. utility industry conveys the image of a beached whale. The industry is locked in a morass of federal, state, and local regulations. Its profits are limited, its return on investment is stifled--it is squeezed to the point where its ability to invest in the future is an order of magnitude below that of other high-technology industries. Its relations with the public are in a shambles. It has invested for three decades in the nuclear technology that is vital to our future. Through fault of its own, through fault of the regulators, and through its inability to help the people and the media understand the role of nuclear power, the electric power industry is witnessing what anti-nuclear forces optimistically describe as the "end of the first nuclear age."

We continue to rely heavily on oil for transportation, oil and gas for space heating and industrial processing, and we are looking more to coal for our electric supply. Yet the use of these fuels, burned with oxygen and emitting CO_2, continues to cause concerned groups to raise questions about possible widespread environmental and economic damage; questions, for example, on the effect of carbon-dioxide emissions on weather patterns, and on the location of precipitation relative to farming, population centers, and the preservation of our rain forests. Should this growing concern lead to restrictions on burning hydrocarbon-based fuels? What are our options?

In some countries and for differing reasons, nuclear energy sources are the answer. For us, at present, that option is a bleak one. Recent events characterized in the public's mind by Three Mile Island, have caused an effective moratorium not only on the use of the present technology, but also on the development of new technologies based on nuclear energy. Costs of very similar plants vary by factors of two to three, depending on who manages the design and construction. We are swiftly losing our domestic nuclear industry. There is a national need to explore this technology anew. Using what we have learned from the past, universities and the industry, working together could develop a nuclear cycle and associated technology that meet safety standards and environmental constraints that are acceptable to our people. Today such standards are hard to find if they exist at all. Here is a problem fraught with the exigencies of political, regulatory, and technological constraints. But it is a problem we must face, with or without government leadership. We must provide the technological leadership and the people to meet this need.

The reductions in growth of electric load are an opportunity! An opportunity to review the way we have done business in the past. To review the way we have designed the system from megawatt of generation to watt of load. We should review our choices for a nuclear fuel cycle; from raw material in, to waste material out. We should review the concept of central-station power plants with economies of scale as their justification, and look again at the potential benefits of mass-produced small generation units, with the benefits of cogeneration and the economies of large numbers. We should resist the tendency to be reactionary when it is suggested that we review the boiling water reactor. It should not be viewed as an admission of choosing the wrong technology or of having made mistakes in the past. Our history as engineers is in some sense one of making mistakes. Science and technology is a living process, one in which we learn from errors in judgment or understanding, for it is our business to be working at the frontiers of the unknown. In engineering and science the only constant is change.

The vacuum tube was not a mistake; it was replaced with the transistor, but it was part of the evolution of electronics. The differential analyzer was not a mistake; it was replaced with the computer, but it was a step in the evolution of computers. Newton's First Law was not a mistake; it was replaced with more complex laws, but it was part of the evolution of the laws of mechanics. Similarly, the present nuclear technology and central-station power plants are not mistakes; they are a part of the evolution of our future electric supply. The utility industry must play a central role in defining the environment and the technology that will shape the form of our electric supply of the future. Otherwise they will not be a part of the future. Philip Sporn said it very well when he described the job for utilities as "inventing the future."

On the educational side, there is much to be done and room for many changes and improvements. Many universities with the firm support of the industry, continue to run programs that are strongly focused on the needs of yesterday's and today's practising utility engineer. Standard fare includes rotating machinery, relaying, system planning and dynamics, transmission, distribution, etc. Many of these skills are needed today, but can they not be boiled down to a thicker broth? Are we training engineers for this industry when we should be educating them? What is the role for an understanding of the new technologies in a potential utility engineer's education? The role of new technology in the future of the utility industry is clear. Without it the industry would, will, and should wither on the vine. Digital-signal-processing, computer-aided design and manufacturing, new materials, integrated circuits, nonlinear programming, artificial intelligence, a host of other disciplines, unconventional from a utility point of view, will be central and essential to the future of the industry and therefore to its professional people. As educators, it is our role to recast the changing technology into fundamentals and to teach them to our students. This is the distinction between training and education. We must prepare our students for a lifetime of contributions–for the long-term. We must provide them with an education that is exchangeable at par in twenty years.

Once a young engineer graduates from our institutions, his or her education has only begun. For the creative engineer and leader, engineering is a commitment to a lifetime of learning. Today technology is changing at a pace that is incomparable with our past experience. An engineer who has been working in the profession for five years is in the midst of changing science and the way it can be utilized for the public good. A four-year or six-year program of education is simply not adequate to prepare for forty years of contributions. Offering subjects *at* the university, for the professional to come back to and take for credit or audit is simply inadequate. We need to work with industry to make the prospect of *lifelong* education a natural extension of the university into the environment of the workplace. It took educators 200 years after the invention of the printing press to accept the printed word as a way to learn. We need to investigate the use of tutored video instruction, the use of computers with interactive graphics, textbooks with floppy disks instead of figures, and any other mode of communication and learning that we can find, to deliver opportunities to the professional engineer and the professional learner in as natural and easily accessible way as possible. We must work together with industry to develop, define, and teach these subjects in ways that integrate the learning process into the professional workplace. This requires initiative and leadership from universities and industry. It needs to be done if we are to assure the healthy utility system that is so central to our nation's future. It needs to be done and together we must do it!

Earlier I painted a very bleak picture with respect to our economic future as a nation,—but there *are* many signs for and reasons for hope. Positive changes are beginning to occur in the automobile industry. New management and process-control techniques are being introduced. Labor is working more closely with management to produce a higher quality, competitive product. The computer, semiconductor, and electronic industries are rising to the challenges of foreign competition. Changes, many seemingly small, but positive changes are occurring that will soon synergize and show the true mettle of the American people. Our electric supply is critical to implementing new technological contributions to our productivity. Where would computers and electronics be without a reliable, economic electric supply? Will we be prepared to meet the demand on our electric supply brought on by new growth and new technology on the time scale required? What will be the national response if we cannot? The industry must provide the leadership *now* to develop the electric technology for the future. We need an activism to bring the changes that must occur if the utility industry is to have the flexibility to act. New regulations are needed. We must redesign our entire nuclear technology system—from the mining of the fuel to its disposal. The days of one-of-a-kind, nuclear plants have passed. We must design smaller units, with the constraints of siting, safety, and fuel disposal included from the beginning. We must qualify and license these plants and their supporting structure generically before they leave the manufacturer. We must work to utilize all of our conversion technology in an integrated fashion, including coal-gas separation, conversion, and higher efficiency combustion systems. We must re-examine old practices. Do large-scale centralized generating plants based on economies of scale still make sense? Or do very small, consumer-located co-generation units which utilize existing fuel and electric distribution systems based on economies of large numbers, make more sense? Will it require the injection of small residential co-generation units into our marketplace by the Europeans or the Japanese for us to consider this prospect seriously? The use of new materials brings higher efficiencies to the conversion processes. New load-control techniques using microtechnology can displace much of the need for spinning reserve. This requires not only the application of existing technology. It requires creative approaches to regulation and new perspectives on the role of utilities. It requires major changes in our regulatory practices. It requires an aggressive, leadership stance of the industry, *aggressive* not passive.

The industry must take the initiative in defining for the public the need for these changes.

Is there a future for electric utility research in our universities? The problems exist—there is inventing to be done. There are opportunities to review the entire premise of central-station power—is there a future in small distributed power plants that use waste heat for space heating and cooling—plants that are designed, operated, and monitored by our electric utilities? There are opportunities in the design of electric machines. Electric machines that operate in simple applications such as pumps, fans, and compressors consume 25 percent of all electric energy in the U.S. Cutbacks in fuel use that would come from the development of high-efficiency electrical machine systems could save billions of dollars per year. The typical efficiency improvement for electrical machines is the addition of more copper. But the development of variable-speed electric machine systems using machines, inverters, and smart electronic control algorithms designed—integrated—to work together, offers far better improvements. A more effective interface between the utility and the user can offer opportunities for spot pricing, for smart controls that bring the benefits of choice of time and use and therefore cost to the consumer, and brings the use of load control as an alternative to spinning reserve to the utility. These are only examples of a host of opportunities for the utility industry to lead, to develop long-range research that can lead our nation to an electric utility system that is aggressive, visionary, and ready for the future. The problems are waiting to be formed and solved.

Is there a future for electric utility research in universities? Today our students, especially students in electrical engineering have a myriad of wonderful choices for their future. Opportunities to work in new and innovative environments where change and challenge and excitement are commonplace. They have opportunities to work in industries where four-and-one-half to six percent of sales is spent on R&D, not four-tenths of one percent!! This translates directly into the environment students experience on the campus. Industrial research support on the campuses of leading universities has grown markedly in the past five years. The academic environment has strong similarities with the marketplace. The competition is among career paths rather than among products. But the mechanisms are the same. If good students are to enter the utility field they must see the opportunities, vitality, and fresh approaches that the industry offers through the visionary and exciting research it supports on our nation's campuses. The Electric Power Research Institute should add a major component to its expenditures to support long-term, visionary university research that will attract our best students. It should do so realizing the fundamental role of university research in affecting students' choices. It should do so realizing that fundamental research is endemic to the university and essential to the future of the industry. They should do so realizing that they must play a role on our campuses, otherwise they will have lost out in the most fundamental competition of all—the competition for talent.

Should we steer our students into this field? Is there a future for talented young people in this field? Should we continue in the next decades to operate laboratories and hire faculty whose focus is electric-power technology? The answers to these questions for us largely depend on the industry's initiatives, its leadership, and its vision for the future of the electric power system of this country. It is not a simple task, the utilities do not have the free options that are available to the free-enterprise system. But they are taking the beating from the regulators and the public too passively. Together, working together, we can change the environment for our nation's electric supply. We can work together to change the image and develop a vision. The legacies of Thomas Edison and Philip Sporn for invention and inventing stand as a reminder that if we are not part of the solution, then we are part of the problem.

THE ROLE OF THE DEPARTMENT OF ENERGY IN ENERGY RESEARCH

Alvin W. Trivelpiece

Alvin W. Trivelpiece received the B.S. degree from California Polytechnic State University in 1953, and the M.S. and Ph.D. degrees from California Institute of Technology in 1955 and 1958, respectively. From 1959 through 1966 he was an Assistant Professor and then Associate Professor of Electrical Engineering at the University of California at Berkeley. From 1966-76 he was Professor of Physics at the University of Maryland and, while on leave from the University in 1973-75, served as Assistant Director of the AEC Division of Research on Controlled Thermonuclear Reactions. In 1976 he became Vice President for Engineering and Research at Maxwell Laboratories, Inc. in San Diego, California. In 1978 he moved to Science Applications, Inc. in La Jolla, California, as Corporate Vice President for Research, a post he held until 1981 when he was appointed Director of the Office of Energy Research of the U. S. Department of Energy, his present position. Dr. Trivelpiece holds several patents in the area of physics and is the author and coauthor of over 100 technical reports. He is the author and coauthor of two textbooks on Plasma Physics. He is a Fellow of the American Physical Society, IEEE, and AAAS, and is a member of the New York Academy of Sciences, the American Nuclear Society, and several related organizations.

Editor's Note: In May, 1987, Dr. Trivelpiece became the Executive Officer of the American Association for the Advancement of Science (AAAS). In the same year he received the Secretarial Gold Medal of the U.S. Department of Energy, and the California Institute of Technology Distinguished Alumnus Award.

I would like to preface my remarks with the comment that celebrating the Centennial of electrical engineering education at Cornell through a nationwide lecture series is an unusual approach to an auspicious occasion. Cornell and electrical science are intertwined in our history. Ben Franklin, in the 1750's, was probably our first electrical engineer. Eighty years later, in the 1830's, Joseph Henry was the second. It took 50 years more for the technology of electricity to really flourish. This was an extraordinary change in human history. The year 1885, which we now commemorate, follows Thomas Edison's first demonstration of the incandescent lamp by six years and precedes Nikola Tesla's invention of the induction motor by three. Since that time the technology and science associated with electricity have rapidly come to be central to the development of industrial societies and to our understanding of the laws of nature.

In 1882 Edison started the first U.S. central station power plant with 59 customers. Just down the road from Cornell, Charles Steinmetz was bringing modern electrotechnology from infancy to adolescence. It is remarkable that in 1885 Cornell began the education of the young in this embryonic field and has been doing so ever since. Nearly all occupied homes and our entire industrial and commercial enterprise are served by a network of over 3,000 electric power companies. An enormous array of appliances and devices has totally transformed the way people live, while the computer, microelectronics and robotics revolutions are in the process of changing how we work, think, communicate and even organize our society. The founding fathers could hardly have imagined the entire nation able to watch the Congress on C-span, or that electronic mail would routinely replace sending mail by messenger along the Boston Post Road.

For some forty of the 100 years that we are celebrating, the Department of Energy (DOE) and its predecessors have been involved in stimulating the technology and the science leading to energy for the future. The Department has also been the main focal point, within the government, for the Nation's investment in fundamental research in the physical sciences. These roles are closely interrelated and much of the new energy technology that holds promise for the future depends directly on our past investments in understanding how nature works. For that reason I will just sketch for you the general role of the DOE in Energy R&D, since more detailed discussions have already been presented today by the experts in some of these areas. I will then say a few things about the long-range fundamental research programs and why they deserve your attention.

The energy industry is perhaps the largest commercial entity in the nation. It is so crucial to the effective functioning of our society that some parts of it are regulated to ensure that the public is adequately served. Along with the progress that has been made toward deregulation of energy and other industries in recent years, there has come a clarification of the purpose of federal involvement in R&D.

In particular, the federal role in energy research and development is concentrated in areas where the incentives for and availability of private investment are severely limited or nonexistent. In these areas, competitive rates of return on investment cannot be realized by private firms, either because they are difficult to predict or highly risky, or because they would require an exceptionally long time to be realized. The success of federal research and development programs in providing options for private investment and commercial development depends on effective technology transfer. The federal government does not have responsibility for the adoption of new energy technologies by the private sector and often is not a customer for such technologies. Its contribution to the development of future energy resources lies in the prompt and effective transfer of results from federally funded generic research and development activities to private entrepreneurs, innovators, businesses, and industry, and in facilitating transfers of technology within the private sector.

The relationship between government and industry is particularly complex as regards the electric utilities. The electricity industry is unique in several aspects. Design, licensing, and construction lead times for new base-load generating capacity are 10 to 14 years. The introduction of a successful new technology to the point of reasonable market penetration requires 20 to 30 years. Economically sized units of capacity are often very large, so the prototypes of production scale are large, and usually necessary to bring the financial risks to a reasonable level to assure reliability. Regulation on a state-by-state basis makes it difficult for the individual firm or the industry as a whole to take adequate account for long-term needs in mounting R&D programs.

R&D related to electricity supply tends to involve long lead times and very large expenditures for individual projects as a technology reaches the pilot plant and demonstration stages. Such research and development has long been supported both by the private sector (utilities and their suppliers) and by government. In general, the private sector has concentrated most of its efforts on relatively near-term technology (initial use within 10 years), much of it related to transmission, distribution, storage, and utilization. Industry, however, also has applied significant funding to intermediate-term technology (initial use 10-25 years away).

The Federal Government has historically supported both intermediate-term and very long-term development, and relevant basic and supporting research. Thus, the two sectors' approaches have been complementary, with the intermediate-term work in which both have been engaged being carefully coordinated, often involving jointly funded projects. Because of the very long lead time and large investment required to bring new electric-supply technologies to fruition, financial support by the Federal Government has been of particular importance to the Nation's long-range electricity supply.

One consequence of this sharing of responsibilities is that the Federal R&D role is predominant in areas aimed at electricity supply such as magnetic fusion, breeder reactors, other advanced reactors, the nuclear fuel cycle and in longer-range renewable technologies such as geothermal and photovoltaic energy conversion. The principal complement to the Federal effort is the Electric Power Research Institute (EPRI), funded primarily by the investor-owned utilities. Working out the balance of activities between the DOE, EPRI, and individual firms requires continued attention, but it has largely been an effective partnership.

R&D that leads to the production of combustible fuels is dominated in the near term by the very broad-based private industry competence that provides most of our conventional energy. The fossil fuels (petroleum, natural gas, and coal) are our primary resources and will remain so for years to come. But the basic fact that has made energy such an issue over the past decade is that our society is organized largely around petroleum for many uses, particularly transportation, and it is petroleum that is both most convenient and least available.

Continued long-term production of our domestic fossil resources will depend on the interplay between advances in extraction technologies, the difficulties of developing increasingly marginal known deposits, and discovery of presently unknown reserves. In the case of oil and gas, long-term production from conventional sources is likely to decline, certainly within the next century. In the case of coal, the known U.S. resource base – the world's largest known reserves – will be adequate for several centuries at current and projected rates of demand. Actual long-term development will depend importantly on the ability of private and public-sector research groups to develop acceptable environmental protection technologies: adequate technologies for efficient use of coal, both directly and by conversion, and on the availability of adequate economical coal transportation facilities. Federal research and development efforts in areas such as advanced coal cleaning, coal conversion, and enhanced oil recovery are intended to ensure economic development and use of domestic fossil resources over the long term.

It is likely that over the longer term the lowest-cost energy supply, as well as our national economic and security interests, will dictate the substitution of synthetic liquids and gas for present supplies of oil and natural gas. The same changes that have led to revision of energy projections have resulted in currently diminished incentives for private investment in synthetic fuels. In this connection, the Department of Energy is conducting research and development programs related to synthetic fuels, and the Synthetic Fuels Corporation is implementing a strategy of support for the most promising synthetic-fuel technologies using a number of economic incentives.

The net conclusion, then, is that longer-range research oriented toward advanced technologies, particularly for the clean use of coal, and toward novel uses of fuels, remains primarily a federal activity with near-term efforts being a private sector responsibility.

Of course, it is not energy that is important to society but the use of energy to provide needed services. The technologies that use energy are, simply, everything we buy, do, use, make, etc. Efficiency in the many ways we use energy obviously confers some long-term benefits on society as a whole, and the encouragement of improved end-use efficiency has been a generic goal that is widely recognized. But, since economic benefits go hand-in-hand with such efficiency improvements, there is not a strong rationale for Federal R&D that would improve industrial processes generally or develop vehicles or design buildings. The R&D activities of private industry, however, must depend on the existence of a sound base of technology and scientific understanding. It is in this role that DOE, in the long run, will contribute to improved end-use efficiency.

At this point let me turn more generally to the DOE programs that extend our scientific understanding. The advancement of science produces information that strengthens the Nation's technological competitiveness. Conversely, improvements in technology provide the enabling capability for new scientific advances. I'm talking about being able to measure, to count, to compute, to detect, in a variety of increasingly complex and subtle ways. DOE programs have been in the forefront of advancement of this knowledge.

The research programs of DOE are inherently long-term in character and high risk in terms of being able to identify the practical consequences of any individual project. It has therefore been long established that the Federal Government must play a primary role in the support of the important research programs that contribute to new understandings and to expansion of the technology base. These programs focus on exploring and understanding the fundamental laws of nature, and on the properties of matter and living systems. In fact, the Federal Government does support most of the basic research in the United States. The private sector generally concentrates on applied research and development, as in most cases it would be difficult to recover expenditures on technology-based research.

Perhaps our programs on High Energy and Nuclear Physics are the most fundamental in character and seemingly most removed from practical experience. These programs are a direct outgrowth of the sequence of inquiry that led to the discovery of electricity, of electromagnetic radiation, of the particle nature of matter, of atomic structure, and of the atomic nucleus; knowledge that has brought many practical benefits.

The High Energy and Nuclear Physics Programs seek a deeper understanding of the nature of matter and energy and the basic forces that exist between the fundamental constituents of matter. Approximately 90 percent of the support for research conducted in the United States in High Energy Physics and over 80 percent in Nuclear Physics comes from these programs. As a result of this continuing research, we are expanding our knowledge about the universe, the fundamental constituents of matter, and the laws of nature that underlie all physical processes.

The Nuclear Physics Program investigates the structure and properties of the atomic nucleus and the forces governing its structure, interactions, and properties. The High Energy Physics Program focuses on obtaining an understanding of the ultimate structure of all matter and energy through the study of properties of elementary particles, which are a consequence of all the basic forces.

The knowledge, techniques, and instruments developed in the High Energy and Nuclear Physics Programs have been applied in many areas. Besides contributing to the United States economic and National defense capabilities, these applications include contributions to therapeutic and diagnostic medical practice, nuclear electricity generation, and an array of procedures in industry and other scientific areas.

Progress in these fields is very much dependent upon large-scale facilities and specialized instruments, particularly particle accelerators. At present, the major DOE program facilities in Particle Physics include three large accelerator centers at Brookhaven National Laboratory (BNL), Stanford Linear Accelerator Center (SLAC), and Fermi National Accelerator Laboratory (Fermilab). Each of these centers contributes unique research capabilities to the National program. The three National centers are utilized for experimental work by a broad base of user groups from universities spread throughout this country and abroad.

For the future, initial planning and R&D is underway that could lead to a new machine that would be the largest research facility ever constructed. Called the Superconducting SuperCollider (SSC), this device would accelerate and guide counter-rotating beams of protons to produce collisions with an effective energy 20 times greater than we will be able to attain in the upgrading of Fermilab that is now underway. Such a machine might cost $6 billion or more to construct and equip and might take six to eight years to assemble after all the R&D is done. Even its design is a major scientific and technical challenge. But each small step in this field has moved us toward new science *and* new technology. If built, the SSC will be the centerpiece of the U.S. High-Energy Physics effort for decades to come.

Since large and complex accelerator facilities are the essential tools of the nuclear scientist, the success of the Nuclear Physics Program depends upon their existence and their ability to operate effectively. Seven accelerators are now operated as National facilities with beam time made available to all the

Nation's scientists on the basis of scientific merit and technical feasibility. In the judgment of its practitioners, however, the field of Nuclear Physics depends strongly upon two complementary proposed machines: first, a continuous electron-beam accelerator (R&D is underway for this machine and a site has been selected), and second, a relativistic heavy-ion collider that would explore new states of nuclear matter.

The world of our common experience is made up of things that are as they are due to the electronic structure of atoms and the ways that atoms can continue into molecules, condense into liquids, and solidify into materials. The nature of these transformations and structures has been under investigation for centuries. The part of DOE that is devoted to these lower-energy phenomena is called Basic Energy Sciences (BES).

The goal of the BES Program is to develop basic scientific and engineering knowledge that can be used by the Government and the private sector to develop new concepts, products, materials and processes. Although the principal focus of the BES Program relates to energy development and use, a number of other important national goals are also supported. U.S. leadership in science and technology, economic growth, and training for tomorrow's scientists are all beneficiaries of properly conceived programs of basic research. While contributions from basic research cannot always be identified immediately or directly, the knowledge that flows from this research forms the foundation on which new ideas and concepts can build in each of these areas.

The BES Program annually supports approximately 1200 individual research projects in the physical, biological and mathematical sciences. These projects are selected on the basis of scientific excellence, relevance to the long-term goals of BES, and their contributions toward a balanced research program. This research is carried out largely by scientists at universities and National Laboratories across the country. An important benefit that results from supporting university researchers is the technology transfer that occurs when graduate students carry their new knowledge into the private sector after completing their studies.

In addition to contacts with universities and laboratories, BES also maintains ties with industry through a variety of mechanisms. These include participation in workshops and the use by industry of the specialized research facilities supported by BES and conducted in seven areas: materials sciences, chemical sciences, nuclear sciences, mathematical sciences, engineering and geosciences, biological sciences, and advanced energy projects. Research in these basic scientific disciplines is carried out by several other Federal agencies and by private industry as well. The U.S. is unique in that we have a strategic triad of university, national laboratory and industry effort that can produce excellent research that ultimately has practical results.

To illustrate my point, some 25 years ago the materials program supported the initial research on metallic glasses. These glassy metals have an atomic structure that is different from the metals with which we are familiar. We now know that metallic glasses are useful in magnetic applications because they can be magnetized and demagnetized with less energy loss than occurs in ordinary structured metals. We also know that they are potentially the strongest, most corrosion-resistant materials known, and that they are economical to manufacture because the production process uses less energy than is consumed by conventional processes.

As a result of that original research and subsequent basic and developmental research done by industry and universities, glassy metals have made a significant impact in the commercial areas of communication and lighting; for example, they are being used in the recording and playback heads of tape recorders and in the starting circuits in fluorescent lights. Research in the private sector continues on production techniques for glassy metals. One potential additional use might be the substitution of glassy metals for steel in electric power transformers. Losses in magnetic energy in these transformers amount to more than $600 million a year, most of which might be saved with the use of glassy metals.

An important point to note is that a wide range of different applications has been found based on a generic improvement in the understanding of fundamental phenomena. Some of the applications relate to energy, some to other significant and important uses. Some of the science that DOE supports will turn out to have such payoffs – and some will turn out to be more a matter of cleaning up a dusty corner of the knowledge bin. The paradox is that if we knew in advance what the outcome would be, we wouldn't have to do the research!

In addition to its physical-science-oriented programs the DOE has a strong and continued effort in the life sciences that has contributed uniquely to improved understanding of the environment and the consequences of energy use for our human health. The Biological and Environmental Research (BER) Program represents the only long-term research effort in the Department and in the Nation that is specifically focused on energy-related health and environmental issues. The goals of this research program are (1) to provide sound, scientific data, through basic and applied research, required to reduce the health and environmental uncertainties associated with energy technology and weapons development for the Nation's future energy and National security needs; and (2) to develop new and improved instrumentation and methodologies for use in diagnosis and treatment of human disease.

Information obtained as a result of this research program has provided the basis for developing cost-effective protection standards for the nuclear industry, the Government's case in litigations on the effects of fallout from weapons testing at the Nevada Test Site and the Marshall Islands, and the development of strategies for control of emissions from processes used for conversion of coal to synthetic oil and gas.

Health and environmental concerns are associated with many of DOE's programs. Examples include issues that arise from the combustion of fossil fuels (*e.g.*, acid ran and carbon dioxide) and from exposure to low levels of radiation of complex chemical mixtures such as those associated with many conventional and synthetic fuels. Reducing these health and environmental uncertainties is not a simple task. The majority of exposures about which we are concerned are exposures to low concentrations of energy-related substances over long periods of time, that is, over months or years. As a result, effects on humans and ecosystems are expected to be subtle and to occur long after the initial exposure. In addition, the health and environmental effects caused by chronic, low-level exposures often are not unlike those that occur spontaneously from natural causes or those that result from the action of toxic agents common to natural and man-made environments, such as sunlight, smoke and medical x-rays. Thus, it is difficult to determine or measure the health or environmental effects of chronic, low-level exposures against the "noisy" background of similar effects that can arise from other causes.

Research in these areas is complex, subtle and interdisciplinary. Human and environmental health are clearly overriding concerns and the necessary protection is specifically a Federal responsibility. DOE programs fulfill a unique function in building the knowledge that will keep us from inadvertently poisoning ourselves or our planet.

There are many other DOE program areas, as addressed by others today, that have a strong science content, particularly the magnetic-fusion program. Conservation, solar, nuclear and fossil programs are all dependent for success on the underpinning science that supports their objectives. A broad and continuing science effort should transcend the short-term popular swings of enthusiasm for the various forms of energy production.

The DOE role in science is fully as needed as the DOE role in energy. The two roles are, in fact, inseparable. It is this strong science base that made it possible for the Atomic Energy Commission to be transformed, in a few short years, into the Department of Energy with concerns and capabilities in all areas of energy research and technology. It is the science base that provides for facilities and knowledge that can later be applied to other national needs, including defense and various areas of industrial productivity. It is this science base that has also provided part of the information on which the new biotechnology depends.

The unique building blocks of DOE scientific research are those institutions that we call National Laboratories. These institutions can combine disciplines to attack a problem, have the continuity of effort and permanence that transcends short-term fads and fancies, and have the tradition of excellence that is essential for attracting, training and motivating talent. In one sense, the role of DOE is simply to provide an effective work environment for some of the world's best minds, agree with them on general direction and motivation, and help the Nation to reap the benefits.

Universities, as well, play an important part in the success of our programs. Good people are the most precious of resources. Our support of universities assists in the education of more than 5,000 graduate students at any one time in addition to providing high-quality research results. Perhaps as important is the fact that university scientists are free to follow their own sense of direction and to swim against the current of prevailing views. This attitude provides science and society with a vital capacity for self-criticism that helps to keep us on the road to excellence.

Cornell University has been a leader in many of the sciences that DOE supports. The fact that much of the Cornell support in these areas comes from the National Science Foundation is one of the strengths of our approach to science in this country. There have been many interactions between Cornell and DOE. Of particular note is that Robert R. Wilson, after a wartime stint at Los Alamos, became a Cornell professor and designed outstanding particle accelerators until he was lured away to head Fermilab in the late 60's. We were delighted to present Bob Wilson with the Enrico Fermi Award last year. And now we have captured Boyce McDaniel (part-time) and Maury Tigner to help us deal with the questions surrounding the Superconducting SuperCollider.

Closer to the daily bread of electrical engineers, the principal elements of DOE's Energy Systems and Storage program were developed by Sam Linke when the program was still at NSF. Peter Auer organized many of the studies and workshops that helped usher in the Government's response to the energy crisis. Bob Hughes now heads Associated Universities Inc., which operates the Brookhaven National Laboratory, and Ravi Sudan successfully blends support from several sources, including DOE, in what continues to be one of the most impressive programs in my own field of plasma physics.

So the ties between DOE and Cornell have been close – because that's the way things work best in this country. Free interchange of people, of ideas and sometimes some healthy rivalry is good for science and is good for the country. Let's keep it that way.

FUTURE DIRECTIONS IN ELECTRICAL ENGINEERING

SOCIETAL IMPACTS AND FUTURE DIRECTIONS

PROCEEDINGS of the

Cornell Electrical Engineering Centennial Symposium

at Ithaca, New York—June 12, 1985

Simpson Linke, editor

Promethean Press
Ithaca, New York

PROCEEDINGS OF THE
CORNELL ELECTRICAL ENGINEERING CENTENNIAL SYMPOSIUM:
FUTURE DIRECTIONS IN ELECTRICAL ENGINEERING-
SOCIETAL IMPACTS AND FUTURE DIRECTIONS

Copyright 1988
by the School of Electrical Engineering
Cornell University, Ithaca, New York
All rights reserved

Published by Promethean Press
Ithaca, New York

Printed in the United States of America

November, 1988

CONTENTS

PREFACE

The celebration of the Centennial Year of the School of Electrical Engineering at Cornell University featured a series of symposia based on the central theme of "Future Directions in Electrical Engineering". Each seminar concentrated on a major facet of the discipline, with individual topics presented by outstanding specialists in each field. The symposia were held at sites around the country that are widely known as centers of activity for each particular subject area, thereby providing significant updates on pertinent electrical engineering technologies at appropriate locations, and allowing a large number of Cornell Electrical Engineering alumni to participate in the celebration in a meaningful way.

The ultimate success of the Symposium Series must be credited to the dedicated efforts of the faculty, symposia, and local alumni chairmen who were responsible for planning the programs, engaging the participants, and attending to local arrangements within specific time constraints. The following schedule of events attests to the overall magnitude of the task and also displays the broad range of topics that were presented:

March 15, 1985—Microelectronics—Stanford, Calif.
April 1, 1985—Computers—E. Windsor, N. J.
April 10, 1985—Communications—Boston, Mass.
April 17, 1985—Atmospheric and Space Sciences—St. Louis, Mo.
April 25, 1985—Energy for the Future—Washington, D. C.
June 12, 1985—Societal Impacts and Future Directions—Ithaca, N. Y.

Each session was videotaped to facilitate preparation of the proceedings of each meeting and for possible use of the individual presentations in classroom seminars and elsewhere. The audio portions of the tapes were transcribed into preliminary text and coordinated with figures obtained from the video images and from hard copy provided by some of the authors. Each presentation was edited with a view toward achieving continuity and coherence without

compromising the natural style of the author. In several instances it was necessary to introduce substantial changes in format and phrasing in order to achieve this goal; in others, the hard copy supplied by the authors made the task easier. In all cases, the final versions were sent to the authors for their criticisms, corrections, and final approval. I am grateful to all of the contributors for their excellent cooperation and assistance in bringing their presentations to this present form.

I wish to thank Professor Edward D. Wolf for his enthusiastic interest in this project and for authorizing financial support from the National Research and Resource Facility for Submicron Structures to assist in the publication of the Symposia Series *Proceedings.* Additional funding from the College of Engineering and the School of Electrical Engineering is also gratefully acknowledged.

It would not have been possible to produce this work without the skillful secretarial assistance of Catherine Kuhl, Deborah Overslaugh, and Linda Struzinsky who performed the difficult task of transcribing the tapes into typed versions of the unedited manuscripts. The expert guidance provided by Mrs. Diana Nathanielsz on format and production of this volume is very much appreciated.

I am particularly grateful to Catherine Kuhl for her diligence and care in forming the camera-ready copies of several of the manuscripts on that versatile new instrument that is so illustrative of "future directions in electrical engineering" — the laser printer.

Simpson Linke

Ithaca, New York

INTRODUCTION

John A. Nation

Director, School of Electrical Engineering

The celebration of our Centennial Year affords a unique opportunity to review past accomplishments, to evaluate our present circumstances, and to make some reasonable estimates of prospects for the coming years. A glance at the history of the University reveals that the Cornell *Register* for 1883 contained a "Special Notice" from President Andrew Dickson White announcing the establishment of a "Course in Electrical Engineering." President White had convinced the University's Trustee Executive Committee to create the new program "provided such course occasions no additional expense."

The early curriculum was rigorous and in addition to the traditional engineering topics of mathematics, physics, mechanics, mechanical drawing, and shop work, required rhetoric, French and German. The senior year also included courses in several Cornell specialties such as photometry and electric lamps, telegraph instruments, lines and cables, the dynamo, and electric motors. The first baccalaureate degrees in electrical engineering in the United States were awarded by Cornell (as well as by M.I.T.) in 1885, and the first Ph.D. in the new discipline was conferred by President White at Cornell in the same year. Thus we are truly celebrating 100 years of electrical engineering education.

One of the major figures in the E.E. School of those days was Professor Harris J. Ryan who ran the Electrical Engineering Department essentially as a one-man faculty for 17 years and is credited with having trained almost 1000 electrical engineers during that period. Ryan left Cornell in 1905 to become head of the Electrical Engineering Department at Stanford. It is interesting to note that 12 of the first 13 members of the Stanford faculty were educated at Cornell.

This brief summary of the beginning of electrical engineering instruction at Cornell would not be complete without mention of the internationally known Professor Vladimir Karapetoff who joined the faculty following Professor Ryan's departure and made many contributions to early electrical engineering theory and analysis in several textbooks and many technical articles. Something of the flavor of electrical engineering studies of the time is suggested by the following excerpt from Karapetoff's *Electrical Laboratory Notes*, published in 1906:

> "In coming to the laboratory, bring with you a slide rule, an inch rule or tape, a speed counter, a screw

> driver and a pair of plyers [sic]. This will save you time and trouble of looking for them or borrowing them. Do not forget to have a pocket knife for skinning off wire; a bicycle wrench is also sometimes very handy to have."

In these days of hand calculators and personal computers, it is rare indeed to find a student who has even seen a slide rule, much less to have used one.

Classical electrical engineering education at Cornell was devoted principally to electric machinery, power systems, telecommunications, and early electronics. As the technology developed from the 20's through the 40's, the curriculum was, of course, continually upgraded as new and advanced concepts were introduced, but the basic format remained essentially unchanged until the mid-50's when research began to play a central role in the E.E. School, coincident with the move from Franklin Hall to the newly constructed Phillips Hall. The following figure among the highlights of our research development over the years:

> An early solar telescope project initiated an era of Radio Astronomy research that culminated in the establishment of the world's largest radio-radar telescope at Arecibo, Puerto Rico, now known as the National Astronomy and Ionosphere Center.
>
> A research program in the new field of Physical Electronics gradually expanded into the present National Research and Resource Facility for Submicron Structures housed in the Knight Laboratory adjacent to Phillips Hall.
>
> The Old High Voltage Laboratory on Mitchell Street, converted to an Energy Laboratory in 1960, served as the staging area for the Cornell Extra-High-Voltage Cable Test Program and housed E.E. research on high-voltage vacuum switches and arc physics until the formation of the interdisciplinary Laboratory of Plasma Studies in 1966. This facility, together with adjacent installations, now contains the largest particle-beam program in any university in the world and has made many important contributions to the understanding of intense electron and ion beams, particularly as they relate to inertial confinement thermonuclear fusion experiments.
>
> The national energy crisis in the early 70's created renewed interest in power and energy research within the School leading to the combination of power-system elements, control theory, and new computer facilities, that now constitutes the unique Kettering Power Systems

Laboratory for the analysis of modern complex power-generation and transmission networks.

Two recent events attest to the high regard that the University and the E.E. School enjoy in the national research community:

1. The Program in Microscience and Technology was established in 1982 and receives about $2 million annually through the Center of Excellence Program of the Semiconductor Research Corporation.

2. The NSF-sponsored Cornell Center for Theory and Simulation in Science and Engineering, with its associated supercomputer production facility, will augment and expand research capability in the School and will encourage greater interdisciplinary activities with other scientific units on campus.

One of our strengths in recent years, as the Electrical Engineering School has evolved from a purely teaching organization into a teaching and research institution, has been our ability to attract good new faculty. A positive measure of this quality is in our receipt of Presidential Young Investigator Awards from the National Science Foundation in the first year of the establishment of these awards. The E.E. School received 5 of the total of 26 awards made available to electrical engineering departments in the nation. Our success in attracting these high-quality young people bodes well for the continued future growth of the School.

Remarkable changes in electrical engineering education have occurred in the School within the past 25 years as the influence and impact of new faculty members and their research have brought modern concepts to curriculum and laboratories that had been essentially identical to those of World War II days. The advent of computers has been particularly significant. We have been very fortunate in the last decade in the support that we have received from industrial corporations. Their generous donations of equipment including computers, work stations, and state-of-the-art measurement devices have been of great value in upgrading our laboratories to present-day standards.

On the other hand, the current huge interest in computer engineering and electrical engineering has introduced major enrollment problems in the School. Using 1965 as a year of reference, the size of the faculty has remained fixed at about 40 members whereas the number of undergraduate students has tripled. Our senior class in 1985-86 numbered 186 students. We have 265 graduates students in the School with about 145-150 MS/PhD students and the remainder in the Master of Engineering Program, the latter being the second largest professional degree program in the University. This year we graduated

30 MS/PhD students and 88 Master of Engineering degree candidates. An encouraging factor is that 12 of these advanced-degree students were women. Today about 15 percent of the students in the E.E. School are women and 25 percent are minority students.

The increase in student enrollment is coupled with a research effort that is growing at a compound rate in excess of 10 percent per year in constant dollars. Clearly we are approaching the limits of our present capability. It is hard to realize that within 30 years of our initial occupancy of Phillips Hall we are now bursting at the seams. Further, the building which was then state-of-the-art, is now an antiquated facility in terms of present-day technology, with little in the way of the clean-air environments, air conditioning, and other amenities that are necessary for the proper functioning of modern equipment.

Plans are in progress to build a new engineering complex on campus in which electrical engineering will have substantially more space than in the past. When these new facilities are completed, our faculty will be able to work at its full potential. We expect to see a relatively modest growth in the number of faculty members. Since a faculty lifetime post is forty years, and student-interest time cycles are rather shorter than that, the response to large student increases in enrollments and interest in electrical engineering cannot be followed instantly by new faculty positions. Attempts will be made to maintain our undergraduate size at about its present level with some growth allowed in the number of our graduate students. It is certain that there will be many more interactions with industry in the near future, and these will take on a much broader form as the years go by. For example, an exchange program may be arranged with industrial laboratories so that their scientists will come to work in the School for a period while a faculty member will have the opportunity to spend corresponding time in industrial environments.

This brief perspective of past, present, and future encourages us to begin our second century with a true sense of excitement. We may be justly proud of the important contributions that our School of Electrical Engineering has made to many of the developments that are reported in these *Proceedings*. As we celebrate these past achievements, we can look forward with confidence to producing equally significant additions to our challenging discipline.

PART 6

SOCIETAL IMPACTS AND FUTURE DIRECTIONS

PROCEEDINGS of the

Cornell Electrical Engineering Centennial Symposium

at Ithaca, New York—June 12, 1985

"I have the pleasant task of welcoming you to this Symposium on "Electrical Engineering: Its Societal Impact and Future Directions". It is the final program in the series of cross-country symposia on major topics in electrical engineering that have been conducted this Spring in celebration of the Centennial Year of the Cornell School of Electrical Engineering. Extrapolating from past accomplishments in research, development, and education in this field, we can predict with confidence that our future will be influenced significantly by new electrical technology. Our panelists today are primarily concerned with the coupling between that technology and human needs, and speak from the vantage point of outstanding accomplishment in their respective fields. They will address the complex issues of what society requires, and perhaps what it will receive, from new approaches to energy production, communications, information systems, computers, and engineering education.

I take this opportunity to express my thanks and appreciation to Dale Corson, our Symposium Chairman, and to the other members of my Committee, for their dedicated efforts in assembling this distinguished panel. I also acknowledge and thank, John Belina, our Arrangements Chairman, for his excellent work in the management of the local facilities for this Symposium.

Our Chairman, Dale Corson, obtained the A.B. degree at Emporia College in 1934, the M.A. degree at the University of Kansas in 1935, and the Ph.D. degree at the University of California in 1938. Following two years as a Research Fellow at the University of California, he was a member of the faculty at the University of Missouri for the next five years with interim periods of service at the MIT Radiation Laboratory and as a Technical Adviser to the War Department. In 1945 he joined the staff of the Los Alamos Scientific Laboratory. In 1946 he came to Cornell as an Assistant Professor, became Associate Professor in 1947, Professor in 1952, and Chairman of the Department of Physics in 1956. Starting in 1959, he served a four-year term as Dean of Engineering, a six-year term as Provost, an eight-year term as President, and a two-year term as Chancellor of the University. He became President Emeritus in 1979. He is a member of the National Academy of Engineering (NAE) and a Fellow of the American Physical Society, the American Academy of Arts and Sciences, and the New York Academy of Sciences. He has been awarded honorary degrees from six universities. Overall, Dale has given 39 years of distinguished service to science and education at Cornell"

– Professor Herbert J. Carlin, Symposium Faculty Chairman
From Welcoming Remarks, June 12, 1985
Ithaca, New York

Editor's Note: Dr. Corson received the Public Welfare Medal of the National Academy of Sciences in 1987 and the Arthur M. Bueche Award of the NAE in 1988, both for his contributions to science policy.

"As General Traffic Manager of this Symposium, it is appropriate to open these proceedings by reading from the Cornell *General Catalog*, of 1883:

> The rapid development of the application of electricity has created a demand for thoroughly trained engineers conversant with electrical science. Recognizing this demand, the trustees of Cornell University receive students desiring to fit themselves to enter this constantly changing and extending field. The special studies embrace the theory of electricity and of machines and the methods of electrical transmission and of electrical measurements. The university possesses a very extensive collection of electrical apparatus for measurements and other accessory apparatus. In the new laboratory, every facility is provided for the use of electrical apparatus under the most favorable conditions, and a workshop provides for the construction of special instruments for investigation.

This statement contains the first mention of electrical science in the *General Catalog*, and we are here today to celebrate the 100th anniversary of the graduation of those first students. We are faced with a time of enormous problems and enormous opportunities, and today we are going to examine those problems and those opportunities.

"I will comment briefly on two fields: the first is energy, a subject that causes me a great deal of concern. I have spent a fair amount of time in that field, including consideration of exotic forms of energy. I was chairman of a two-year study by the National Research Council on the solar-satellite power system, and I know a lot of ways not to get energy. Our oil supply is going to be gone, not in my lifetime certainly, but in the lifetime of our students. This morning we will hear from Professor Bethe about fission power, a prospect that appears to be closed to us by political action that has resulted from considerable popular pressure. If we are going to have fission power in amounts sufficient to last several hundred years, we must have breeder reactors, but there has been no serious attempt to develop them in this country. Fusion is uncertain. People I know who work in the field tell me that we will light no light bulbs for the next forty years. What will happen after that is somewhat uncertain, but I hope the outlook is optimistic. Thus we are left with using our coal supply that has sufficient reserves to last for several hundred years. Carbon dioxide, however, poses a serious pollution problem. In recent months I have had the pleasure of examining research in that area in some detail. There is no doubt that the CO_2 in the atmosphere, along with some other gases of human origin such as methane, are rising at a rather frightening rate. Together these gases can produce serious heating of the atmosphere. The magnitude of this worrisome problem is not understood and a great deal of research is required.

"The other field is information handling, which will be given considerable attention today. I have a couple of pertinent observations on that topic. The number of technical journals in the world has grown with a remarkably constant doubling time of 15 years, from the mid-17th century until the World War II era, starting in 1660 with the *Philosophical Transactions* of the Royal Society of London. Although I am not certain of the exact figure, that doubling time is shorter now, perhaps somewhere around eight years. The actual number of technical journals is not important. Depending on how you define the technical journal, there may be 200,000, perhaps more, perhaps less. It does not matter how many there are because in eight years there will be twice as many. Some 2,000 or 3,000 pages of text come off the printing presses of the world every minute. How are we going to deal with this publication explosion? In addition, we are in a period when the Africans, the Asians, and the South Americans are not producing much. When they start producing information at the same rate as in the Western world, we will literally be up to our necks in information. What does all this mean? What are we going to do about it? How can we deal with it? What does it mean for our educational system? We are going to discuss these questions today."

–Professor Dale R. Corson, Symposium Faculty Chairman
From Introductory Remarks, June 12, 1985
Ithaca, New York

NUCLEAR POWER IN THE UNITED STATES AND ABROAD

Hans A. Bethe

Hans A. Bethe received the Ph.D. degree from the University of Munich, Germany in 1928. He was an Instructor in Theoretical Physics in Germany until 1933 when he moved to England and became a Lecturer at the Universities of Manchester and Bristol. In 1935 he joined the Physics Department at Cornell University as an Assistant Professor and was promoted to Professor in 1937. In 1975 he became the John Wendell Anderson Professor of Physics Emeritus. Dr. Bethe received the Nobel Prize in Physics in 1967. In addition to his five honorary degrees, he has received many other honors and awards including the National Medal of Science in 1976, the Fermi Award from the U.S. Atomic Energy Commission in 1961, and the U.S. National Medal of Merit in 1946 for his wartime contributions at the Los Alamos National Laboratory. In 1985 he received the National Science Board's Vannevar Bush Award. He was a technical adviser to the U.S. delegation at the International Conference on Peaceful Uses of Atomic Energy, Geneva, Switzerland in 1955, and has served as a consultant to many industrial and government laboratories. He is a member of the Royal Society of London, the American Philosophical Society, the National Academy of Sciences, the American Physical Society, and the American Astronomical Society. Dr. Bethe's research work on nuclear physics, stellar evolution, and most recently on the peaceful uses of atomic energy, are justly celebrated. He has been a moving force in the efforts to institute international control of nuclear armaments.

"Professor Bethe came to Cornell 50 years ago and has been here ever since. He holds every honor there is: Nobel Prize, National Medal of Merit, Fermi Award, National Medal of Science, and one, recently awarded, that I like particularly is the Vannevar Bush Award from the National Science Foundation. This latest award is presented to people who have made important contributions to public-science policy in this country. Professor Bethe has had a remarkable research career, a teaching career of great significance, and an important public-policy career."

—Dale Corson

My subject is one of the topics that Professor Corson mentioned in the beginning, namely how do we obtain our electric power. For the last year, approximately, the newspapers have been full of very depressing stories about nuclear power. For some time there was hardly a week in which there was not some news report that this nuclear power plant has been cancelled after having been 90% complete, and that other plants, although fully complete, have not been licensed for operation. Still other news items stated that the nuclear power industry is in a very bad way financially because of the uncertainties about completion and licensing of plants. It appears that in very many cases the problem of completing a nuclear power plant depended more on electric-utility finances than on any regulatory activity. So generally it has been bad news all around in our nuclear power program.

Figure 1 shows some of the promise and some of the bad news that I will discuss in a few minutes. Let us turn for a moment to the good news. The existing nuclear power plants still produce power rather cheaply at 3.5 cents per kilowatt hour, on average the same as coal plants. The new power plants are more expensive, as I will show in more detail in a minute, but note that the price of electric power from oil-fired plants is roughly twice as great as that for nuclear power. I think it will be best if I turn to Figure 2 which gives the electric generation from various power sources. The good news is shown by the data in the figure. The total electric-power production is interesting to consider from the viewpoint of Electrical Engineering. The total U.S. electric-energy generation from 1973 to 1984 increased roughly 30%, from 1800 billion kilowatt hours to 2400 billion kilowatt hours. There were three years of recession in which electric power did not increase at all. The gross national product decreased in that slump but electric power stayed constant. It is well known that electric power is much more lively than the economy as a whole. Of this total power production, coal accounted for an increasing fraction starting from 46% and rising to 56%. Nuclear power had a similar increase, relatively speaking a much bigger increase, from 4.5% to 13.5% of total production. Natural gas and hydro-electric power stayed approximately constant in total production. In fact, hydropower cannot increase very much because we have already used practically all of our good sources of hydropower, so that percentage went down a bit. Comparing these average percentages we see that the great achievement is that electric-power production by oil has gone down from 17% to 5% and this was achieved by the increase of power production by nuclear and by coal. Therefore nuclear power has not been quite as unsuccessful as might be assumed from the newspaper reports. One always gets the reports of trouble but very seldom the reports of success.

Returning to Figure 1 we see that the bad news is the cost of investment in nuclear power plants. How much does it cost to build one kilowatt of

The Good News:

Existing plants provide cheap electricity:

Plant type	Cost in Cents/kWhr
Nuclear	3.5
Coal	3.5
Oil	6.9

Percent of Total U. S. Electric Generation in 1984

Source	Percent
Nuclear	13
Gas, Hydro	13
Oil	5

The Bad News:

U.S. Investment Costs per kW_e:

Actual costs	$1200 to $4500
Median costs	$1800

Building time: 6 to 15 years

Delays are now often caused by problems of utility financing

Figure 1. Present Nuclear-Power Comparison

Year	Total (GWhr)	Coal (%)	Nuclear (%)	Gas, Hydro (%)	Oil (%)
1973	1860	46	4.5	16.5	17.0
1979	2250	48	11.4	13.5	13.5
1982	2240	53	12.6	13.7	6.6
1984	2410	56	13.5	12.8	5.0

Figure 2. Electric-Power Generation from Various Energy Sources

power? That number rose from about $400 a kilowatt ten years ago to anything between $1200 and $4500 today. The famous power plant of the Long Island Lighting Company cost $4500 a kilowatt. Plants in Florida and Arizona cost $1200 per kilowatt. The median is about $1800. For a large plant that is a lot of money. At the time the first order for a nuclear plant was placed, the building time was about five years. It has now increased to anything between six and fifteen years. The utilities are reluctant to place new orders, not only because of the high cost, but even more because the cost is unpredictable. It may be anything between $1200, which is perfectly tolerable, to $4500 which is completely intolerable and cannot be recovered by a reasonable price for electricity. So that's the sad story of nuclear power in the United States. There have not been any new orders for nuclear power plants for five years or more, and most of the previous orders have been cancelled rather than new orders given.

Figure 3 shows that conditions are very different in France. Total nuclear power installed in the United States at the moment is 77 gigawatts electric, that is 77 thousand megawatts of electric power. In France it is half as much but the population of France is just about 20% of the population of the United States. Consequently, nuclear power plants in France produce nearly 60% of their electric power, whereas in the United States it is only 14%. The cost in dollars per kilowatt of newly-built plants was about the same in France and in the United States in 1974, at $400 per kilowatt in each country. It is now $800 in France. That does not represent an increase because these are current dollars and there has been that much inflation and more, so that the price in constant dollars in France has decreased rather than increased. Note that the median price has increased in the United States to more than twice what it is in France. The important number in Figure 3 is the ratio of the cost of

Nuclear Plants		France	United States
Power Installed (GW)		38	77
Percent of Electric Generation		58	14
Installation costs:	year		
($ per KW)	1974	400	400
	1984	800	1800
Nuclear/Coal Investment Ratio		1.4	2.4

Energy-Production Costs		(Cents/kWhr)	
Source	France	U.S. (old)	U.S. (new)
Nuclear	1.4	3.5	6
Coal	2.4	3.5	-
Oil	7.0	7.0	-

Figure 3. Comparison of French and U.S. Nuclear-Plant Parameters

building a nuclear power plant to building a coal power plant. That ratio is 1.4 per kilowatt in France but 2.4 in the United States. If the ratio is 1.4 the difference can be easily recovered since nuclear fuel is very much cheaper than coal. If the ratio is 2.4, however, the two are about even at best. Figure 3 also shows that the cost per kilowatt hour of nuclear power is 1.4 cents in France and 3.5 cents in the United States. The cost for coal power is much higher in France than it is for nuclear power. In the United States, however, the costs are the same for existing plants with previously low building costs. One may wonder why even coal power is more expensive in the United States than in France. The reason for that I think is a matter of bookkeeping. The French charge themselves relatively little interest, taxes, etc., while we have many such charges. Oil, of course, in each country is very much more expensive than either nuclear or coal.

Why is it that the French can do it so well and we do it so badly? The answer is detailed in Figure 4. The French just are much more efficient in building nuclear power plants. Their efficiency is shown, for instance, in the amount of materials they use, amounting to about 2/3 of what we use for the same size nuclear power plant. When we compare man-hours of labor, the French use is only 1/3 of ours and in terms of qualified labor, such as engineers and draftsmen, we require 8 times the number needed in France. I have all of this information from Pierre Zaleski, one of the leading people in

Reactor-Construction Components	Cost Ratio: France/U.S.
Materials	2/3
Labor (Man hours)	1/3
Engineers and Draftsmen	1/8

Manhours Used in the United States (Millions)

	1967	1983
Total	3.5	30
In the Field	1.3	22

Construction Time (years)

	1974	1983
France	6.2	5
United States	5.5	7-15

Figure 4. Nuclear Plant Construction Comparison

the French nuclear power establishment, who gave an excellent talk last November in Florida at a meeting organized by the University of Miami. In the United States we use millions of man-hours to build a nuclear power plant of about a million kilowatts capacity. In 1967 we used a total of 3.5 million man-hours, in 1983 we used 30 million. And it is even worse when we compare the man-hours in the field. That went from 1.3 million in 1967 to 22 million, increasing by a factor of 17. We must therefore conclude that something is wrong about the way we build nuclear power plants. It can be done much better and it is being done much better in France. Back in 1974, for nuclear plants that were completed in that year, construction time took longer in France than in the United States. Since that year the time for construction in France has decreased from 6 years to 5 years, while in the United States it has increased from 5.5 to anything from 7 to 15 years. Most U.S. plants require 10 to 15 years with very few below 10 years.

What additional factors contribute to the French success? The information in Figure 5 was presented to us by another Frenchman, Dr. Vendryes, who received the Fermi Prize from the Department of Energy last year and who is one of the leading nuclear engineers in France. He is largely responsible

1. One Vendor, One User:
   a. Country is the right size
   b. Country has a centralized government
   c. Key decisions are made by <10 people
2. Standardization:
   a. Families of plants fixed at 900 to 1400 MW_e
   b. Original design was the Westinghouse pressurized-water reactor (PWR).
3. Need for Nuclear Power:

   Country has no oil, little coal, and little hydropower
4. Political Support from all Parties
5. Country also Leads in Breeder-Reactor Development

Figure 5. Contributions to Nuclear Success in France

for the design, at least in principle, of the French breeder reactor Phénix. Again why can the French do it so well? First of all is the point that they have only one vendor and one user. There is only one firm which designs and builds nuclear power plants in France and that is the French government. There is only one user, one utility, and that is Electricité de France, which is also run by the government. Now that does not mean that I would advocate that everything be done by the government, but we certainly have made a mistake by fragmenting the design of nuclear power plants, even more, fragmenting the construction among many architect/engineers, even more, fragmenting the users, the utilities, of which we have well over a hundred. In this respect we would be much better off if we followed the advice of Alvin Weinberg, long-time director of the Oak Ridge National Laboratory, who has said many times that the production of power from nuclear fuel should be concentrated in a few companies which then would sell their power to the present electric utilities, organizations that are eminently suited for distributing the power to the customers. Nuclear power plants have different problems than power plants based on fossil fuels. Therefore it would be perfectly reasonable to have a different type of company to run them.

A second point, which was made by Vendryes, is that France is a country that is just the right size for nuclear power. It is big enough so that it requires a large amount of power; it is small enough so that there is no need for fragmenting the manufacturers of the power plants and the operators of the plants among many companies. France profits from having a centralized government which goes back at least four hundred years, so that means that a very small number of people, less than ten, in fact, can decide what is to be done. The main thing however is standardization. The French Electric Power Company, the company that designs and builds the plants, has developed a number of standard power plants, about four families, one after the other. At first they designed a plant for 900 megawatts. Now their plants are at the 1400-megawatt level. In each family they design many plants to the same standards whether they are built in Northern France or Southern France. By building many plants they obtain experience and find out what's wrong with each plant. Then they can proceed with the next family. In the United States we have omitted this standardization. The original design of the French plants came from this country. It was the Westinghouse design of a pressurized water reactor.

The third point in which France excels is the need for nuclear power and that of course is decisive. The French have no oil, we have quite a lot of oil. They have no natural gas, we have a lot of that. They have little coal, just about enough for metallurgical purposes, and we have lots and lots of coal. Therefore the United States doesn't really need nuclear power but in France it is the only sensible power source that they have. This fact was recognized at a very early time so the whole country got behind it. There is political support from all parties and it has never been a big issue as it has been in the United States. There have been occasional protests, but in general there has been very firm support. France is also leading in the development of breeder reactors.

From France let me go to other countries abroad. Figure 6 shows the way nuclear power is distributed at the moment and how it will be distributed in 1990. I have only listed twelve of the more important countries. Note that Belgium gets 50% of its power from nuclear power, Sweden gets 40%, the United Kingdom nearly 20%, the United States only 14%. The underlined countries are the most important, that is they have the largest nuclear-power establishments. France expects to build many more power plants by 1990 at

	1984			1990	
Country	Nuclear capacity (GW_e)	Percent of total capacity (%)	Percent of total generation (%)	Nuclear capacity (GW_e)	Percent of total generation (%)
Belgium	3.5	27	49	5.4	40
Canada	8.6	10	12	13	12
France	38	39	57	56	54
W. Germany	14	15	23	23	25
Japan	19	14	20	34	19
S. Korea	2.0	13		7.6	
Spain	4.9		8.7		
Sweden	9.5	24	41	9.5	
Taiwan	4.0	32		4.9	30
USSR	24			71	
UK	6.6	11	18	12.4	
USA	77	11	14	100	13
Total	211		13	337	

Figure 6. Present and Projected Nuclear Capacity in Various Countries

which time they want to derive 75% of their electric generation from nuclear power. Japan is building furiously, but as yet they don't have as large a percentage, while in the United States the building is rather slower. It is likely that after 1990 there will be very few additional plants in this country because there are very few orders at the moment.

Finally, in Figure 7, I have the projected nuclear-power generation capacity in the future. At present, nuclear power provides about 11% of the world-wide power generation. Nuclear power generation is expected to double by 1990 and again to double by 2000, at which time it will be about one fourth of total power generation. I think it is obvious from all this that nuclear power is alive and well everywhere, or nearly everywhere, except in the United States.

The question that arises is what can we do to improve its status in the United States? What can we learn from other countries? Evidently, the United States is still blessed with lots of other fuels, especially coal. Evidently, therefore, the need in the United States is not as great as it is in countries like France or Japan or even Germany and England. On one point, however, we could learn from the others to standardize our plants. That suggestion has been mentioned over and over again but it has not been done. We would profit from that both in reliability and in cost. Of course it is not quite as easy to standardize in the United States as it is in France because of the great climatic variations in this country. The climate in Florida and the climate in Minnesota are rather different. So complete standardization is not possible but much more could be done than has been done.

Year	Nuclear Capacity (GW_e)	Nuclear Generation (Billions of kWhr)
1983	190 (= 9% of total)	960 (= 11% of total)
1990	390 ± 25	2250 ± 100
1995	500 ± 60	3000 ± 330
2000	640 ± 120 (= 16 ± 2% of total)	4000 ± 650 (= 24 ± 2% of total)

Figure 7. Projected Worldwide Nuclear Power Capacity and Generation

Let me now go to another subject, namely the safety of nuclear power. The first big study of safety was finished in 1975. It is the so-called Rasmussen Report, followed by many other investigations that have vindicated the methodology used in that study. The Rasmussen Report, however, did not have a very good data base for its conclusions because the investigators estimated the probability of trouble from general industrial experience which was the only information available. In the meantime a much better data base, the so-called licensee-event reports, have been developed, since every electric utility has to report every anomalous event that occurs in their reactors. The enemies of nuclear power give a great deal of publicity to the number of these reports. It is as if every car owner were to report whenever his car doesn't start or whenever there is trouble in any part of it, such as a problem with the lights. Rather, we should be proud of the number of these reports because it shows that every anomalous behavior is reported and one can learn from it.

On the basis of these reports the Oak Ridge National Laboratory made new safety studies about two years ago and concluded that the Rasmussen Report was too optimistic about the status of nuclear power safety when it was first written. In the meantime, the half-dozen or so bad reactors have been retrofitted and brought up to normal standards, so by now the safety of nuclear power plants is better than that presented in the Rasmussen Report. The Three Mile Island event has made nuclear power much safer because the utilities were willing to retrofit their reactors that were substandard. The meltdown probability is about 1 in 20,000 years but the most important change is the so-called "source" term. If there is a meltdown, which, of course, is a very rare event, then the important question is how much radioactivity will be released to the public. That amount was grossly overestimated by the Rasmussen Report because again they had only guesses based on experiments. Three Mile Island has helped a great deal here because it was found that the amount of iodine released in that accident was much much less than expected. Iodine is the worst substance for giving radioactive exposure to humans. It has a lifetime of about 8 days, and it can be biologically absorbed in the thyroid. In the laboratory it is found that cesium and iodine form a compound very easily. The resulting cesium-iodide is not volatile and therefore is not ejected from the reactor. As long as there is water in the reactor it remains safely contained in that water. There have been many estimates of the source term, including a very extensive one sponsored by the American Physical Society under Richard Wilson. Anyone who is interested in this subject will learn a great deal from that study.

Another point which has worried many people is waste disposal. Contrary to what everybody says, and particularly what was said in the three-hour broadcast by ABC on the 6th of June, we *do* know how to dispose of waste. We *do* have safe waste-disposal methods. There is just no technical doubt that we can do these things. Under the sponsorship of the National Academy of Sciences a very good committee was appointed to discuss waste disposal. The committee was made up of geologists, radiochemists, and reactor experts. That committee and also everybody familiar with the problem agrees that the way to dispose of high-level waste is to place it deep down in the earth. As indicated in Figure 8, a deep shaft is dug by normal mining methods. Containers filled with nuclear waste are taken down to the bottom of the shaft where there are many tunnels extending in every direction. Along each tunnel there are side drifts which are just long enough to take one container of nuclear-waste material. The committee members set themselves the task of finding the maximum exposure of any person to radioactivity if such burial is performed. They found that at one place in the United States, namely in the area of Hanford, Washington, there is an almost ideal site where, after about 100,000 years of burial, the maximum exposure would be about 1 microrad per year. The exposure from natural sources is approximately a hundred millirads plus an equal amount from diagnostic medical sources. So the worst exposure would be less than 1 part in 100,000 of the natural radiation experienced by a person. For the first thousand years or so, the container and its contents of radioactive material of glass composition can be expected to remain intact. After a couple of thousand years, disintegration will occur and the barrier preventing the radioactivity from coming to the surface becomes the earth

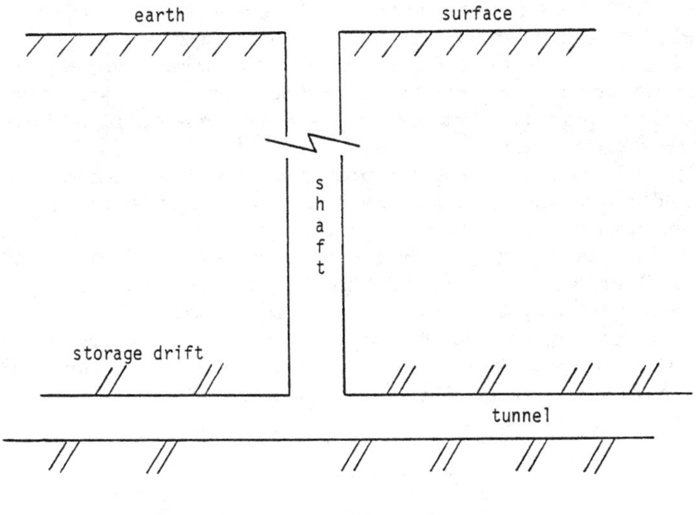

Figure 8. Plan of Deep-Earth Waste Disposal

itself. And, of course, the earth is a marvelous filter, as everybody knows. Where does one get the purest water? By drinking from a spring where the water comes right out of the earth and where it has been filtered by the earth. The presence of ground water will, of course, transport some of the radioactivity but at a very very slow pace. The previously stated exposure rates are based on these conditions. In a different region where the main underground material is granite rather than basalt, the exposure will be 10 microrads rather than 1 microrad, still a negligible amount.

Let me now discuss the breeder reactor. There is a good breeder reactor that is being used in France and a better one is being built. We have fallen behind very badly in this area but there is some cheerful news to report. A completely new type of breeder reactor is being developed by the Argonne National Laboratory. In most nuclear power plants the fuel is in the form of uranium or plutonium-oxide which have the very good property of being resistant to the released neutrons that might destroy the fuel material. But Argonne has found that a metallic alloy of uranium and plutonium with a bit of zirconium, is just as resistant and that fact was not known 20 years ago when we started out in the breeder business. The Argonne people have had an experimental breeder reactor in Idaho going for about 20 years, and now they are studying the metal-fuel-element breeder and how to process the material which it releases. For a metal, entirely different methods can be used than for oxide, namely chemical and electrolytical processing at high temperatures of the order of a thousand degrees Celsius. When some chloride is added, the metal becomes a molten salt that can be electrolyzed, resulting in fuel reprocessing that may be much simpler and hopefully cheaper than the present methods. It can be done in small batches, and no plutonium ever needs to leave or enter the plant. So I think this new process offers great hope for the United States breeder program.

Finally, since this discussion is, after all, about Electrical Engineering, we should consider the need for electricity. At the moment we have surplus capacity of electric power plants. If we assume an economic growth of 3% per year, Dr. Chauncey Starr, of the Electric Power Research Institute in Palo Alto, California, has calculated that by 1991 we shall have used up our surplus capacity. By the year 2000, if we retire the truly obsolete plants and do not order any new ones, then we shall have only 60% of the required electric-power capacity. So it is high time for the utilities to order new plants, and perhaps by 1991 some of them may again be nuclear plants.

[The following discussion resulted from some questions raised by members of the audience.]

Q.

A. On the question of the status of nuclear fusion, it looks quite good in the sense that the Princeton magnetic-confinement fusion experiment is going and maybe in a year or two from now they may reach the stage of "break even" where as much energy comes out as is put in. Recall that in order to start a fusion reaction it is necessary to inject energy into the reactor and so an energy balance must be obtained. Scientifically, I think it is going well and will continue to improve if the government continues to support it. I think very little of inertial-confinement fusion. It is a fairly hopeless undertaking by comparison, and the government rightly has decreased the support for inertial fusion, while keeping the support for magnetic confinement about constant.

So much for the technical side. The engineering side is far less promising. The best and most recent estimates I have seen say that fusion will cost approximately twice as much as nuclear power from fission, so I believe that for our lifetimes (I don't know how many young people are here) it will be nuclear power from fission and not from fusion. Now when we have the breeder, then there will be fuel for nuclear power from fission essentially indefinitely; certainly in this country for ten thousand years and probably longer than that. Therefore fusion will not be needed for a long time. We should keep at it, however, and we should continue to develop it. I know, in fact, a very beneficial use for fusion. Since fusion produces a lot of neutrons per unit energy, these neutrons could be absorbed in some material like thorium or natural uranium. In this way, one could make large quantities of fissionable isotopes, U-233 or Pu-239, respectively. This method could replace the breeder, and rough estimates indicate that it may cost about the same.

Q.

A. The questioner states that it would seem from the literature that progress from the inertial procedure has perhaps not matched that of the tokamak but shows considerable promise for a fraction of the investment. He then asks why I consider the inertial approach to be hopeless. In response I have two reasons: One is it hurts my intuitive sense to take an apparatus the size of this 400-seat auditorium and focus its energy on a tiny pellet and then try to get large amounts of energy from such very small bits of material. The other problem is that the way to produce the necessary high temperature and high compression is commonly by use of lasers. The laser which has been mostly used, namely the neodymium laser, is unsuitable for any high-duty-cycle operation so it is necessary to go to some type of gas laser. The most promising is the rare-gas halogen laser which is still in its early development. How well it will do we don't know. While the Livermore experiments have been quite successful, to a large extent they have been done with a laser which has no growth possibility to steady operation. Further, the pellet must be compressed to extremely high density before it is heated and that is a big problem. Another difficult problem that must be solved is having to keep the pellet completely round while it is being heated.

Q.

A. On the question on the coolant in the breeder reactor, it will continue to be sodium. One reason is that sodium is very compatible with metal; it does not attack metal while water attacks many metals. The second reason is that sodium is a marvelous transporter and conductor of heat so that its use is also very good to ensure the breeder against accidents. The trouble with sodium, of course, is that ultimately it is necessary to transfer the heat to water, and sodium and water do not get along with each other very well.

Q.

A. In response to the question on the contributions to the energy problem of photovoltaic generation and hydrogen-oxygen fuel cells, I know very little about the latter so I prefer not to comment on that subject. I believe photovoltaic generation is improving somewhat because higher efficiencies are now being obtained. I think, however, that it will have a hard time competing economically with existing sources. I think it still will be a long time before photovoltaics can compete with either fission or coal.

Professor Corson: I might make a comment about that. In my study of the satellite power system a great deal of attention went into photovoltaics and solar-energy generation. The general problem with this technology is the extremely difficult matter of power distribution. Sunshine is plentiful in Arizona but to get that sunshine transported in useful form to Ithaca, New York is a very difficult job. There are things called surge impedances in transmission lines that limit the length of these lines. Cryogenic power transmission may help but it is still a long way off. The possibilities are there but the most we can hope for in our lifetime is perhaps to see 20% of our energy provided in this way.

Q.

A. The question is could one get a breeder program that uses thorium as the raw material and uranium 233 as the fissionable material, and could existing reactors be used by backfitting them to this process. I don't think a breeder can be obtained in this way. I do think, however, that a reactor might be developed which uses very little uranium. One such reactor is being designed at the moment by Dr. Radkowsky who used to be a technical assistant to

Admiral Rickover in the Navy program. It looks fairly promising and has the very great advantage of possible backfitting into existing reactors. In fact that is what he has very much in mind. But it should not be called a breeder reactor. A better name would be: material-conserving reactor.

TELECOMMUNICATIONS REGULATION: A CASE STUDY OF THE IMPACT OF A TECHNOLOGY ON SOCIAL INSTITUTIONS

Alfred E. Kahn

Alfred E. Kahn received the A.B. and M.A. degrees from New York University in 1936 and 1937, respectively, and the Ph.D. degree from Yale University in 1942. In 1945 he was appointed Assistant Professor and Chairman of the Department of Economics at Ripon College and remained there until 1947 when he joined the faculty at Cornell University as an Assistant Professor of Economics. He was promoted to Associate Professor and Professor in 1950 and 1955, respectively, and was Chairman of the Department of Economics from 1958 to 1963. In 1967 he became the Robert Julius Thorne Professor of Economics, and served as Dean of the College of Arts and Sciences at Cornell from 1969 to 1974. Dr. Kahn is the recipient of five honorary degrees and has held a number of important government posts including service as the Chairman of the New York State Public Service Commission from 1974 -1977, Chairman of the U.S. Civil Aeronautics Board from 1977-1978, and Chairman of the President's Council on Wage and Price Stability from 1978-1980. He is the author or co-author of four textbooks on Economics, and was a member of the Board of Editors of the *American Economic Review* from 1961-1964. He is a member of the American Academy of Arts and Sciences, the American Economic Association, the National Association of Regulatory Utility Commissions, and the National Governing Board of Common Cause. He is also a member of National Economic Research Associates (NERA).

"I've forgotten what year Fred came to Cornell. It was a few years after the war. He was a professor of Economics and served as Chairman of the Department. He became Dean of the College of Arts and Sciences while I was President, and, of course, I did not give him adequate funds to allow him to do his job properly. He went to Albany to be Chairman of the Public Service Commission. From there he went to Washington to be the Chairman of the Civil Aeronautics Board where he deregulated himself out of a job. He went on to be President Carter's Inflation Czar; and as soon as he solved that problem, he came back to Cornell. Now there is one thing I want to say, because I don't want Fred to say it. He got a lot of mileage in Washington with an aphorism that was really mine. I think I told it to him in the first place, and I'm going to stop him from saying it this time because *I'm* going to say it. The statement is that a dean is to his faculty as a fire hydrant is to a dog."

—Dale Corson

I wasn't intending to use that one today, Dale. I did want to make one observation, however, about my success in combating inflation, despite all appearances to the contrary: Those of you who are not economists may not understand that sometimes policy changes have their effect after only a lag of three or four years.

I'm somewhat intimidated by the prospect of participating in this symposium, because I take it my role is one, in effect, of attempting to construct a bridge between what C. P. Snow referred to several decades ago as the "two cultures"—that is, the culture of science and technology on the one side and of our social and political institutions on the other. Unfortunately, I am the most perfect exemplar of his thesis about the deplorable rift between the two, and I'm particularly distressed that you have me following Hans Bethe, whom I have always regarded with awe as its leading refutation.

Last week I attempted, equally foolhardily, to fulfill the role of rapporteur and grand summarizer of a two-day conference on communications technology and public policy, which began with a fascinating talk by Dr. Arno Penzias, Vice President for Research at Bell Labs, and a Nobel Laureate. I felt exactly the way I had felt when my wife and I went to the Comédie Française in Paris almost a year ago. I understood thousands of words and not one single sentence. I well remember sitting there mystified at the news that UNIX are, or is, very popular on college campuses, and you're never going to convince me that NYNEX is anything but an over-the-counter drug for the terminal stages of Pacific Telesis.

I list these disqualifications of mine with dismay. Let me turn, however, to what I think I do know and understand: Until recently, the industry that supplied us with telecommunications services was a monopoly protected by exclusive franchises from competition and tightly, thoroughly regulated like other public utility monopolies. The field of telecommunications, more broadly defined, consisted of a series of sheltered preserves with rigid, legally impenetrable barriers erected between them. Each, for the most part, was the similarly restricted exclusive domain of a licensed, certificated, regulated, and putatively monopolistic supplier. For example, record or written communications, that is to say typed or teletyped communication, was the exclusive preserve of Western Union, while voice communications was the exclusive domain of the Bell System and the 1500 interconnected independent telephone companies, each confined to its own, exclusive market boundaries, which the others were forbidden to cross. In fact, AT&T was at one point forced to sell its TWX (or was it its TELEX?) to Western Union so as to maintain that historic distinction between record and voice.

Similarly, satellite transmission was for years the exclusive domain of COMSAT: AT&T was forbidden to enter, being confined instead to terrestrial communications facilities, from which, in turn, COMSAT was excluded.

Similarly, video was designated as the preserve of Cable TV companies, who have been hooking up a large percentage of the homes in America with coaxial cable—even today telephone companies are, for the most part, not permitted to go into that business; and cable companies were excluded from providing switched telephone service. Finally, there was a similar separation between the supply of computers and computing services, the storage, processing, and retrieval of data, on the one hand, which remained generally unregulated and competitive, and the offer, on the other hand, of telecommunication services, with the suppliers of each strictly forbidden to trespass into the market preserve of the other—AT&T, under the terms of the 1956 antitrust consent decree, was confined to the provision of regulated monopoly services.

Now even a technological idiot like me can understand that those historic distinctions have become technologically meaningless. Coaxial cable, microwave, satellite, fiber optics, the transistor and microprocessor, the super computer (observe how lightly those words trip off my tongue without ever passing through my brain) have literally made thousands of companies capable of supplying telecommunications services—indeed made it logical to perform telecommunication service as the link between their other operations—and made the telephone companies correspondingly capable of supplying all those other services: the modern telephone exchange is itself a computer. And worse, those artificial legal boundaries, whatever their historic justification, became, and to the extent that they are still there, remain, technologically and economically mischievous, obstructive of efficiency and of the process of innovation itself. In effect, they try to draw legal institutional lines through a body of water—worse, through a turbulently-running, technological stream that is changing all the time. Those lines are, in the long run, doomed to obliteration.

I think it's only fair to say that the old system served us very well. We routinely encountered and sometimes uttered the opinion that ours is the best telephone system in the world. I'm not aware of any effort systematically to investigate the validity of that opinion, but we've had enough experience with enough telephone systems elsewhere to accept it at least as a plausible judgment. I also think it is possible that some of the legal barriers served a kind of infant industry purpose: we may have developed satellite communications more rapidly precisely because we entrusted it at the outset to a COMSAT, and kept AT&T out, in view of the fact that AT&T had its preponderant financial stake in terrestrial facilities and therefore might have had inadequate incentive to develop the less capital-intensive satellite facilities. As a regulated utility, its profits varied with the size of its rate base: so—to put it overly simply—it was not in its interest to shift from terrestrial to satellite transmission facilities. In short, a distortion of company incentives introduced by rate base/rate-of-return regulation may in turn have required another distortion—an artificial barrier to AT&T entry. Perhaps, similarly, we developed cable TV, the wiring of America, more rapidly because we entrusted it to independent entrepreneurs and kept out the telephone companies, who might have been tempted to hold it back for the good static reason that there's no need to have those two separate sets of connections into the house.

Finally, I feel it is worth pointing out that we still don't know, even today, to what extent the telephone business, both long-distance and local-exchange, may still really be natural monopolies—to what extent, that is, service might still be provided most efficiently by a single supplier, planning and operating an integrated national network and in a position to exploit to the fullest extent the economies of scale and of networking. There really is a strong possibility, which we have yet to test, that the entry of competitors like MCI, Sprint, Satellite Business Systems and the like into the long-distance business might have been precluded by the telephone companies if they had not been subjected to regulatory constraints on their ability to compete. Arbitragers and resellers of long-distance service have, similarly, been able to enter into local competition in large measure merely because of the artificial structure of non-cost-related prices that regulators have imposed upon the telephone companies; and the same has certainly been true in important measure of the many companies that have been bypassing the facilities of the local telephone companies.

The most prominent instance of those artificial regulatory inducements to competitive entry is the system under which we have progressively, over the decades, loaded billions of dollars of costs onto long-distance calling. I refer here to approximately 10 billion dollars of annual cost associated merely with hooking a hundred million or so subscribers to the telephone system, giving them a dial tone when they pick up the phone. These costs are generally conceded to be non-traffic sensitive: they are not increased one iota by the placing of additional long-distance calls. But over the years, for political reasons, the state regulators, with the cooperation of the Federal Communications Commission, transferred more and more of those costs over to the interstate jurisdiction (as well as to intrastate toll service), with the result that the price of long-distance calling appears to be inflated by something like 200 percent over the marginal cost of providing it. It is entirely possible therefore that the entry of the MCIs into these markets was simply the artificial consequence of an arbitrary, regulatorily imposed system of pricing. It is a very actively contested question today whether these competitors can survive if and when AT&T is finally permitted to compete fairly and freely with them. And of course there is even less chance of survival if AT&T is free to compete unfairly.

The other side of the coin of these ever-larger transfers of costs to long-distance calling is that it enabled state legislators and regulators to avoid the embarrassment of having to increase local basic rates as much as they would otherwise have had to do. As a result, between 1940 and 1980 the average basic monthly charge for telephone service rose something like 55 percent less than the Consumer Price Index, i.e., that rate fell something like 55 percent in real terms; and this contributed to the virtual universality of telephone subscription today. In 1940, only 37 percent of the households in the United States had telephone service; in the 1980 census, it was about 93 percent.

The basic charge, in most jurisdictions, includes unlimited local calling. Here is another economic absurdity—a zero marginal charge for a service that has positive marginal costs, financed by charging long-distance service far more than its marginal costs. The point is that this kind of cross subsidy was possible only under monopoly: once competitors were free to enter the long-distance business, they threatened to eliminate the monopoly surcharging of that service that generated the subsidies unless some way were found to levy the surcharges on them as well.

A final consideration on the side of the possibility that telephone service may really be a natural monopoly and that the present phenomenon of competition is merely an artificial construct of selective and, in some measure, economically ridiculous regulation itself will be apparent to anyone who has puzzled over the new telephone bills, or has had the experience of calling the telephone company when the phone doesn't work, only to be asked: "Are you sure it's our line?", or "Could it be your interior wiring?", or "Could it be your jack?", or "Could it be AT&T's or Sears Roebuck's telephone equipment?" Many of these people will not need to be convinced of the benefits of giving one company end-to-end responsibility for the quality and continuity of service.

Having recognized these possibilities, I remained convinced that the dynamic development of telecommunications and computing technology ultimately made inevitable the destruction of that tidy, neatly compartmentalized, noncompetitive, cartelized, compulsively regulated, historic structure of the telecommunications industry. Once, for example, we had microwave transmission, with its very low capital costs, it was inevitable that we would eventually permit large users of telecommunications services to construct their own private systems. That change marks the first step in the industry's deregulation—when the Federal Communication Commission, in 1959, opened up the above-890 megahertz spectrum to these private microwave systems. Once that door was opened, I think it was inevitable that we would permit the MCIs and the Sprints and the others to come in to supply the same service on a common carrier basis, as the FCC did in 1969 and 1971. And once that happened, I think it was likewise inevitable that we would eventually permit those competitors to supply switched toll and WATS service to the public at large in direct competition with the Bell System, as the United States Circuit Court of Appeals forced an unwilling FCC to do in its Execunet decision in 1977. One of the reasons I emphasize all these dates is to observe that the last of them was seven years before the divestiture of AT&T. The point is that deregulation and the breakup of the Bell System were not the same thing; the former—which clearly traces back, at least, to the Execunet decision in 1977 and really to the above-890 megahertz decision of 1959—did not necessarily require the latter.

In short I believe that in the face of the burgeoning of these technologies and the potential proliferation of services, both phenomena cutting across all the traditional legal and regulatory boundaries, we would have been unwilling, indefinitely, to rely on a single monopolist supplier, however efficient and however large the number of Nobel laureates on its payroll. The whole rationale of deregulation of the airlines was the conception that government regulators are incapable of visualizing or anticipating what the optimum structure of an industry is, who is best adapted to supply what services at what price, and that is, of course, the virtue of leaving such determinations to the competitive market. Given the far more complicated and versatile technological possibilities in telecommunications, this particular case for deregulation must be even stronger there.

In the case of communications, the adaptation of our public policies to these imperatives, as well as to the new competitive order, is still highly incomplete. The political resistances continue to be intense. But I think it is already possible to discern the principal consequences of deregulation from our experience to date in this industry, along with the others—airlines, trucking, railroading, financial markets—that have in recent years been subject to a similar revolutionary transformation.

1. It has produced a sharp increase in the intensity of competition.
2. As a consequence—except in the case of the railroads—prices have come to be much more closely aligned with costs. That's the way competition works: competitors tend to enter the markets in which prices have been held artificially above cost, and in so doing, dry up such monopoly revenues as may have previously been used to subsidize the provision of other services below cost.
3. Customers are being offered the choice among a much richer variety of price-quality options.
4. With their newly received freedom to structure their operations as they please, and subjected to sharply intensified competitive pressures, suppliers have achieved great improvements in efficiency. All of these industries have experienced an upsurge of productivity.

5. The new competition has exerted powerful downward pressure on previously inflated—in many cases outrageously inflated—wages: it is not genuinely appreciated to what extent the monopoly profits made possible by governmentally-enforced cartelization were appropriated by the teamsters, airline pilots, machinists, and the rest. The unions were not foolish to have opposed deregulation.
6. Except in the case of the railroads, competition has produced declines in average prices, in real terms, and, finally,
7. On average, an improvement in the quality in service. (I wish there were time to document these categorical assertions for the benefit of the skeptical among you.)

This quick summary doesn't do justice to the exceptions, and certainly not to the turmoil and the pain that has been inflicted on previously protected, insulated, privileged businesses and groups of employees. But it is nevertheless, I believe, broadly and fairly descriptive of what has happened. For the most part, we see evidence of all these same consequences in communications as well:

1. First of all, there has been a sharp increase in the intensity of competition. AT&T introduced an exhibit in an FCC proceeding over a year ago, showing that as of the end of 1983, one or more of its long-distance competitors was already offering service in counties embracing 75 percent of the U.S. population, and that percentage is undoubtedly higher today. It also showed—and this I found even more striking—that among its business customers with bills of $150 or more a month, who accounted for about 80 percent of its total business billings, no less than 41 percent of the companies in the entire country, and no less than 56 percent in the particular counties where one or more of its competitors were represented—MCI, Sprint, Satellite Business Systems, Western Union, US Tel—were already subscribing to their service. These percentages will doubtless increase very rapidly, so that the availability of competitive long-distance services must be getting very close to universal.

   The Bell operating companies are now required, under the terms of the antitrust consent decree, to extend to those competitors access to the local exchange network for the initiation and termination of calls equal in quality—for example, via one-plus dialing—to the access heretofore provided only to AT&T Long Lines; and as that goal is reached—by the end of 1986 in most markets—all customers are being required to opt for their preferred long-distance supplier—the one they will reach just by dialing one. And a recent FCC decision decrees that those customers who do not opt for a preferred supplier will be distributed among the competitive suppliers in the same proportion as the ones who do.

Similarly, the Bell System's share of the customer equipment market has plummeted. AT&T's share in the sale of PBX's, new switchboards, is below 30 percent. Western Electric's share in the supply of telephonic equipment has declined as well. Scarcely a day passes without the announcement of some new telecommunications venture by companies setting up their own systems. Banks, financial houses, industrials, real estate developers, railroads, utility companies engage in this activity, with many of them selling their surplus capacity to the general public. New York Tel, for example, alone, supplied to the New York Public Service Commission a partial list of just a few examples in New York State. Bankers Trust, Brooklyn Union Gas, and Citibank all use private microwave systems in New York City to bypass, totally, New York Telephone facilities and connect directly either with customers or branch offices of their own or with long-distance carriers. Crocker National Bank, General Electric, IBM, J.C. Penney, and the *New York Times* use private satellite networks. Niagara Mohawk in Syracuse, Grumman Corporation on Long Island, *Reader's Digest* in Pleasantville, and the M&T Bank in Buffalo all use fiber-optics by-pass systems, again totally circumventing the New York Telephone facilities. The State of New York uses a private microwave system between New York and Albany and soon will have one between Albany and Buffalo. Manhattan Cable TV offers a data-communications network for several large corporations, mainly banks in New York. Merrill Lynch is laying or planning to lay fiber-optics cable right down Wall Street, to connect directly with a teleport on Staten Island, thereby offering an integrated cable/satellite joint venture transmission network totally circumventing the local telephone company. Olympia & York, the biggest real-estate developer, is constructing a number of "smart" commercial buildings and complexes, with their own computing, computer-access, teleconferencing, telephone systems, asking the telephone company merely to put a trunk at their door. Digital-termination service, a packetized microwave technique that requires only a small roof-top antenna, is capable of distributing data over a 360-degree circle with a seven-mile radius, and so of connecting directly with an interexchange carrier, once again bypassing the telephone company. These various developments are so numerous and valuable that there is even a new periodical devoted to them; its title: *By-Pass News.*

2. In consequence, as in other deregulated industries, competition, where it has been permitted to operate, has driven prices closer to cost. The qualification is, of course, a very important one. State public utility commissions, intent upon protecting the basic monthly charge to residential subscribers (who are numerous, vocal, and vote), are strenuously resisting the squeezing out of the excessive charges on the long-distance business by incorporating them in their charges to the long-distance companies for access to the local network. The inflation of those access charges, of course, merely encourages more by-pass by those carriers; since divestiture, AT&T, itself, is now the biggest potential by-passer of all. Ultimately, I suggest, the logic of competition will therefore succeed in squeezing out that subsidy, and drive the prices of the various communications services—both the prices artificially inflated and the prices artificially suppressed under present regulatory polices—to their respective economic costs.
3. Communications exemplifies, even more so than the other recently deregulated industries, the efficacy of competition in ensuring a multiplication of the options available to consumers. All you have to do is read the daily newspapers to follow the explosion of communication services, standardized and tailor made, offered to the market. One of the other things that Arno Penzias pointed out to us last week, which I thought very illuminating, was that in what he referred to as the old days of "technological scarcity," it was the vendor who devised and designed the communications system; the big customer would go to the vendor and ask what services were available. Today, in the era of what he characterized as "technological abundance," it is the user who defines the architecture, who tells the vendor what he needs, and demands a system designed to satisfy those needs.

   A telecommunications monopoly could modulate and moderate the pace at which it went from analog to digital, or introduced wide-band high-speed data-transmission facilities. A telephone company subject to intense competition, in contrast, has no choice but to press ahead with such innovations. One interesting and difficult consequence of the introduction of competition has therefore been to leave the telephone companies with an enormous investment in outmoded facilities, inadequately depreciated under regulatory policies that, framed under the assumption of monopoly and an ability to control the rate of innovation, provided for writing off investments over very long periods of time. The consequence of the introduction of competition, with its more rapid and uncontrollable embodiment of the latest technologies in new facilities, has been to depress the market value of the telephone companies' assets far below the value at which, insufficiently depreciated, they are listed in the books. AT&T recognized this discrepancy, upon entering the deregulated era, by simply writing off something like six billion dollars of its overvalued assets. In the case of the operating companies, we are talking apparently about an even greater aggregate discrepancy, an even greater amount by which the book value of their assets, and therefore of their equity, exaggerates their true earning capacity in the new competitive era—which raises the same fascinating question as is raised by the multi-billion dollar investments by electric companies in failed nuclear plants: who is going to swallow those losses, reflecting—in the telephone case—the historic underrecovery of investments via depreciation charges to ratepayers?

4. On the quality of service consequent on deregulation of communications, I think that there are two phenomena to be observed. On the one side, the proliferation of services and combinations of services, the acceleration of innovation, which have clearly been a plus. On the other side, there have been the consequences of the dissolution of AT&T—the complaints among ordinary users about the loss of a single responsible supplier and complaints from business users about delays in getting service instituted.

As I have already observed, the deregulation of the communications industry is not synonymous with the break up of AT&T, which became effective on January 1, 1984. The opening of this industry to the forces of competition goes back 25 years.

At the same time, the two changes were and are not unrelated. The previous structure of AT&T, and perhaps even more the policies that it chose to adopt in response to the threat of competition, stood as a very powerful obstacle to the emergence of effective competition, critically because the company controlled the access of its long-distance competitors to the local-exchange networks, to the hundred million or so lines to ultimate placers and receivers of calls, and exercised that control in such a way as to deny its rivals a fair opportunity to compete with it—for example, by refusing to provide them with interconnections of the same quality as its Long Lines enjoyed.

It seemed to the people involved in the antitrust proceedings against AT&T, therefore, that the cleanest solution would be simply to separate the Bell companies' control over the bottleneck facilities from their operations in the competitive areas—the manufacture of equipment and the provision of long-distance service—so as to eliminate both the incentive and the ability to handicap rivals in these ways. Acceptance of this separation looked like a sensible thing from AT&T's point of view, too, because they were faced with the strong possibility that they would be convicted of violation of the antitrust laws, which would have exposed them to billions of dollars of damages resulting from private suits by aggrieved parties. It also had the enormous attraction of promising to release Long Lines (along with Bell Labs and Western Electric) from the constraints of an earlier antitrust consent settlement, which had confined them to the provision of regulated communications services, freeing them instead to exploit their scientific and technological talents, free of regulation, wherever they saw an opportunity for profit.

The concept was not ridiculous. The local telephone companies, now severed from AT&T, left in control of the local-access network and prohibited from competing with the companies dependent on them for access to that network, would no longer have any incentive to discriminate among different long-distance companies or terminal equipment suppliers. On the contrary, the separated local telephone companies would presumably have every incentive to encourage all such companies to develop uses of and attachments to its network—burglar-alarm, fire-alarm, time-of-day services, meter reading, electronic publishing, and all kinds of equipment attached to the customer's line—because the result would be higher revenues all around. (Remember, whenever you place a call to someone with a recording machine, there are no uncompleted, no-charge calls!) Developers of these services, in turn, free of the fear that they would be competing with the company on which they were dependent for access to the facilities, would emerge more willingly and in greater numbers in exploiting the possibility of these technologies. It seems to me that conception was not crazy.

And AT&T, separated from the controllers of those bottlenecks, could in principle safely be freed to compete anywhere and everywhere it chose.

Unfortunately, however, the conception was simplistic. The operating companies—according to that conception, natural monopolists—have instead, as I've already pointed out, been subject to intensifying competition by bypassers. At the other end, AT&T Communications remains overwhelmingly dominant in the provision of long-distance communications: there is still genuine reason for uncertainty about how much monopoly power it may still have and whether its much smaller rivals are really in a position to compete effectively with it, and whether, or in what respect, it may therefore safely be deregulated.

Moreover, as I've likewise pointed out, regulators and legislators are still eager to have their cake and eat it too. That is to say, to demand the benefits of competition but at the same time to try to retain the benefits of cross-subsidization, in order to save them from the unpleasant necessity of having to increase the basic monthly charge. And so they continue to insist on collecting billions of dollars a year in economically illegitimate charges to long-distance carriers for access to the local networks: some 75 percent of AT&T's total costs consist of such access charges.

Finally, so long as AT&T and the operating companies retain monopoly power over some customers, in some markets, and prices to those captive customers are regulated in traditional fashion, regulators quite properly worry about letting them price their services subject to competition or bypass as they choose, for fear that they will be tempted to cross-subsidize the competitive operations by transferring costs to the monopoly services. And so they have been resisting the deregulation of the competitive prices and limiting the ability of the Bell successor companies to meet competition.

There is no way in which I can do justice to the complicated maze of interrelated problems involved in trying to manage a system part competitive and part regulated. One thing must be clear: the introduction of competition and the attempt to preserve a regime of prices grossly out of line with cost are simply incompatible; and in the long run, I believe this means the latter effort must give way, because I don't think we can ever unscramble those competitive eggs. So the big question of social policy is how do we reconcile competition with preserving some of the social goals, particularly the one of universality of service.

Every economist knows the answer: the way to take care of the five percent or so, at most, of the 93 percent of American households with telephone service who might be tempted by increasing basic monthly charges to drop off the system is not to hold down the price for all 93 percent. The way to help poor people is to help poor people, with some sort of means-tested life-line rate, financed, if need be, by a marginally higher basic monthly charge to the other 88 percent.

There are a host of related questions, to which I really can't do justice. Can we go the whole way to deregulation of AT&T? When? What about the complaints of the MCIs and Sprints that they are still too small to compete with it? Ironically, the advent of equal-quality access at cost-based rates—the denial of which has been one of their major historic grievances—may take away their main basis for survival until now—the ability to offer inferior service at a very substantial cost discount: conceivably once they have to pay the same price for access as AT&T, they are doomed. What do we do about their protestations? Is our only hope for preserving competition to protect competitors from competition? For how long? Meanwhile, of course, AT&T and the operating companies keep clamoring for "level playing fields," a legitimate contention if one truly wants competition. But of course the Christians and the lions in the Coliseum also competed on a level playing field.

Meanwhile, and very close to finally, under the terms of the Modified Final Judgment's confinement of the operating companies to the provision of local (strictly intra-LATA) plain old telephone service under comprehensive regulation continues to conflict with the dictates of technology. In this respect, the decree is trying to perpetuate the old legal-jurisdictional boundaries in a rapidly moving liquid; but the barriers work in only one direction: others are free to enter the Bell companies' markets and compete with them. It is only the operating companies who cannot move outside of their boxes—in effect, the very box in which AT&T had previously been confined by the 1956 decree. Restrictions like these are in an important sense inhibitive of competition itself: industries are limited in their ability to follow their technology wherever it leads.

Are there ultimate solutions to this puzzle of running a system that is both competitive and regulated? It seems to me the logic of free entry and free pricing for some demands total deregulation. Not necessarily right away—there is a reasonable case for transitional protection of customers who lack the protection of competitive alternatives. But competition is ultimately incompatible with asymmetrical regulation that ties the hands of one of the competitors in order to protect the others. Ultimately, I think, we must dissolve the structural restrictions on the ability of companies—all companies—to follow their technology where they please, and protect monopoly customers from the burden of cross-subsidization by effecting a *financial* separation of the revenue requirements associated with the regulated services from whatever revenues are received from the competitive services. Once you break the financial link you don't have to break the operational link.

Unless and until we do so, I think there is a non-zero possibility we could end up in this country with two communication systems—a shiny, bright, efficient, technologically advanced, high-performance, low-cost system for the big users, who cannot be denied access to the full benefits of modern technology; and an antiquated, decrepit, high-cost, low-performance system for the smaller users, provided by financially deprived local telephone companies. I think there is a non-negligible danger that the operating companies will go the way of the railroads if legislators and regulators try to cling to the old system, if they fail in their obligation to find new and more efficient ways of satisfying the social imperative of something like universality of service, if they continue to obstruct the new technology, if they force telephone companies to continue indefinitely to stick to plain old telephone service and operate within the old closed compartments.

Now I don't really think that is going to happen. But my principal reason for this comparative optimism is the central theme of this convocation—namely, that the imperatives of technology and competition must ultimately triumph. The reason, an economist will recognize, is that we are not talking about a zero-sum game. There is a larger national pie available to us to the extent that we price efficiently, to the extent that we rely on competition, to the extent that we actively probe the potentialities of these exuberant technologies. Ultimately that will force political leaders and regulators to adapt to these inescapable changes, find alternative ways of achieving the social goals, and welcome the vast benefits that they promise all of us.

Dr. Penzias said something like this at the end of his keynote talk: that technology does not impose imperatives on us, it does not dictate what we do, it provides us rather with new enriched options. "Technology," he said, "is not a driver but a facilitator." I think he was being excessively modest. I think, instead, that our experiences tell us two other things as well: number one, as Thorstein Veblen observed almost a hundred years ago, it is not only true that necessity is the mother of invention, it is equally true that invention is the mother of necessity. There is no clearer illustration of this proposition than the telephone itself. What was once a toy has, by virtue of its having been invented and made available, become an absolute necessity. And today the pressures are simply irresistible on all businesses, if they are to survive, to avail themselves with the utmost intensity of whatever technology is making available to them.

Second, that it is technology that will ultimately determine the public policies that we adopt, from the above-890 megahertz decision in 1959, to—shortly, I believe—a dissolution of the constraints imposed on the Bell telephone companies by the Modified Final Judgment. I believe that it will not take us another 25 years to remove those constraints as it did to let AT&T out of the box in which it was confined by the 1956 decree. Ten years from now I expect we will look back and wonder what all the fuss was about.

WHITHER ELECTRICAL ENGINEERING?

Maurice G. A. Bernard

Maurice G. A. Bernard graduated from the École Polytechnique in Palaiseau, France in 1948 and from École Nationale Superieure des Télécommunications (ENST) in Paris, France in 1953. He was awarded the degrees of Dr. ès Sci., and lic. ès Sci (Dr. of Sciences and Graduate in Sciences) in 1958. From 1957-1970 he was a professor at the ENST College of Physics and Chemistry and also was a lecturer at École Polytechnique until 1973 when he became a professor at that institution. In 1961 he was a Visiting Professor at Stanford University in the United States. Following his return to École Polytechnique he was responsible for the Department of Electronics from 1969 -1975, and served as Deputy Executive Manager of that department until 1978 when he became Director of the French National Center for Telecommunications Research, a post he held until 1981. His present position is Director of the École Polytechnique, the prestigious university of science and engineering that was founded by Napoleon in 1794. where he is responsible for all educational and research programs. Dr. Bernard is the author of two textbooks on semiconductor theory and has many publications in the fields of solid-state physics and quantum electronics. His awards include the Chevalier Legion d' Honneur, the Chevalier Ordre Nationale du Merit, the Chevalier O. Palmes Academy, and the Blondel Medal (1962).

Editor's Note: Dr. Bernard's current position is Dean of Scientific Studies at École Polytechnique, Palaiseau.

"Dr. Bernard, a world citizen who has been a Visiting Professor in a number of institutions in this country, will now present his thoughts on the future of Electrical Engineering."

—Dale Corson

The good thing about international conferences is that you travel; and when you travel, you learn a lot. For instance, this morning I learned from Professor Hans Bethe that we Frenchmen, after all, are not so bad. I also found in his fascinating lecture a lesson on what I may call some very special relativity. Let me explain: I arrive here in a capitalist country and learn how good, how efficient, a centralized governmental institution can be. Now two days from now I will be back in France, a country run by Socialists as far as I know; and I will be told by everybody again that the entrepreneurial spirit is exactly what we lack in France.

I also learned something very important from Fred Kahn when he mentioned in his marvelous speech his experience with the Comédie Française, where he could understand thousands of words but did not catch the story. Then I realized how different I am from an economist because I did not catch a single word of his lecture, but I guess I understood what he meant.

To celebrate today, the centennial anniversary of the first American Ph.D. in Electrical Engineering, granted here at Cornell University, offers a two-fold opportunity:

- First, to acknowledge your University as one of the most famous in this country, both for the remarkable foresight of its Founding Fathers and for the high level of excellence reached today by your School of Electrical Engineering.

- Second, to glance back at the recent past to see if it can help us in our present tasks by providing some ideas and guidelines about our future and how to prepare for it.

This is what I will try to do.

A TOUCH OF HISTORY

Although electrical phenomena have been known for a long time, their presence did not keep people in earlier days from believing that the world was made of fire, water, earth, and air. Before the end of the 17th century, electrical experiments remained curiosities. During the 18th century, some key discoveries, both in Western Europe and in the USA, enlarged the field of knowledge in electrophysics. Electrical Engineering emerged as a separate field of study in the 19th century, and throughout the 20th century has progressed at an ever increasing pace. The discipline has become multiform, and pervades almost every scientific and technical domain, and is certain to affect our society as a whole.

The history of electricity is clearly bound to the history of modern times. It shows that major progress in a scientific field proceeds through qualitative jumps that are superimposed on the constant accumulation of knowledge.

Each jump is the result of an encounter, of a collision. I will limit myself to two very different examples:

1) Figure 1 depicts the first electrostatic machine as built by Otto Von Guericke at the end of the 17th century. A sulphur sphere rotates about an axis and electric charges appear as a result of friction. The Leyde bottle, shown in Figure 2, was discovered soon afterwards. The 18th century was a time of intense scientific curiosity, and the phenomena associated with the "electric fluid" caused many "experiments" of the kind displayed in Figure 3 to take place in many of the "salons" of the large cities of Western Europe and of the east coast of the USA. Around 1750, Benjamin Franklin made the assumption that lightning appearing during storms was similar to the flashes that occurred during experiments on the "electric fluid." Two years later, Dr. Franklin and his son demonstrated the validity of this assumption by means of a very dangerous experiment that used a kite flown with a conducting cord. Fortunately, as reported in the volume shown in Figure 4, the experiment was conducted without harm to either man. At about the same time, Dalibard, the parish priest of Marly, near Versailles, was making a similar experiment with a long metallic pole. For the first time, Franklin established the link between lightning and the discharge phenomena observed in electrostatic machines.

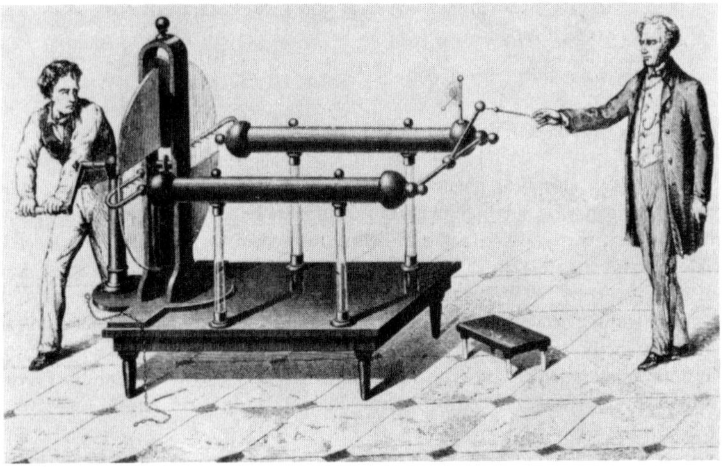

Figure 1. The Electrostatic Machine of Otto von Guericke

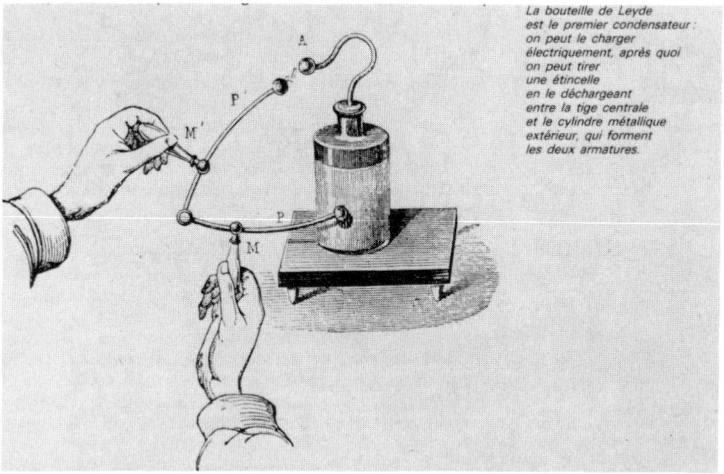

Figure 2. The Leyden Jar

Figure 3. An Experiment on the Electric Fluid: The Knight's Armor

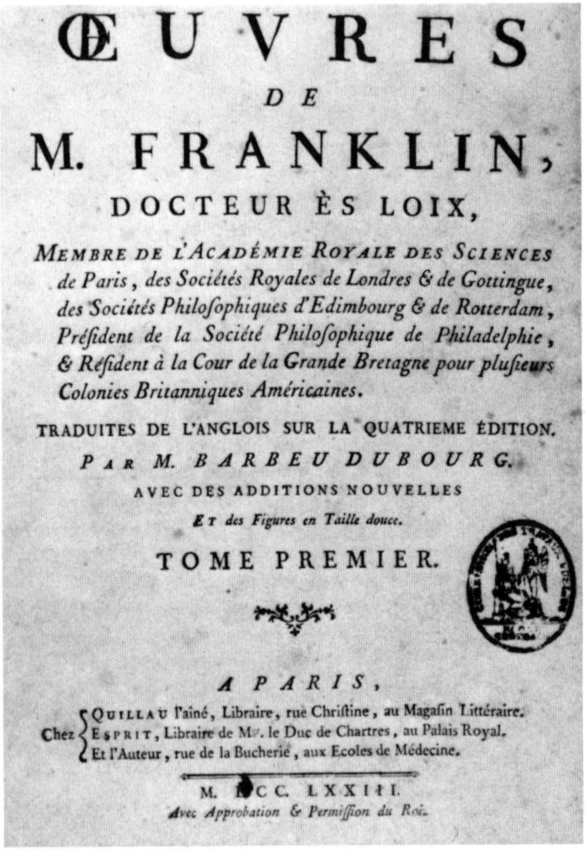

ŒUVRES

DE

M. FRANKLIN,

DOCTEUR ÈS LOIX,

Membre de l'Académie Royale des Sciences de Paris, des Sociétés Royales de Londres & de Gottingue, des Sociétés Philosophiques d'Edimbourg & de Rotterdam, Président de la Société Philosophique de Philadelphie, & Résident à la Cour de la Grande Bretagne pour plusieurs Colonies Britanniques Américaines.

TRADUITES DE L'ANGLOIS SUR LA QUATRIEME ÉDITION.

Par M. Barbeu Dubourg.

AVEC DES ADDITIONS NOUVELLES

Et des Figures en Taille douce.

TOME PREMIER.

A PARIS,

Chez { Quillau l'aîné, Libraire, rue Christine, au Magasin Littéraire. Esprit, Libraire de M. le Duc de Chartres, au Palais Royal. Et l'Auteur, rue de la Bucherie, aux Ecoles de Médecine.

M. DCC. LXXIII.

Avec Approbation & Permission du Roi.

Figure 4. The Works of Benjamin Franklin

2) By the turn of the 20th century, Maxwell's equations were already 20 years old and firmly established. At the same time, pure mechanics, summarized by dynamic equations, had produced a succession of major achievements for several decades. The equations of Figure 5 precipitated a major scientific conflict between the two different disciplines that was finally resolved when Albert Einstein looked at both sets with new eyes, unveiled special relativity, matched the two disciplines, and transformed the conflict into a powerful new theory.

CREATIVE COLLISIONS

The large majority of scientific discoveries are the result of what I shall call a "creative collision" between two pre-existing elements of knowledge. These encounters may appear under various circumstances:

- Collision (as in Figure 5) between two basic elements of knowledge that have been expressed in terms of concepts and models. Fundamental research usually progresses as a result of such encounters.
- Fertilization of basic knowledge to produce a new technology, or vice versa, as indicated in Figures 6 and 7.
- Combination of two technologies into a new one, as indicated in Figure 8.

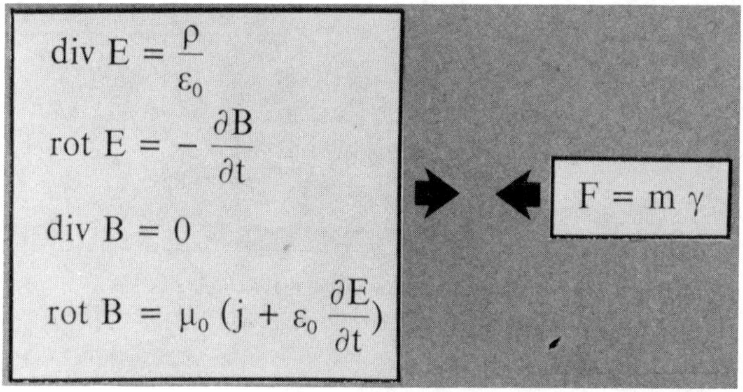

Figure 5. Maxwell's Equations and the Law of Dynamics

Figure 6. The Prime Numbers Produce Cryptology

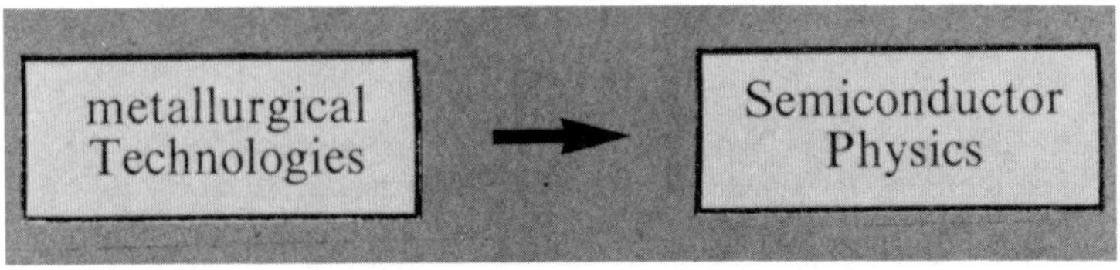

Figure 7. Metallurgical Technologies Produce Semiconductor Physics

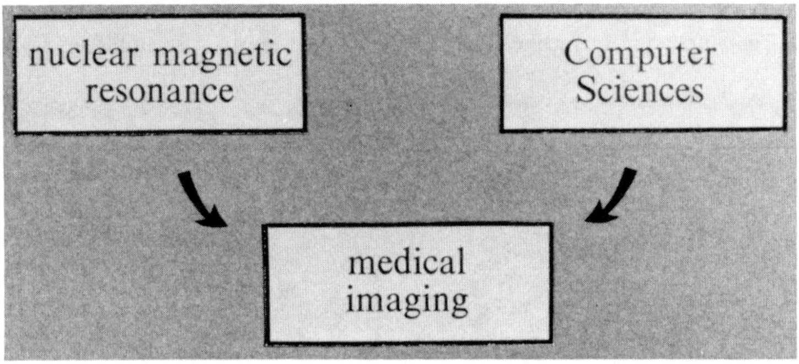

Figure 8. Nuclear Magnetic Resonance and Computer Science Combine to Produce Medical Imaging

Innovation relies on similar encounters. Innovation occurs when an invention acquires a social dimension, when it pervades the whole society as a product or as an attitude. We can consider this as an encounter between a discovery and society.

Which laws do these creative collisions obey? The answer is still obscure. Instead of waiting for a few geniuses to make discoveries why not deliberately organize these creative collisions? Specialists in creativity have thought about the problem, but they can only suggest that we cross each concept with all the others. Thus we are more likely to invent the square wheel than to take a significant step towards the Great Unified Theory.

For an invention as well as for an innovation to occur, some very special conditions must be present. A maturation must take place. As long as such conditions are not fulfilled, nothing will happen. The invention waits for the inventor, the innovation keeps sleeping. But when these conditions are satisfied, discoveries and innovations arise rapidly. The present outburst of interest in the biological sciences, the sciences of life, is a good example to which I shall return later. Unfortunately, favorable conditions are identified after the event, not before. From this point of view, it would be wise to give more attention to the history of science and technology as will be seen below. All of these statements sound familiar to any scientific mind. It might be interesting, however, to look carefully at some of their consequences.

EXPONENTIAL KNOWLEDGE

The most remarkable consequence is that the number of discoveries to be made during a certain period of time is proportional to the amount of knowledge already accumulated. If enough inventors are available, one must expect scientific progress to develop exponentially. The part of the world population contributing to the advancement of science continues to grow. So does the welfare of those countries that can devote a sustained part, or an increasing part, of their gross national product (GNP) to research. As a result, the number of scientists is increasing rapidly. Alan Toffler, in 1970, estimated that 90 percent of all the world scientists were still alive!

One must expect a fantastic acceleration of Man's knowledge and this perspective can but make us wonder . . . How will the world look only a few generations from now? Curiously, the man in the street is very excited about what he learns concerning his remote ancestors a million years ago (see Figure 9) but does not seem very anxious to know what his descendents will look like a thousand years from now, as projected in Figure 10. Of course in the first case he estimates that some facts are firmly established while in the second case, although a thousand times closer to us, there is nothing we can really predict. We may even retrogress to the conditions of Figure 11. It is truer than ever to state that "Now even the future is no longer what it used to be."

Although predictions are difficult, especially when they deal with the future, I shall come back to the two main domains where Man's knowledge is exploding right now:

The first one is information in a broad sense. For the last twenty years it has been said that we were entering the Information Age. Thanks to continuous progress in silicon semiconductors, then in III-V compounds, and now with the aid of optical fibers, the computer-hardware substratum is constantly improving in capacity, speed, reliability, cost effectiveness, etc. At the same time, computer software becomes more complex, more powerful, more intelligent. Thus information technologies progressively affect all jobs, all forms of knowledge, and probably will even affect our ability at comprehension and the way we learn.

Figure 9. Our Ancestor of a Million Years Ago

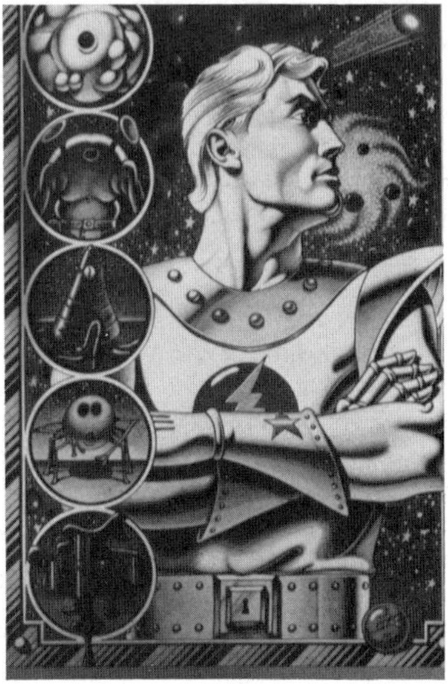

Figure 10. Our Descendent in 1000 Years

The second domain, even more important, is the huge field of Life Sciences that, at present, is expanding in all directions. Since the discovery of the DNA structure by Watson and Crick thirty years ago, what a change has occurred in the last few years! One of my colleagues, a medical professor, returning from a recent cancer conference, told me that half the papers presented at the conference would have not been understood five years earlier!

BACK TO SCHOOL

People responsible for education are faced with a major problem: how to adjust instruction at institutions of higher learning to the needs of the future. How are we to manage the evolution of Electrical Engineering? More precisely, what are we going to teach and how?

A century ago, electrical engineering instruction was concerned with explaining the principles of electric machinery, as illustrated in Figure 12. The subsequent development of electrical communications required application of the Telegrapher's Equation (Figure 13) to transmission lines. Electric and electronic circuitry (Figure 14) followed soon thereafter; and in modern times,

Figure 11. Perhaps in Another 1000 Years We Will Start Over Again!

we see the influence of semiconductor physics (Figure 15), quantum optics, microelectronics, etc. Tomorrow shall we teach molecular biology?

At École Polytechnique, Palaiseau, I am responsible for teaching and research, and I have a practical problem. The students graduated from École Polytechnique are among the best in France with scientific training. They get jobs in pure and applied research, technical projects, and various industrial and public corporations. Should they have a minimum but effective knowledge of molecular biology? Life Sciences are deeply affected by the various disciplines that are usually taught in an electrical engineering department, such as electronics, microelectronics, computer sciences, signal processing, communication, etc. How will the young who are being trained in those disciplines be able to fertilize life sciences if they have no background in molecular biology? But if we decide to teach a new discipline, are we going to teach less mathematics, or less physics, or less electronics? Over a relatively short period of time, teaching requirements have gone through great changes. We should expect that in the near future our teaching objectives will continue to change.

It is not enough, however, to learn something different. Today one must learn more. This evolution will be made easier due to various factors: science advancement results from successive synthesis; some particular forms of knowledge incorporated in more general presentations may be temporarily

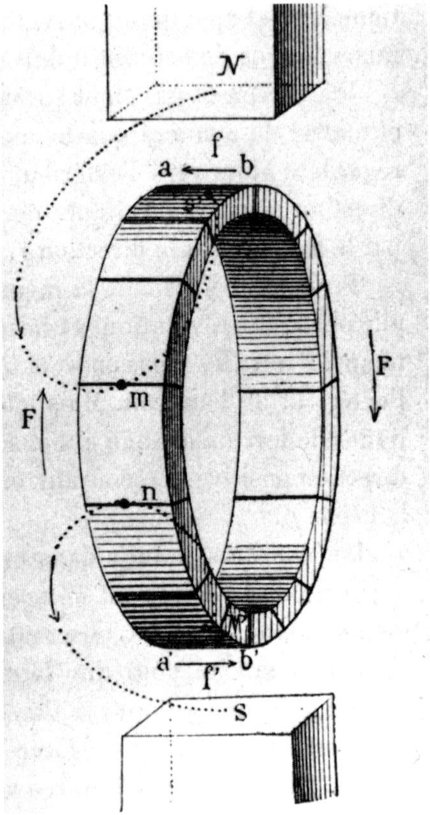

Figure 12. The Rotor of an Electric Machine

$$\frac{\partial^2 U}{\partial x^2} - \frac{1}{V^2}\frac{\partial^2 U}{\partial t^2} - \frac{2\beta}{V}\frac{\partial U}{\partial t} - \frac{\beta}{\tau V^2}U = 0$$

Figure 13. The Telegrapher's Equation

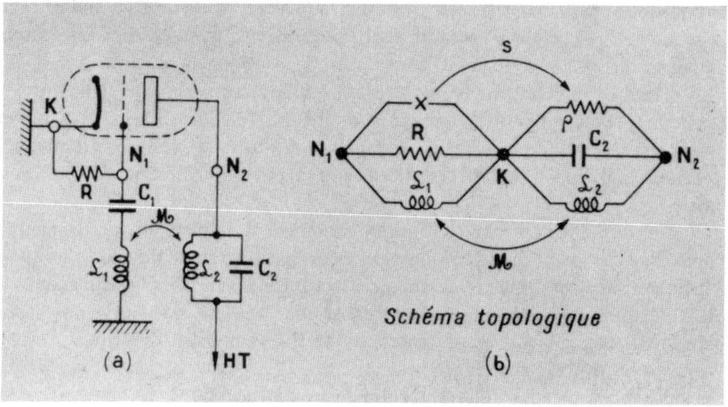

Figure 14. A Vacuum-Tube Electronic Circuit

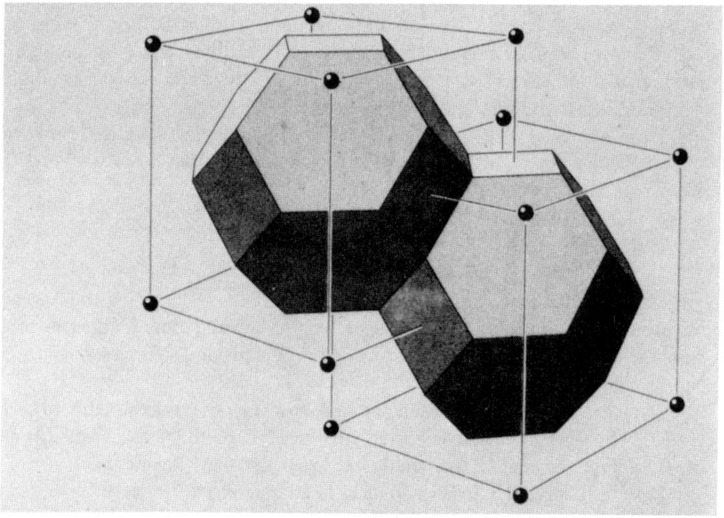

Figure 15. The Brillouin Zones

overlooked; data bases grow in size, availability, accessibility, and in cost effectiveness. But there is cause for preoccupation: it is never easy to cope with an increasing exponential function.

Learning more, however, could require longer learning time. One must be extremely cautious about this point: in the USA and in France, it has been observed recently that corporate recruiters tend to favor younger graduates, a phenomenon that may not necessarily stem from present economic conditions alone.

One can predict that continuing education will play an increasing role and must be given greater attention. In fact, students at the early levels of their university education should recognize that their acquisition of knowledge is incomplete and must realize that without post-university study they will become even more ignorant. At the very least, they must develop awareness of their potential shortcomings. This relativisation of knowledge belongs to a somewhat different philosophy of education than the usual one–a modern form of skepticism that must favor research and the spirit of innovation by challenging accepted ideas and by encouraging the creative collisions that were mentioned earlier. It seems to me that the History of Sciences and Techniques must be developed. Epistemology is not a new discipline. Teaching it, however, would probably stimulate a sense of intellectual adventure and of entrepreneurial spirit and help to question accepted ideas.

It is clear also that we have come into an age of shared knowledge: access to data bases will become easier and cheaper, and work will become more collective and more interactive. The winners will be those who will be able to associate with others having complementary knowledge and skills. The ability of individuals to interact and communicate with each other will become more and more appreciated by society at large. Engineering education, however, must develop these qualities among students to a level that is well beyond the present standard.

CONCLUSIONS

Mankind has gone through three very different ages. During the first age, for about 1 or 2 million years, Man, in very much the same manner as other animals, wandered through his natural environment and survived by hunting and foraging of plant life.

During the second age, Man learned how to raise animals and to grow plants. This agricultural age started between 10,000 to 20,000 years ago and caused a large segment of the population to inhabit fixed homes.

The third age, the industrial one, is characterized by large amounts of energy at man's disposal. It is only 100 or 200 years old.

Now we are entering a new age, the fourth one. As I said earlier, there are two domains in which the "explosion" of knowledge is obvious: information and life. These two domains are becoming more and more closely connected. The main issue of the human adventure is the understanding of man by man, that is to say the understanding of his own brain. When the understanding of the basic processes that take place in our brains reaches a certain threshold, it is very likely that human and social sciences will be deeply affected. We are probably quite a distance from this point, and one can still dream of several possibilities that may occur. The most important consequence should be the action of man on his own intelligence. The brain of a baby is like a book, either large or small, whose pages are still almost empty. Apparently the brain itself selects from the environment what is to be written and determines the way it is written. The number of parameters that control these processes is enormous and their action is extremely complex. But one day many of these parameters will be identified and many of their actions will be understood. Then human and social sciences will be revolutionized and unimaginable possibilities will appear. Teaching will be deeply affected, and the efficiency of the educational process will begin to improve significantly.

Various kinds of dangers, however, will be associated with these new developments; but we will not be facing a new situation: human life has always been dangerous—it is only the nature of the danger that is changing. As unpredictable dangers from the natural environment disappear, new dangers arise that issue from knowledge itself. Man never was more responsible for his own destiny.

I wish to acknowledge useful discussions with Claude Maury and Alfred Fontenilles.

ELECTRICAL ENGINEERING: FROM SOLDERING IRONS TO SIMULATORS

Lewis M. Branscomb

Lewis M. Branscomb received the A.B. degree from Duke University in 1945, and the M.S. and Ph.D. degrees from Harvard University in 1947 and 1949, respectively. He was an Instructor in Physics at Harvard from 1950-51, and a Lecturer in Physics at the University of Maryland from 1952 until 1954 when he was appointed as Chief of the Atomic Physics Section of the National Bureau of Standards (NBS) in Washington, D.C. In 1960 he became Chief of the Atomic Physics Division, then transferred to NBS in Boulder Colorado as Chief of the Laboratory of the Astrophysics Division, a post he held until 1969 when he was appointed Director of NBS. From 1962 to 1969 he was also a Professor of Physics at the University of Colorado. In 1972 he became Vice President and Chief Scientist of IBM Corporation, his present position. In addition to his many honorary degrees, Dr. Branscomb received the Proctor Prize of the Research Society of America in 1972, and five awards from the U. S. Department of Commerce for distinguished service. He became a member of the National Science Board in 1978 and served as the Chairman of that body from 1980-1984. He has been the U. S. representative on a number of international councils and has served on many national and international committees. Dr. Branscomb is a Fellow of the American Physical Society, AAAS, and the American Academy of Arts and Sciences. He is a member of the National Academy of Sciences, the National Academy of Engineering, the National Academy of Public Administration, the American Philosophical Society, and several societies concerned with Astronomy and Geophysics.

"Someone like Lew Branscomb who is the Chief Scientist at IBM and who has all of these other responsibilities, surely has some words of wisdom for us."

—Dale Corson

Electrical engineering has evolved in the last century in ways that could never have been envisioned by Cornell's EE faculty back in 1885. Even so, important elements of the field were well established even then, and persist today. But electrical engineering is changing very fast, and in the future will be not only quite a different field, but one of greatly expanded importance to mankind.

The primary changes are increased systems complexity, both technologically and in dependence on behavioral and cognitive human requirements. This afternoon, I want to explore those two directions, which I believe will profoundly affect both the teaching and the practice of electrical engineering.

The products of electrical engineering are shrinking the world, creating an extraordinary diversity of human communities, promoting creativity in previously rigid human organizations, and allowing the fruits of knowledge to plant their seeds in any land that reaches out for them.

Communications and information-processing hardware cannot, alone, produce these effects. The ability to replicate, to adapt, and to interconnect the extraordinary diversity of systems required by the even more diverse needs of the world's population, depends on systems technology and software and ever more powerful computation. Success in these new endeavors depends on a much more profound understanding by engineers of the human individuals and societies for whom the systems are designed.

These two facets of the new engineering—engineering for human use, and higher-level abstraction through simulation and modeling—will be the important research tasks of the next century. It is not too early to get started.

EE A CENTURY AGO

Even 100 years ago, EE at Cornell was—like all fields of engineering, then and now—rooted in three areas: science, systems, and applications.

The curriculum dealt directly with some of the most important applications of that day—motors and dynamos, electric lamps, and the telegraph. Systems issues were included from the very beginning: design of electric lighting plants, including protection against fire; construction of and fault isolation in telegraph lines.

By the '20s, the scope of applications embraced wireless telegraphy, central-station electric power, and electric railways. Engineering, then as now, was understood to enable applications of science for the needs of mankind; but explicit recognition of the need to study the human side of that equation was limited, 100 years ago, to six lectures on hygiene and two hours a week of military drill.

As the science that underlies engineering changes, so do the tools, the systems thinking, and the nature of the benefits and challenges for society.

Engineers 100 years ago had already discovered three of the enduring main themes of electrical engineering: power, light, and communications; but the only devices available were switches, transformers, and passive elements. No logical devices or even amplifiers were available. Information could be communicated digitally (in Morse code), but it could not be stored, retrieved, or processed electrically.

EE TODAY

One hundred years later, the number of EE courses at Cornell has grown by a factor of 15. The original three themes–energy, radiation, and communications–are still mainstream interests:

- Electric power is still supplied by rotating machines, but now also by fuel cells, solar-energy systems, ion engines, and someday, fusion plasmas.
- The electric lamp has evolved from a carbon filament to the laser.
- Communications, already digital at its birth in the telegraph key and Samuel Morse's bit code, has evolved through wireless telegraphy and telephony to deep-space communications, computer networks, and optical communications. Now a fourth theme is added: information processing, with its associated applications in control systems–a profound change indeed, as we shall see.

The tools of engineering have also changed. When I was in school, circuit diagrams were hand drawn using plastic templates for the symbols for individual circuit components. If you wanted a copy of the completed drawing, you made it on an Ozalid machine. Circuits were assembled manually with clumsy soldering irons that today would be used to repair leaks in roof gutters. A highly complex circuit might have several dozen circuit elements.

Engineers were very adept at assembly and disassembly of such large amounts of circuitry. A favorite pastime of EE students at Duke University in the early '40s was to assemble a broadcast-band radio transmitter in a dormitory room, go on the air and say a few impertinent voice-over words on one of the local Durham AM channels. They would hold their hot soldering irons at the ready until they heard the siren on the FCC truck. Then, in a cloud of smoke from the rosin-cored solder, they would rip the transmitter down to its discrete elements, finishing just before the rap on the door. Today's hackers are no match for those EE students of the 1940s; they usually get caught red-handed with their commercially-built hardware still intact.

The engineer today is separated by additional levels of abstraction from his hardware, compared to those days. Computer modeling and simulation facilitate the exploration of systems ideas that would be expensive and time consuming to build in prototype.

I wonder how many of today's younger students appreciate the literal meaning behind the phrase "to breadboard," which means to prototype? Those who have never screwed tube sockets, resistors, capacitors, and switches down to a wooden breadboard probably also have never experienced the pleasure of eating the homemade bread that was kneaded and shaped on such boards in my part of the country.

SIMULATION FOR PROTOTYPING: THREE REASONS WHY

There are two primary reasons for substituting simulation for prototyping. First, one may wish to build a system that is so complex that it cannot be prototyped with enough reliability to test it. Second, one may not be able to wait the two to three extra years in development that is entailed in going through a full prototype design in current technology, followed by remapping in new technology. By then, the design concept might be obsolete or noncompetitive when finally manufactured.

The only way to manufacture a complex system that must be state-of-the-art when it is built, is to incorporate new component technology and new design ideas concurrently. Unfortunately, system design and component specifications are interdependent. If one uses the current generation of electronic components to prototype a computer of novel architecture, one cannot predict with confidence that it will work when implemented in new technology. The answer is to simulate both the design and the components concurrently and test them both together.

Electrical engineering has already made giant strides toward the automation of the engineering process in many fields of application. Let me take a few minutes to illustrate that by describing, in some detail, the process of designing and building a very large, modern computer such as the IBM 3090, the successor to the multi-million dollar 3084s being installed here in Cornell's Computer Center.

The electrical engineers who design today's computers certainly do not do it by sitting down with a textbook and a hand calculator and solving transforms. Far from it. They are very heavily engaged in the development of tools to automate the task of design. IBM may well have as many EEs *developing* computer tools to design, test, and manufacture our systems as *using* those tools to build computers. This wasn't always so, even in the computer industry.

Fred More, who helped design IBM's Stretch Machine, that powerful computer of 1960 vintage, tells me Stretch was developed with virtually no help from computers at all. It was all slide-rule circuit analysis and other manual procedures. The primitive transistors of that era were what they were, and engineers had to live with them. Today, all the EEs in the world literally could not design our largest new computers without the help of the largest predecessor computers then available on which to run the simulations and test programs.

DESIGN AND PROCESS AUTOMATION IN THE COMPUTER INDUSTRY

Let me illustrate that procedure with a brief review of the steps in the development of a large machine such as the new IBM 3090. A new system begins, obviously, with a set of requirements defining its function and architecture, its desired performance, its reliability and serviceability goals, and the expected manufacturing costs and volumes. From an engineering point of view, the manufacturing requirements are more important than most people realize because it's one thing to design a big system that is to be built at a rate of one per month, and quite another when you're going to build several each day. Typically, a large computer will comprise several subsystems: for example, the central processing unit or CPU, a storage subsystem, the channels, and a control processor. A multiprocessing system like the IBM 3084 will require four times as many subsystems plus the multiprocessing control hardware. Over a million circuits will be required for each uniprocessor.

For each of these subsystems, there is a set of electronic technologies to be specified and developed. There are the silicon chips, the first and second levels of packaging for the chips, the power, cooling and control elements, the mechanical frames, and the service processor. I'll describe the design and production control for the component technology first then return to the problem of building a computer out of this technology. In reality, these two activities proceed concurrently.

Just designing the basic silicon devices involves a host of decisions about implant materials, insulators, metallurgies, and the like. These design choices for building the basic circuit elements are made with the help of three-dimensional device modeling programs, whose power (given a large enough computer to run them on) enables a great variety of transistor designs to be explored with a minimum of laboratory prototyping and testing.

Another group of engineers takes these device designs and begins to design circuits using a program called ASTAP (A Simulation and Test Analysis Program) to do dc and ac modeling of logic elements, memory arrays, and passive devices such as resistors, capacitors, and wires. They consider wirability, interconnections, manufacturability and intrinsic reliability, and again there are programs to help—for yield analysis, to check whether all design rules have been adhered to, and so on.

Manufacturability of a new chip technology involves things like turnaround time for a new design, throughput in production, part number and engineering-change control, and ultimately, the release to manufacturing of hundreds of megabytes of computer data for each LSI chip designed, defining everything it takes to produce the chip. For submicron chips, we estimate that over a gigabyte of data will be fed to the direct-write electron-beam lithography tool for each mask level used in the manufacturing process for each part number.

Designing the ceramic multi-chip modules with 36 wiring planes, through which more than a hundred chips can be interconnected, is not so different from designing in silicon technology, except that the materials (ceramics, polymers, a variety of metal alloys) are quite different. The available software modeling tools are not quite so sophisticated as for the silicon chips, but substantial progress is also being made in automating this area of design.

The electric fields of the complex wiring patterns inside the multi-layer ceramic modules are modeled—using ASTAP again—to investigate the signals that will emerge when each wire is energized, and importantly, how much signal information gets coupled to adjacent wires. Still other programs deal with wirability, power distribution, and chip-attachment modeling to anticipate the thermal expansion of the module when each chip is powered up according to its function.

The ceramic-module manufacturing process requires, just as chip production does, the part-number control, release data, yield data, turnaround time, and throughput analysis—all done with queuing models, flow-control modeling, and huge data bases.

The circuit boards into which each of the thermal-conduction-cooled ceramic modules is inserted are still another design automation problem, for these boards must interconnect 1,200 chips divided among nine modules, each of which has 1,800 pins. But once again, the electrical parameters are altered since the materials system changes, this time to organic and glass materials and copper, instead of ceramics and paste metallurgy.

Power technology, however, is different—much more like traditional electrical engineering—with hundreds of amperes of current at low voltages. Power supplies are modeled, but usually prototypes are built to prove out the designs. Here, physical space and cooling are always a problem, both for the power supplies themselves, which take up about half the space in a modern computer, and for the rigid power-bus lines that carry power to the circuits. So there is a lot of computer-aided design and analysis that looks at mechanical designs, fits and tolerances, and heat-flow modeling of the devices and thermal paths. Indeed, throughout each component of the system, device junction temperatures and thermal paths are modeled so that reliability won't be compromised in the design.

Just as it takes a lot of space to get the power in, it takes a lot of space to get the heat out. We use an enhanced version of the CADAM program—and "3-D" solid modeling as well—to design not only the cooling paths but all other mechanical aspects of a machine like the 3090, from mechanical loadings to cable routing and access for service personnel. The control of the entire power assembly—sequencing the 50 different voltages of the 3090 when the system is turned on, for example—and the self-test routines to be sure all thermal limits are met, requires a simulation and software control system.

Procedures that monitor hundreds of sense points to isolate what's wrong in the event of a problem are also simulated to keep the machine from self-destruction, and to make recovery as quick and painless as possible. In the 3090, remote serviceability is a key availability element. In most cases, the machine does a self-diagnosis and calls up the parts depot, ordering specific part numbers to be sent. The customer engineer may well receive the replacement part even before he knows the machine has experienced a fault.

As each of these elements of the building blocks of technology are being simulated, designed, and tested, another entire team of engineers is developing the system structure. Having settled on the architecture for the machine (for example, its instruction set), they must now design the machine organization. For even with compatible architectures, a line of machines of different speed and cost must have different specific designs in order to make best use of the appropriate technologies. These engineers define the data paths and how various operations will be implemented, the location of the registers, what kind of buffers are needed, how much pipelining is required, etc.

The immediate result of their efforts is what we call a first-pass "technology-independent system design"—one for which the detailed specifications of the technology are not yet taken into account. The purpose of this first-pass design is to verify whether the design does, functionally, what it is supposed to do, assuming the technology will work when it is ready.

Meanwhile, still other engineers are devising a whole family of test cases by taking the architecture of the new machine's systems software—like MVS-XA, our large system-control program. These segments of the existing programs, kernels, and previous cases that have caused problems, are then run against the technology independent design, using simulation. Until recently this procedure has involved a *software* simulation on an existing, large-scale computer; but since a 10-MIP machine can only simulate about one instruction a minute, this has been very time consuming. Tens of thousands of hours of simulation are not unusual to verify that all the instructions work, in all the sequences in which they can arrive, with all the interrupt conditions that can occur. Clearly, the compute-load for simulating the new higher levels of system complexity, as higher levels of circuit integration are achieved, is rising faster than the computing power of the machines that are being developed.

To break that potential barrier to the ability to design ever faster machines, IBM has developed special-purpose *hardware* simulators such as EVE, our Engineering Verification Engine. EVE, a special-purpose supercomputer with 256 processors, is thousands of times faster, and able to handle logical models hundreds of times larger, than software simulators on the largest general-purpose machine.

Today, science is getting the most attention to its needs for super-fast computing, but engineers will be the biggest MIP consumers. Of course, EEs will design the machines for both groups. While the functional simulation is going on, engineers are taking the technology-independent design and partitioning it—on chips, modules, boards and frames, translating the ANDs, ORs, and bits into physical circuit elements. Logic design of the thousands of chips for the 3080 series of computers was a major engineering effort. About half of the higher-speed chips in the 3090 were designed automatically, using logic-synthesis tools. Logic design may, in time, in fact must, in time, become an obsolete specialty.

When the chips are designed, we take the resulting technology-*dependent* design and do a Boolean comparison with the technology-*independent* design, which we have tested for logical completeness. We iterate this process over and over until the two are compatible. At this point, it is time to start generating the Release Interface Tapes, or RITs, that contain all the data needed by Manufacturing to build each component on their automated lines. RITs for chips go to the Quick-Turn-Around-Time line in the East Fishkill Plant, where electron-beam pattern-making bypasses the time-consuming need to build photomasks for chip pattern making. Other RITs go to Fishkill's Thermal Conduction Module pilot line and to Endicott for first models of the Clark circuit boards. Finally, many thousands of test patterns—tests for opens and shorts, dc and ac tests, functional tests, stuck-fault tests, etc.—are generated automatically by various programs to ensure that the components going into pilot manufacturing are, in fact, testable.

Now for the first time in the process, physical hardware components—chips, TCMs, boards, cables, power supplies, etc.—begin arriving from pilot lines and plants. Teams of engineers get their first chance to assemble an actual computer and run real-life functional and physical tests on all its subassemblies. These assemblies are then combined and testing starts all over again, but now with more complex tests that simulate real systems programming.

If the system is a multiprocessor—for example, a four-way machine like the 3084—it must be tested again in this configuration. When it passes, and the development engineers are satisfied, the Product Assurance team, an independent testing group, goes through the whole process again, testing for function, performance, and especially for reliability and serviceability. They do these tests with the system in an environmental chamber in which the machine is cooked, cooled, steamed, and dried to determine its failure limits. When the system exits this product test, it is then installed in selected internal and customer computer rooms and very carefully observed as it is put into service. The 3090 is in that Early-Ship Program now. When volume shipment begins later this year, an extraordinarily reliable and serviceable computer of very considerable power will be put in mass production.

DESIGNING LARGE SYSTEMS IN THE FUTURE

Obviously, it takes many very large computers, indeed some special-purpose supercomputers, to design a new big machine in a reasonable length of time. We could not build a modern VLSI computer like the 3090 without powerful software simulators and simulation engines. An attempt to do so would result in some serious problems:

- Circuits would probably not work without device and circuit-modeling tools.
- Designs could never be completed in time without logic synthesizers to wire the chips automatically.
- Knowledge of the validity of the design could never be available in advance without automatic tools to generate thousands of test patterns.

Even if we could design the machine and its components and verify the correctness of their design, we could never build them in quantity without the vast data bases and the process automation that automatically transfers designs into manufactured units.

FUTURE DESIGN AND PRODUCTION TECHNOLOGY

Is this the end of the line? Certainly not. As mathematicians improve the algorithms in these tools, and the machines on which they run become more powerful, it will be very important to integrate these separate processes into a much smaller number of steps, with shared data bases. Eventually, it should be possible to start with a description of the machine organization in a formal design language and produce automatically ("compile," if you like) the system structure, the component designs, and the test routines.

I can imagine a small team of engineers at workstations in a design "war-room" overseeing the automatic generation of the system and physical designs—pretty much untouched by human hands except for guidance. This could greatly reduce design time and cost. Progress down this path has already caused the largest machines to come down in cost per unit of computing power by 15 to 25 percent per year for many years.

Surely this trend is to be welcomed, for it gives the electrical engineer much more power in the practice of his craft. Just as the engineer of the 1940s did not have to make his own resistors and tubes (as his predecessor in the 1930s had to do) it should be possible to generate new machines comprising millions of circuits and package and test them all automatically in the 1990s without having to do it manually.

But one does pay a price. The system builder no longer builds from the best parts that his semiconductor design colleagues can make in the laboratory. He builds from the parts that are specified in the design rules, which describe allowed tolerances in the physical characteristics of the technology. Some degree of opportunity for optimization is lost in the quest for productivity and the ability to handle complexity levels that would otherwise be prohibitive.

INTUITION IN A WORLD OF SIMULATION

If the engineer is isolated in this manner from the physical artifacts with which he works, does he lose some element of physical intuition? Is some instinctive feeling for physical reality—born of touching and feeling the hardware—necessary inspiration for invention? Many of us—with backgrounds in hands-on experimental work—worry that modern engineering is getting too theoretical. I don't know, but I suspect that 20 years from now we will look back on these days when people building small machines often spent hours with wire-wrap machines assembling useful products from standard vendor components and say, "What a primitive way to design. How could they get a feeling from one or two physical prototypes for the range of function in the product if they could not try 100 hypothetical designs against a thousand possible tasks, enriching their intuitive judgment to impart the best qualities into the artifacts they build?"

If the entire process of electrical engineering design of computers today collapses into a single automated process, starting with the systems architecture and functional specifications, what will be left for engineers to do? Plenty. For I contend that the most important, and the most complex, part of the design of intelligent devices and tools for mankind—the "front end" of the design process—has hardly been touched at all. That is the user-specific, human-centered elements of design—the second great challenge for electrical engineering.

HUMAN CENTERED DESIGN

It is quite extraordinary that most computers—and indeed most other electrical systems for human use--are designed as though the computer were merely an abstract device capable of a variety of mathematical manipulations. Alan Turing, indeed, laid the foundations of modern computer science by exploring just that notion, but *computer science* is the only scientific field built around the study of an artifact of man; and *electrical engineering* is the field that takes that science and produces useful systems that satisfy mankind's needs.

But how can you build an intelligent machine to help an intelligent person if you know a great deal about the machine but almost nothing about the person? I have looked at many Ph.D. curricula in computer science and have yet to find one that requires formal graduate study in behavioral science. I note that Cornell's electrical engineering program offers a strong option in bioengineering, but this is much more oriented to physiology and prosthetics than to cognitive and behavioral characterization.

Interestingly, computer scientists often appeal to aesthetics to guide them in the absence of a quantitative and predictive theory for the behavior of information systems as perceived by humans. Thus—for want of verified principles with which to test system and software designs, or even to judge the superiority of one design over another—the information industry tends to grasp at the arbitrary and rather meaningless concept of "user friendly."

Academics, on the other hand, tend to define an equally arbitrary artificial model of what a human being might *expect* from a useful machine—which they call artificial intelligence (AI). Both academics (hankering to be entrepreneurs) and industrialists (eager to follow academic fads) cover their bets by asserting that only with artificial-intelligence techniques can a truly user-friendly machine be built. In the meantime, there are some brave souls, out in universities and industry, trying to build a rigorous behavioral science to underpin computer science and engineering.

Today's AI research is aimed at a very useful but limited goal. We focus on giving the machine the capability to manipulate its stored information in a more sophisticated manner, mimicking a caricature of a rational model of human reasoning which is the basis for "expert systems" research; but we must also try to understand the behavior that the machine must exhibit if we are to find it a congenial information assistant.

ARTIFICIAL PERSONALITY

There is a famous bit of doggerel written by a student on the bulletin board of the University of Wisconsin computer room:

"I'm sick and tired of this machine;
I wish that they would sell it.
It never does just what I want,
But only what I tell it."

I call this a plea for *artificial personality*, rather than *artificial intelligence*. We want the computer to respond in a predictable, or at least understandable way. We want what it does to have integrity, within the concept of the task we have assigned it. We hope the machine will be smart, but we insist that it should be well-behaved.

AI works on the easy problems, for its paradigm of a human being is complete as far as sensory physiology is concerned. Great progress has been made in speech synthesis and recognition, in pattern recognition and image processing, and in robotics; and as for the reasoning contained in "knowledge-based" systems, there *are* problems human beings address in the form of inference rules and facts. But very few people are working on the hard, important problem of developing system-design principles that permit computers and other intelligent systems to serve ordinary people in a way that is compatible with people's needs.

The field of human factors is, of course, an old one and is particularly well developed in the design of aircraft and spacecraft cockpits. During and since World War II, anthropologists and psychologists did excellent work on instrument and control designs for fighter pilots and astronauts. But this tradition of designing interfaces for highly-trained specialists in perfect physical trim is an inappropriate base for design of computer systems for ordinary people. After all, has NASA ever tried to design a cockpit so that any passenger can come up front and fly the airplane? That is the problem electrical engineers will be called upon to solve when we put computer networks at the fingertips of untrained members of the general public. We haven't the luxury of designing only for astronauts.

EFFECTS ON ORGANIZATIONS

It will not be sufficient to consider only those system characteristics that relate to the needs of a single person. The systems designed by electrical engineers, especially the communication networks, are likely to have a profound effect on organizations and society as a whole. Careful study of these effects can—perhaps I should say *must*—lead to new principles for good systems design. At the same time, engineers can use those networks to fulfill their own professional needs.

Let me cite an example. Experience with ARPANET led to the somewhat surprising discovery that electronic mail, not host processor load-sharing, was the important application. The two most important consequences of such networking are:

(a) The spontaneous creation of communities of individuals with shared interests, and

(b) The possibility of remote collaboration, sometimes among individuals who have never met face-to-face.

IBM has a great deal of experience with these benefits, for our internal network, called VNET, interconnects more than 1,800 host computers, accessible by over 200,000 people in 75 states and countries throughout the world. It has given a great boost to creativity, tends to remove middle management as a bottleneck to exploiting that creativity, and generally bypasses the formal authority structure for those activities where that structure is inappropriate.

Universities around the world are realizing the same benefits as similar networks grow. In the U.S., 119 campuses and other research facilities are connected through BITNET, which in turn is connected to ARPANET and CSNET. One hundred and twenty-two universities and national laboratories, in almost all the countries of Europe, are connected up through the European Academic Research Network, or EARN, which also reaches Kuwait, Israel, and Egypt. Twenty Canadian universities are interconnected through NETNORTH. In Japan, the Science University of Tokyo is connected across the Pacific. All of these 262 institutions are interconnected through transborder-leased services paid for by IBM.

The EARN directors have suggested that this global assembly of peer-connected university systems be coordinated under the name World University Network, or WUNET. I predict that WUNET will grow like wildfire, as indeed VNET did without the need for management stimulation. In addition to Japan, universities in Hong Kong, Singapore, and Australia may soon be added. All this is happening with a minimum of formal government support or management, save the necessity for post and telephone authorization to lease the communications lines. The freedom to share information, to work together, that this system provides will be a powerful antidote to protectionism, isolationism, and other restraints on intellectual mobility.

UNIVERSITIES AS TOOL BUILDERS

Today tool-building, of the kind I have described, goes on primarily in industry, but universities are also making important contributions, especially in chip design methods. In the future the heart of electrical engineering practice will be so built into the software tools and the special engines on which they run that substantial academic energy will have to be devoted to this technology.

Fortunately, the networks these tools have built will enable this kind of effort to be shared among schools and with industry. Most important, it can be shared between the school and its graduates, providing a continuing linkage between professionals and their home institutions. Since these tools will have substantial commercial value, some of the resource problems of engineering education might find relief from this activity.

SOCIAL SCIENCE, HUMANISM, AND ENGINEERING

I am confident that electrical engineering in the future will enjoy powerful capabilities to use large computers to simulate complex systems. There is, of course, no more complex system than human intelligence and personality. Will it ever be possible to build a reasonable simulation of man—at least one sufficient to provide quantitative design guidance to builders of intelligent machines?

Again, I don't know. I doubt that either computer science or behavioral science are really sciences yet. I have heard a science defined as "a field of study in which untrue statements can be shown to be false." But the effort must be made; otherwise there is no rigorous guide to systems evolution except the market place. That pragmatic guide may be quite adequate for most industrial purposes, but I do not think universities or leading companies should be satisfied with it. The requirement for simulation of the hardware and software designs will bring us powerful programs and very powerful hardware, but some solid research in cognitive and behavioral science will have to make those capabilities useful.

Where can we start? We might at least take the six lectures in hygiene that each electrical engineer was required to take at Cornell in 1885 and replace them with six lectures about the human as seen from the perspective of modern science. That description of the bounds of our knowledge about man might help make us more modest about our claims for man-built intelligent systems and point the way to a broader base of participation of other disciplines, along with electrical engineers, in this kind of research.

SHAPING THE INFORMATION AGE: THE ROLE OF ELECTRICAL ENGINEERING

S. J. Buchsbaum

Solomon J. Buchsbaum received the B.S. and M.S. degrees from McGill University in 1952 and 1953, respectively, and the Ph.D. degree from MIT in 1957. He became a member of the Technical Staff of Bell Laboratories in 1958, and except for a three-year period from 1968-71 when he was Vice President in Charge of Research at Sandia Laboratories (a subsidiary of the Western Electric Company) he has held various technical and administrative positions in that organization. In 1979 he became Executive Vice President of Customer Systems at AT&T Bell Laboratories in Holmdel, New Jersey, his present position. He has served on many technical advisory boards including two U.S. Presidential Advisory Committees in the mid 1970s, as well as on National Advisory Boards for the AEC, ERDA, NSF, DOE, and the U.S. Navy. He is the co-author of a book on waves in plasmas, has published over 50 research papers in the areas of microwave plasmas in magnetic fields and in solid-state plasmas, and has been issued eight patents. In 1977 he received the Secretary of Defense Medal for Outstanding Public Service, and the Anne Molson Gold Medal. Dr. Buchsbaum is a member of the National Academy of Sciences and the National Academy of Engineering. He is a Fellow of the American Physical Society, the American Academy of Arts and Sciences, IEEE, and AAAS.

"Because Sol is involved with so many public activities I sometimes wonder when he has time to work for Bell Labs. One of his concerns is the governance of the National Academy of Sciences and the National Academy of Engineering. Since I spend most of my time these days working for these two organizations, he is my boss, in a sense, and I am required to report to him once a year– but I was called to report *four* times last year. Most of those meetings were at Wood's Hole, however, so it wasn't too bad."

–Dale Corson

Someone once said, "Celebrating a centennial provides an excellent opportunity to extol the past and exaggerate the future"—or was it "exaggerate the past and extol the future"? In any case, I would like to speak about both today, with emphasis on the future and do so by examining the roles that electrical engineering—and electrical engineers—have played and will be playing in shaping the information age.

Before discussing those roles, however, perhaps it is worth defining the information age—or at least putting it in some perspective. For example, what really distinguishes the information age? What sets it apart from the industrial age, the space age, the nuclear age, the age of anxiety, the age of enlightment, or any of those many other ages the world is supposed to have lived through or be living in? How will it be different, and how might it contain elements of each of these other "ages"?

First, let me say what the information age will *not* be. Professors Corson and Bernard referred today to the fact that the world is experiencing an "information explosion". I do not think that the technologies of the information age will provide a panacea for that profound event. In fact, many of the same technologies that help us deal with information are helping us discover new things and thus add to our storehouse of data and knowledge. So we can expect the information age to be an increasingly complex, a somewhat confusing, but certainly a more challenging, era. Later in this discussion I will have more to say about how we will cope with these developments.

Secondly, an information age will *not* be a time when everyone will be either programming or operating a computer, or working at a McDonalds. Some pundits are projecting this kind of two-tier society. While it is forecast that a greater number of people will be in service jobs, and many of those will be handling information, much more of the physically and mentally tedious work of today will be automated, and people will have opportunities to do many more creative things.

As a result, we do face a vast restructuring of employment along with a major effort to upgrade education and retraining. This transformation will require ongoing attention for decades because we cannot even begin to imagine the new industries and career opportunities that information-age technologies will spawn—any more than we could anticipate the enormous economic and social changes—along with new jobs—brought about by such innovations as electric power, the automobile, or the telephone.

Finally, I do *not* think we can view the information age as a "*post-*industrial age"—in the sense that many people misconstrue that expression. When Daniel Bell coined it in the early 1970s, he meant it to denote the time when only a small percentage of the population would be directly engaged in heavy industry and manufacturing.

Because of such statements, too many people began to believe that in an information age our manufacturing industries would somehow become unimportant. As we are rapidly learning, however, nothing could be further from the truth. We are finding that information is becoming an important partner of industry rather than its replacement—an essential component that is making all industry more productive—and in particular it is revitalizing mature industries and turning them into high-tech operations. We need to take advantage of this trend—we are starting to do so—and not a bit too soon in view of our worldwide industrial competition.

So, while the information age will not bring us utopia, I believe that the technologies behind it will do much to ameliorate, if not solve, many of the problems building up today.

- They will help us deal with the growing complexity of the world with its trillions of human transactions per day;
- They will improve our productivity, in both industry and services, so as to not only keep up with human growth but provide a higher standard of living for more people;
- They will help us make the most of our energy and natural resources, and concomitantly reduce waste and environmental impacts; and
- They will give more people the opportunity to be both more productive and creative, at all stages of their lives, by providing us with the technologies to process and convert enormous amounts of raw data into useful information and higher levels of knowledge.

Overall, the information age may not be a panacea to all of the world's problems, but it is going to go a long way toward providing us with the new tools and opportunities we will urgently need in the years ahead.

Now, what do these comments have to do with the centennial we are commemorating today and with the field of electrical engineering? The answer, in a word, is *everything*. For it has been, in good measure, the past work of electrical engineers that has made this age possible, and it will also be largely their role in the future to assure that the new possibilities of the information age are realized. Thus, I can see no more exciting field to be in at this time than electrical engineering.

My excitement is not just due to the fact that so many of the technologies that electrical engineers will be working *on* will be central to the progress of the times. Rather, it is because they will be working quite closely *with* so many of the other disciplines and professions that are also at the frontier of new technological and social change. This trend has already been going on in several fields, including the one that I am representing today.

Of course to many of you in electrical engineering, such involvement is nothing new nor has it been so at my institution, AT&T Bell Laboratories, where over the years our electrical engineers have worked closely, not only with their brethren in other engineering disciplines, but also with professions ranging from artists to astronauts, cosmologists to chemists to coin-makers, medical men to movie producers, physicists to psychologists, and, I am sorry to say, more than our fair share of lawyers and regulators.

It is simply amazing how many disciplines are used, and have been created, to solve what a former chairman of AT&T called, "the two great problems of electrical communications". One problem is getting from here to there–*transmission*. The other is connecting anyone to anyone else–*switching*. Those problems were difficult enough to solve in voice communications, but we did so, creating a national network that includes some 22,000 switching centers and over a billion miles of transmission paths. The trillions of interrelated parts of this system, all designed to interact compatibly and reliably, can in a few seconds complete any one of 6×10^{15} possible connections. In other words, anyone with a telephone can reach anyone else anywhere with a telephone (almost all of the time!).

Today those problems of transmission and switching are compounded by new challenges, those of processing and moving information in the forms of voice, data, and image–interactively–and connecting not only people to people but any combination of machines and people. Now, considering those two great problems in the context of today's needs, how are we going to achieve the information age? How are we going to link people and computers into one global, intelligent, interactive network that is essentially as easy to use and as reliable as today's telephone system?

I would like to devote the rest of my talk to answering these questions–to describing the technologies and their progress that will allow us to get from here to there. Then I will conclude with some thoughts on how they might all be put together–systems engineered–under an architecture that will universalize information services essentially to the same extent to which we have brought universal telephone service.

One key to accomplishing this communications level lies in advancing three underlying technologies. They are microelectronics, photonics, and software. Microelectronics provides the ability to process and store large amounts of digital information. Photonics is basic to the transmission of very large amounts of such information. Software makes it possible to customize this hardware, to allow its operation and control by people, to provide it with "intelligence."

As my boss, Ian Ross, President of Bell Labs, is fond of pointing out, these three technologies exemplify our involvement in the microscopic and the complex. It is our role as engineers to make the complex simple–to engineer these complex technologies into elegant solutions to people's problems.

Certainly nothing illustrates our involvement in the microscopic and the complex more than microelectronics. Here we have the technological challenge of making integrated circuits with components of ever decreasing size and placing an increasing number of them on to a silicon chip smaller than a postage stamp. To accomplish this feat requires reducing the line size of such components until they are now approaching, and will soon slip below, the wavelength of visible light.

How well are we doing these tasks, and how much further can we go? Today we have the capability of putting one million components on a chip. At AT&T, we plan to have a one-megabit random-access memory (RAM)—which contains two million components—coming off our factory production line by the end of this year. This is just another landmark, however, on the way to reaching today's known limits of the technology: at least 100 million components per chip. We should be there by the end of this century—which we must keep reminding ourselves is less than 15 years away. Considering other frontier research going on in microelectronics today, including that being performed here at Cornell, this goal may turn out to be far from the end of the line.

Now, what do these advances in microelectronics accomplish for us? Among other things, they have been exponentially decreasing the cost per circuit, while increasing the speed and reliability of those circuits in similar measure. Their major accomplishment has been the increase in the processing power of computers. We take advantage of that fact, of course, in telecommunications. But just as important for us is to employ the power of microelectronics to reduce large communications subsystems—even whole systems—to a single chip of silicon, or at worst to a few chips. In our jargon, we put a "system on a chip". For example, not too many years ago echo cancellation in telephony required a box the size of a refrigerator. Today, echo cancellation is performed by a single chip of silicon especially designed to perform that function. That chip is but one of some two hundred such custom chips—each a "system on a chip"—that we will design and put into manufacture this year.

Beyond this achievement, the availability of ever more powerful single-chip computers and networking of such single-chip computers will be needed if we are going to achieve the kind of systems in the future that can understand and respond to natural language, do real-time translation, provide robots with visual abilities similar to that of people, and a host of other things that today remain in the realm of science fiction.

Moving on to photonics, or lightwave technology, we see a technology of unusual flexibility and utility and of vast transmission capability. Lightwave systems use lasers or light-emitting diodes to send pulses of light, in digital format, through ultrapure optical fiber. Here, at last, we have a technology, unlike any previous technology, that can be used efficiently and economically just about anywhere information has to be sent from place to place. It is already being used today as optical data links within computers and switching systems; as intrapremises and intracampus information-distribution systems; as connecting links between customer premises and local telephone company switching hubs; as trunks between switching hubs within cities and metropolitan areas; and as long-haul transmission systems connecting distant cities and indeed, very shortly, continents.

The overall capability of such systems has been doubling each year, a trend that should continue for many years. We measure these advances in terms of increases in both the distances that pulses of light can be sent unamplified, and in the rate of pulse transmission. Thus the product rate, measured in millions of bits per second, and distance, measured in miles or kilometers, is a convenient criterion for our progress in lightwave technology.

Lightwave technology advances constitute something like an international olympics, with records being broken periodically, and the lead changing hands among principal competitors—at this time mainly the U.S. and Japan. In recent laboratory experiments, Bell Labs set a "distance record" by transmitting 420 million bits per second for 203 kilometers. In another experiment, we sent two billion bits per second for 130 kilometers on a single fiber without amplification. That pulse rate, by the way, could transmit the entire text of the 30-volume *Encyclopaedia Britannica* in less than a second. (Now if we could only teach our kids to read that fast —).

Do these accomplishments mean that we are approaching the limits of lightwave technology? No, the theoretical limit is several orders of magnitude away. It should be possible to span a continent or an ocean without a repeater and still not reach the theoretical limit.

The third principal technology of the information age is software. While of the three, software may be the technology—or science or art—that seems to be farthest removed from electrical engineering, its progress is critical to the entire success of digital communication and information management. In the United States today, software alone represents an industry with annual revenues now exceeding $10 billion. Approximately 10,000 companies are developing software for sale, and the estimated number of programmers in U.S. industry ranges from 500,000 to one million. At AT&T, more than 40 percent of our technical people are involved in software development. We currently employ about 300 Ph.D.s in software research.

Why do we have such involvement in software? There are two major reasons. One is the huge demand for software in the communications business. The other is because of the complexity of software, which, ironically is used to achieve simplicity. It is essentially through software that we can take extremely complex technological systems, tailor them to customers' needs, and make them "friendly" and reliable. At the same time, all this technology is, in effect, "transparent" to the user.

The difficulty with software lies in its production: the equivalent of one programmer-year is required to write from 1,000 to 10,000 lines of advanced code, depending on the type of systems application. The tremendous demand for software in the information age just begins to be realized when one considers its need in our U.S. telephone system. Here is a system in which an untrained user with a simple 12-button keyboard has access to a complex network, one which is actually an interconnection of thousands of computers that control the network. That network operates on 40 million lines of code, and the telephone that accesses it relies on switches using 1.6 million lines of code.

These numbers demonstrate that over the years we have learned to create big software systems, developing them to meet cost, performance, and schedule objectives; but the software challenge of the future remains enormous when one considers the forthcoming needs of a network that will manage integrated voice, data, and video communications and information processing with the same efficiency and facility that we have come to expect of the telephone system.

Furthermore, we see such a future information-age system as one that is composed of a myriad of interconnected networks whose functions can be defined and redefined by users to meet their individual and changing needs. Such networks will be, for the most part, "software defined" by those users, so that they will be able to access, arrange, and use, when and where they want them, only those services or combination of services that are needed to run their offices, factories, professions, and lives.

How are we going to put together the technologies I have described in order to create such an incredible system? Let me outline a few thoughts we have at AT&T that center principally on some of our plans for the systems engineering and network architecture that will be needed to accomplish this objective. I think these observations will point out many of the major challenges that lie ahead for electrical engineering–for those of you already in the profession and for those here at Cornell and at other universities who are training the next generation of electrical engineers.

As we look ahead to creating a telecommunications network that truly meets the needs of an information age, certain criteria and objectives come into focus. First of all, we see such a network as encompassing major integrated roles for a public network, private networks, and customer-premises equipment; and we believe that a great amount of computer intelligence belongs in all three. In fact, a key to our architectural plan is the extensive use of distributed intelligence–and particularly the synergistic linking of intelligent premises equipment with intelligent networks.

The systems-engineering goals for an evolving network architecture are broad. The network must be responsive to instantaneous input from users. It must continually and rapidly incorporate leading-edge technology. It must be evolvable and robust, avoiding dead-end systems to protect customers investments, and it must provide high-quality, reliable performance as well as cost-effective services.

Our major architectural planning at AT&T includes five specific objectives by which we intend to achieve the broad systems engineering goals I have just mentioned. The objectives are open, standard interfaces; integrated service and network management; distributed intelligence; consistent user interfaces; and universal connectivity. I will say a few words about each of these aims.

First, achieving "open, standard interfaces" is critical, especially in data communications. It is crucial to have both strong participation in and support of international and domestic standard bodies. This requirement represents a major challenge to the electrical engineering profession, nationally and worldwide. In addition, architectures must support major equipment vendor interfaces as well as formal standards, procedures that are necessary in order to protect customer investments and to give customers freedom of choice in selecting vendors.

Second, achieving "integrated service and network management" is vital to customers. Today, people are depending more and more on telecommunications and information technology for the core functions of their businesses. Because of this dependence, they are demanding the capability to define, monitor, and modify their services in a flexible manner.

Achieving the third objective of "distributed intelligence" is crucial to the evolution of network architecture. Computer technology is used extensively in the communications industry, with applications ranging from workstations to sophisticated switching machines, and including advanced transmission and operating support systems.

Distributed network intelligence requires robust, flexible, signaling systems to provide the means for connecting various intelligent systems. One such signaling system, common-channel signaling, is deployed today to link computer-controlled switching machines in the network. Such arrangements have already made many innovative services possible. The next major step in distributed intelligence is the linking of intelligent customer-premises equipment with the intelligent network, thereby providing the flexible message-based signaling capability of the so-called Integrated Services Digital Network (ISDN). I will have more comments about ISDN later.

A major challenge is posed by the fourth architectural objective of "consistent user interfaces" that must be both logical and uniform. These interfaces are vital to making it easy to use our ever more sophisticated products and services. The challenge here is to provide these consistent user interfaces across families of products and services, and to do so without stifling creativity in their conception and development. This broad concern ranges from the basic level of consistency in the use of the touch-tone telephone pad to the levels of consistency for common tones, announcements, displays, keyboards, menus, directories, and the like. Even the common touch-tone pad can be intimidating, for example, when unfamiliar procedures are encountered as one tries to make calls from different hotels or from dissimilar telephones. Needless to say, there are a number of technological opportunities for achieving consistent as well as user-friendly interfaces--including, for example, techniques such as voice recognition.

Our fifth and final architectural objective is achieving "universal connectivity". By this, I mean connectivity at all levels. We must be able to interconnect various private and public networks. We need a full range of standard protocols, and we must have connectivity across applications such as the electronic mail and messaging systems of different vendors.

To cite one major example, voice networks today have achieved universal connectivity. They are basically transparent to the user and are relatively simple to interconnect. In contrast, many computer networks are private, inflexible, and isolated by protocol incompatibilities. As noted earlier, achieving the objective of open, standard interfaces is critical to solving this problem. Maintaining privacy and security is a significant challenge here.

Much of the industry, AT&T included of course, is embracing this architectural approach to the evolution of the world-wide network. The first step in this approach is the previously mentioned Integrated Services Digital Network (ISDN). The name ISDN embodies two notions. The first is "digital network": whatever connection is established from any place to any place shall be a digital connection. The second notion is that of "integrated services". The digital connection shall serve any mixture of bundled services: voice, data, facsimile, or whatever. That is easier said than done because the various services require greatly different treatments for, say, encoding, formatting, and, very importantly, signaling.

At AT&T we are aggressively supporting, developing, and implementing the concept and standards of ISDN, which are rapidly becoming recognized worldwide. The infrastructure for achieving our network architectural objectives and ISDN is largely in place. For example, product families of switching and transmission systems are currently being deployed to support ISDN. An abundance of processing and data bases exist in the form of computer-controlled switching machines and associated operations support systems. Extensive interprocessor communications are also available *via* common-channel signaling.

End-to-end digital connectivity is very important for ISDN. This process, I am pleased to say, is becoming a reality with a rapidity that I would not have expected a decade ago. The reason for this growth is simple since with microelectronics, lightguides, and software, it is cheaper to handle, process, or transmit analog information, such as voice, digitally rather than in its native analog form. Therefore, digital systems are being deployed at an ever faster pace in the telecommunications network to handle voice as well as being readily available to handle data.

Thus, advanced digital access and distribution systems are being deployed widely and rapidly. Local exchange carriers are deploying a variety of digital loop systems, including lightwave communications links. In fact, what was once a wire network, is becoming very sophisticated, with advanced capabilities for reconfiguration and bandwidth management. Digital transport capabilities are also being deployed in both exchange and interexchange networks. The prime vehicle, of course, is lightwave communications, but digital radio systems are also playing an important role; and we are devoting significant attention and effort to developing control and management systems.

All this brings me logically to the question of our broader vision of the future. That vision seems to evolve out of both our advancing technology and the needs and wants of people and business as we enter the information age. For example, the wide-ranging set of goals and needs I described earlier will no doubt be augmented by sophisticated additions–as in the area of wideband communications. New needs that actually do emerge will be determined partially by people's expanded perceptions of what is possible as a result of new and evolving technologies. This dynamic process is what helps make our business so exciting.

As a result, there will be three broad areas of increasing benefits. First, there will be a continually broadening range of customer service and product choices, and these will have common characteristics of more and more power and capability, of better and better economics, and of virtually limitless options for customization. Second, data, image, and video communications will eventually become as routine and transparent as voice is today. Third, customers will have constantly expanding flexibility in their capabilities to manage communications resources.

Now, what kind of network evolution, along with what capabilities, do we envision beyond the achievement of the near-term goal of basic ISDN? We see not only the achievement but also the extension of the previously mentioned architectural objectives into entirely new service capabilities. More specifically, we see capabilities for a smooth–indeed, a seamless–evolution from both the installed bases of network and of customer-premises equipment. Adherence to internationally accepted standards will enable compatibility at all levels. We will see, on a widespread basis, simplified as well as integrated-service and network-management capabilities, and the integrated-access capabilities of ISDN will be extended both to integrated-transport and to integrated-service applications.

Moreover, network service flexibility will continually increase. Specifically, there will be a practically limitless menu of distributed software-based services. These include services using software-defined networks—instantaneously set up to the customer's specifications and under the customer's control. Network capabilities and resources such as bandwidth will be available to customers on demand, to give them dynamic access to exactly the resources needed for any specific application, including the ability to economically obtain a wideband channel for video or for advanced distributed computing applications. Importantly, families of advanced services will be available to all customers—ranging from large-business customers to small-business and home customers. Finally, we will see the achievement of universal connectivity, linking and uniting the nations of the world.

All of these factors add up to how we envision the network of the future—a truly global system of information movement and management, one that will permit the universalizing of the information age. Clearly, we have a long way to go to achieve this vision. We face many technological challenges—not to mention the political, economic, and social challenges that invariably accompany such technological change.

This is where your second century here at the Cornell University School of Electrical Engineering begins—with these technologies in hand, this vision ahead, and these challenges yet to be met. To these I am sure each of you can add many more, as you see where and how your profession and your specific interests can contribute to our country's and the world's needs.

As I tried to indicate at the beginning of my remarks, we are entering an era in which we can gain the ability to use information as an ever more powerful force for human productivity and the fulfillment of human needs. Thus, we have a great opportunity to shape the information age to meet those needs.

- We can make it a new industrial age, with more and better material goods available to and affordable by more people.

- We can make it a new age of enlightenment, one that provides more people with a better education and new knowledge.

- We can make it a new age of opportunity—opportunity for better health, for a greater understanding of the world and its people, perhaps for a more secure and peaceful world.

I hope that these will be among our larger goals as we shape the information age, and I know that you here at Cornell share them, not only in your fine School of Electrical Engineering but throughout all the schools and departments of this outstanding university. Therefore, as you celebrate this centennial, I am sure you are already well on your way to a successful second century—one as rich in accomplishments and contributions as your first hundred years.

Thank you for letting me share my thoughts with you on this momentous and happy occasion.